ACTIVE AND PASSIVE ANALOG

An Introduction

McGraw-Hill Series in Electrical and Computer Engineering

Senior Consulting Editor

Stephen W. Director, *Carnegie Mellon University*

Circuits and Systems
Communications and Signal Processing
Computer Engineering
Control Theory
Electromagnetics
Electronics and VLSI Circuits
Introductory
Power and Energy
Radar and Antennas

Previous Consulting Editors

Ronald N. Bracewell, Colin Cherry, James F. Gibbons, Willis W. Harman, Hubert Heffner, Edward W. Herold, John G. Linvill, Simon Ramo, Ronald, A. Rohrer, Anthony E. Siegman, Charles Susskind, Frederick E. Terman, John G. Truxal, Ernst Weber, and John R. Whinnery

Electronics and VLSI Circuits

Senior Consulting Editor

Stephen W. Director, *Carnegie Mellon University*

Consulting Editor

Richard C. Jaeger, *Auburn University*

Colclaser and Diehl-Nagle: *Materials and Devices for Electrical Engineers and Physicists*
Elliott: *Microlithography: Process Technology for IC Fabrication*
Fabricius: *Introduction to VLSI Design*
Ferendeci: *Physical Foundations of Solid State and Electron Devices*
Franco: *Design with Operational Amplifiers and Analog Integrated Circuits*
Geiger, Allen, and Strader: *VLSI Design Techniques for Analog and Digital Circuits*
Grinich and Jackson: *Introduction to Integrated Circuits*
Givone and Roesser: *Microprocessors/Microcomputers: Introduction*
Hamilton and Howard: *Basic Integrated Circuits Engineering*
Hodges and Jackson: *Analysis and Design of Digital Integrated Circuits*
Huelsman: *Active and Passive Analog Filter Design: An Introduction*
Long and Butner: *Gallium Arsenide Digital Integrated Circuit Design*
Millman and Halkias: *Integrated Electronics: Analog, Digital Circuits, and Systems*
Millman and Grabel: *Microelectronics*
Millman and Taub: *Pulse, Digital, and Switching Waveforms*
Offen: *VLSI Image Processing*
Roulston: *Bipolar Semiconductor Devices*
Ruska: *Microelectronic Processing: An Introduction to the Manufacture of Integrated Circuits*
Schilling and Belove: *Electronic Circuits: Discrete and Integrated*
Seraphim: *Principles of Electronic Packaging*
Singh: *Physics of Semiconductors and Their Heterostructures*
Smith: *Modern Communication Circuits*
Sze: *VLSI Technology*
Taub: *Digital Circuits and Microprocessors*
Taub and Schilling: *Digital Integrated Electronics*
Tsividis: *Operations and Modeling of the MOS Transistor*
Wait, Huelsman, and Korn: *Introduction to Operational and Amplifier Theory Applications*
Walsh: *Choosing and Using CMOS*
Yang: *Microelectronic Devices*
Zambuto: *Semiconductor Devices*

ACTIVE AND PASSIVE ANALOG FILTER DESIGN

An Introduction

Lawrence P. Huelsman

University of Arizona

McGraw-Hill, Inc.

New York St. Louis San Francisco Auckland Bogotá
Caracas Lisbon London Madrid Mexico Milan Montreal
New Delhi Paris San Juan Singapore Sydney Tokyo Toronto

This book was set in Times Roman.
The editors were Anne T. Brown and Eleanor Castellano;
the production supervisor was Janelle S. Travers.
The cover was designed by Carol Couch.
R. R. Donnelley & Sons Company was printer and binder.

ACTIVE AND PASSIVE ANALOG FILTER DESIGN
An Introduction

1 2 3 4 5 6 7 8 9 0 DOC DOC 9 0 9 8 7 6 5 4 3

ISBN 0-07-030860-8

Library of Congress Cataloging-in-Publication Data

Huelsman, Lawrence P.
Active and passive analog filter design: an introduction / Lawrence P. Huelsman.
p. cm. — (McGraw-Hill series in electrical and computer engineering. Computer engineering)
Includes index.
ISBN 0-07-030860-8
1. Electric filters, Active—Design. 2. Electric filters, Passive—Design. I. Title. II. Series.
TK7872.F5H797 1993 92-40111
621.3815′324—dc20

ABOUT THE AUTHOR

Lawrence P. Huelsman received the BSEE degree from Case Institute of Technology and the MSEE and Ph.D. degrees from the University of California at Berkeley. He is a Life Fellow of the *Institute of Electrical and Electronic Engineers*. He currently holds an appointment as Professor Emeritus of Electrical and Computer Engineering at the University of Arizona.

Dr. Huelsman is the author of seventeen books including: *Basic Circuit Theory, 3rd Ed.* published by Prentice-Hall, Inc., and *Introduction to the Theory and Design of Active Filters*, published by McGraw-Hill, Inc. Japanese, German, Korean, Spanish, and Russian translations have been made of several of his books. He has also published many papers in the area of active circuit theory.

He has served as Associate Editor of the *IEEE Transactions on Circuit and System Theory* and the *IEEE Transactions on Education* and was technical chairman of the IEEE Region Six Annual Conference. He is a member of the steering committee for the Midwest Symposium on Circuits and Systems. He is a member of several scientific, engineering, and honorary societies, including Tau Beta Pi, Phi Beta Kappa, Eta Kappa Nu, and Sigma Xi. He has received the Anderson Prize of the College of Engineering and Mines of the University of Arizona for his contributions to education.

To my wife
Jo

CONTENTS

PREFACE

OVERVIEW

This book is designed to provide the basic material for an introductory course in the theory and design of active and passive analog filters. Although the book has some of its origins in an earlier work (*Introduction to the Theory and Design of Active Filters*, L. P. Huelsman and P. E. Allen, McGraw-Hill Book Company, 1980), it has been rewritten to emphasize modern trends and applications and to provide greater flexibility for the instructor and increased clarity for the students. Among the changes in content are the inclusion of a chapter on basic passive filters. This addition is especially important since many modern active filter techniques use passive structures as a prototype. Several changes have been made in the presentation of the active filter materials to eliminate the coverage of older seldom-used techniques and to add material on new design methods. Among these latter are the inclusion of material on switched capacitor filters and on filters that use operational transconductance amplifiers. To provide easier usage by both the instructor and the students, most of the examples have been rewritten, and many new ones have been added. Numerous solved drill problems are given throughout the text. The problems at the end of the chapters have been completely revised. In its present form, the book is well suited for a self-contained one-semester introductory course in analog filters for seniors or first-year graduate students. Sufficient material is included to provide the instructor with considerable flexibility in choosing the depth of coverage he or she desires for the various topics presented. The book is also suitable for a one-year course. In this case, all of the material would normally be covered.

CONTENTS BY CHAPTER

The first chapter of the book provides introductory material that outlines the scope and purpose of the filtering problem. Basic circuit theory concepts such as the role of the network function and the complex-frequency variable s are reviewed. Their relation to sinusoidal steady-state performance characteristics are illustrated. The all-important techniques of frequency and impedance denormalization, which will be used throughout the book, are presented. Examples are given to illustrate to the student how filters are used in several applications.

In Chapter 2, the subject of approximation is presented. The treatment includes not only the usual Butterworth, Chebyshev, and Thomson characteristics but also the very important inverse-Chebyshev and elliptic ones. A treatment is given of the use of phase and delay approximation and their role in determining time-domain filter performance. Extensive tables are given of pole-zero locations and polynomial coefficients for the most commonly encountered filter specifications. Quadratic factors are tabulated to provide ready application to cascade active filter design methods. Frequency transformations are covered in depth.

The important subject of sensitivity is introduced in Chapter 3. Algebraic relations for the relative sensitivity function are developed that are used extensively to facilitate following sensitivity analysis applications. The limitations placed on sensitivity measure as a result of its first-order approximation are illustrated. The most frequently used types of sensitivity are defined and interrelations between them are developed. A new section on parasitic sensitivity has been included.

Passive filters are introduced in Chapter 4. The treatment begins with a development of the general properties of driving-point functions and uses these properties to develop methods of *LC* driving-point synthesis. A treatment of the general properties of transfer functions is next developed. The constraints imposed by ladder topologies are derived and a development is made of single- and double-resistance-terminated *LC* ladder filters. The use of passive filter tables and the methods for applying source conversion, reciprocity, change of variables, and duality to extend them is described. Extensive passive filter realization tables are given in an appendix. The use of coupled resonator configurations as an alternate to band-pass filter configurations developed through transformation methods is introduced. A treatment of the sensitivity of passive filters is included.

Single-amplifier active *RC* filter configurations are introduced in Chapter 5. The fundamental role of feedback and the use of a standard passive network are defined. Application of these concepts are made to the basic Sallen and Key configuration and to the more versatile infinite-gain structure. The extension of this structure to the general biquadratic Friend, Harris, and Hilberman (STAR) circuit is presented. Sensitivity considerations for these filter types are illustrated. The effects of operational amplifier gain bandwidth on these configurations is discussed. Compensation methods for gain bandwidth are described.

Multiple-amplifier active *RC* filters are presented in Chapter 6. Treatments are given of the popular state-variable (KHN) filter, the Tow-Thomas (resonator) configuration, and the Mossberg-Åkerberg modification. The details of typical universal active filter packages are described. Extensions of the basic multiple-amplifier filters to realize biquadratic functions are presented. Sensitivity considerations for these filter types are discussed. Design equations for the various realizations are given. The effects of operational amplifier gain bandwidth on the various realizations and the use of compensation methods are given. An introductory treatment is given of the use of multiple operational transconductance amplifier filters.

Direct active *RC* filter realization methods are covered in Chapter 7. The properties of generalized-impedance converters are developed. The use of these converters in realizing synthetic inductors and FDNRs (frequency-dependent negative resistors) to eliminate the requirement for inductors in passive filters is described. Other direct realizations that are described include the leapfrog passive network structure and the primary resonator block. The sensitivities of all these direct realizations are discussed. An introductory treatment of switched capacitor filters is given.

Several appendices are included to provide more extensive resources than are given in the body of the text. These include tables for passive filter realizations for various approximations, listings of network functions for various types of elliptic characteristics, and a description of the properties of operational amplifiers.

SUGGESTED OUTLINE FOR ONE-SEMESTER COURSE

A one-semester course using this book as a text might be structured as follows:

Chapter 1 Assign this chapter as material to be reviewed by the student.

Chapter 2 Cover the "all-pole" magnitude approximation for the Butterworth and Chebyshev cases in Secs. 2.1 and 2.2 thoroughly. Give an introduction to the elliptic approximation in Sec. 2.4. Cover Sec. 2.5 on frequency transformations in depth. Give an introduction to the phase, delay, and time-response material in Secs. 2.6 and 2.7.

Chapter 3 Cover the material on relative sensitivity and function sensitivity in Secs. 3.1 and 3.2 in depth. Give an introduction to the Q and ω_n sensitivity in Sec. 3.5.

Chapter 4 Give an introduction to the material on energy functions in Sec. 4.1. Cover the material on the properties and synthesis of *LC* driving-point functions in Secs. 4.2 and 4.3 in depth. Give an introduction to the theoretical material on transfer functions in Secs. 4.4 and 4.5. Cover the material on the single-resistance-terminated lossless ladder synthesis in Sec. 4.6 in depth. The material on the use of tables in Sec. 4.9 should be covered thoroughly. The other sections in this chapter may be omitted without loss of continuity.

Chapter 5 The material in this chapter provides the theory of the single-amplifier active filter. It should be covered completely. If time is limited, Sec. 5.5 may be omitted without loss of continuity.

Chapter 6 The material on the state-variable and Tow-Thomas filters in Secs. 6.1 and 6.2 should be covered thoroughly. The remaining sections can be omitted without loss of continuity. Section 6.6 on filters using operational transconductance amplifiers should be included if possible.

Chapter 7 Cover the material on the generalized impedance converter and its use in realizing synthetic inductors in Secs. 7.1 and 7.2 in depth. The treatment of the leapfrog filter in Sec. 7.4 is also very important. The other sections may be eliminated without loss of continuity, although Sec. 7.6 on switched capacitor filters should be included if at all possible.

ACKNOWLEDGMENTS

The material presented in this book has been tested in the classroom in note form in a senior/graduate course at the University of Arizona. The course has also been offered through the National Technological University.

The author would like to express his appreciation to the many persons who assisted and encouraged him in the preparation of this book. He is especially grateful to Kai Chang, Texas A&M University; Jaime Ramirez-Angulo, New Mexico State University; James A. Resh, Michigan State University; Thomas M. Scott, Iowa State University; Martin Snelgrove, University of Toronto; Bang-Sup Song, University of Illinois; F. William Stephenson, Virginia Polytechnic Institute and State University; and David J. Willis, Oregon Institute of Technology for their most helpful definitive reviews of the entire manuscript. Special thanks are also due to all the students who responded enthusiastically to the offer of "one point for every error you find." In this group, Shirley Theriot, Edward M. Ochoa, and Anita Singh were the outstanding winners. Finally, my thanks go to my wife, Jo, for her patience, understanding, and support throughout the project.

Lawrence P. Huelsman

ACTIVE AND PASSIVE ANALOG FILTER DESIGN
An Introduction

CHAPTER 1

INTRODUCTION

The study of almost any engineering subject, be it electrical, mechanical, hydraulic, thermal, etc., can always be divided into two parts, analysis and synthesis. In *analysis* we are concerned with finding the characteristics or properties of some existing system. The process is illustrated by the flowchart shown in Fig. 1.1(a). This may be read, "Given a system, find its properties." Frequently, of course, the system may "exist" only as a schematic showing the interconnection of idealized elements. In that case, the schematic defines a *model* of the system, and the analysis then gives the properties of the model. If the system (or model) is completely specified, its properties are, of course, unique. Thus, in an analysis problem, there is *only one solution*.

In *synthesis* (or *design*), on the other hand, the starting point is a desired set of properties, and the goal is to find a system in actual or (more usually) in modeled form that has those properties. The process is illustrated by the flowchart shown in Fig. 1.1(b). This may be read, "Given a set of properties, find a system possessing them." In general, as indicated in the figure, there is usually more than one such system. Thus, in a synthesis problem, *the solution is rarely unique*. Because of this nonuniqueness, a final step in the synthesis process is usually required, namely, the evaluation of several different systems, all of which have the desired original set of properties, to find out which one is best. Before this can be done, we must define what is meant by the word *best*. Another way of looking at this is that an additional property or properties must be added to those originally specified in order that a unique choice may be made from among the systems that have been found. Obviously, the synthesis process is considerably more complicated (and challenging) than the analysis one!

In this book our goal is to investigate the design of a specific class of systems called filters. These may be thought of as frequency-selective signal transmission devices in which certain ranges of frequencies (the *passbands*) are passed from the input to the output, while other ranges (the *stopbands*) are rejected. Filters occur widely in the

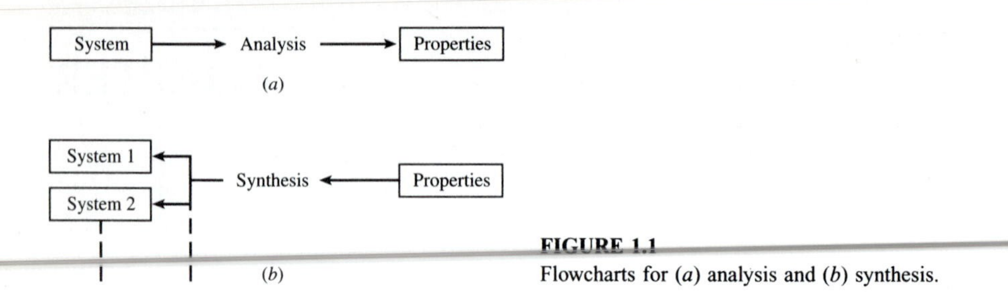

FIGURE 1.1
Flowcharts for (*a*) analysis and (*b*) synthesis.

electronic circuits associated with modern audio, communications, and signal processing fields. Audio systems use them for preamplification, equalization, and tone control. In communication circuits they are used for the tuning of specific frequencies and the elimination of others. In telephony, filters are used to decode the tone frequencies used for dialing. Digital signal processing systems use filters to prevent aliasing. Indeed, it is difficult to find any moderately complex electronic device that does not contain one or more filters. In our study of filters we will be concerned with finding ways of interconnecting network elements so as to realize circuits that have specific filtering characteristics. To achieve this goal, we will need to consider several topics. One of these is the determination of methods for expressing or approximating the properties of filters in such a way as to facilitate their synthesis. This topic is called *approximation*. Another important topic is the development of methods for expressing the ways in which the properties of a given filter realization change as a result of variations of the values of the elements from their design specifications. The variations can be the result of aging, temperature changes, component tolerance, and so forth. This topic is called *sensitivity*. It provides valuable techniques for comparing alternate realizations of the same set of design specifications.

Filter implementations have evolved drastically over the years as a result of changes in technology. The original electrical filter was realized as a collection of discrete *RLC* elements. When good quality active components, namely, operational amplifiers, became available, they provided a means for eliminating the inductor, thus leading to *active RC filters*. As suitable batch-processed thin-film technologies emerged, these active filters realized significant advantages of economy, size reduction, and versatility. Most recently, the field of filters has witnessed the introduction of sampled-data signal processing methods that have made it possible to eliminate resistors through the use of metal-oxide semiconductor (MOS) switches. The result is the *switched capacitor filter* that provides even greater application potential. In the following chapters these topics and others will be presented in such a way as to give the reader a comprehensive and unified treatment of modern filter theory.

1.1 NETWORK FUNCTIONS

In this section we briefly review some of the basic concepts of network theory. Additional discussions of these concepts may be found in the texts listed in the Bibliography.

Transformed Variables

The usual variables that we associate with an electronic circuit are the voltages and currents measured at various points in the circuit. These are "physical," or "real-world," variables in the sense that they can be measured with meters, can be displayed on an oscilloscope, and if they are large enough, can even provide us with a tangible indication of their presence such as a spark or a shock. In most filtering situations, these variables are not constant but have values that change with time. Thus, we write them in the form $v(t)$ and $i(t)$ and refer to them as being in the *time domain*. The time-domain determination of the filtering characteristics for a specific application represents a specification of how some response (or output) variable $r(t)$, which may be a voltage or a current, is produced as the result of some excitation (or input) variable $e(t)$, which may also be either a voltage or a current. Determining the relations between the input and the output variables of time, however, gives little insight into the way in which the various elements of the network enter into the frequency-selection process of the filter. Instead, for circuits with linear time-invariant elements, we usually use the Laplace transform to create new (transformed) variables, namely, $R(s)$ for the response variable and $E(s)$ for the excitation variable. These are related to $r(t)$ and $e(t)$ by the Laplace transform

$$R(s) = \mathcal{L}\{r(t)\} = \int_0^\infty r(t)e^{-st}\,dt \qquad E(s) = \mathcal{L}\{e(t)\} = \int_0^\infty e(t)e^{-st}\,dt \tag{1}$$

where $\mathcal{L}\{\ \}$ may be read "the Laplace transform of." In the above integrands, the exponent in the term e^{-st} must be dimensionless. Therefore, we conclude that the quantity s must have the dimensions of reciprocal time, namely, frequency. Thus the variables $R(s)$ and $E(s)$ are said to be in the *frequency domain*, and s is called the *complex-frequency variable*. Its values are frequently displayed on a two-dimensional *complex-frequency plane*. The relation between $R(s)$ and $E(s)$ is in general given by defining a *network function* $N(s)$ as the ratio of response to excitation for the case where all network initial conditions are zero. Thus

$$N(s) = \frac{R(s)}{E(s)} \tag{2}$$

Example 1.1-1 Network function for a third-order network. As an example of a network function, consider the third-order network shown in Fig. 1.1-1. If we define the response variable $R(s) = V_2(s)$ and the excitation variable $E(s) = V_1(s)$, by routine circuit

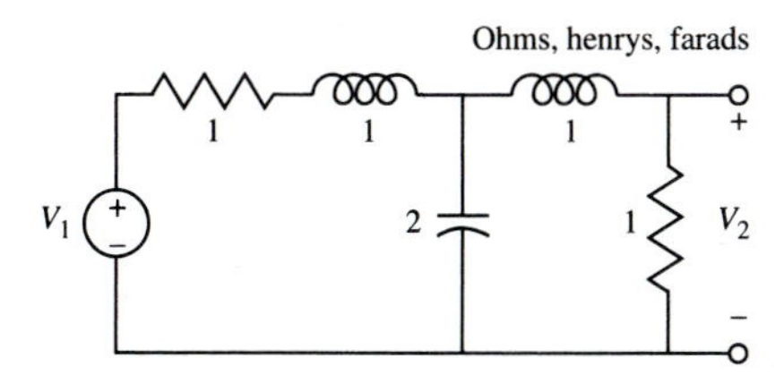

FIGURE 1.1-1
A third-order network.

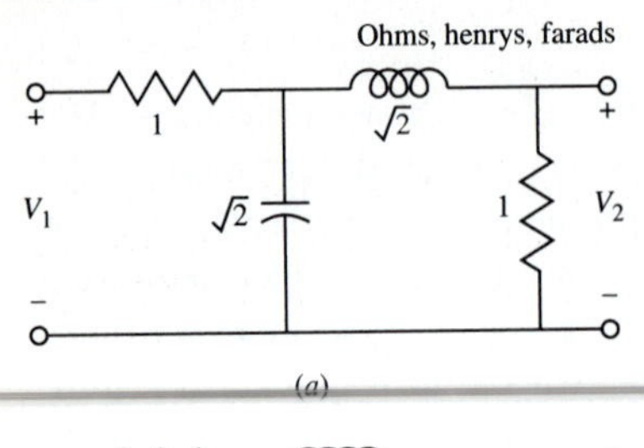

(a)

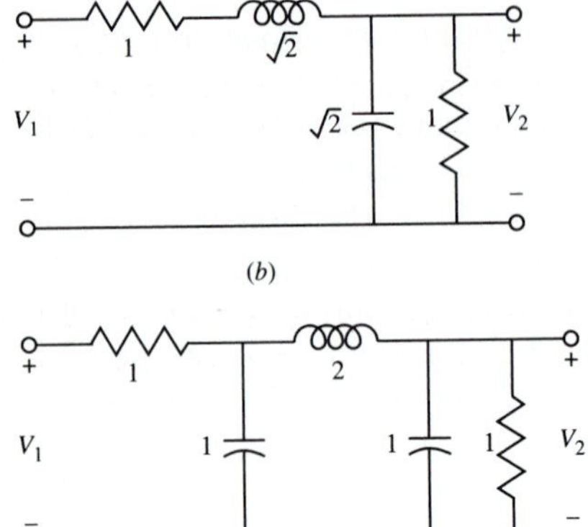

(b)

(c)

FIGURE 1.1-2
Example networks.

analysis, we find the network function $N(s)$ is

$$N(s) = \frac{V_2(s)}{V_1(s)} = \frac{0.5}{s^3 + 2s^2 + 2s + 1}$$

Exercise 1.1-1. For each of the networks shown in Fig. 1.1-2, find the network function $N(s) = V_2(s)/V_1(s)$.

Figure	Answers—network functions
1.1-2(*a*)	$N(s) = \frac{V_2(s)}{V_1(s)} = \frac{0.5}{s^2 + \sqrt{2}s + 1}$
1.1-2(*b*)	Same as for Figure 1.1-2(*a*)
1.1-2(*c*)	$N(s) = \frac{V_2(s)}{V_1(s)} = \frac{0.5}{s^3 + 2s^2 + 2s + 1}$

Sinusoidal Steady State

Complex frequency-domain expressions of the type given for $N(s)$ in Example 1.1-1 are completely general in the sense that they apply to almost any type of time-domain excitation function. Most filtering requirements, however, are based on the *sinusoidal steady-state* behavior of a network. Thus they assume that the excitation has the form

$$e(t) = \sqrt{2}E_0 \cos(\omega t + \alpha) \tag{3}$$

where E_0 is the root-mean-square (rms) value of $e(t)$, ω is the frequency in radians per second, and α is the phase in radians. For such an excitation, assuming $N(s)$ is stable,[1] after the transient time-domain components of $r(t)$ have decayed to the point where they are negligible, i.e., after a "steady state" has been reached, $r(t)$ will have the form

$$r(t) = \sqrt{2}R_0 \cos(\omega t + \beta) \tag{4}$$

where R_0 is the rms value of $r(t)$ and β is its phase. The relation between $e(t)$ and $r(t)$ under these conditions is readily found by using complex numbers called *phasors* to represent the sinusoidally varying quantities. For $e(t)$ and $r(t)$ we will use $\mathscr{E}$ and $\mathscr{R}$ to represent the phasors. These are defined as

$$\mathscr{E} = E_0 e^{j\alpha} \qquad \mathscr{R} = R_0 e^{j\beta} \tag{5}$$

The relationship between the phasors $\mathscr{E}$ and $\mathscr{R}$ is directly obtained by replacing the variable s in the network function $N(s)$ in (2) with the variable $j\omega$. Thus we obtain

$$N(j\omega) = \frac{\mathscr{R}}{\mathscr{E}} = \frac{R_0 e^{j\beta}}{E_0 e^{j\alpha}} \tag{6}$$

Frequently we are interested only in the way the rms magnitudes of the excitation and response sinusoids are related. In such a case, taking the magnitude of (6), we obtain

$$|N(j\omega)| = \left|\frac{\mathscr{R}}{\mathscr{E}}\right| = \frac{|\mathscr{R}|}{|\mathscr{E}|} = \frac{R_0}{E_0} \tag{7}$$

This magnitude may of course be expressed in logarithmic measure (decibels) by taking $20 \log |N(j\omega)|$. Alternately, we may be interested in the phase difference between the excitation and response sinusoids. In this case, taking the argument of (6), we may write

$$\arg N(j\omega) = \arg \frac{\mathscr{R}}{\mathscr{E}} = \arg \mathscr{R} - \arg \mathscr{E} = \beta - \alpha \tag{8}$$

Example 1.1-2 Magnitude and phase of a network function. As an example, consider the network function given in Example 1.1-1. For $s = j\omega$ we obtain

$$N(j\omega) = \frac{0.5}{(j\omega)^3 + 2(j\omega)^2 + 2j\omega + 1}$$

The magnitude and phase functions defined in (7) and (8) are

$$|N(j\omega)| = \frac{0.5}{[(1 - 2\omega^2)^2 + (2\omega - \omega^3)^2]^{1/2}}$$

$$\arg N(j\omega) = -\tan^{-1} \frac{2\omega - \omega^3}{1 - 2\omega^2}$$

Plots of these functions are shown in Fig. 1.1-3.

---

[1] By *stable* we mean that for any bounded excitation $e(t)$, the response $r(t)$ will also be bounded.

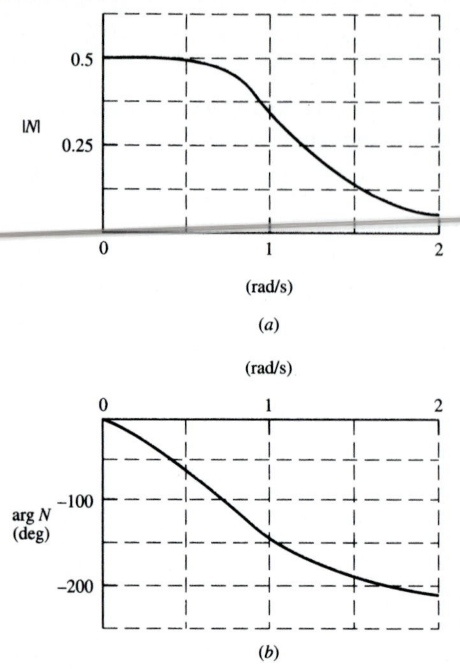

FIGURE 1.1-3
Magnitude and phase plots for the network of Fig. 1.1-1.

Exercise 1.1-2. Find the expressions for the magnitude and phase of the network functions defined in Exercise 1.1-1.

Figure	Answers	
1.1-2(*a*)	$\lvert N(j\omega)\rvert = \dfrac{0.5}{[(1-\omega^2)^2 + 2\omega^2]^{1/2}}$	$\arg N(j\omega) = -\tan^{-1}\dfrac{\sqrt{2}\omega}{1-\omega^2}$
1.1-2(*b*)	Same as for Fig. 1.1-2(*a*)	
1.1-2(*c*)	Same as for Example 1.1-2	

From the preceding example, we see that the network function $N(s)$ defined in (2) has a very important property, namely, at all frequencies it determines the sinusoidal steady-state response of the network for which it is specified. Almost all filter synthesis methods use the network function as a starting point.

1.2 PROPERTIES OF NETWORK FUNCTIONS

Driving-Point Functions

In the preceding section we introduced the concept of a network function. Such functions may be defined in various ways. The simplest situation occurs for the case where the voltage and current variables of a network are only of interest at a single pair

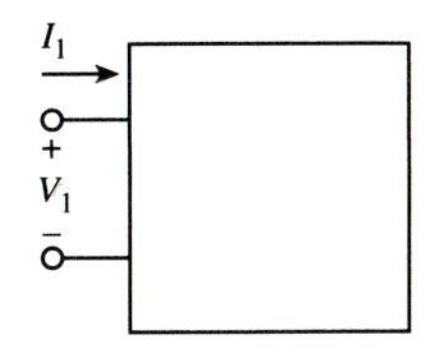

FIGURE 1.2-1
A one-port network.

of terminals. By convention, the relative reference polarities of the variables are arranged as shown in Fig. 1.2-1, the network is referred to as a *one-port network*, and the network functions are called *driving-point functions*. Only two types of driving-point network functions are defined for a one-port network. If current is treated as the excitation variable, then the network function is defined as a *driving-point impedance* $Z(s) = V_1(s)/I_1(s)$; while if voltage is considered as the excitation variable, the network function is defined as a *driving-point admittance* $Y(s) = I_1(s)/V_1(s)$. Obviously $Y(s) = 1/Z(s)$.

Transfer Functions

The situation that is of more interest in practical filtering applications is the one where there are four external terminals brought out from the network. These are arranged into pairs called *ports* so as to define four variables, two of voltage and two of current. By convention, the relative reference polarities are arranged as shown in Fig. 1.2-2, and the network is called a *two-port network*. The lower terminals of the two ports are frequently common, in which case the network is sometimes referred to as a *three-terminal network*. In addition to the driving-point functions that may be defined for such a two-port network, any network function that involves variables from both of the two ports is called a *transfer function*. Such functions can be transfer impedances and admittances as well as dimensionless transfer voltage functions such as $V_2(s)/V_1(s)$ or $V_1(s)/V_2(s)$ or dimensionless transfer current functions such as $I_2(s)/I_1(s)$ or $I_1(s)/I_2(s)$. The latter require that a path be available for the flow of the response current. This is usually provided by putting a short circuit or a load across the terminals of the response port.

Poles and Zeros

In general, a network function $N(s)$ has the form of a ratio of a numerator polynomial $A(s)$ to a denominator polynomial $B(s)$. Thus

$$N(s) = \frac{A(s)}{B(s)} \tag{1}$$

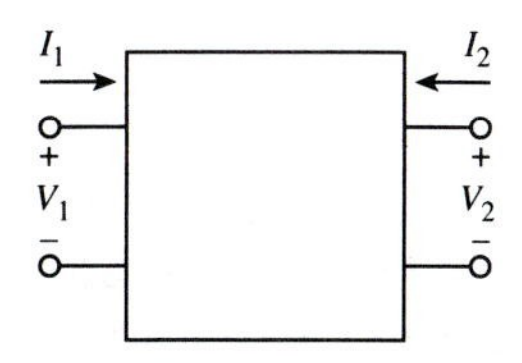

FIGURE 1.2-2
A two-port network.

The zeros of the numerator polynomial $A(s)$ are referred to as the *zeros of the network function* $N(s)$; that is, they are the values of s, also referred to as the *locations on the complex-frequency plane*, where the magnitude of $N(s)$ is zero. Zeros of $N(s)$ also occur at infinity when the degree of the numerator polynomial is less than the degree of the denominator one. The zeros of the denominator polynomial $B(s)$, on the other hand, are referred to as the *poles of the network function* $N(s)$. They are the values of s, or the locations on the complex-frequency plane, where the magnitude of $N(s)$ is infinite. The location of these poles is directly related to the filtering properties of a given network. For example, for the network to be stable, the poles of its network function must be in the left-half plane, or if on the $j\omega$ axis, they must be simple, i.e., only of first order. Networks whose network functions have right-half-plane poles, or have $j\omega$-axis poles of greater than first order, are unstable; i.e., they have an unbounded response for a bounded input, and thus they do not represent a physically useful situation.

Example 1.2-1 Poles and zeros of a network function. As an example of pole and zero locations, consider the network function

$$N(s) = \frac{9s}{(s+1)^2(s^2+2s+2)}$$

This function has a second-order pole at $s = -1$, simple poles at $s = -1 \pm j1$, and a zero at $s = 0$. Since the denominator is of degree 4 while the numerator is only of degree 1, the function will also have a third-order zero at $s = \infty$.

Exercise 1.2-1. Determine the poles and zeros (including the zeros at infinity) of the following network functions:

	Answers	
Function	**Poles**	**Zeros**
$\frac{3}{s^2+s+1}$	$-0.5 \pm j\sqrt{3}/2$	2 at ∞
$\frac{7s}{s^2+2s+2}$	$-1 \pm j1$	$0, \infty$
$\frac{s^2+1}{s^2+2s+5}$	$-1 \pm j2$	$\pm j1$

Poles and Zeros and Magnitude Response

Certain configurations of poles and zeros in a network function $N(s)$ produce easily recognized effects in the magnitude function $|N(j\omega)|$. Some examples follow.

Example 1.2-2 Poles close to the $j\omega$ axis. As an example of the relation between the poles of a network function and its magnitude characteristic, consider

$$N_a(s) = \frac{1}{s^2 + 0.6s + 1} \qquad \text{poles at } -0.3 \pm j0.9539$$

$$N_b(s) = \frac{1}{s^2 + 0.25s + 1} \qquad \text{poles at } -0.125 \pm j0.9922$$

The corresponding plots of $|N_a(j\omega)|$ and $|N_b(j\omega)|$ are shown in Fig. 1.2-3. Note that as the poles get closer to the $j\omega$ axis, the "resonant peak" of the magnitude characteristic becomes more prominent.

Example 1.2-3 Zeros on the $j\omega$ axis. As an example of the relation between the zeros of a network function and its magnitude characteristic, consider

$$N(s) = \frac{s^2 + 2}{s^2 + 0.25s + 1} \qquad \text{zeros at } \pm j\sqrt{2}$$

The corresponding magnitude plot of $|N(j\omega)|$ is shown in Fig. 1.2-4. Note that the zeros on the $j\omega$ axis produce a zero in the magnitude characteristic.

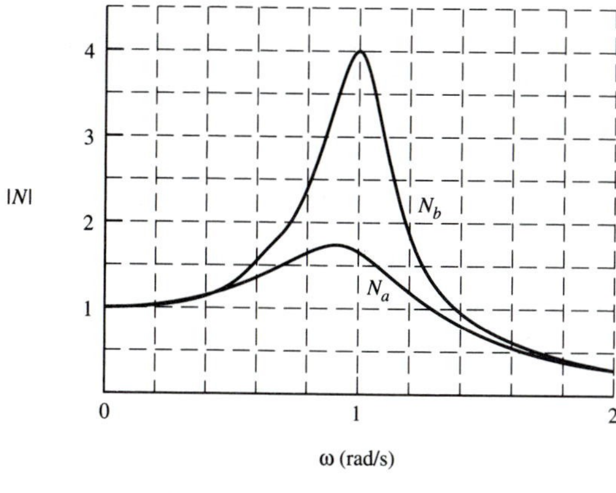

FIGURE 1.2-3
Effect of poles close to $j\omega$ axis.

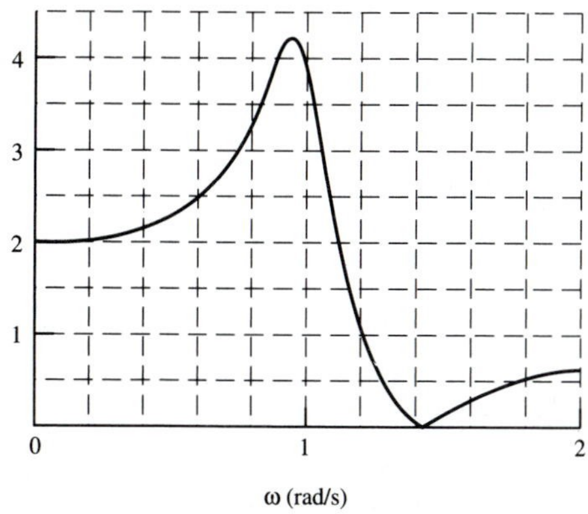

FIGURE 1.2-4
Effect of zeros on $j\omega$ axis.

Effect of Element Types

Now let us consider how the form of a network function for a given circuit is affected by the elements of which the filter is comprised. In general for lumped, linear, finite networks, $A(s)$ and $B(s)$ will be polynomials with real coefficients. Thus the network function will have the form

$$N(s) = \frac{A(s)}{B(s)} = \frac{a_0 + a_1 s + a_2 s^2 + \cdots}{b_0 + b_1 s + b_2 s^2 + \cdots} \tag{2}$$

In this case $N(s)$ is called a (real) *rational function*, i.e., a ratio of polynomials. For a three-terminal network containing only the passive elements resistors, capacitors, and inductors (an *RLC* network), the coefficients b_i of (2) will all be nonnegative and the poles of $N(s)$ will always be in the left-half plane and/or be simple on the $j\omega$ axis. If active elements such as operational amplifiers are present, however, the coefficients b_i may also be negative and the poles of $N(s)$ may be in the right half of the complex-frequency plane. In this case, of course, the network will be unstable.

A further classification of the properties of the poles of $N(s)$ may be made considering the location of these poles for networks comprised of various classes of elements. For example, for networks with *RLC* elements, the poles may be anywhere in the left half of the complex-frequency plane and/or on the $j\omega$ axis. In the latter case they must be simple. For *LC* elements the poles will lie only on the $j\omega$ axis (and be simple). For *RC* or *RL* elements the poles will be only on the negative real axis or at the origin. However, for networks consisting of *RC* elements and operational amplifiers, the poles may be anywhere in the complex plane. Filters realized by such networks are referred to as *active RC filters*. They basically provide all the advantages of *RLC* networks, in that they may have any desired left-half-plane pole locations, but they do not have the disadvantages of size, weight, nonlinearity, and nonintegrability associated with the inductive element. The properties of the poles for networks comprised of various classes of elements are summarized in Table 1.2-1.

TABLE 1.2-1
Characteristics of poles for various classes of network elements

Elements				
R	***C***	***L***	**Operational amplifier**	**Order and location of poles**
X	X			Simple on the negative real axis and at the origin
X		X		Simple on the negative real axis and at the origin
	X	X		Simple on the $j\omega$ axis and at the origin
X	X	X		Any order in the left-half plane and simple on the $j\omega$ axis
X	X		X	Any order anywhere in the complex plane

1.3 TYPES OF FILTERS

In this section we briefly introduce the major types of filters that we shall be concerned with in this book. A further treatment of the various types is given in Chap. 2.

Low-Pass Filters

The first filter type that we shall consider is the *low-pass* one. By definition, such a filter has the properties that low-frequency excitation signal components, down to and including direct current, are transmitted, while high-frequency components, up to and including infinite ones, are blocked. Thus the magnitude of a low-pass network function has ideally the appearance shown in Fig. 1.3-1(*a*). The range of low frequencies that are passed is called the *passband* or the *bandwidth* of the filter. As shown in the figure, it is equal to the value of the highest frequency ω_c transmitted. This frequency is also called the *cutoff frequency*. In practice, the ideal magnitude characteristic shown can only be approximated. Several such approximations are shown in Figs. 1.3-1(*b*) through (*d*). Note that in every case, the magnitude of the network function reaches zero only at infinite frequency. For these characteristics, the bandwidth is defined in terms of some maximum excursion of the magnitude characteristic from its peak value. The peak value may occur at zero frequency, as shown in Fig. 1.1-3(*b*) or at some intermediate frequency or frequencies, as shown in Figs. 1.3-1(*c*) and (*d*). The excursion or ripple is frequently specified as 3 dB, but other values may also be chosen. Magnitude characteristics in which all the maximum peak and the minimum valleys in a given range of frequencies have the same value, as shown in Figs. 1.3-1(*c*) and (*d*) for the passband and (*e*) for the passband and the stopband, are called *equal-ripple* in these regions. A characteristic in which the derivative of the magnitude does not change sign over a given range of frequencies is called *monotonic*. For example, the characteristic shown in Fig. 1.3-1(*b*) is completely monotonic. As other examples, the characteristics shown in Figs.

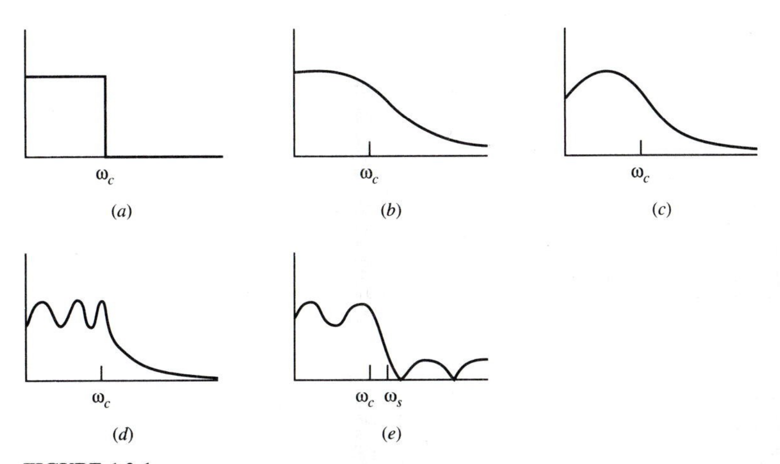

FIGURE 1.3-1
Some magnitude functions for low-pass filters.

1.3-1(c) and (d) may be described as being equal ripple in the passband and monotonic outside of the passband. The low-pass network functions whose magnitudes are shown in Figs. 1.3-1(b) through (d) most commonly have all their zeros located at infinity and are called *all-pole functions*. Their numerator polynomials are of zero degree, i.e., a constant, independent of the degree of the denominator polynomial. Their general form is

$$N(s) = \frac{H}{B(s)} \tag{1}$$

where H is not a function of s and the form of the polynomial $B(s)$ depends on the elements of the network.

Example 1.3-1 Low-pass network functions. As examples of network functions for low-pass networks, consider the voltage transfer functions for the second-order networks shown in Fig. 1.3-2 [the triangle symbol in Fig. 1.3-2(b) represents an ideal voltage-controlled voltage source of gain 1.586]. We find

$$N_a(s) = \frac{V_{2a}(s)}{V_{1a}(s)} = \frac{0.5}{s^2 + \sqrt{2}s + 1} \qquad N_b(s) = \frac{V_{2b}(s)}{V_{1b}(s)} = \frac{1.586 \times 10^8}{s^2 + \sqrt{2} \times 10^4 s + 10^8}$$

The plots of the magnitude characteristics are similar to the one shown in Fig. 1.3-1(b). Note that at direct current, the voltage transfer function for the network shown in Fig. 1.3-2(a) must have a value of 0.5, and the one for the network of Fig. 1.3-2(b) a value of 1.586 (the capacitors behave as open circuits and the inductor as a short circuit). These values are readily verified in the network functions given above by letting $s = 0$. An example of a third-order network function is given in Example 1.1-1. The methods used to synthesize these filters and the other ones used as examples in this chapter will be the subject of much of the material of the following chapters in this book.

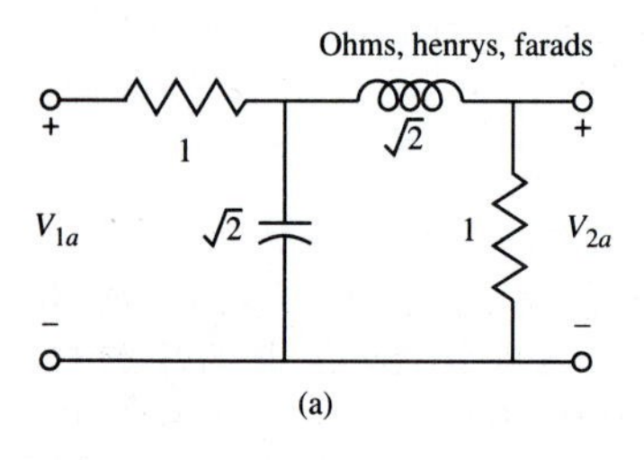

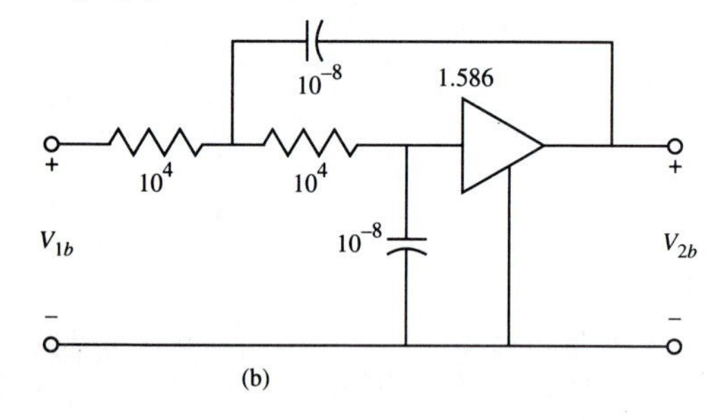

FIGURE 1.3-2
Low-pass filters.

The magnitude characteristic of a low-pass filter has a nonzero positive value at zero frequency. Thus, the phase characteristic for all low-pass filters whose network functions have the form shown in (1) starts at zero (assuming $H > 0$) and decreases for increasing frequency. If the network function is rational, the maximum phase is $-90n°$ at infinite frequency, where n is the order of the denominator polynomial $B(s)$. Thus, the maximum phase of the filter shown in Fig. 1.1-1 is $-270°$.

A modification of the basic low-pass magnitude characteristic occurs when equal-ripple behavior is present in both the passband and in a stopband having a range $\omega_s \leq \omega < \infty$, as shown in Fig. 1.3-1(e). Such a characteristic is called an *elliptic* one. The infinite frequency magnitude can be zero or (as shown) nonzero. To have an elliptic characteristic, the network function must have zeros on the $j\omega$ axis of the complex-frequency plane. Thus elliptic network functions have the form

$$N(s) = H \frac{\prod_i (s^2 + \omega_i^2)}{B(s)} \tag{2}$$

where the zeros of the numerator are on the $j\omega$ axis and the zeros of $B(s)$ are in the left-half plane.

Example 1.3-2 Elliptic network function. As an example of a network function for a low-pass elliptic network, consider the filter shown in Fig. 1.3-3. The voltage transfer function is

$$\frac{V_2(s)}{V_1(s)} = \frac{0.176524(s^2 + 5.153209)}{(s + 1.116765)(s^2 + 0.763717s + 1.629108)}$$

The plot of the magnitude characteristic is similar to the one shown in Fig. 1.3-1(e), except that the first peak occurs at zero frequency, there is only one finite null in the stopband, and a null occurs at infinity. Note that at direct current, from Fig. 1.3-3, the voltage transfer function must have a value of 0.5. This is readily verified by letting $s = 0$ in the network function given above.

High-Pass Filters

The second type of filter that we shall consider in this section is the *high-pass filter*. It has the property that low frequencies (the stopband) are blocked while high frequencies (the passband) are transmitted. Several magnitude characteristics for high-pass network functions are shown in Fig. 1.3-4. The terminology to describe these is similar to that

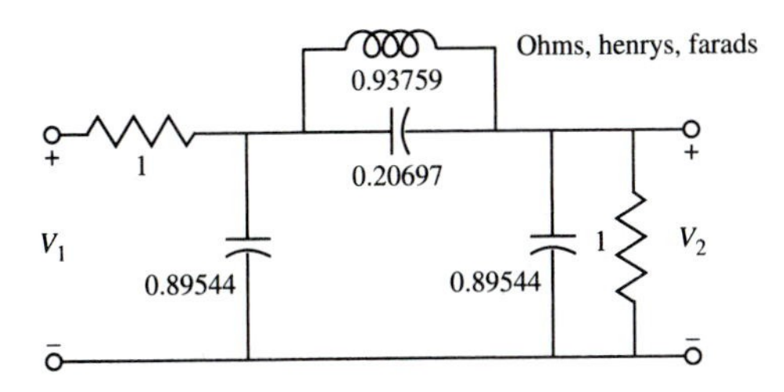

FIGURE 1.3-3
Elliptic low-pass filter.

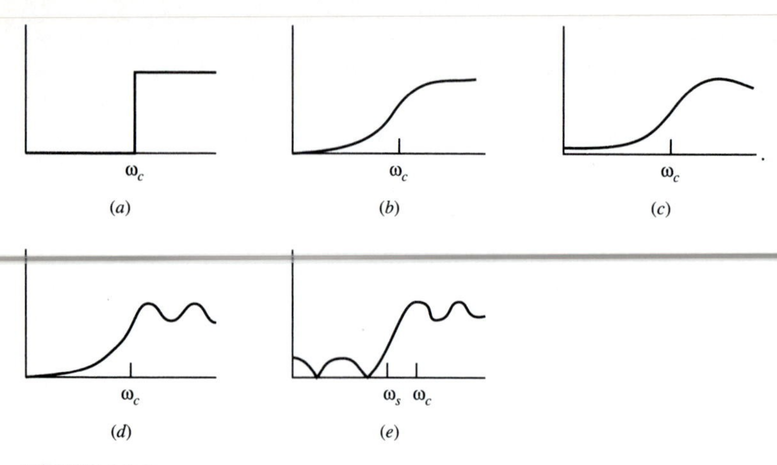

FIGURE 1.3-4
Some magnitude functions for high-pass filters.

used for the low-pass ones. Figure 1.3-4(*a*) shows an ideal (not realizable) characteristic. Figure 1.3-4(*b*) shows a completely monotonic characteristic. Figures 1.3-4(*c*) and (*d*) show characteristics that are equal ripple in the passband and monotonic outside the passband. Figure 1.3-4(*e*) shows a characteristic that is equal ripple in the passband and in the stopband $0 \le \omega \le \omega_s$, that is, an elliptic function. Its zero-frequency magnitude may be zero or (as shown) nonzero. For high-pass filters the passband is of infinite width, since in theory it extends to infinite frequency. As a result, rather than specifying the bandwidth, we more meaningfully specify the *cutoff frequency*, which is shown as ω_c in the figures. The high-pass functions with magnitude characteristics shown in Figs. 1.3-4(*a*) through (*d*) most commonly have all their zeros located at the origin of the complex-frequency plane. Thus, for rational functions they have the form

$$N(s) = \frac{Hs^n}{B(s)} \tag{3}$$

where H is a constant and n is the degree of the denominator polynomial $B(s)$.

Example 1.3-3 High-pass network function. As an example of a network function for a high-pass network, consider the filter shown in Fig. 1.3-5. The voltage transfer function is

$$\frac{V_2(s)}{V_1(s)} = \frac{0.5s^3}{s^3 + 2s^2 + 2s + 1}$$

The plot of the magnitude characteristic is similar to the one shown in Fig. 1.3-4(*b*). Note that at infinite frequency, from Fig. 1.3-5, the voltage transfer function must have a value of 0.5 (the capacitors behave as short circuits and the inductor as an open circuit). This is readily verified by taking the limit as s approaches infinity in the network function given above.

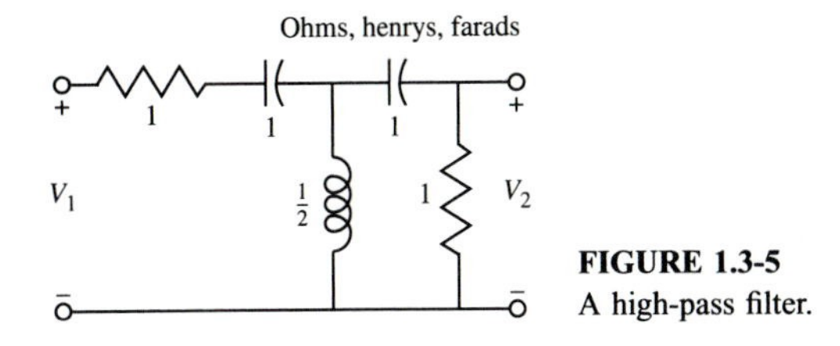

FIGURE 1.3-5
A high-pass filter.

The phase characteristic for a rational high-pass network function having the form given in (3) starts at $+90n^\circ$ at zero frequency and decreases to zero degrees at infinite frequency.

Band-Pass Filters

The third general filter type that we shall consider here is the *band-pass filter*. It has the property that one band of frequencies (the passband) is transmitted, while two bands of frequencies, namely, those below and above the passband, are blocked (the stopbands). Several network function magnitude characteristics for band-pass filters are shown in Fig. 1.3-6. The range of frequencies that is passed is called the *bandwidth* (BW) and is defined as the difference between the frequencies that define the edges of the passband. Using ω_1 and ω_2 as shown in the figures to define the passband edges, we obtain

$$\mathrm{BW} = \omega_2 - \omega_1 \tag{4}$$

The *center frequency* ω_0 of the passband is defined as the geometric mean of the band-edge frequencies. Thus

$$\omega_0 = \sqrt{\omega_1 \omega_2} \tag{5}$$

Figure 1.3-6(a) shows an ideal (not realizable) band-pass characteristic. Figure 1.3-6(b) shows a characteristic that is monotonic in the sense that on either side of the center

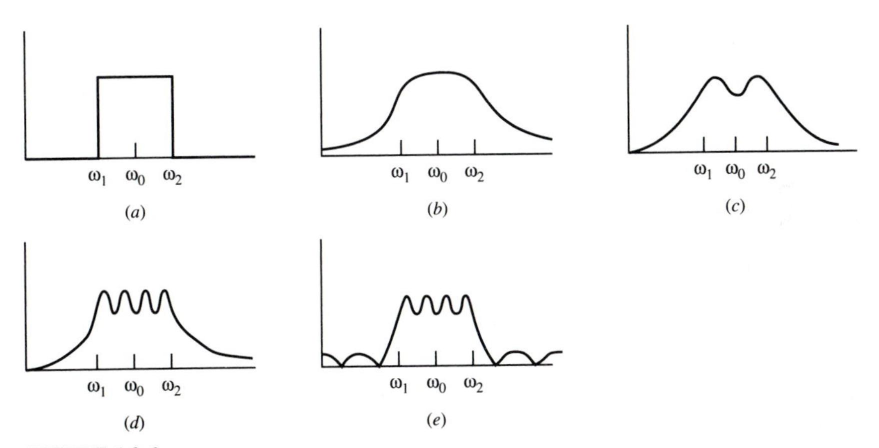

FIGURE 1.3-6
Some magnitude functions for band-pass filters.

frequency ω_0 the derivative of its magnitude characteristic does not change sign. Similarly, in Figs. 1.3-6(*c*) and (*d*), in the passband the behavior is equal ripple, while outside the passband the behavior is monotonic. Finally, in Fig. 1.3-6(*e*) an elliptic band-pass characteristic is shown. Its magnitude at zero and infinite frequencies may be zero, or (as shown) nonzero. All the band-pass magnitude characteristics shown in Figs. 1.3-6(*a*) through (*d*) have network functions that most commonly have half their zeros at the origin and the other half at infinity. Thus, for rational functions they have the form

$$N(s) = \frac{Hs^{n/2}}{B(s)} \tag{6}$$

where H is a constant and n is the degree of the denominator polynomial $B(s)$ and is always even. The phase characteristic for such a function starts at $(+90n/2)^\circ$ at zero frequency and decreases to $(-90n/2)^\circ$ at infinite frequency. It is zero at the center frequency ω_0.

Example 1.3-4 Band-pass network function. As an example of a network function for a band-pass network, consider the filter shown in Fig. 1.3-7. The voltage transfer function is

$$\frac{V_2(s)}{V_1(s)} = \frac{0.5s^2}{s^4 + \sqrt{2}s^3 + 3s^2 + \sqrt{2}s + 1}$$

The plot of the magnitude characteristic is similar to the one shown in Fig. 1.3-6(*b*). The center frequency $\omega_0 = 1$ rad/s and the bandwidth of 3-dB down frequencies is 1 rad/s. The phase characteristic ranges from $+180^\circ$ at zero frequency to -180° at infinite frequency. Note that at the center frequency, from Fig. 1.3-7, the voltage transfer function has a value of 0.5 (the series inductor and capacitor behave as a short circuit and the shunt inductor and capacitor behave as an open circuit). This may be verified by letting $s = j1$ in the network function given above.

Band-Reject Filters

The fourth general filter type that we shall consider here is the *band-reject filter*, also called a *band-elimination filter* or a *notch filter*. It has the property that one band of frequencies (the stopband) is blocked, while two bands of frequencies, namely, those below and above the stopband, are passed (the passbands). Several network function magnitude characteristics for band-reject filters are shown in Fig. 1.3-8. The range of

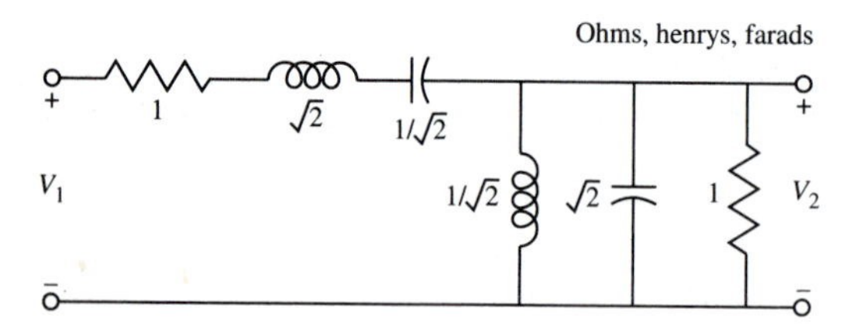

FIGURE 1.3-7
A band-pass filter.

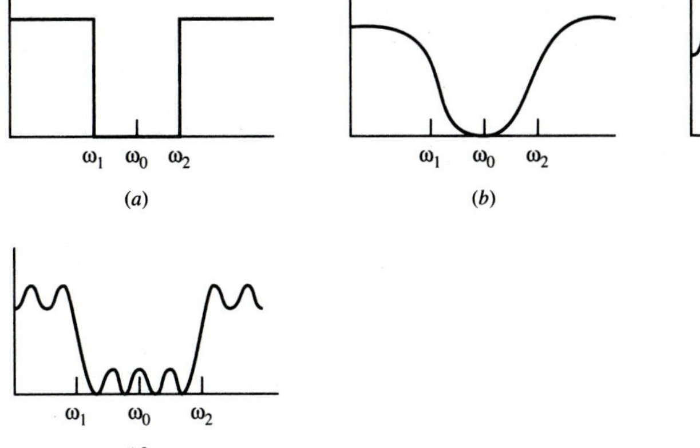

FIGURE 1.3-8
Some magnitude functions for band-reject filters.

frequencies that is blocked is called the *bandwidth* (BW) and is defined as the difference between the frequencies that define the edges of the stopband. Using ω_1 and ω_2 as shown in the figures to define the stopband edges, we obtain

$$\mathrm{BW} = \omega_2 - \omega_1 \tag{7}$$

The *center frequency* ω_0 of the stopband is defined as the geometric mean of the band-edge frequencies. Thus

$$\omega_0 = \sqrt{\omega_1 \omega_2} \tag{8}$$

Figure 1.3-8(*a*) shows an ideal (not realizable) band-reject characteristic. Figure 1.3-8(*b*) shows a monotonic characteristic. In Fig. 1.3-8(*c*) the characteristic is equal ripple in the passbands and monotonic in the stopband. Finally, in Fig. 1.3-8(*d*), an elliptic characteristic that has equal-ripple behavior in the stopband and the passbands is shown. All the band-reject magnitude characteristics shown in Fig. 1.3-8 have network functions that have zeros on the $j\omega$ axis. They are of the form

$$N(s) = \frac{H \prod_i (s^2 + \omega_i^2)}{B(s)} \tag{9}$$

where H is a constant and the zeros of $B(s)$ are in the left-half plane. The phase characteristic for such a function has a 180° positive discontinuity at each (sinusoidal) frequency corresponding with the location of the $j\omega$-axis zeros.

Example 1.3-5 Band-reject network function. As an example of a network function for a band-reject filter, consider the network shown in Fig. 1.3-9. The voltage transfer function is

$$\frac{V_2(s)}{V_1(s)} = \frac{0.5(s^2+1)^2}{s^4 + \sqrt{2}s^3 + 3s^2 + \sqrt{2}s + 1}$$

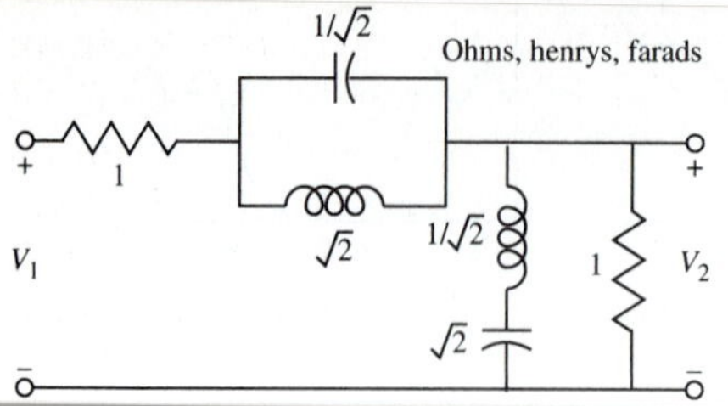

FIGURE 1.3-9
A band-reject filter.

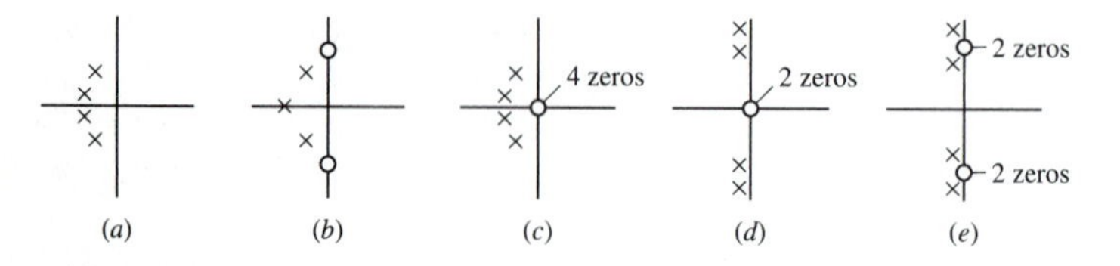

FIGURE 1.3-10
Typical pole-zero plots for filter functions: (*a*) low pass (all pole); (*b*) elliptic; (*c*) high pass; (*d*) band pass; (*e*) band reject.

The plot of the magnitude characteristic is similar to the one shown in Fig. 1.3-8(*b*). The center frequency $\omega_0 = 1$ rad/s. The bandwidth of 3-dB down (measured from the zero frequency and infinite frequency values of 0.5) frequencies is 1 rad/s, and the bandwidth of 20-dB frequencies is 0.31702 rad/s. The second-order zeros at $s = \pm j1$ produce a 360° discontinuity in the phase characteristic at 1 rad/s. Note that at the center frequency, the shunt inductor and capacitor in Fig. 1.3-9 act as an open circuit and the series inductor and capacitor act as a short circuit. Thus, signals at this frequency are effectively blocked. At zero and infinite frequencies, the circuit acts like a two-resistor voltage divider, and the voltage transfer function has a value of 0.5. This value may be verified in the network function given above by letting $s = 0$ and $s = \infty$.

Each of the filter types described in this section has a distinctive pole-zero configuration. Some examples are given in Fig. 1.3-10. Another useful filter function is the all-pass one. We will defer a treatment of this until later in the text.

1.4 FREQUENCY AND IMPEDANCE DENORMALIZATIONS

In this section we discuss the use of *denormalizations* in filter design.

Frequency Denormalization

The specifications usually given for the design of filters involve frequencies that have values of thousands of cycles per second. Synthesis calculations, however, are most easily done using frequencies of a few hertz or radians per second, since the numerical computations are simplified (and mistakes are minimized) by not having to carry along various powers of 10. Design tables for various filter characteristics also use similar convenient frequency values. Such values are usually referred to as *normalized frequency* values. To convert such normalized values to the "real-world" frequencies actually required in a given filter application, we use a process called *frequency denormalization*. This involves a change of complex-frequency variable. If we consider p as the normalized complex-frequency variable and s as the denormalized one, then the frequency-denormalization process is defined by the relation

$$s = \Omega_n p \tag{1}$$

where Ω_n is called the *frequency-denormalization constant*. For example, consider an inductor with a frequency-normalized value of L henrys and a corresponding frequency-normalized impedance $Z_n(p) = pL$. The denormalized impedance $Z(s)$ is found from (1) to be $Z(s) = sL/\Omega_n$; thus, it represents an inductance of value L/Ω_n. Similarly a denormalized capacitor will have a value of C/Ω_n. Additional relations are readily developed for the way in which other network elements are affected by frequency denormalization. These are summarized in the first line of Table 1.4-1. The corresponding effects on network functions are given in the first line of Table 1.4-2. In using these relations, it is helpful to remember that elements such as resistors and controlled sources, whose network functions are not functions of s, are left invariant by a frequency denormalization; whereas reactive elements, such as inductors and capacitors, which are characterized by network functions in which s appears, change value. For such reactive elements, the change in their value is in inverse proportion to the change in frequency.

Example 1.4-1 Frequency denormalization of a band-pass network. As an example of frequency denormalization, consider the (normalized) network shown in Fig. 1.4-1(*a*). The voltage transfer function for this, using p as the complex-frequency variable, is readily shown to be

$$\frac{V_2(p)}{V_1(p)} = N_n(p) = \frac{0.6p}{p^2 + 0.6p + 2} \tag{2}$$

The poles are at $p = -0.3 \pm j1.38203$. From the results given in Sec. 1.3 we may identify this as a band-pass characteristic. It may be shown to have a center frequency of $\sqrt{2}$ rad/s (the square root of the zero degree coefficient of the denominator). The frequency denormalization of (1) may be applied to (2) to develop a magnitude characteristic having exactly the same shape but with a center frequency of 10 kHz by specifying the frequency-normalization constant $\Omega_n = (2\pi/\sqrt{2}) \times 10^4 = 4.4429 \times 10^4$. Substituting $s/(4.4429 \times 10^4)$ for p in (2), we obtain the denormalized network function $N(s)$,

$$\frac{V_2(s)}{V_1(s)} = N(s) = \frac{0.6 \times 4.4429 \times 10^4 s}{s^2 + 0.6 \times 4.4429 \times 10^4 s + 2 \times (4.4429 \times 10^4)^2} \tag{3}$$

which has poles at $s = -1.33287 \times 10^4 \pm j6.14019 \times 10^4$ and is readily shown to have a center frequency of 10^4 Hz. Applying the same frequency denormalizations to the elements of the network shown in Fig. 1.4-1(*a*), we obtain the frequency-denormalized network shown in Fig. 1.4-1(*b*), which realizes (3).

TABLE 1.4-1
Effect of frequency and impedance denormalization on network elements

Denormalization	R	C	L	VCVS gain = α	ICIS gain = β	VCIS gain = g	ICVS gain = r	Ideal transformer, turns ratio = N	FDNR†
$s = \Omega_n p$ (frequency)	R	$\frac{C}{\Omega_n}$	$\frac{L}{\Omega_n}$	α	β	g	r	N	$\frac{D}{\Omega_n^2}$
$Z = z_n Z_n$ (impedance)	$z_n R$	$\frac{C}{z_n}$	$z_n L$	α	β	$\frac{g}{z_n}$	$z_n r$	N	$\frac{D}{z_n}$
Frequency and impedance	$z_n R$	$\frac{C}{\Omega_n z_n}$	$\frac{z_n L}{\Omega_n}$	α	β	$\frac{g}{z_n}$	$z_n r$	N	$\frac{D}{\Omega_n^2 z_n}$

† This element is introduced in Chap. 7.

TABLE 1.4-2
Effect of frequency and impedance denormalization on network functions

Denormalization	Poles p_i	Zeros z_i	Voltage transfer function $V_2/V_1 = A(p)$	Current transfer function $I_2/I_1 = B(p)$	Transfer impedance $V_2/I_1 = Z(p)$	Transfer admittance $I_2/V_1 = Y(p)$
$s = \Omega_n p$ (frequency)	$\Omega_n p_i$	$\Omega_n z_i$	$A\left(\frac{s}{\Omega_n}\right)$	$B\left(\frac{s}{\Omega_n}\right)$	$Z\left(\frac{s}{\Omega_n}\right)$	$Y\left(\frac{s}{\Omega_n}\right)$
$Z = z_n Z_n$ (impedance)	p_i	z_i	$A(p)$	$B(p)$	$z_n Z(p)$	$\frac{Y(p)}{z_n}$
Frequency and impedance	$\Omega_n p_i$	$\Omega_n z_i$	$A\left(\frac{s}{\Omega_n}\right)$	$B\left(\frac{s}{\Omega_n}\right)$	$z_n Z\left(\frac{s}{\Omega_n}\right)$	$Y\left(\frac{s}{\Omega_n}\right)/z_n$

The role of frequency normalization in synthesis can be further visualized by reference to Fig. 1.4-2. Either of two paths may be followed to synthesize a denormalized filter (the block at the right) starting from a normalized network function (the block at the left). In the upper path the synthesis operation is applied first, followed by a frequency denormalization. In the lower path the order of the two operations is reversed. Following either path produces the same result; however, the upper one is recommended since in it the numerical computations are greatly simplified.

Impedance Denormalization

A second type of denormalization frequently used to simplify the numerical computations made during a synthesis operation is called an *impedance denormalization.* It permits the use of elements with numerical values in a range close to unity, rather than

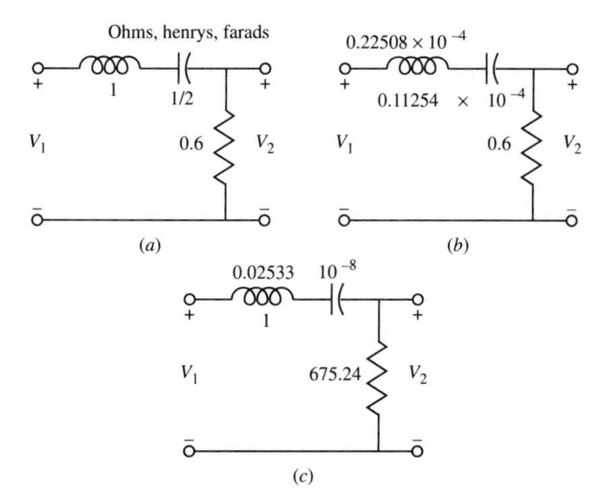

FIGURE 1.4-1
Frequency and impedance denormalization of a filter.

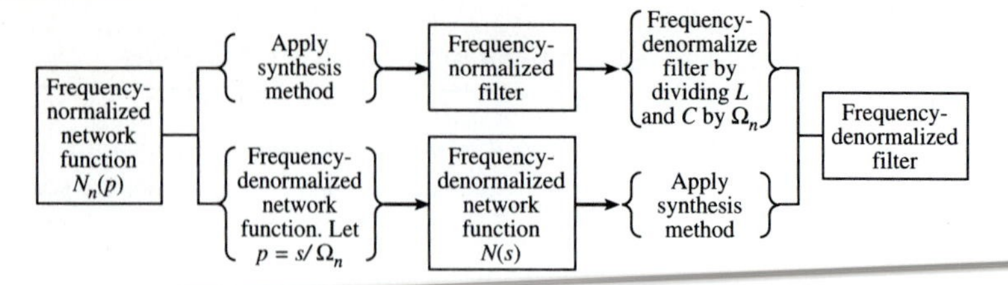

FIGURE 1.4-2
Flowchart for performing synthesis and frequency denormalization.

ones with real-world multipliers of 10^3, 10^6, 10^{-6}, 10^{-12}, etc. A normalized impedance $Z_n(s)$ can be denormalized to the impedance $Z(s)$, that is, converted to practical values, by the relation

$$Z(s) = z_n Z_n(s) \tag{4}$$

where z_n is called the *impedance-denormalization constant*. For example, consider an inductor with an impedance-normalized value of L henrys. Its impedance-normalized impedance is $Z_n(s) = sL$, and the denormalized (practical) impedance is $Z(s) = sz_nL$, representing an inductor of value z_nL. On the other hand, for a capacitor with an impedance-normalized value of C farads, the impedance-normalized impedance is $Z_n(s) = 1/sC$, and the denormalized impedance is $Z(s) = z_n/sC$, representing a capacitor of value C/z_n. Thus we see that when an impedance denormalization is made, the values of inductors (in henrys) and capacitors (in farads) move in opposite directions. Similar relations are readily developed for other types of network elements and for network functions. They are summarized in the second lines of Tables 1.4-1 and 1.4-2.

Example 1.4-2 Impedance denormalization of a band-pass network. As an example of an impedance denormalization, let us choose z_n such that the capacitor in the network shown in Fig. 1.4-1(*b*) is changed to a value of 10^{-8} F. From (4) we find that $1/s10^{-8} = z_n/(s \times 0.11254 \times 10^{-4})$. Thus $z_n = 1125.4$. The denormalized values of the network elements are given in Fig. 1.4-1(*c*). Since the network function of (3) for the network in Fig. 1.4-1(*b*) is dimensionless, it is unaffected by the impedance transformation; thus, it also applies to the network shown in Fig. 1.4-1(*c*). It should be noted that, as indicated in the third lines of Tables 1.4-1 and 1.4-2, the operations of frequency and impedance denormalization are commutative: they may be applied in either order, and the results obtained will be the same.

In practice, impedance denormalization frequently permits more effective use of a given implementation technology by minimizing the total required area and thus the cost of a circuit configuration.

1.5 EXAMPLES OF FILTER APPLICATIONS

In this section we present some examples of filter applications. In the remaining chapters of the book we will introduce methods for designing filters for applications similar to these.

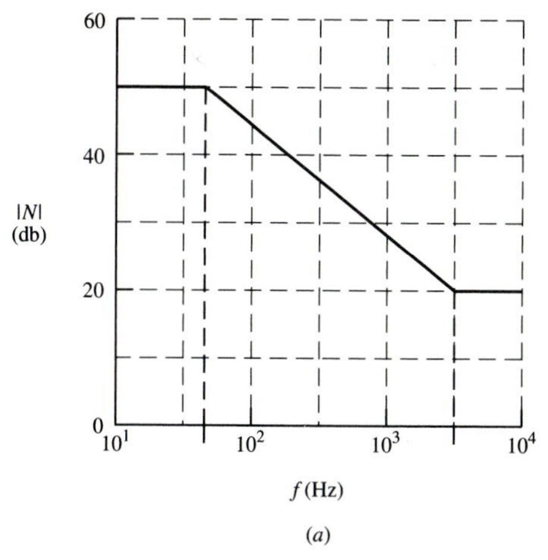

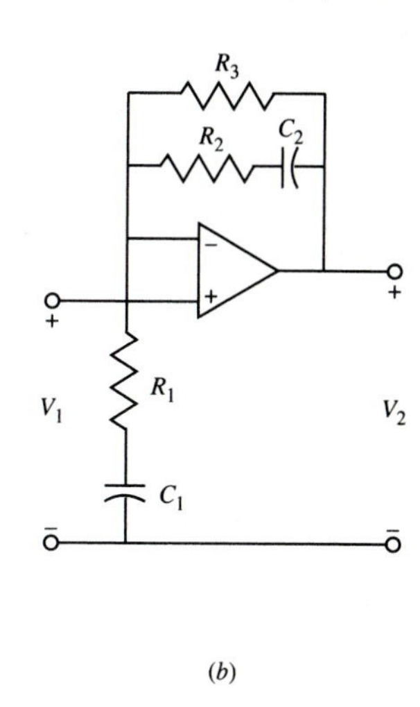

FIGURE 1.5-1
Magnitude specification and filter for tape preamplifier.

Tape Preamplifier

An interesting filter application occurs in high-fidelity audio systems that use a signal from a tape head as an input. Such an input signal must be filtered by a tape preamplifier before entering the main audio amplifier. The preamplifier must provide frequency-selective gain for the signal. Specifically, lower frequency signals must be amplified more than higher frequency ones, since they have a lower velocity as they pass the tape head and thus induce a lower voltage. A standard NAB (National Association of Broadcasters) response curve for the magnitude of the voltage transfer function for a tape preamplifier is shown in Fig. 1.5-1(*a*). A circuit that realizes this characteristic is shown in Fig. 1.5-1(*b*). To simplify the analysis of this circuit, we will assume that the value of C_1 is chosen so that in the frequency range of interest, the magnitude of its impedance is much less than the value of R_1, namely,

$$|Z_{C_1}| = \left| \frac{1}{j\omega C_1} \right| \ll R_1 \tag{1}$$

For this condition, the voltage transfer function is

$$\frac{V_2(s)}{V_1(s)} = 1 + \frac{R_3}{R_1} \frac{1 + sC_2R_2}{1 + sC_2(R_2 + R_3)} \tag{2}$$

The desired performance is obtained by choosing the element values so that

$$\frac{1}{C_2R_2} = 2\pi \times 3183 \qquad \frac{1}{C_2(R_2 + R_3)} = 2\pi \times 50 \qquad 20 \log_{10} \frac{R_3}{R_1} = 55 \tag{3}$$

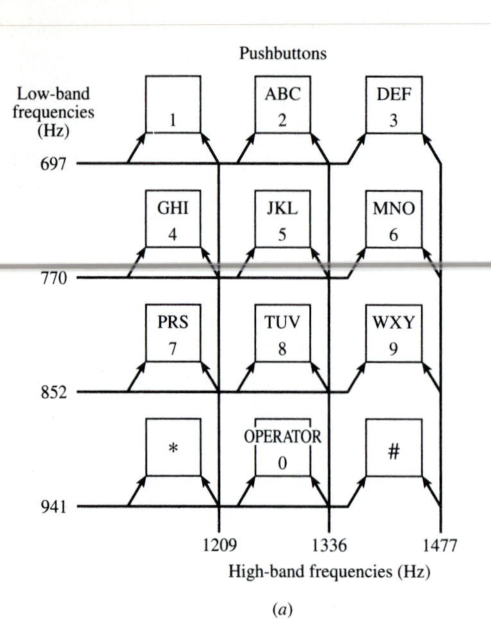

(a)

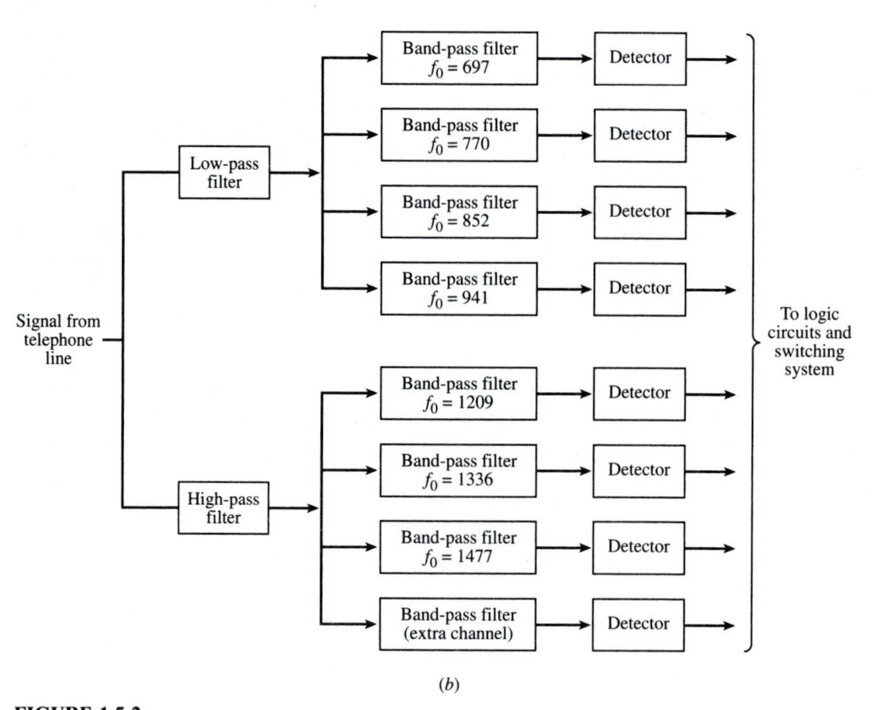

(b)

FIGURE 1.5-2
Touch-tone telephone dialing system.

Touch-tone Dialing

Another interesting example of how filters are used in a real-world application is the Touch-tone dialing system used in most telephone installations. In such a system, the telephone handset has a group of 12 pushbuttons. When depressed, each of these pushbuttons energizes oscillators that simultaneously generate and put on the line a low-band and a high-band audio frequency signal. The frequency assignments and the arrangement of the pushbuttons are shown in Fig. 1.5-2(*a*). For example, the 3 pushbutton generates 697 and 1477 Hz. Thus, each pushbutton is identified by a two-tone signal code. At the central telephone office, these signals are decoded using a set of filters arranged as shown in Fig. 1.5-2(*b*). Low-pass and high-pass filters are used to separate the low-band and high-band groups of frequencies, and band-pass filters are used to identify specific frequencies in each band. Detectors monitor the output of the band-pass filters and, through the use of simple logic circuits, permit identification of the pushbutton pressed. This system permits far more rapid signaling than is possible with the older rotary-dial system, in which individual dc pulses are counted to identify each digit. Obviously, filters play an important role in the functioning of the Touch-tone system.

Monolithic Pulse Code Modulation Filter

An example of a high-technology filter that embodies many of the ideas that will be presented in this book is the monolithic pulse code modulation (PCM) filter. It is used with an encoder that samples an analog input signal and converts it to (typically) 8-bit digital PCM data at 8000 samples per second. The filter consists of a transmitting section that prepares the data for encoding and transmission and a receiving section that reconstructs transmitted data. Each of these sections usually contains several individual analog filters. A block diagram of a typical transmitting section is shown in Fig. 1.5-3. The first filter is an antialiasing filter. It is required because the filter sections that follow it are usually realized using switched capacitor (sampled data) designs. It consists of a second-order low-pass filter with a cutoff frequency of 12 kHz. Its magnitude characteristic has the general form shown in Fig. 1.3-1(*b*). The second filter is used to eliminate any 60-Hz signal that occurs from power line coupling. It consists of a band-reject filter with two complex poles and a pair of zeros on the $j\omega$ axis. Its magnitude characteristic has the form shown in Fig. 1.3-8(*b*). The third filter is called a transmit low-pass filter. It is used to prevent any signal component above 4 kHz from entering the encoder. It requires a very sharp cutoff and is usually implemented by an elliptic function with a magnitude characteristic similar to the one shown in Fig. 1.3-1(*e*). The fourth filter is an output smoothing one that is used to prevent broadband noise components from entering the encoder. Its properties are similar to those of the antialiasing filter.

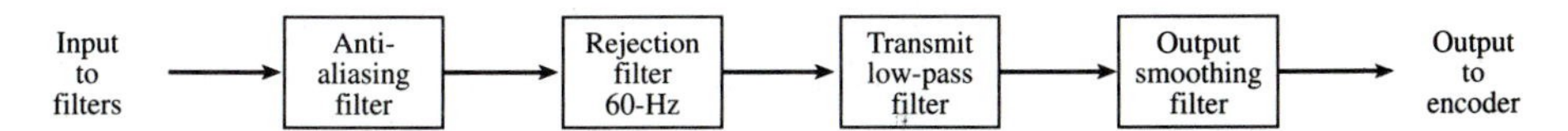

FIGURE 1.5-3
Components of the transmitting section of a PCM filter.

Modern technology makes it possible to integrate the entire collection of transmitting and receiving filters on a single monolithic substrate that is encapsulated in a single 16-pin dual in-line package (DIP). An example is the Intel 2912 PCM filter.

PROBLEMS

Section 1.1

1. Derive the voltage transfer function given in Example 1.1-1 for the network shown in Fig. 1.1-1.
2. For the network shown in Fig. 1.1-2(*a*), let the excitation $e(t) = v_1(t) = \cos t$. Find the sinusoidal steady-state response $r(t) = v_2(t)$.

Section 1.2

1. (*a*) Find the voltage transfer functions for each of the networks shown in Fig. P1.2-1.
   (*b*) Find expressions for the magnitude and phase of the voltage transfer functions of each of the networks.
   (*c*) Show that the pole locations of each of the network functions agree with the cases presented in Table 1.2-1.

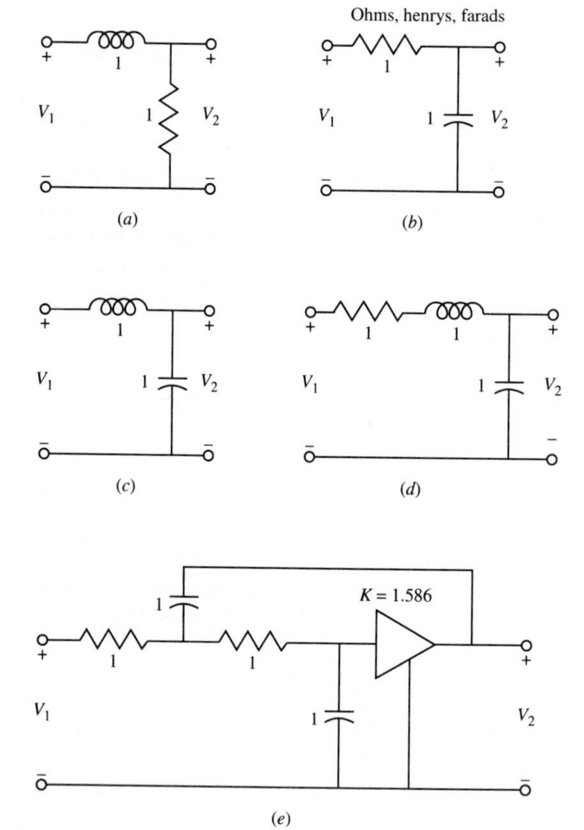

FIGURE P1.2-1

2. Find the pole and zero locations for each of the following network functions:

$$N_a(s) = \frac{H}{s^3 + 2s^2 + 2s + 1} \qquad N_b(s) = \frac{Hs}{s^3 + 2s^2 + 2s + 1}$$

$$N_c(s) = \frac{Hs^2}{s^3 + 2s^2 + 2s + 1} \qquad N_d(s) = \frac{Hs^3}{s^3 + 2s^2 + 2s + 1}$$

3. (*a*) Draw a phase characteristic for the network function $N_a(s)$ given in Example 1.2-2.
   (*b*) Repeat part (*a*) for the network function $N_b(s)$.
4. Draw a phase characteristic for the network function $N(s)$ given in Example 1.2-3.
5. Show that the driving-point admittance at port 1 [defined at the terminal pair where $V_1(s)$ is defined] for the network shown in Fig. 1.1-1 has poles that are the same as those found for the voltage transfer function in Example 1.1-1.

Section 1.3

1. Derive the voltage transfer function given in Example 1.3-1 for the network shown in Fig. 1.3-2(*b*).
2. Derive the voltage transfer function given in Example 1.3-2 for the network shown in Fig. 1.3-3.
3. Derive the voltage transfer function given in Example 1.3-3 for the network shown in Fig. 1.3-5.
4. Derive the voltage transfer function given in Example 1.3-4 for the network shown in Fig. 1.3-7.
5. Derive the voltage transfer function given in Example 1.3-5 for the network shown in Fig. 1.3-9.
6. Magnitude characteristics of filter functions are sometimes illustrated using units of attenuation rather than gain for the plot ordinate scale. Assuming such a convention, identify the ideal magnitude plots shown in Fig. P1.3-6 as low pass, high pass, band pass, or band reject.

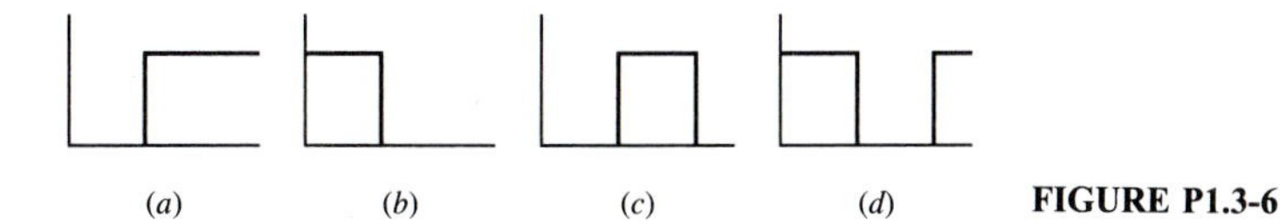

FIGURE P1.3-6

Section 1.4

1. (*a*) The low-pass network shown in Fig. 1.1-1 has a bandwidth of 1 rad/s for the specified element values. Find the element values that will result when a frequency denormalization is made such that the bandwidth is changed to 1 kHz.
   (*b*) Find the voltage transfer function for the frequency-denormalized network found in part (*a*).
   (*c*) Make an impedance denormalization on the network found in part (*a*) such that the terminating resistor has a value of 10 kΩ
   (*d*) Find the voltage transfer function for the impedance- and frequency-denormalized network found in part (*c*).

2. (*a*) For the frequency denormalization and network defined in part (*a*) of Prob. 1, find the driving-point admittance at the input (left) port of the network. Note that the driving-point admittance for the normalized network was found in Prob. 5 of Sec. 1.2.

(*b*) Find the driving-point admittance at the input (left) port for the frequency- and impedance-denormalized network defined in part (*c*) of Prob. 1.

3. (*a*) The high-pass network shown in Fig. 1.3-5 has a cutoff frequency of 1 rad/s for the specified element values. Find the element values that will result when a frequency denormalization is made such that the cutoff frequency is changed to 1 kHz.

(*b*) Find the voltage transfer function for the frequency-denormalized network found in part (*a*).

(*c*) Make an impedance denormalization on the network found in part (*a*) such that the terminating resistor has a value of 10 kΩ.

(*d*) Find the voltage transfer function for the impedance- and frequency-denormalized network found in part (*c*).

4. (*a*) The band-pass network shown in Fig. P1.4-4 has a center frequency of 1 rad/s and a bandwidth of 0.1 rad/s for the specified element values. Find the element values that will result when a frequency denormalization is made such that the center frequency is changed to 1 kHz.

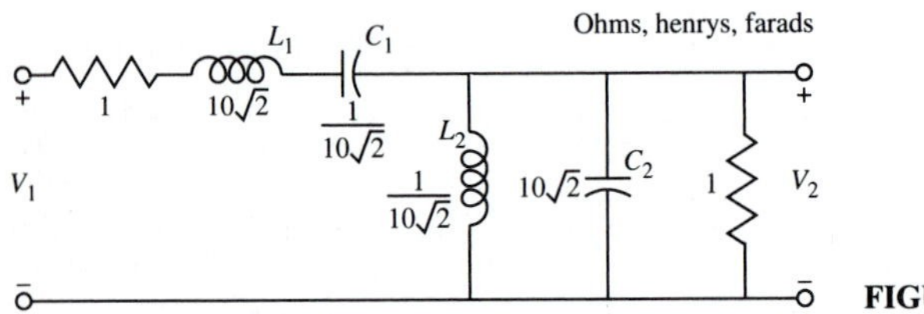

FIGURE P1.4-4

(*b*) What is the bandwidth of the frequency denormalized network?

(*c*) Make an impedance denormalization on the network found in part (*a*) such that the terminating resistor has a value of 1 kΩ.

CHAPTER 2

APPROXIMATION

The majority of practical filter specifications are based on sinusoidal steady-state performance requirements. These are usually given for magnitude and/or phase characteristics as a function of a real frequency variable ω (radians per second) or f (hertz). The actual synthesis techniques used to find active or passive filter realizations, however, invariably use the network function, a ratio of polynomials in the complex frequency variable s, as a starting point. The process of relating sinusoidal steady-state characteristics to a network function is called *approximation*. Figure 2.1 illustrates the difference between synthesis and approximation.

2.1 MAGNITUDE APPROXIMATION—THE MAXIMALLY FLAT CHARACTERISTIC

One of the most frequently used types of approximation is that relating the magnitude $|N(j\omega)|$, specified either by a mathematical expression, a set of data values, or a plotted waveshape, to a rational function $F(s)$, so that in some specified sense $|F(j\omega)|$ approximates $|N(j\omega)|$. Ideally, of course, we would like the two magnitude functions to be identical, and in many cases we shall find that this is true. Magnitude specifications are usually given either in decimal (linear) or logarithmic measure. In the latter case decibels [$20 \log |N(j\omega)|$], abbreviated dB, are used. The left two columns of Table 2.1-1 give some examples of linear and logarithmic magnitude values.

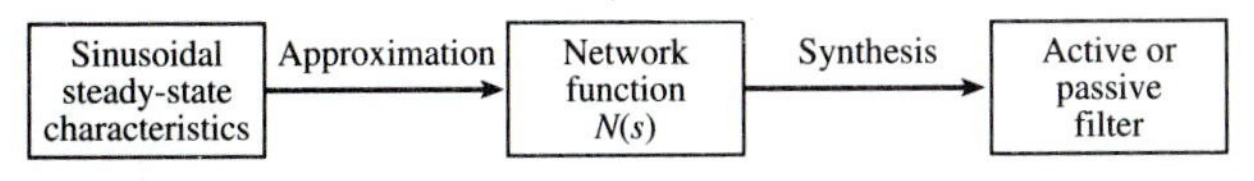

FIGURE 2.1
Difference between approximation and synthesis.

TABLE 2.1-1
Relations for $|N(j\omega)|$ and ε†

$\|N(j\omega)\|$			
Decimal	**dB**	ε^2	ε
0.999000	−0.008690	0.00200	0.04475
0.990000	−0.087296	0.02030	0.14249
0.950000	−0.445528	0.10803	0.32868
0.900000	−0.915150	0.23457	0.48432
0.707107	−3.010300	1.00000	1.00000
0.500000	−6.020600	3.00000	1.73205
0.200000	−13.979400	24.00000	4.89898
0.100000	−20.000000	99.00000	9.94987
0.050000	−26.020600	399.00000	19.97498
0.010000	−40.000000	9999.00000	99.99500

dB	**Decimal**	ε^2	ε
−0.01000	0.998849	0.00231	0.04801
−0.10000	0.988553	0.02329	0.15262
−0.50000	0.944061	0.12202	0.34931
−1.00000	0.891251	0.25893	0.50885
−2.00000	0.794328	0.58489	0.76478
−3.01030	0.707107	1.00000	1.00000
−5.00000	0.562341	2.16228	1.47047
−10.00000	0.316228	9.00000	3.00000
−20.00000	0.100000	99.00000	9.94987
−40.00000	0.010000	9999.00000	99.99500

† The quantity ε is used in Sec. 2.2.

Necessary Properties of $|N(j\omega)|^2$

We begin our study of magnitude approximation by considering the necessary properties that a magnitude function must have. It is more convenient to actually consider the square of the magnitude function. Thus, we may write

$$|N(j\omega)|^2 = N(j\omega)N^*(j\omega) = N(j\omega)N(-j\omega) \tag{1}$$

where the superscript asterisk indicates the complex conjugate and where the justification for the right member of the equation is that, for rational functions with real coefficients, the conjugate of the function is found by replacing the *variable* by its conjugate, that is, by replacing $j\omega$ with $-j\omega$. Now let $N(s)$ have the form

$$N(s) = \frac{c_0 + c_1 s + c_2 s^2 + c_3 s^3 + c_4 s^4 + \cdots}{d_0 + d_1 s + d_2 s^2 + d_3 s^3 + d_4 s^4 + \cdots} \tag{2}$$

The term $N(j\omega)$ will have the form

$$N(j\omega) = \frac{c_0 - c_2\omega^2 + c_4\omega^4 - \cdots + j(c_1\omega - c_3\omega^3 + \cdots)}{d_0 - d_2\omega^2 + d_4\omega^4 - \cdots + j(d_1\omega - d_3\omega^3 + \cdots)} \tag{3}$$

Inserting this relation in the right member of (1), we readily see a first property of $|N(j\omega)|^2$, that it will be a ratio of even polynomials.

Example 2.1-1 $|N(j\omega)|^2$ as a ratio of even polynomials. For the case where $N(s)$ has the form

$$N(s) = \frac{3(s+1)}{s^2 + 0.1s + 1}$$

if we substitute $j\omega$ for s, we obtain

$$N(j\omega) = \frac{3(j\omega + 1)}{1 - \omega^2 + 0.1j\omega}$$

From (1) we find

$$|N(j\omega)|^2 = \frac{3(j\omega+1)}{1-\omega^2+0.1j\omega}\,\frac{3(-j\omega+1)}{1-\omega^2-0.1j\omega} = \frac{9(1+\omega^2)}{1-1.99\omega^2+\omega^4}$$

This is a ratio of polynomials with only even powers of the variable ω.

If we now evaluate (1) by letting $\omega = s/j$, we may define a function $T(s^2)$ as

$$T(s^2) = |N(j\omega)|^2 \Big|_{\omega=s/j} = N(s)N(-s) \tag{4}$$

In the right member of (4), the substitution of $-s$ for s as the argument of $N(s)$ simply reflects the pole and zero positions of $N(s)$ through the origin of the s plane. As a result, the product $N(s)N(-s)$ has poles and zeros that are symmetrically located with respect to both the real and the imaginary axes. We call this *quadrantal symmetry.* In general, the numerator and denominator polynomials of $T(s^2)$ can have three types of factors:

1. $s^4 + as^2 + b$ where a and b may be positive or negative
2. $as^2 + b$ where a and b have opposite signs
3. $as^2 + b$ where a and b have the same signs

The first and second types have the necessary quadrantal symmetry but the third type does not unless it has even multiplicity, that is, unless it appears as $(as^2+b)^2$, $(as^2+b)^4$, and so on. In this case the resulting even-order $j\omega$-axis zeros have the necessary symmetry.

Example 2.1-2 Factors of numerator and denominator polynomials of $T(s^2)$. The numerator and denominator polynomials of $T(s^2)$ may be expressed in the form $P(s^2) = A(s)A(-s)$, where $P(\cdot)$ and $A(\cdot)$ are polynomials with real coefficients. For example (type 1), consider $P(s^2) = s^4 + \sqrt{2}s^2 + 1$, which has roots at

$$s^2 = -0.7071 \pm j0.7071 = 1e^{j135^\circ},\ 1e^{-j135^\circ},\ 1e^{j(135+360^\circ)},\ 1e^{-j(135+360^\circ)},$$

If we take the square root of the various elements in the preceding equation, we obtain

$$s = \sqrt{-0.7071 \pm j0.7071} = 1e^{j67.5^\circ},\ 1e^{-j67.5^\circ},\ 1e^{j247.5^\circ},\ 1e^{-j247.5^\circ}$$
$$= \pm 0.3827 \pm j0.9239$$

If we assign the left-half-plane roots to $A(s)$ and the right-half-plane ones to $A(-s)$, we obtain

$$P(s^2) = s^4 + \sqrt{2}s^2 + 1 = (s^2 + 0.7654s + 1)(s^2 - 0.7654s + 1) = A(s)A(-s)$$

As an example of the second type of decomposition,

$$P(s^2) = -s^2 + 1 = (s + 1)(-s + 1) = A(s)A(-s)$$

As an example of the third type,

$$P(s^2) = (s^2 + 1)^2 = (s^2 + 1)(s^2 + 1) = A(s)A(-s)$$

The results given above may be summarized as follows.

Summary 2.1-1 Properties of $|N(j\omega)|^2$. For a given $|N(j\omega)|^2$ to be the magnitude-squared function of some rational function $N(s)$, it is necessary and sufficient that:

1. The function $|N(j\omega)|^2$ be a ratio of even polynomials in ω
2. In the function $T(s^2)$ defined in (4), any poles or zeros on the $j\omega$ axis be of even order.

The sufficiency of the two conditions given in Summary 2.1-1 is readily demonstrated by factoring $T(s^2)$ into the product $N(s)N(-s)$, taking the left-half-plane poles and half of any even-order $j\omega$-axis pole pairs from $T(s^2)$ as the poles of $N(s)$ and similarly assigning either right- or left-half-plane zeros and half of any even-order $j\omega$-axis zeros from $T(s^2)$ as the zeros of $N(s)$. The restriction of using only the *left-half-plane* poles from $T(s^2)$ is of course simply a stability consideration. An example follows.

Example 2.1-3 Finding a network function from its magnitude specification. As an example of the sufficiency argument given above, consider the function

$$|N(j\omega)|^2 = \frac{\omega^2 + 1}{\omega^4 + 1} \tag{5}$$

The related function $T(s^2)$ is easily seen to be

$$T(s^2) = \frac{-s^2 + 1}{s^4 + 1} = \frac{(s + 1)(-s + 1)}{(s^2 + \sqrt{2}s + 1)(s^2 - \sqrt{2}s + 1)} \tag{6}$$

Comparing this with (4), we see that there are two network functions $N(s)$ that satisfy (5). These are

$$N_1(s) = \frac{s + 1}{s^2 + \sqrt{2}s + 1} \qquad \text{and} \qquad N_2(s) = \frac{s - 1}{s^2 + \sqrt{2}s + 1} \tag{7}$$

where, in the expression for $N_2(s)$, we have reversed the signs in the numerator, since such an operation has no effect on $|N(j\omega)|^2$.

Maximally Flat Magnitude Function

The necessary and sufficient conditions developed above for magnitude-squared functions in general are readily applied to specific filter characteristics. As an example, consider the determination of a magnitude-squared function that, in the low-frequency range starting at zero, has as flat a characteristic as possible. One way of obtaining such a flatness is to set as many derivatives of the function as possible to zero at $\omega =$ 0 rad/s. Such a function is called *maximally flat*. To see how this can be done, we may write an expression for a general magnitude-squared function $|N(j\omega)|^2$ as follows:

$$|N(j\omega)|^2 = H^2 \frac{1 + b_1\omega^2 + b_2\omega^4 + \cdots}{1 + a_1\omega^2 + a_2\omega^4 + \cdots} \tag{8}$$

If we now divide the denominator into the numerator, we obtain

$$|N(j\omega)|^2 = H^2[1 + (b_1 - a_1)\omega^2 + (b_2 - a_2 + a_1^2 - a_1b_1)\omega^4 + \cdots] \tag{9}$$

Now consider a general MacLaurin series, i.e., a Taylor series expansion at the origin, of an arbitrary function $F(\omega)$. This has the form

$$F(\omega) = F(0) + \frac{F^{(1)}(0)}{1!}\omega + \frac{F^{(2)}(0)}{2!}\omega^2 + \frac{F^{(3)}(0)}{3!}\omega^3 + \frac{F^{(4)}(0)}{4!}\omega^4 \cdots \tag{10}$$

where $F^{(i)}(0)$ is the ith derivative of $F(\omega)$ evaluated at $\omega = 0$. Comparing this expression with the expansion for $|N(j\omega)|^2$ given in (9) and recalling that such an expansion must be unique, we see that due to the even nature of $|N(j\omega)|^2$, all its odd-ordered derivatives are already zero. In addition, for the second derivative to be zero, we require that the coefficients a_1 and b_1 be equal. Similarly, for the fourth derivative to also be zero requires that, in addition, a_2 equal b_2, etc. Thus, the general maximally flat magnitude-squared function $|N(j\omega)|^2$ in (8) is characterized by the restriction that

$$a_i = b_i \tag{11}$$

for as many coefficients as possible.

Example 2.1-4 Function with a maximally flat magnitude. Consider the function

$$N(s) = 3\frac{2s + 1}{s^2 + ks + 1}$$

To determine the value of the coefficient k required to make the magnitude (at the origin) maximally flat, we determine

$$|N(j\omega)|^2 = \frac{|3(j2\omega + 1)|^2}{|1 - \omega^2 + jk\omega|^2} = 9\frac{1 + 4\omega^2}{1 + (k^2 - 2)\omega^2 + \omega^4}$$

Equating the numerator and denominator coefficients of ω^2, we obtain $4 = k^2 - 2$ or $k = \sqrt{6}$.

Butterworth Function

As an example of the application of the maximally flat magnitude criterion, consider the magnitude function for a *low-pass network.* The plot of such a function will show a flat characteristic at low frequencies and a drop-off to some low value at high frequencies. Thus, ideally we might conceive that it would appear as shown in Fig. 2.1-1. The characteristic shown in the figure, however, is not realizable.[1] As a more practical approach to finding a low-pass function, let us try to approximate it by choosing a magnitude-squared function $|N(j\omega)|^2$ that satisfies the maximally flat criteria at $\omega = 0$. This should generate the desired flatness of the curve, at least for low frequencies. Next, to provide the eventual drop-off of the characteristic at higher frequencies, we will locate all the transmission zeros of the function at infinity; thus, the numerator of $N(j\omega)$ will simply be a constant and all the coefficients b_i of (8) will be zero. For a maximally flat characteristic from (11), the coefficients a_i must also be set to zero, except of course for the highest order one. The resulting magnitude-squared function has the form

$$|N(j\omega)|^2 = \frac{H^2}{1 + \omega^{2n}} \tag{12}$$

where we have chosen the coefficient multiplying the ω^{2n} term as unity in order to provide a frequency normalization. This function is called a *Butterworth function.*[2]

Summary 2.1-2 Properties of Butterworth functions. A low-pass Butterworth function having the form given in (12) has the following properties:

1. The range of frequencies $0 \le \omega \le 1$ rad/s is called the *passband.*
2. The range of frequencies $\omega \ge 1$ rad/s is called the *stopband.*
3. At $\omega = 1$ rad/s, $|N(j\omega)| = H/\sqrt{1 + 1^{2n}} = H/\sqrt{2} = 0.7071H$, independent of the value of n.
4. At $\omega = 1$ rad/s, the slope of $|N(j\omega)|^2$ is proportional to $-\frac{1}{2}n$.
5. The function $|N(j\omega)|$ is a *monotonic* (continually decreasing) function of ω.

The function defined in (12) is usually referred to as a *normalized* Butterworth function. Since $20 \log [|N(j1)|/|N(j0)|] = 20 \log 0.70711 = -3.0103$ dB, the frequency of 1 rad/s is usually referred to as the *−3-dB frequency* or the *3-dB down frequency.*

---

[1] A general necessary and sufficient condition for realizability is the Paley-Wiener criterion, which requires

$$\int_{-\infty}^{\infty} \left| \frac{\ln |N(j\omega)|}{1 + \omega^2} \right| d\omega < \infty$$

where $|N(j\omega)|$ is the magnitude characteristic being tested.

[2] S. Butterworth was a British engineer who described this type of response in connection with electronic amplifiers in his paper "On the Theory of Filter Amplifiers," *Wireless Eng.*, vol. 7, 1930, pp. 536–541. Over a decade later, V. D. Landon applied the phrase *maximally flat* in his paper "Cascade Amplifiers with Maximal Flatness," *RCA Rev.*, vol. 5, 1941, pp. 347–362.

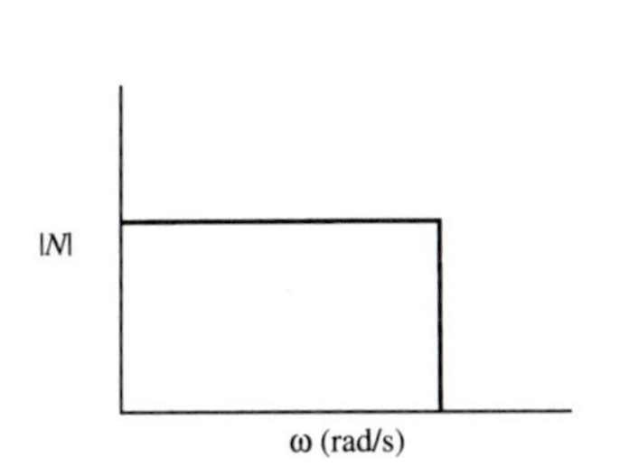

FIGURE 2.1-1
An ideal low-pass magnitude characteristic.

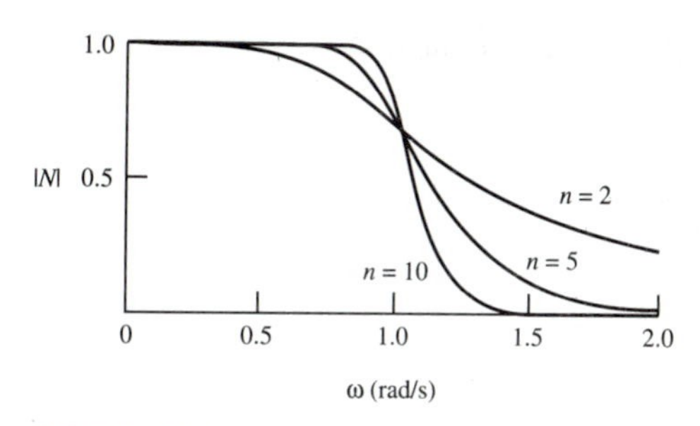

FIGURE 2.1-2
The magnitude of Butterworth functions of various orders.

Plots of the magnitude of the Butterworth function for various values of n are shown in Fig. 2.1-2. Numerical values of Butterworth functions are given in decimal and decibel form in Table 2.1-2. For sufficiently large values of frequency the attenuation is $20n$ dB per decade of frequency beyond $\omega = 1$ rad/s, where n is the degree of the function.

Pole Locations

The locations of the poles of a network function $N(s)$ that has a Butterworth magnitude characteristic may be found using (4) and (12). Thus we obtain

$$N(s)N(-s) = \left.\frac{H^2}{1+\omega^{2n}}\right|_{\omega^2=-s^2} = \frac{H^2}{1+(-1)^n s^{2n}} \tag{13}$$

Setting the denominator polynomial of (13) to zero, we find that the poles are located at the values of s that satisfy the relation

$$s = [-(-1)^n]^{1/2n} \tag{14}$$

TABLE 2.1-2
Values of the magnitude of maximally flat magnitude (Butterworth) functions with normalized ($\omega = 1$ rad/s) passband

	ω = 0.7 rad/s		ω = 0.8 rad/s		ω = 0.9 rad/s		ω = 1.1 rad/s		ω = 1.2 rad/s		ω = 1.5 rad/s	
n	**Mag**	**dB**	**Mag**	**dB**	**Mag**	**dB**	**Mag**	**dB**	**Mag**	**dB**	**Mag**	**dB**
2	0.8980	−0.93	0.8423	−1.49	0.7771	−2.19	0.6370	−3.92	0.5704	−4.88	0.4061	−7.83
3	0.9459	−0.48	0.8901	−1.01	0.8081	−1.85	0.6007	−4.43	0.5009	−6.01	0.2841	−10.93
4	0.9724	−0.24	0.9254	−0.67	0.8361	−1.55	0.5640	−4.97	0.4344	−7.24	0.1938	−14.25
5	0.9862	−0.12	0.9503	−0.44	0.8611	−1.30	0.5275	−5.56	0.3729	−8.57	0.1306	−17.68
6	0.9932	−0.06	0.9673	−0.29	0.8830	−1.08	0.4916	−6.17	0.3176	−9.96	0.0875	−21.16
7	0.9966	−0.03	0.9787	−0.19	0.9021	−0.89	0.4566	−6.81	0.2688	−11.41	0.0584	−24.67
8	0.9983	−0.01	0.9862	−0.12	0.9185	−0.74	0.4228	−7.48	0.2265	−12.90	0.0390	−28.18
9	0.9992	−0.01	0.9911	−0.08	0.9325	−0.61	0.3904	−8.17	0.1903	−14.41	0.0260	−31.70
10	0.9996	−0.00	0.9943	−0.05	0.9442	−0.50	0.3597	−8.88	0.1594	−15.95	0.0173	−35.22

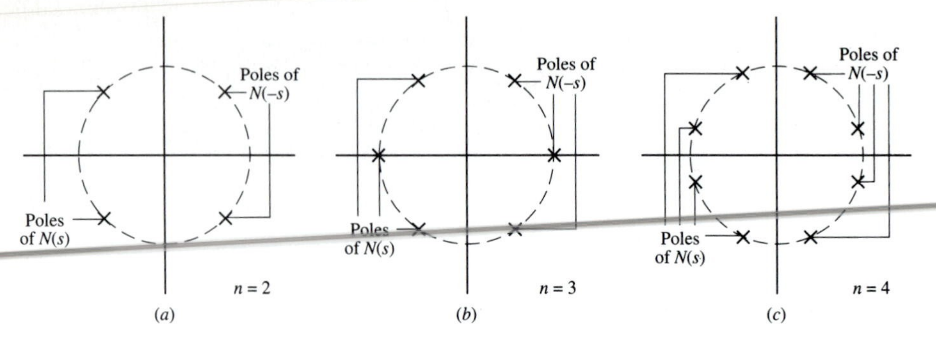

FIGURE 2.1-3
Poles of $N(s)N(-s)$ of (13).

Thus, for n even, $s=(-1)^{1/2n}=e^{j\pi k/2n}$ ($k=1, 3, 5, \ldots, 4n-1$), and for n odd, $s=(1)^{1/2n}=e^{j\pi k/2n}$ ($k=0, 2, 4, \ldots, 4n-2$). From these relations we see that the poles of $N(s)N(-s)$ are equiangularly spaced around the unit circle, as shown in Fig. 2.1-3. Retaining only the left-half-plane singularities, we find that the poles of $N(s)$ are given as $p_k=\sigma_k+j\omega_k$, where

$$\sigma_k=-\sin\frac{2k-1}{2n}\pi \qquad \omega_k=\cos\frac{2k-1}{2n}\pi \qquad k=1, 2, 3, \ldots, n \tag{15}$$

The denominator polynomials characterized by these roots are called *Butterworth polynomials*. The values of the polynomial coefficients for a polynomial $P(s)$, where

$$P(s)=a_0+a_1s+a_2s^2+\cdots+a_ns^n \tag{16}$$

are given by noting that, since all the poles are located on the unit circle, $a_0=1$. The other coefficients are determined by the iterative relation

$$a_k=\frac{\cos\,[(k-1)\pi/2n]}{\sin\,(k\pi/2n)}a_{k-1} \qquad k=1, 2, \ldots, n \tag{17}$$

Values of the pole locations, quadratic factors, Q's,[3] and coefficients for these polynomials are given in Table 2.1-3. Note that the coefficients are symmetric, so that

$$\begin{aligned} a_0 &= a_n = 1 \\ a_1 &= a_{n-1} \\ a_2 &= a_{n-2} \\ &\vdots \end{aligned}$$

Passive filter realizations of Butterworth functions are given in App. A.

---

[3] For a quadratic factor of the form s^2+a_1s+1, the quality factor $Q=1/a_1$. The use of Q will be covered in more detail in Chap. 5.

TABLE 2.1-3*a*

Denominator coefficients of maximally flat magnitude (Butterworth) functions of the form $s^n + a_1 s^{n-1} + a_2 s^{n-2} + \cdots + a_2 s^2 + a_1 s + 1$ with passband 0 to 1 rad/s

n	a_1	a_2	a_3	a_4	a_5
2	1.414214				
3	2.000000				
4	2.613126	3.414214			
5	3.236068	5.236068			
6	3.863703	7.464102	9.141620		
7	4.493959	10.097835	14.591794		
8	5.125831	13.137071	21.846151	25.688356	
9	5.758770	16.581719	31.163437	41.986386	
10	6.392453	20.431729	42.802061	64.882396	74.233429

TABLE 2.1-3*b*

Pole locations, quadratic factors ($s^2 + a_1 s + 1$), and Q's† of maximally flat magnitude (Butterworth) functions with passband 0 to 1 rad/s

n	Poles	a_1	Q
2	$-0.70711 \pm j0.70711$	1.41421	0.70711
3	$-0.50000 \pm j0.86603$	1.00000	1.00000
4	$-0.38268 \pm j0.92388$	0.76537	1.30656
	$-0.92388 \pm j0.38268$	1.84776	0.54120
5	$-0.30902 \pm j0.95106$	0.61803	1.61803
	$-0.80902 \pm j0.58779$	1.61803	0.61803
6	$-0.25882 \pm j0.96593$	0.51764	1.93185
	$-0.70711 \pm j0.70711$	1.41421	0.70711
	$-0.96593 \pm j0.25882$	1.93185	0.51764
7	$-0.22252 \pm j0.97493$	0.44504	2.24698
	$-0.62349 \pm j0.78183$	1.24698	0.80194
	$-0.90097 \pm j0.43388$	1.80194	0.55496
8	$-0.19509 \pm j0.98079$	0.39018	2.56292
	$-0.55557 \pm j0.83147$	1.11114	0.89998
	$-0.83147 \pm j0.55557$	1.66294	0.60134
	$-0.98079 \pm j0.19509$	1.96157	0.50980
9	$-0.17365 \pm j0.98481$	0.34730	2.87939
	$-0.50000 \pm j0.86603$	1.00000	1.00000
	$-0.76604 \pm j0.64279$	1.53209	0.65270
	$-0.93969 \pm j0.34202$	1.87939	0.53209
10	$-0.15643 \pm j0.98769$	0.31287	3.19623
	$-0.45399 \pm j0.89101$	0.90798	1.10134
	$-0.70711 \pm j0.70711$	1.41421	0.70711
	$-0.89101 \pm j0.45399$	1.78201	0.56116
	$-0.98769 \pm j0.15643$	1.97538	0.50623

† All odd-order functions also have a pole at $s = -1$.

Example 2.1-5 Pole locations of a Butterworth function. For a second-order ($n = 2$) Butterworth function, we may use (15) to find the following pole locations:

$$\text{For } k = 1 \qquad \sigma_1 = -\sin\frac{\pi}{4} = -0.7071 \qquad \omega_1 = \cos\frac{\pi}{4} = 0.7071$$

$$\text{For } k = 2 \qquad \sigma_2 = -\sin\frac{3\pi}{4} = -0.7071 \qquad \omega_2 = \cos\frac{3\pi}{4} = -0.7071$$

The network function is

$$N(s) = \frac{H}{(s + 0.7071 - j0.7071)(s + 0.7071 + j0.7071)} = \frac{H}{s^2 + \sqrt{2}s + 1}$$

The first-degree coefficient in the denominator can also be found directly from (17):

$$a_1 = \frac{\cos(0\pi/4)}{\sin(\pi/4)}1 = \frac{1}{1/\sqrt{2}} = \sqrt{2}$$

Exercise 2.1-5. Use (15) and (17) to find the pole locations and the network function for:

1. A third-order Butterworth function
2. A fourth-order Butterworth function
3. A fifth-order Butterworth function

Answers. The answers may be found in Table 2.1-3.

Determination of Order

A fundamental problem in the design of filters is the determination of the order of filter function required to meet a set of filtering specifications. The specifications usually consist of a passband set and a stopband set. They have the following form:

1. *Passband.* In the passband $0 \le \omega \le \omega_p$ radians per second ($0 \le f \le f_p$ hertz), the maximum deviation of the magnitude characteristic is K_p dB.
2. *Stopband.* In the stopband $\omega \ge \omega_s$ radians per second ($f \ge f_s$ hertz), the minimum attenuation of the magnitude characteristic is K_s dB. This attenuation is measured from the greatest passband value of the characteristic.

A plot illustrating the quantities defined above is shown in Fig. 2.1-4. To meet the specifications, the actual magnitude characteristic must lie within the unshaded area. For a Butterworth characteristic, the monotonicity property ensures that if the stopband specification K_s is met at the stopband frequency ω_s, it will be satisfied for all $\omega \ge \omega_s$. Note that K_p and K_s are specified as positive constants.

To determine the order of function required to meet a given set of specifications, we define

$$\Omega = \frac{\omega_s}{\omega_p} = \frac{f_s}{f_p} \qquad M = \sqrt{\frac{10^{0.1K_s} - 1}{10^{0.1K_p} - 1}} \tag{18}$$

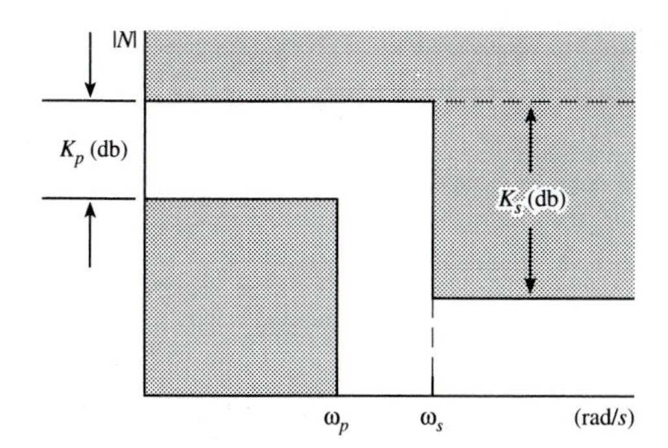

FIGURE 2.1-4
Low-pass filter specifications.

For the Butterworth case the required order n_B is given as

$$n_B = \frac{\log M}{\log \Omega} = \frac{\ln M}{\ln \Omega} \tag{19}$$

The derivation of (18) and (19) is left as an exercise for the reader. The required order of the function is the next highest integer greater than n_B.

Example 2.1-6 Order of a Butterworth function. A Butterworth function must meet the following specifications:

$$f_p = 1000 \text{ Hz} \qquad K_p = 3.0103 \text{ dB}$$
$$f_s = 1300 \text{ Hz} \qquad K_s = 22 \text{ dB}$$

From (18) and (19) we find

$$\Omega = \frac{1300}{1000} = 1.3 \qquad M = \sqrt{\frac{10^{2.2} - 1}{10^{0.30103} - 1}} = 12.5495$$

$$n_B = \frac{\ln 12.5495}{\ln 1.3} = \frac{2.5297}{0.2624} = 9.6419$$

A 10th-order function will meet the specifications.

Exercise 2.1-6. For each of the following sets of specifications for a Butterworth function, determine the required order.

Passband		Stopband		Answers
ω_p (or f_p)	K_p, dB	ω_s (or f_s)	K_s, dB	n_B
1	3.0103	2	25	4.1501
1	3.0103	1.5	25	7.0947
1	3.0103	2	15	2.4683
1	3.0103	1.5	15	4.2195
1	1	2	25	5.1248
1	1	1.5	25	8.7609
1000	1	1100	25	37.2705

Frequency Denormalization

The normalized Butterworth function defined in (12) has a passband specified by $\omega_p = 1$ rad/s and $K_p = 3.0103$ dB. The pole locations and polynomials given in Table 2.1-3 for these specifications may readily be converted to other values of K_p by making a frequency denormalization. To do this, we solve (12) for the value of frequency ω_{KP} at which the function has the (desired) K_p attenuation. Thus we find

$$20 \log \frac{|N(j0)|}{|N(j\omega_{KP})|} = 20 \log\sqrt{1 + \omega_{KP}^{2n}} = K_p \tag{20}$$

This may be solved to obtain

$$\omega_{KP} = (10^{0.1K_p} - 1)^{1/2n} \tag{21}$$

The frequency-denormalization constant is $1/\omega_{KP}$.

Example 2.1-7 Frequency denormalized Butterworth function. To find the pole locations for a third-order Butterworth function that has 2 dB attenuation at 1 rad/s, from (21) we find

$$\omega_{2\text{dB}} = (10^{0.2} - 1)^{1/6} = 0.9145 \text{ rad/s}$$

The frequency denormalization constant is $\Omega_n = 1/0.9145 = 1.0935$. Using Table 2.1-3, the pole locations are

$$-1.0935 \qquad -0.5468 \pm j0.9470$$

The resulting denormalized Butterworth function is

$$N(s) = \frac{H}{s^3 + 2.1870s^2 + 2.3915s + 1.3076}$$

Note that the coefficients are no longer symmetric.

Exercise 2.1-7. For each of the sets of values for n, ω_p, and K_p given below, find the pole locations of the denormalized Butterworth function.

			Answers	
n	ω_p, rad/s	K_p, dB	ω_{KP}	Poles
3	1	1	0.79835	-1.253, $-0.626 \pm j1.085$
4	1	1	0.84459	$-0.453 \pm j1.094$, $-1.094 \pm j0.453$
4	1	2	0.93516	$-0.409 \pm j0.988$, $-0.988 \pm j0.409$
5	1	1	0.87361	-1.145, $-0.354 \pm j1.089$, $-0.926 \pm j0.673$
5	1	2	0.94778	-1.055, $-0.326 \pm j1.003$, $-0.854 \pm j0.620$

2.2 MAGNITUDE APPROXIMATION—THE EQUAL-RIPPLE CHARACTERISTIC

In the preceding section we introduced one kind of magnitude approximation, namely, the maximally flat type. This particular approximation was shown to be characterized by the fact that the derivatives of the magnitude-squared function are set to zero at zero frequency. Thus, the approximating effect is concentrated at a single frequency, i.e., zero. One result of this is that the transition from passband to stopband is not as sharp as is needed for many applications. In this section we describe a different type of approximation, one in which the approximating effect is spread over the entire passband. Such an approximation is said to have an *equal-ripple* characteristic.

Chebyshev Polynomials

The normalized low-pass equal-ripple magnitude approximation may be developed by writing the magnitude-squared function $|N(j\omega)|^2$ in the form

$$|N(j\omega)|^2 = \frac{H^2}{1 + \varepsilon^2 C_n^2(\omega)} \tag{1}$$

where $C_n(\omega)$ is a polynomial of order n. If these polynomials have the properties

$$\begin{aligned} 0 \le C_n^2(\omega) \le 1 \qquad & \text{for } 0 \le \omega \le 1 \\ C_n^2(\omega) \ge 1 \qquad & \text{for } \omega \ge 1 \end{aligned} \tag{2}$$

then the permitted values for $C_n^2(\omega)$ and $|N(j\omega)|$ lie in the unshaded areas shown in Fig. 2.2-1. The value of ε determines the limits of variation in the passband $0 \le \omega \le 1$ rad/s. For example, for $\varepsilon = 1$, the range of passband variation is 3.0103 dB. Other pairs of values for $|N(j\omega)|$ and ε are given in Table 2.1-1. The stopband is defined for $\omega \ge 1$ rad/s.

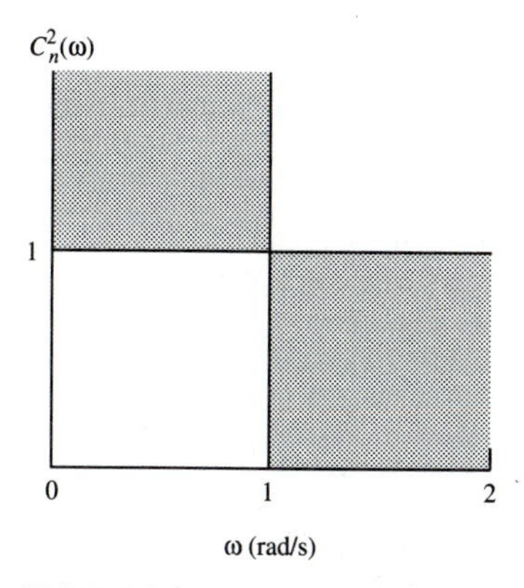

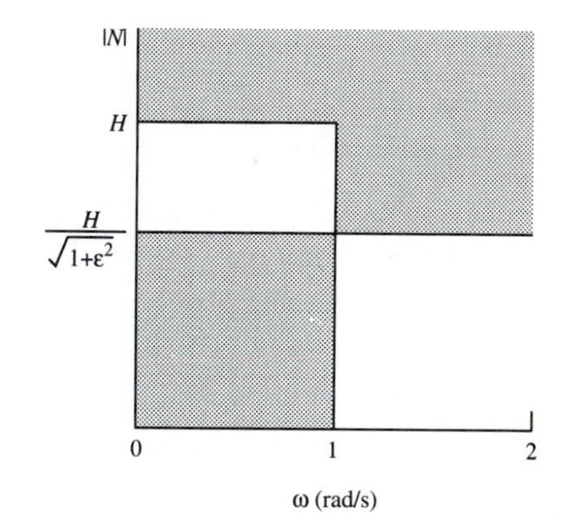

FIGURE 2.2-1
Permitted areas for $C_n^2(\omega)$ and $|N(j\omega)|$ in (1).

A set of polynomials $C_n(\omega)$ that have the properties specified above are the *Chebyshev polynomials*.[1] These are defined as follows:

$$\begin{aligned} C_1(\omega) &= \omega \\ C_2(\omega) &= 2\omega^2 - 1 \\ C_3(\omega) &= 4\omega^3 - 3\omega \\ &\vdots \\ C_{n+1}(\omega) &= 2\omega C_n(\omega) - C_{n-1}(\omega) \end{aligned} \tag{3}$$

where the last expression is valid for all $n > 1$. The Chebyshev polynomials may also be written using the expressions

$$C_n(\omega) = \cos\,(n \cos^{-1} \omega) \qquad 0 \le \omega \le 1 \tag{4a}$$

$$C_n(\omega) = \cosh\,(n \cosh^{-1} \omega) \qquad \omega \ge 1 \tag{4b}$$

$$C_n(\omega) = \tfrac{1}{2}[(\omega + \sqrt{\omega^2 - 1})^n + (\omega + \sqrt{\omega^2 - 1})^{-n}] \qquad \omega \ge 1 \tag{4c}$$

where the expression in (4*c*) provides an alternative to the cosh function in (4*b*). Plots of the values of some of the polynomials $C_n^2(\omega)$ are shown in Fig. 2.2-2. It is readily apparent that they have the desired properties.

Example 2.2-1 Values of Chebyshev polynomials. To find the values of a fifth-order Chebyshev polynomial at $\omega = 0.7$ and $\omega = 1.2$ rad/s, we may use (3) to determine

$$C_5(\omega) = 16\omega^5 - 20\omega^3 + 5\omega \qquad C_5(0.7) = -0.67088 \qquad C_5(1.2) = 11.25312$$

Alternately, from (4*a*), we obtain

$$C_5(0.7) = \cos\,(5 \cos^{-1} 0.7) = \cos\,(5 \times 0.79540) = -0.67088$$

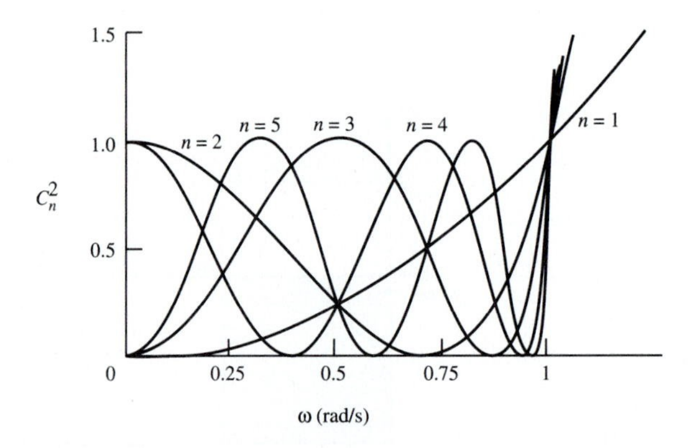

FIGURE 2.2-2
The value of some Chebyshev polynomials $C_n^2(\omega)$.

[1] These polynomials were originally used in studying the construction of steam engines by P. L. Chebyshev in the paper "Théorie des mécanismes connus sous le nom de parallelogrammes," *Oeuvres*, vol. I, St. Petersburg, 1899. The spellings *Tschebyscheff* and *Tchebysheff* for the author's name also appear frequently in the literature. All these spellings are transliterations of the Russian name Чебышёв.

Similarly, from (4*b*), we find

$$C_5(1.2) = \cosh\,(5 \cosh^{-1} 1.2) = \cosh\,(5 \times 0.62236) = 11.25312$$

Functions having the form of (1) in which the quantities $C_n(\omega)$ are Chebyshev polynomials are called *Chebyshev functions* or *equal-ripple functions*.

Summary 2.2-1 Properties of Chebyshev functions. A low-pass Chebyshev function having the form given in (1) has the following properties:

1. The range of frequencies $0 \le \omega \le 1$ rad/s is called the *passband*.
2. The magnitude characteristic in the passband is equal ripple.
3. The range of frequencies $\omega \ge 1$ rad/s is called the *stopband*.
4. The magnitude characteristic in the stopband is monotonic.
5. For n odd, $|N(j0)| = H$, and for n even, $|N(j0)| = H/\sqrt{1+\varepsilon^2}$ for all n.
6. At $\omega = 1$ rad/s, $|N(j1)| = H/\sqrt{1+\varepsilon^2}$ independent of the value of n.
7. The passband peaks occur at the frequencies at which $C_n^2(\omega) = \cos^2(n \cos^{-1} \omega) = 0$. These are defined by $\omega_{\text{peak}} = \cos\,(k\pi/2n)$ for $k = 1, 3, 5, \ldots$.

Plots of the magnitude of equal-ripple functions for $\varepsilon = 1$ and $n = 2, 5, 10$ are shown in Fig. 2.2-3. A comparison of the plots shown in Fig. 2.2-3 with the corresponding ones for the maximally flat magnitude case given in Fig. 2.1-2 shows that the stopband attenuation in the vicinity of the cutoff frequency is considerably higher for an equal-ripple characteristic of any given order. For sufficiently high values of frequency, of course, the attenuation will be $20n$ decibels per decade of frequency past $\omega = 1$ rad/s, just as was the case for the maximally flat magnitude characteristic. Some specific values of $|N(j\omega)|$ are given in Table 2.2-1. The -3-dB frequencies for various ripples and orders are given in Table 2.2-2. These frequencies are determined by the relation

$$\omega_{3\text{dB}} = \cosh\left(\frac{1}{n}\cosh^{-1}\frac{1}{\varepsilon}\right) \qquad (\varepsilon \le 1) \tag{5}$$

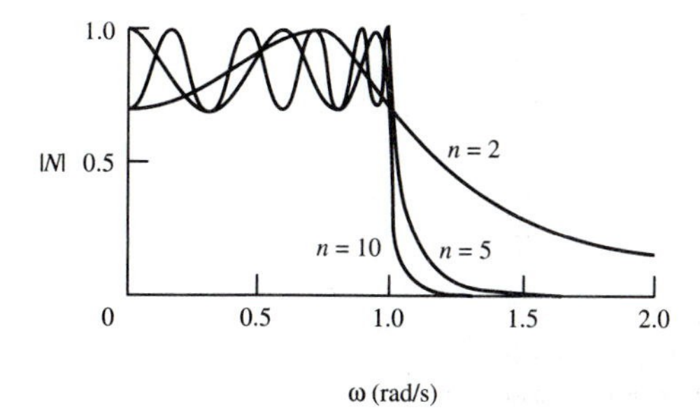

FIGURE 2.2-3
The magnitude of equal-ripple (3-dB) functions of various orders.

TABLE 2.2-1
Values of the magnitude of equal-ripple (Chebyshev) low-pass functions with normalized passband 0 to 1 rad/s

		ω = 1.1 rad/s		ω = 1.2 rad/s		ω = 1.5 rad/s	
Ripple	***n***	**Mag**	**dB**	**Mag**	**dB**	**Mag**	**dB**
$\frac{1}{2}$ dB	2	0.8958	−0.96	0.8359	−1.56	0.6331	−3.97
	3	0.8165	−1.76	0.6539	−3.69	0.3031	−10.37
	4	0.6864	−3.27	0.4266	−7.40	0.1209	−18.35
	5	0.5244	−5.61	0.2465	−12.16	0.0465	−26.65
	6	0.3698	−8.64	0.1355	−17.36	0.0178	−35.00
	7	0.2481	−12.11	0.0732	−22.71	0.0068	−43.36
	8	0.1624	−15.79	0.0394	−28.10	0.0026	−51.72
	9	0.1051	−19.57	0.0211	−33.50	0.0010	−60.08
	10	0.0677	−23.39	0.0113	−38.90	0.0004	−68.44
1 dB	2	0.8105	−1.82	0.7226	−2.82	0.4896	−6.20
	3	0.6966	−3.14	0.5103	−5.84	0.2133	−13.42
	4	0.5438	−5.29	0.3081	−10.23	0.0833	−21.58
	5	0.3894	−8.19	0.1720	−15.29	0.0319	−29.91
	6	0.2635	−11.58	0.0934	−20.59	0.0122	−38.27
	7	0.1732	−15.23	0.0503	−25.96	0.0047	−46.63
	8	0.1123	−19.00	0.0270	−31.36	0.0018	−54.99
	9	0.0723	−22.81	0.0145	−36.76	0.0007	−63.35
	10	0.0465	−26.65	0.0078	−42.17	0.0003	−71.71

TABLE 2.2-2
Half-power (−3-dB) frequencies for equal-ripple magnitude (Chebyshev) low-pass functions with passband 0 to 1 rad/s for *n* = 2, . . . , 10

	Frequency (rad/s)								
Ripple (dB)	**2**	**3**	**4**	**5**	**6**	**7**	**8**	**9**	**10**
0.010000	3.303615	1.877180	1.466904	1.291217	1.199412	1.145268	1.110609	1.087064	1.070331
0.100000	1.943219	1.388995	1.213099	1.134718	1.092931	1.068001	1.051927	1.040955	1.033131
0.200000	1.674270	1.283455	1.156346	1.099154	1.068517	1.050188	1.038351	1.030262	1.024489
0.500000	1.389744	1.167485	1.093102	1.059259	1.041030	1.030090	1.023011	1.018167	1.014707
1.000000	1.217626	1.094868	1.053002	1.033815	1.023442	1.017205	1.013164	1.010396	1.008418

Pole Locations

We may now use the expression for $|N(j\omega)|^2$ given in (1) to determine the pole locations for an equal-ripple network function. Following the development given for a general magnitude-squared network function in the preceding section, we may write

$$N(s)N(-s) = \left| N(j\omega) \right|^2_{\omega=s/j} = \frac{H^2}{1+\varepsilon^2 C_n^2(s/j)} \tag{6}$$

Thus, the poles of the product $N(s)N(-s)$ are the roots of $C_n^2(s/j) = -1/\varepsilon^2$ or $C_n(s/j) = \pm j/\varepsilon$. Using the trigonometric form for $C_n(\omega)$ given in (4), we may write

$$C_n\left(\frac{s}{j}\right) = \cos\left(n \cos^{-1}\frac{s}{j}\right) = \pm\frac{j}{\varepsilon} \tag{7}$$

To solve this equation, we first define a complex function as

$$w = u + jv = \cos^{-1}\frac{s}{j} \tag{8}$$

Substituting this expression in (7), we obtain

$$\cos n(u + jv) = \cos nu \cosh nv - j \sin nu \sinh nv = \pm\frac{j}{\varepsilon} \tag{9}$$

Equating the real parts of the second and third members of this relation gives $\cos nu \cosh nv = 0$. Since $\cosh nv \geq 1$ for all values of nv, this equality requires $\cos nu = 0$. This may be written in the form

$$u_k = \frac{2k-1}{2n}\pi \qquad k = 1, 2, 3, \ldots, 2n \tag{10}$$

Equating the imaginary parts of (9) and recognizing that for the values of u defined by (10), $\sin nu = \pm 1$, we obtain

$$v = \frac{1}{n}\sinh^{-1}\frac{1}{\varepsilon} \tag{11}$$

where we have retained only the positive value for v. Equation (8) may now be put in the form

$$s = j\cos(u_k + jv) = \sin u_k \sinh v + j\cos u_k \cosh v \tag{12}$$

This relation specifies the poles of the product $N(s)N(-s)$. The left-half-plane poles are assigned to $N(s)$ to complete the determination of the network function. Thus we see that the poles of $N(s)$ will be at $p_k = \sigma_k + j\omega_k$, where

$$\sigma_k = -\sin u_k \sinh v \qquad \omega_k = \cos u_k \cosh v \qquad k = 1, 2, \ldots, n \tag{13}$$

and where u_k and v are defined in (10) and (11).

Example 2.2-2 Determination of a Chebyshev function. As an example of the use of (10), (11), and (13), consider the determination of the poles of a second-order equal-ripple low-pass network function having a 3-dB ripple in the passband. In this case $n = 2$, and from (10), $u_1 = \pi/4$ and $u_2 = 3\pi/4$. For a 3-dB ripple, $\varepsilon = 1$, and from (11), $v = 0.44069$. Using

(13), we now obtain

$$\sigma_1 = -\sin\frac{\pi}{4}\sinh 0.44069 = -0.32180$$

$$\omega_1 = \cos\frac{\pi}{4}\cosh 0.44069 = 0.77689$$

$$\sigma_2 = -\sin\frac{3\pi}{4}\sinh 0.44069 = -0.32180$$

$$\omega_2 = \cos\frac{3\pi}{4}\cosh 0.44069 = -0.77689$$

Thus the network function is

$$N(s) = \frac{H}{(s + 0.32180 + j0.77689)(s + 0.32180 - j0.77689)}$$

$$= \frac{H}{s^2 + 0.64359s + 0.70711}$$

Exercise 2.2-2. For each of the following sets of specifications, find the poles of a low-pass normalized Chebyshev function:

n	K_p, dB	Answers (pole locations)
2	0.1	$-1.18618 \pm j1.38095$
3	0.1	$-0.48470 \pm j1.20616,\ -0.96941$
2	1.5	$-0.46109 \pm j0.84416$
3	1.5	$-0.21006 \pm j0.93935,\ -0.42011$
2	4.0	$-0.26878 \pm j0.75647$
3	4.0	$-0.12510 \pm j0.89272,\ -0.25021$

Passive network realizations of equal-ripple functions are given in App. A. Values of the pole locations, quadratic factors and their Q's and ω_n's, and the coefficients of the denominator polynomials for values of n from 2 to 10 and values of ripple of 0.5 and 1.0 dB are given in Table 2.2-3.[2] The locus on which the poles lie may be determined by starting with the basic trigonometric relation $\sin^2 u_k + \cos^2 u_k = 1$. Inserting the relations for $\sin^2 u_k$ and $\cos^2 u_k$ from (13) we obtain

$$\frac{\sigma_k^2}{\sinh^2 v} + \frac{\omega_k^2}{\cosh^2 v} = 1 \tag{14}$$

This equation represents an ellipse centered at the origin of the p_k plane with an ordinate

[2] For a quadratic factor of the form $s^2 + a_1 s + a_0$, the quality factor $Q = \sqrt{a_0}/a_1$ and the undamped natural frequency $\omega_n = \sqrt{a_0}$. The use of these quantities will be covered in more depth in Chap. 5.

TABLE 2.2-3a
Denominator coefficients of equal-ripple magnitude (Chebyshev) functions of the form $a_0 + a_1 s + a_2 s^2 + \cdots + a_{n-1} s^{n-1} + s^n$ with passband 0 to 1 rad/s

n	a_0	a_1	a_2	a_3	a_4	a_5	a_6	a_7	a_8	a_9
						$\frac{1}{2}$-dB ripple				
2	1.516203	1.425625								
3	0.715694	1.534895	1.252913							
4	0.379051	1.025455	1.716866	1.197386						
5	0.178923	0.752518	1.309575	1.937367	1.172491					
6	0.094763	0.432367	1.171861	1.589763	2.171845	1.159176				
7	0.044731	0.282072	0.755651	1.647903	1.869408	2.412651	1.151218			
8	0.023691	0.152544	0.573560	1.148589	2.184015	2.149217	2.656750	1.146080		
9	0.011183	0.094120	0.340819	0.983620	1.611388	2.781499	2.429330	2.902734	1.142571	
10	0.005923	0.049285	0.237269	0.626969	1.527431	2.144237	3.440927	2.709741	3.149876	1.140066
						1-dB ripple				
2	1.102510	1.097734								
3	0.491307	1.238409	0.988341							
4	0.275628	0.742619	1.453925	0.952811						
5	0.122827	0.580534	0.974396	1.688816	0.936820					
6	0.068907	0.307081	0.939346	1.202140	1.930825	0.928251				
7	0.030707	0.213671	0.548620	1.357545	1.428794	2.176078	0.923123			
8	0.017227	0.107345	0.447826	0.846824	1.836902	1.655156	2.423026	0.919811		
9	0.007677	0.070605	0.244186	0.786311	1.201607	2.378119	1.881480	2.670947	0.917548	
10	0.004307	0.034497	0.182451	0.455389	1.244491	1.612986	2.981509	2.107852	2.919466	0.915932

semiaxis of length $\cosh v$ and an abscissa semiaxis of length $\sinh v$, as shown in Fig. 2.2-4.

Determination of Order

The determination of the order of a Chebyshev function follows the procedure outlined for the Butterworth function in Sec. 2.1 and the quantities shown in Fig. 2.1-4. The specifications are:

1. *Passband.* For $0 \le \omega \le \omega_p$ radians per second ($0 \le f \le f_p$ hertz), the maximum ripple of the magnitude characteristic is K_p dB.
2. *Stopband.* For $\omega \ge \omega_s$ radians per second ($f \ge f_s$ hertz), the minimum attenuation of the magnitude characteristic is K_s dB. This attenuation is measured from the greatest passband value of the characteristic.

To determine the order, we repeat the definitions given in (18) of Sec. 2.1:

$$\Omega = \frac{\omega_s}{\omega_p} = \frac{f_s}{f_p} \qquad M = \sqrt{\frac{10^{0.1K_s} - 1}{10^{0.1K_p} - 1}} \tag{15}$$

TABLE 2.2-3b
Pole locations, quadratic factors ($s^2 + a_1 s + a_0$), Q, and ω_n of 0.5-dB equal-ripple magnitude (Chebyshev) low-pass functions with passband 0 to 1 rad/s

n	Poles	a_1	a_0	Q	ω_n
2	$-0.71281 \pm j1.00404$	1.42562	1.51620	0.86372	1.23134
3	$-0.31323 \pm j1.02193$	0.62646	1.14245	1.70619	1.06885
	-0.62646				
4	$-0.17535 \pm j1.01625$	0.35071	1.06352	2.94055	1.03127
	$-0.42334 \pm j0.42095$	0.84668	0.35641	0.70511	0.59700
5	$-0.11196 \pm j1.01156$	0.22393	1.03578	4.54496	1.01773
	$-0.29312 \pm j0.62518$	0.58625	0.47677	1.17781	0.69048
	-0.36232				
6	$-0.07765 \pm j1.00846$	0.15530	1.02302	6.51285	1.01145
	$-0.21214 \pm j0.73824$	0.42429	0.59001	1.81038	0.76812
	$-0.28979 \pm j0.27022$	0.57959	0.15700	0.68364	0.39623
7	$-0.05700 \pm j1.00641$	0.11401	1.01611	8.84180	1.00802
	$-0.15972 \pm j0.80708$	0.31944	0.67688	2.57555	0.82273
	$-0.23080 \pm j0.44789$	0.46160	0.25388	1.09155	0.50386
	-0.25617				
8	$-0.04362 \pm j1.00500$	0.08724	1.01193	11.53079	1.00595
	$-0.12422 \pm j0.85200$	0.24844	0.74133	3.46567	0.86101
	$-0.18591 \pm j0.56929$	0.37182	0.35865	1.61068	0.59887
	$-0.21929 \pm j0.19991$	0.43859	0.08805	0.67657	0.29674
9	$-0.03445 \pm j1.00400$	0.06891	1.00921	14.57933	1.00459
	$-0.09920 \pm j0.88291$	0.19841	0.78936	4.47802	0.88846
	$-0.15199 \pm j0.65532$	0.30397	0.45254	2.21305	0.67271
	$-0.18644 \pm j0.34869$	0.37288	0.15634	1.06040	0.39540
	-0.19841				
10	$-0.02790 \pm j1.00327$	0.05580	1.00734	17.98714	1.00366
	$-0.08097 \pm j0.90507$	0.16193	0.82570	5.61141	0.90868
	$-0.12611 \pm j0.71826$	0.25222	0.53181	2.89134	0.72925
	$-0.15891 \pm j0.46115$	0.31781	0.23791	1.53475	0.48776
	$-0.17615 \pm j0.15890$	0.35230	0.05628	0.67338	0.23723

The Chebyshev order n_C is given as

$$n_C = \frac{\cosh^{-1} M}{\cosh^{-1} \Omega} \tag{16}$$

The derivation of (15) and (16) is left as an exercise for the reader. The required order of the function is the next highest integer greater than n_C.

Example 2.2-3 Order of a Chebyshev function. A Chebyshev function must meet the following specifications (these are the same as were used for a Butterworth function in Example 2.1-6):

$$f_p = 1000 \text{ Hz} \qquad K_p = 3.0103 \text{ dB}$$
$$f_s = 1300 \text{ Hz} \qquad K_s = 22 \text{ dB}$$

TABLE 2.2-3c
Pole locations, quadratic factors $(s^2 + a_1 s + a_0)$, Q, and ω_n of 1.0-dB equal-ripple magnitude (Chebyshev) low-pass functions with passband 0 to 1 rad/s

n	Poles	a_1	a_0	Q	ω_n
2	$-0.54887 \pm j0.89513$	1.09773	1.10251	0.95652	1.05000
3	$-0.24709 \pm j0.96600$	0.49417	0.99420	2.01772	0.99710
	-0.49417				
4	$-0.13954 \pm j0.98338$	0.27907	0.98650	3.55904	0.99323
	$-0.33687 \pm j0.40733$	0.67374	0.27940	0.78455	0.52858
5	$-0.08946 \pm j0.99011$	0.17892	0.98831	5.55644	0.99414
	$-0.23421 \pm j0.61192$	0.46841	0.42930	1.39879	0.65521
	-0.28949				
6	$-0.06218 \pm j0.99341$	0.12436	0.99073	8.00369	0.99536
	$-0.16988 \pm j0.72723$	0.33976	0.55772	2.19802	0.74681
	$-0.23206 \pm j0.26618$	0.46413	0.12471	0.76087	0.35314
7	$-0.04571 \pm j0.99528$	0.09142	0.99268	10.89866	0.99633
	$-0.12807 \pm j0.79816$	0.25615	0.65346	3.15586	0.80837
	$-0.18507 \pm j0.44294$	0.37014	0.23045	1.29693	0.48005
	-0.20541				
8	$-0.03501 \pm j0.99645$	0.07002	0.99414	14.24045	0.99707
	$-0.09970 \pm j0.84475$	0.19939	0.72354	4.26608	0.85061
	$-0.14920 \pm j0.56444$	0.29841	0.34086	1.95649	0.58383
	$-0.17600 \pm j0.19821$	0.35200	0.07026	0.75304	0.26507
9	$-0.02767 \pm j0.99723$	0.05533	0.99523	18.02865	0.99761
	$-0.07967 \pm j0.87695$	0.15933	0.77539	5.52663	0.88056
	$-0.12205 \pm j0.65090$	0.24411	0.43856	2.71289	0.66224
	$-0.14972 \pm j0.34633$	0.29944	0.14236	1.26004	0.37731
	-0.15933				
10	$-0.02241 \pm j0.99778$	0.04483	0.99606	22.26303	0.99803
	$-0.06505 \pm j0.90011$	0.13010	0.81442	6.93669	0.90245
	$-0.10132 \pm j0.71433$	0.20263	0.52053	3.56051	0.72148
	$-0.12767 \pm j0.45863$	0.25533	0.22664	1.86449	0.47606
	$-0.14152 \pm j0.15803$	0.28304	0.04500	0.74950	0.21214

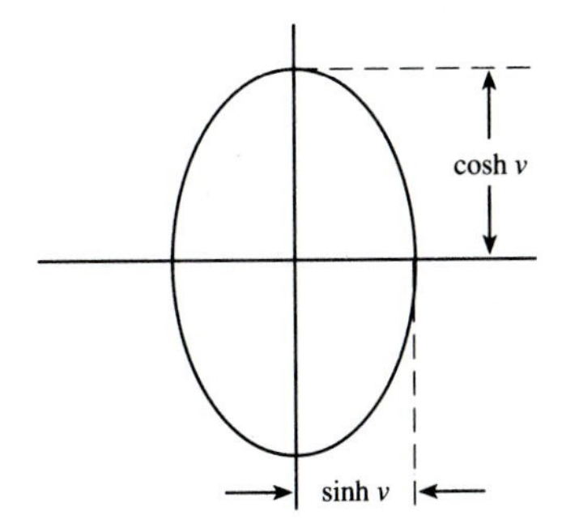

FIGURE 2.2-4
The loci of the poles of equal-ripple magnitude functions.

From Example 2.1-6 we find $\Omega = 1.3$ and $M = 12.5495$. From (16) we find the order

$$n_C = \frac{\cosh^{-1} 12.5495}{\cosh^{-1} 1.3} = \frac{3.2212}{0.7564} = 4.2585$$

A 5th-order function will meet the specifications. Recall that a 10th-order Butterworth function was required. This illustrates the "steeper" cutoff of the Chebyshev function.

Exercise 2.2-3. For each of the following sets of specifications for a Chebyshev function, determine the required order:

Passband		Stopband		Answers
ω_p (or f_p)	K_p, dB	ω_s (or f_s)	K_s, dB	n_C
1	3.0103	2	25	2.7100
1	3.0103	1.5	25	3.7083
1	3.0103	2	15	1.8192
1	3.0103	1.5	15	2.4893
1	1	2	25	3.2235
1	1	1.5	25	4.4109
1000	1	1100	25	9.5706

2.3 MAGNITUDE APPROXIMATION—THE INVERSE-CHEBYSHEV CHARACTERISTIC

In the preceding section we discussed the equal-ripple or Chebyshev magnitude approximation. It is characterized by equal-ripple behavior in the passband and monotonic behavior in the stopband. In this section we introduce a related type of magnitude characteristic, the *inverse Chebyshev.* Its properties are the inverse of those of the equal-ripple approximation, namely, it has a monotonic behavior in its passband and an equal-ripple behavior in its stopband. Its advantage over the equal-ripple approximation is that it has better passband phase characteristics. The effects of phase are discussed in Sec. 2.7. Its main disadvantage is that the filter configurations required to realize this approximation are more complex than those required for the realization of maximally flat magnitude or equal-ripple characteristics.

Inverse-Chebyshev Magnitude Function

To see how the inverse-Chebyshev characteristic may be developed, consider a low-pass equal-ripple function, which, in addition to being frequency normalized for a cutoff frequency of 1 rad/s, has also been normalized so that its peak magnitude is unity. From Sec. 2.2, this has the form

$$|N_{ER}(j\omega)|^2 = \frac{1}{1 + \varepsilon^2 C_n^2(\omega)} \tag{1}$$

where ER stands for "equal ripple" and $C_n(\omega)$ is a Chebyshev polynomial of order n. An example magnitude characteristic for the case where $n = 4$ is shown in Fig. 2.3-1(a). If we subtract (1) from unity, we obtain

$$|N_{1-ER}(j\omega)|^2 = 1 - |N_{ER}(j\omega)|^2 = \frac{\varepsilon^2 C_n^2(\omega)}{1 + \varepsilon^2 C_n^2(\omega)} \tag{2}$$

where $1 - ER$ stands for "1 minus equal ripple." An example magnitude characteristic (for $n = 4$) is shown in Fig. 2.3-1(b). Note that this is a high-pass function with an equal-ripple stopband of range $0 \le \omega \le 1$ rad/s and a monotonic passband for $\omega > 1$ rad/s. The inverse-Chebyshev magnitude characteristic is now found by applying a frequency transformation in which $1/\omega$ is substituted for ω in the right member of (2). Thus, we obtain

$$|N_{IC}(j\omega)|^2 = \frac{\varepsilon^2 C_n^2(1/\omega)}{1 + \varepsilon^2 C_n^2(1/\omega)} \tag{3}$$

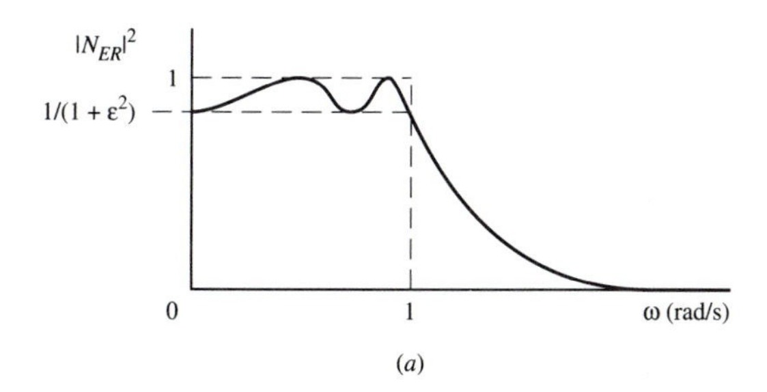

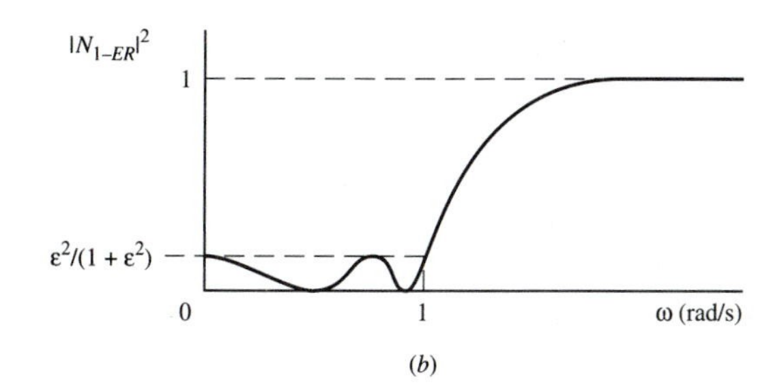

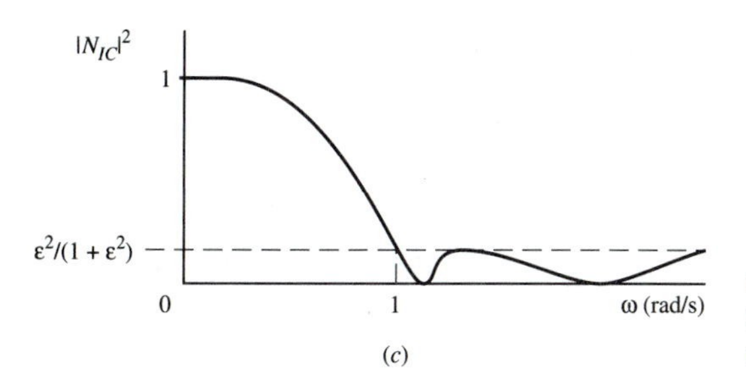

FIGURE 2.3-1
Development of the inverse-Chebyshev magnitude characteristic.

where IC stands for "inverse Chebyshev." The frequency transformation effectively transforms the high-pass function $|N_{1-ER}(j\omega)|^2$ to a low-pass one. The example magnitude characteristic now has the form shown in Fig. 2.3-1(c). Note that this has the desired monotonic passband and equal-ripple stopband behavior.

Order of Inverse-Chebyshev Functions

The properties of the general inverse-Chebyshev magnitude characteristic may be specified in terms of the parameters shown in Fig. 2.3-2. Note that the normalizations chosen for the original equal-ripple function have two effects: (1) the peak magnitude in the passband is unity and (2) the starting frequency for the equal-ripple stopband portion of the characteristic is 1.0 rad/s. The parameters defined in the figure are the passband attenuation K_p (dB), the frequency ω_p (radians per second) at which the passband attenuation is specified, and the stopband attenuation K_s (dB). Specifying these three parameters allows us to determine the values of the quantities n and ε^2 used in (1) to (3), and, as we shall shortly see, to specify the pole and zero locations of the inverse-Chebyshev network function. To determine ε^2, we need only compare (3) and Fig. 2.3-2 at $\omega = 1$ rad/s. Thus, we obtain

$$K_s = 10 \log \left(1 + \frac{1}{\varepsilon^2}\right) \tag{4}$$

Solving for ε^2, we find

$$\varepsilon^2 = \frac{1}{10^{0.1K_s} - 1} \tag{5}$$

To determine n, we begin by comparing (3) and Fig. 2.3-2 at $\omega = \omega_p$. Thus we obtain

$$K_p = 10 \log \left[1 + \frac{1}{\varepsilon^2 C_n^2(1/\omega_p)}\right] \tag{6}$$

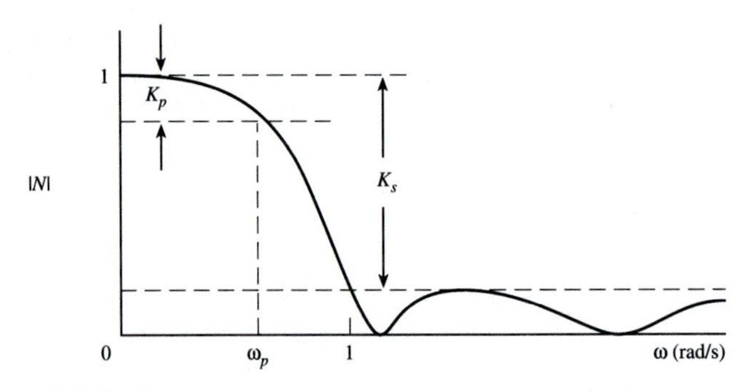

FIGURE 2.3-2
Parameters of an inverse-Chebyshev magnitude characteristic.

Using (5) to eliminate ε^2 and solving for $C_n^2(1/\omega_p)$, we get

$$C_n^2\left(\frac{1}{\omega_p}\right) = \frac{10^{0.1K_s} - 1}{10^{0.1K_p} - 1} \tag{7}$$

Using (4*b*) of Sec. 2.2, we may solve for n to obtain

$$n_{IC} = \frac{\cosh^{-1}\left[(10^{0.1K_s} - 1)/(10^{0.1K_p} - 1)\right]^{1/2}}{\cosh^{-1}(1/\omega_p)} \tag{8}$$

The next highest integer value of n_{IC} determines the required order of the filter.

Now let us compare the order of the filter required in the inverse-Chebyshev approximation with that required in the Chebyshev one. For the latter, using the development of (15) and (16) in Sec. 2.2, we have

$$n_C = \frac{\cosh^{-1} M}{\cosh^{-1} \Omega} = \frac{\cosh^{-1}\left[(10^{0.1K_s^{(C)}} - 1)/(10^{0.1K_p^{(C)}} - 1)\right]^{1/2}}{\cosh^{-1} \omega_s^{(C)}} \tag{9}$$

where $K_s^{(C)}$ is the desired attenuation in dB at the stopband frequency $\omega_s^{(C)}$ and $\omega_p^{(C)} = 1$. We see that the expressions for n_{IC} of the inverse Chebyshev in (8) and for n_C in (9) are the same under the conditions that

$$\omega_s^{(C)} = \frac{1}{\omega_p} \qquad K_s^{(C)} = K_s \qquad K_p^{(C)} = K_p \tag{10}$$

The first of these conditions, however, is exactly the condition realized by frequency denormalizing the equal-ripple specification by $1/\omega_s^{(C)}$ so that its stopband frequency specification is at 1 rad/s, as it is in the inverse-Chebyshev approximation. We conclude that if the other two conditions of (10) are satisfied, the order required for an inverse-Chebyshev specification is exactly the same as that required for an equal-ripple one.

Example 2.3-1 Order of an inverse-Chebyshev function. As an illustration of the determination of the order required for an inverse-Chebyshev characteristic, consider the specification of a magnitude function with $\frac{1}{2}$ dB attenuation at a frequency of 0.5 rad/s in the monotonic passband and a minimum of 18 dB attenuation in the equal-ripple stopband whose range is $1 \le \omega \le \infty$ radians per second. From Fig. 2.3-2 we see that $K_p = 0.5$ dB, $\omega_p = 0.5$ rad/s, and $K_s = 18$ dB. From (10) we find $\omega_s^{(C)} = 1/\omega_p = 2$ rad/s, $K_s^{(C)} = K_s = 18$ dB, and $K_p^{(C)} = K_p = 0.5$ dB. Using the (C) superscript quantities in (15) and (16) of Sec. 2.2, we find

$$M = \sqrt{\frac{10^{0.1\times 18} - 1}{10^{0.1\times 0.5} - 1}} = 22.5589 \qquad \Omega = \tfrac{2}{1}$$

$$n_{IC} = n_C = \frac{\cosh^{-1} 22.5589}{\cosh^{-1} 2} = \frac{3.8088}{1.3170} = 2.8921$$

Exercise 2.3-1. For each of the following sets of specifications for an inverse-Chebyshev function, determine the required order:

Passband		Stopband		Answers
ω_p (or f_p)	K_p, dB	ω_s (or f_s)	K_s, dB	n_{IC}
0.5	3.0103	1	25	2.7100
0.6667	3.0103	1	25	3.7083
0.5	3.0103	1	15	1.8192
0.6667	3.0103	1	15	2.4893
0.5	1	1	25	3.2235
0.6667	1	1	15	4.4109
769.2	1	1000	25	9.5706

Summary 2.3-1 Properties of an inverse-Chebyshev function. A normalized low-pass inverse-Chebyshev function having the form given in (3) and illustrated (for $n = 4$) in Fig. 2.3-2 has the following properties:

1. The range of frequencies $0 \le \omega \le \omega_p$ radians per second is called the *passband*.
2. The magnitude characteristic in the passband is monotonic.
3. The range of frequencies $\omega \ge 1$ rad/s is called the *stopband*.
4. The magnitude characteristic in the stopband is equal ripple.
5. For n even, $|N(j\infty)|$ has K_s dB of attenuation, and for n odd, $|N(j\infty)| = 0$.
6. In the form given in (3), $|N(j0)| = 1$.
7. The order is the same as the order of a Chebyshev function whose parameters are defined by (10).
8. The stopband specification K_s dB is met exactly while the passband specification K_p is usually met conservatively, that is, $20 \log_{10} [|N(0)|/|N(j\omega_p)|] \le K_p$.

Pole and Zero Locations

In the preceding paragraphs we have considered the properties of the magnitude characteristics of the inverse-Chebyshev approximation. Now let us consider the pole and zero locations for the network function. Using the techniques introduced in Sec. 2.1, we may write

$$|N(j\omega)|^2 \Big|_{\omega = s/j} = N(s)N(-s) = \frac{A(s)A(-s)}{B(s)B(-s)} \tag{11}$$

where $N(s)$ is the network function, $A(s)$ is its numerator polynomial, and $B(s)$ is its denominator polynomial. Applying this result to (3), we find that the numerator and denominator polynomials are determined by the relations

$$A(s)A(-s) = \varepsilon^2 C_n^2\left(\frac{j}{s}\right) \tag{12a}$$

$$B(s)B(-s) = 1 + \varepsilon^2 C_n^2\left(\frac{j}{s}\right) \tag{12b}$$

First let us consider the numerator. Setting (12*a*) to zero and using (7) of Sec. 2.2, we obtain

$$C_n\left(\frac{j}{s}\right) = \cos\left(n \cos^{-1}\frac{j}{s}\right) = 0 \tag{13}$$

The techniques for solving this equation closely parallel those used in Sec. 2.2. The details are left to the reader as an exercise. The result is that the zeros $z_k = \alpha_k + j\beta_k$ of the inverse-Chebyshev function are given by

$$\alpha_k = 0 \qquad \beta_k = \frac{1}{\cos u_k} \qquad k = 1, 2, \ldots, n \tag{14}$$

where, as defined in (10) of Sec. 2.2,

$$u_k = \frac{2k-1}{2n}\pi \tag{15}$$

For example, for $n=4$, the zeros are located at $\beta_1 = 1.08239$, $\beta_2 = 2.61313$, $\beta_3 = -2.61313$, and $\beta_4 = -1.08239$. Thus, the numerator polynomial has the form

$$A(s) = (s^2 + 1.17157)(s^2 + 6.82843)$$

where, for convenience, the highest degree coefficient has been assumed to have a value of unity. Values of zero locations for other values of n are given in Table 2.3-1*a*.

Now let us consider the denominator polynomial of the inverse-Chebyshev function. The plots are found by setting (12*b*) to zero. The resulting equation is almost the same as that used in (7) of Sec. 2.2 for finding the poles of a Chebyshev function whose passband has the same attenuation (K_p) as the inverse Chebyshev but is defined for 0 to 1 rad/s and whose stopband has the same attenuation (K_s) as the inverse Chebyshev but starts at $1/\omega_p$ radians per second. The only difference is that the argument

TABLE 2.3-1*a*

Zero locations of inverse-Chebyshev low-pass functions with stopband starting at 1 rad/s

n	Zeros	*n*	Zeros
2	1.41421	8	1.01959
3	1.15470		1.20269
4	1.08239		1.79995
	2.61313		5.12583
5	1.05146	9	1.01543
	1.70130		1.15470
6	1.03528		1.55572
	1.41421		2.92380
	3.86370	10	1.01247
7	1.02572		1.12233
	1.27905		1.41421
	2.30477		2.20269
			6.39245

TABLE 2.3-1*b*

Denominator coefficients of inverse-Chebyshev low-pass functions of the form $a_0 + a_1 s + a_2 s^2 + \cdots + a_{n-1} s^{n-1} + s^n$ with stopband starting at 1 rad/s

n	a_0	a_1	a_2	a_3	a_4	a_5	a_6	a_7	a_8	a_9
					$K_s = 20$ dB					
2	0.20000	0.60000								
3	0.40202	0.94200	1.40538							
4	0.80000	1.71447	2.63713	2.26147						
5	1.60806	3.30466	5.40570	5.14547	3.24707					
6	3.20000	6.42700	11.25412	11.55465	8.81693	4.15618				
7	6.43224	12.74053	23.87419	26.01039	22.69086	13.26607	5.19876			
8	12.80000	25.12519	50.25918	57.46664	55.73028	37.49462	18.97412	6.10805		
9	25.72897	50.19097	106.84537	127.19065	134.81233	100.51592	60.37574	25.38642	7.18269	
10	51.20000	99.43560	224.55721	276.61837	316.78068	255.91794	174.89951	87.56419	33.13797	8.07935
					$K_s = 30$ dB					
2	0.06325	0.34999								
3	0.12655	0.47241	0.97664							
4	0.25298	0.81486	1.56531	1.76483						
5	0.50622	1.51981	2.91421	3.49129	2.64719					
6	1.01193	2.92249	5.73802	7.24909	6.37489	3.56564				
7	2.02487	5.71149	11.59864	15.36356	14.98886	10.20067	4.52221			
8	4.04772	11.24272	23.70902	32.75721	34.70333	26.87293	15.04949	5.48042		
9	8.09948	22.25942	48.81110	70.01983	79.52314	67.79767	44.29240	20.85418	6.46448	
10	16.19086	44.15949	100.69819	149.38105	180.29733	165.95838	121.59780	67.73088	27.70435	7.43698
					$K_s = 40$ dB					
2	0.02000	0.19900								
3	0.04000	0.22709	0.67460							
4	0.08000	0.36528	0.91393	1.35139						
5	0.16001	0.65729	1.55005	2.30828	2.14920					
6	0.32000	1.23932	2.87985	4.38191	4.55189	3.01665				
7	0.64003	2.39096	5.58599	8.73767	9.80490	7.71180	3.92791			
8	1.28000	4.67000	11.07911	17.84609	21.32243	18.76547	11.83453	4.86443		
9	2.56013	9.18939	22.25260	36.90027	46.54164	44.57388	32.30586	16.93122	5.81984	
10	5.12000	18.16376	45.01896	76.76130	101.59852	104.18640	83.70033	51.37418	23.02354	6.78506

TABLE 2.3-1c
Pole locations and quadratic factors ($a_0 + a_1 s + s^2$) of K_s = 20 dB inverse-Chebyshev functions with stopband starting at 1 rad/s

n	Poles	a_0	a_1
2	$-0.30000 \pm j0.33166$	0.20000	0.60000
3	$-0.27597 \pm j0.62840$	0.47105	0.55194
	-0.85345		
4	$-0.20565 \pm j0.78291$	0.65524	0.41129
	$-0.92509 \pm j0.60426$	1.22093	1.85018
5	$-0.15006 \pm j0.86147$	0.76465	0.30011
	$-0.68614 \pm j0.92991$	1.33551	1.37228
	-1.57469		
6	$-0.11182 \pm j0.90476$	0.83110	0.22363
	$-0.47723 \pm j1.03469$	1.29834	0.95446
	$-1.48905 \pm j0.86506$	2.96558	2.97809
7	$-0.08568 \pm j0.93071$	0.87355	0.17135
	$-0.34115 \pm j1.06068$	1.24143	0.68230
	$-1.03813 \pm j1.23957$	2.61425	2.07627
	-2.26884		
8	$-0.06739 \pm j0.94737$	0.90205	0.13478
	$-0.25389 \pm j1.06249$	1.19334	0.50777
	$-0.69933 \pm j1.30663$	2.19633	1.39865
	$-2.03342 \pm j1.13102$	5.41400	4.06684
9	$-0.05424 \pm j0.95868$	0.92200	0.10847
	$-0.19586 \pm j1.05738$	1.15641	0.39173
	$-0.49154 \pm j1.28554$	1.89421	0.98307
	$-1.37368 \pm j1.55835$	4.31544	2.74735
	-2.95207		
10	$-0.04451 \pm j0.96669$	0.93648	0.08902
	$-0.15564 \pm j1.05081$	1.12842	0.31128
	$-0.36270 \pm j1.24771$	1.68833	0.72540
	$-0.90714 \pm j1.59003$	3.35111	1.81428
	$-2.56969 \pm j1.40010$	8.56359	5.13938

of the Chebyshev polynomial $C_n^2(\cdot)$ is inverted and appears as j/s rather than as s/j. As a result, the poles of the inverse-Chebyshev function are simply the reciprocal of the ones found for the equal-ripple case. Thus, they may be specified as

$$p_k = \frac{1}{\sigma_k + j\omega_k} \qquad k = 1, 2, \ldots, n \tag{16}$$

where [repeated from (11) and (13) of Sec. 2.2]

$$v = \frac{1}{n}\sinh^{-1}\frac{1}{\varepsilon} \tag{17a}$$

$$\sigma_k = -\sin u_k \sinh v \qquad \omega_k = \cos u_k \cosh v \qquad k = 1, 2, \ldots, n \tag{17b}$$

and u_k is defined in (15) above.

TABLE 2.3-1*d*
Pole locations and quadratic factors ($a_0 + a_1 s + s^2$) of $K_s = 30$ dB inverse-Chebyshev functions with stopband starting at 1 rad/s

n	Poles	a_0	a_1
2	$-0.17499 \pm j0.18062$	0.06325	0.34998
3	$-0.22043 \pm j0.43315$	0.23621	0.44086
	-0.53578		
4	$-0.19879 \pm j0.61799$	0.42142	0.39757
	$-0.68363 \pm j0.36464$	0.60031	1.36726
5	$-0.16241 \pm j0.73493$	0.56650	0.32482
	$-0.62225 \pm j0.66471$	0.82904	1.24450
	-1.07787		
6	$-0.12969 \pm j0.80842$	0.67036	0.25938
	$-0.49923 \pm j0.83385$	0.94453	0.99846
	$-1.15390 \pm j0.51642$	1.59818	2.30780
7	$-0.10389 \pm j0.85616$	0.74381	0.20777
	$-0.38931 \pm j0.91829$	0.99481	0.77862
	$-0.97130 \pm j0.87987$	1.71759	1.94260
	-1.59322		
8	$-0.08420 \pm j0.88846$	0.79645	0.16839
	$-0.30564 \pm j0.96013$	1.01526	0.61129
	$-0.74807 \pm j1.04916$	1.66035	1.49614
	$-1.60230 \pm j0.66898$	3.01490	3.20460
9	$-0.06920 \pm j0.91116$	0.83501	0.13839
	$-0.24405 \pm j0.98143$	1.02276	0.48809
	$-0.57039 \pm j1.11126$	1.56025	1.14079
	$-1.30089 \pm j1.09934$	2.90086	2.60177
	-2.09543		
10	$-0.05766 \pm j0.92766$	0.86388	0.11532
	$-0.19848 \pm j0.99261$	1.02466	0.39696
	$-0.44234 \pm j1.12717$	1.46618	0.88469
	$-0.97942 \pm j1.27164$	2.57632	1.95884
	$-2.04059 \pm j0.82357$	4.84228	4.08118

Summary 2.3-2 Finding the poles of an inverse-Chebyshev function. The procedure for finding the poles of an inverse-Chebyshev function from specified values of the parameters ω_p, K_p, and K_s given in Fig. 2.3-2 may be summarized as follows:

1. Use (5) to determine the value of ε for the related equal-ripple function from the specified value of K_s.
2. Use (8) [or (10) and (9)] to determine the order of the network function for the specified values of K_s, K_p, and ω_p.
3. Use (17) to find the values of σ_k and ω_k ($k = 1, 2, \ldots, n$), the real and imaginary parts of the pole locations of the related Chebyshev function.
4. The poles of the inverse-Chebyshev function are the reciprocal of those found in step 3 and are computed using (16).

TABLE 2.3-1*e*

Pole locations and quadratic factors ($a_0 + a_1 s + s^2$) of $K_s = 40$ dB inverse-Chebyshev functions with stopband starting at 1 rad/s

n	Poles	a_0	a_1
2	$-0.09950 \pm j0.10050$	0.02000	0.19900
3	$-0.16115 \pm j0.29593$	0.11355	0.32230
	-0.35230		
4	$-0.17116 \pm j0.47610$	0.25597	0.34232
	$-0.50454 \pm j0.24079$	0.31254	1.00907
5	$-0.15592 \pm j0.61087$	0.39747	0.31183
	$-0.52480 \pm j0.48539$	0.51102	1.04960
	-0.78777		
6	$-0.13388 \pm j0.70579$	0.51606	0.26777
	$-0.47103 \pm j0.66535$	0.66456	0.94206
	$-0.90341 \pm j0.34193$	0.93307	1.80682
7	$-0.11268 \pm j0.77234$	0.60920	0.22537
	$-0.39799 \pm j0.78070$	0.76789	0.79598
	$-0.85179 \pm j0.64169$	1.13732	1.70358
	-1.20298		
8	$-0.09456 \pm j0.81975$	0.68094	0.18911
	$-0.33009 \pm j0.85193$	0.83475	0.66019
	$-0.72591 \pm j0.83644$	1.22657	1.45182
	$-1.28166 \pm j0.43964$	1.83593	2.56332
9	$-0.07968 \pm j0.85431$	0.73619	0.15936
	$-0.27374 \pm j0.89634$	0.87835	0.54747
	$-0.59559 \pm j0.94480$	1.24739	1.19119
	$-1.15872 \pm j0.79730$	1.97833	2.31745
	-1.60437		
10	$-0.06764 \pm j0.88008$	0.77911	0.13528
	$-0.22865 \pm j0.92480$	0.90754	0.45731
	$-0.48568 \pm j1.00088$	1.23764	0.97135
	$-0.96184 \pm j1.00996$	1.94516	1.92368
	$-1.64872 \pm j0.53814$	3.00787	3.29744

Example 2.3-2 Determination of an inverse-Chebyshev function. As an example of the use of the results given above, consider the determination of the network function for an inverse-Chebyshev magnitude characteristic having a maximum of 1 dB attenuation at 0.5 rad/s in the passband and a minimum of 20 dB attenuation in the stopband at all frequencies greater than 1 rad/s. Thus, from Fig. 2.3-2, we find $K_p = 1$ dB, $\omega_p = 0.5$ rad/s, and $K_s = 20$ dB. Applying (8), we find $n = 2.78343$; thus a third-order function is required. We may now apply (14) and (15) to find the numerator polynomial. From these relations we get $\beta_1 = 1.154701$, $\beta_2 = 0$, and $\beta_3 = -1.154701$. The numerator polynomial (with unity leading coefficient) is

$$A(s) = s^2 + 1.33333$$

Now let us determine the pole locations. Applying (5), we find $\varepsilon = 0.10050$. Applying the relations of (17), we find the Chebyshev poles are

$$p^{(C)}_{1,3} = -0.58586 \pm j1.33405 \qquad p^{(C)}_2 = -1.17172$$

Taking the reciprocal of these, we find the inverse-Chebyshev pole locations are

$$p_{1,3} = -0.27597 \pm j0.62840 \qquad p_2 = -0.85345$$

Thus, the inverse-Chebyshev network function is

$$N(s) = \frac{0.30151(s^2 + 1.33333)}{s^3 + 1.40538s^2 + 0.94210s + 0.40202}$$

The multiplicative constant has been chosen so as to set the gain at $\omega = 0$ to unity.

Values of the pole and zero locations and the coefficients of the denominator polynomials for different values of n and for some common values of K_s are given in Table 2.3-1. Some passive filter realizations are given in Table 4.7-1.

2.4 MAGNITUDE APPROXIMATION—THE ELLIPTIC CHARACTERISTIC

In the preceding sections of this chapter we discussed two types of magnitude approximation, namely, the maximally flat and the equal ripple. These may both be written in the form

$$|N(j\omega)|^2 = \frac{H^2}{1 + \varepsilon^2 P_n^2(\omega)} \tag{1}$$

where P_n^2 is a polynomial that for the maximally flat case is ω^{2n} and for the equal-ripple case is $C_n^2(\omega)$ (a Chebyshev polynomial). In this section we consider a quite different type of low-pass magnitude characteristic in which the polynomial $P_n^2(\omega)$ is replaced with a *rational function* $R_n^2(\omega)$ having both a numerator and a denominator polynomial. By choosing a specific function called a *Chebyshev rational function*, it is possible to produce a magnitude characteristic that is equal ripple in both the passband and the stopband. An example is shown in Fig. 2.4-1. Note that a *transition band* is defined to

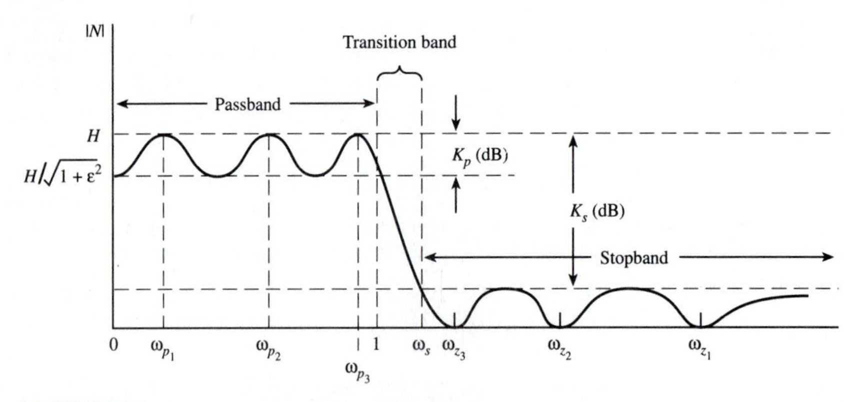

FIGURE 2.4-1
A magnitude function that is equal ripple in both passband and stopband.

link the equal-ripple passband and stopband. For a given order filter, the resulting magnitude characteristic drops off even more steeply than it does in the Chebyshev case, providing the sharpest cutoff of any of the three types of low-pass approximations. The determination of the form of the rational function $R_n(\omega)$ in general requires the use of elliptic functions and elliptic integrals, and the resulting network functions are referred to as *elliptic functions*. Since the original work on such functions was performed by the famous German network theorist W. Cauer, filters with an elliptic characteristic are also called *Cauer filters*.

Chebyshev Rational Functions

The general form of the elliptic magnitude characteristic is specified as

$$|N(j\omega)|^2 = \frac{H^2}{1+\varepsilon^2 R_n^2(\omega)} \tag{2}$$

The Chebyshev rational function $R_n(\omega)$ can be shown to be

$$R_n(\omega) = \begin{cases} M \displaystyle\prod_{i=1}^{n/2} \frac{\omega^2 - \omega_{p_i}^2}{\omega^2 - \omega_{z_i}^2} & \text{for } n \text{ even} \quad (3a) \\ M\omega \displaystyle\prod_{i=1}^{(n-1)/2} \frac{\omega^2 - \omega_{p_i}^2}{\omega^2 - \omega_{z_i}^2} & \text{for } n \text{ odd} \quad (3b) \end{cases}$$

Summary 2.4-1 Properties of the elliptic magnitude characteristic. The expressions given in (2) and (3) have the following properties:

1. The *passband* is defined for $0 \le \omega \le 1$. The constants M are chosen so that in this region $0 \le R_n^2(\omega) \le 1$. As a result, $H \ge |N(j\omega)| \ge H/\sqrt{1+\varepsilon^2}$.
2. The values $\omega = \omega_{p_i}$ at which $R_n^2(\omega) = 0$ represent the *passband peaks* at which $|N(j\omega)| = H$.
3. In the passband, the values of ω at which $R_n^2(\omega) = 1$ correspond with the *passband valleys* at which $|N(j\omega)| = H/\sqrt{1+\varepsilon^2}$.
4. The *stopband* is defined as $\omega \ge \omega_s$. In this region the minimum value of $R_n^2(\omega)$ is R^2_{stopband}, where

$$R^2_{\text{stopband}} \ge \frac{10^{0.1K_s} - 1}{\varepsilon^2} \tag{4}$$

   As a result, $|N(j\omega)|$ has a minimum of K_s decibels of attenuation in this region.
5. In the stopband, the values of $\omega = \omega_{z_i}$ at which $R_n^2(\omega) = \infty$ correspond with the *transmission zeros* at which $|N(j\omega)| = 0$.
6. In the stopband, the values of ω at which $R_n^2(\omega) = R^2_{\text{stopband}}$ correspond with the *stopband peaks* of the magnitude characteristic at which $|N(j\omega)|$ has a minimum attenuation of K_s decibels.
7. The quantities ω_{p_i} and ω_{z_i} have a geometric mean ω_s. Thus, $\sqrt{\omega_{p_i}\omega_{z_i}} = \omega_s$.

A plot of a typical elliptic magnitude characteristic is shown in Fig. 2.4-1.

Example 2.4-1 Chebyshev rational function. As an example of a Chebyshev rational function, consider a third-order function for which $\omega_s = 2$ rad/s[1]:

$$R_n(\omega) = 18.5589 \frac{\omega[\omega^2 - (0.88103)^2]}{\omega^2 - (2.27007)^2}$$

A plot of the function $R_n^2(\omega)$ is shown in Fig. 2.4-2. A logarithmic scale has been used for the ordinate (as a result, the value of zero for the function corresponds with an ordinate value of $-\infty$). The reader should verify the following properties of the function and of their effect on $|N(j\omega)|$:

$R_n^2(0) = 0$	[$\lvert N(j0)\rvert = H$, peak passband value]
$R_n^2(0.527) = 1$	[$\lvert N(j0.527)\rvert = H/\sqrt{1+\varepsilon^2}$, passband valley]
$R_n^2(0.881) = 0$	[$\lvert N(j0.881)\rvert = H$, passband peak]
$R_n^2(1) = 1$	[$\lvert N(j1)\rvert = H/\sqrt{1+\varepsilon^2}$, end of passband]
$R_n^2(2) = 10{,}766$	[$\lvert N(j2)\rvert$ has K_s dB of attenuation, beginning of stopband]
$R_n^2(2.27) = \infty$	[$\lvert N(j2.27)\rvert = 0$, transmission zero]
$R_n^2(3.796) = 10{,}766$	[$\lvert N(j3.796)\rvert$ has K_s dB of attenuation, stopband peak]
$R_n^2(\infty) = \infty$	[$\lvert N(j\infty)\rvert = 0$, transmission zero]

Elliptic Network Functions

We may now determine the general form of elliptic network functions. We first consider the odd case. If the odd form of $R_n(\omega)$ given in (3*b*) is substituted in (2), we may then

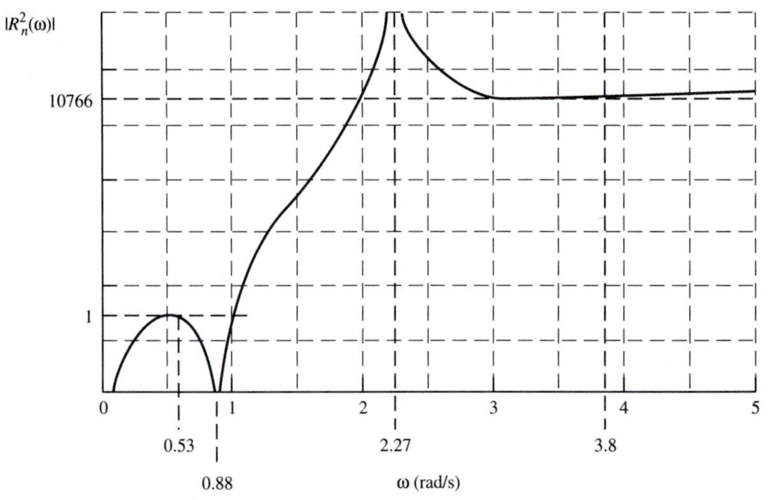

FIGURE 2.4-2
Example Chebyshev rational function.

[1] The numerical values for this function may be found in the references listed in the Bibliography. See, for example, Christian and Eisenman, case CO3, $\theta = 30°$.

obtain the network function by replacing ω with s/j and selecting the left-half-plane poles and half of the $j\omega$-axis zeros (see the discussion in Sec. 2.1). We obtain the following general form for the odd-order elliptic network function:

$$N_o(s) = \frac{H_o \prod_{i=1}^{(n-1)/2} (s^2 + \Omega_i^2)}{a_0 + a_1 s + \cdots + a_{n-1}s^{n-1} + a_n s^n} \tag{5}$$

where the $j\omega$-axis zeros are located at $s = \pm j\Omega_i$. The degree of the denominator polynomial of $N_o(s)$ will be n, while the degree of the numerator polynomial will be $n - 1$. The term $|N_o(j\omega)|$ will have $\frac{1}{2}(n - 1)$ peaks in the passband (plus a peak which occurs at $\omega = 0$), $\frac{1}{2}(n - 1)$ transmission zeros in the stopband, and a zero value at $\omega = \infty$. An example for $n = 5$ is shown in Fig. 2.4-3(*a*).

Now let us consider the case where n is even. Using the even form of $R_n(\omega)$ given in (3*a*), and proceeding as in the above, we obtain a first general form for the even-order elliptic network function:

$$N_a(s) = \frac{H_a \prod_{i=1}^{n/2} (s^2 + \Omega_i^2)}{a_0 + a_1 s + \cdots + a_{n-1}s^{n-1} + a_n s^n} \tag{6}$$

We call this *case A for n even*. The degrees of both the numerator and denominator polynomials are equal to n. The term $|N_a(j\omega)|$ will have $\frac{1}{2}n$ peaks in the passband, $\frac{1}{2}n$ transmission zeros in the stopband, and a nonzero value at $\omega = \infty$. An example is shown for $n = 4$ in Fig. 2.4-3(*b*). A consequence of the fact that the function is finite at $\omega = \infty$ is that a passive *RLC* ladder network realization of the function is only possible if coupled coils, i.e., transformers, are used in the circuit. This is usually undesirable from the standpoint both of cost and of accurate determination of element values. Thus, this function is normally considered only for realization by active *RC* filters of the type that will be introduced in Chaps. 5 through 7. If passive realizations are required, however,

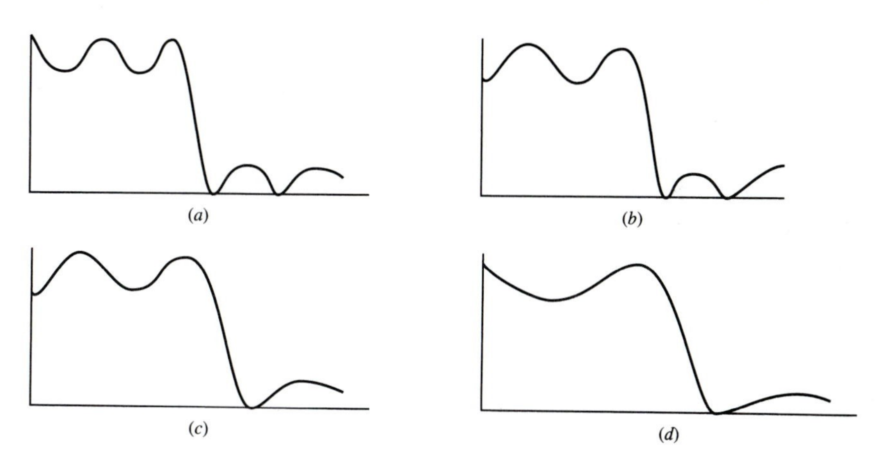

FIGURE 2.4-3
Various forms of elliptic magnitude functions: (*a*) $n = 5$; (*b*) $n = 4$, Case *A*; (*c*) $n = 4$, Case *B*; (*d*) $n = 4$, Case *C*.

the need for transformers may be eliminated by modifying the form of the function $R_n(\omega)$ given in (3*a*). Such a modification effectively consists of a frequency transformation such that the location of the highest frequency infinite-value point is shifted to infinity. If the result is substituted in (2), we obtain a second general form for the even-order elliptic function:

$$N_b(s) = \frac{H_b \prod_{i=2}^{n/2} (s^2 + \Omega_i^2)}{a_0 + a_1 s + \cdots + a_{n-1} s^{n-1} + a_n s^n} \tag{7}$$

We call this *case B for n even*. The quantities Ω_i and a_i, of course, have different values from the ones in (6). The degree of the numerator is $n - 2$ (note that the product index i starts at 2, not 1), while that of the denominator is n. This function may be realized without coupled coils as long as the source and load resistance termination have specific, nonequal values. A price is required, however, for the simplification of the realization, in that for a given order and a given stopband attenuation, the stopband frequency ω_s lies somewhat higher than it does in case *A*. An example is shown for $n = 4$ in Fig. 2.4-3(*c*). The requirement for unequal resistance values is a result of the fact that the function $R_n^2(\omega)$ is finite at $\omega = 0$. If we make another frequency transformation such that the first zero location of $R_n^2(\omega)$ is effectively shifted to the origin, we obtain a network function having the general form

$$N_c(s) = \frac{H_c \prod_{i=2}^{n/2} (s^2 + \Omega_i^2)}{a_0 + a_1 s + \cdots + a_{n-1} s^{n-1} + a_n s^n} \tag{8}$$

This is *case C for n even*. An example of the form of this network function for $n = 4$ is shown in Fig. 2.4-3(*d*). Note that the magnitude at zero frequency is the maximum passband value. The stopband frequency ω_s will be even higher than that of case *B* as a result of the additional requirement on the form of the realization; however, this function can be realized with equal-resistance terminations.

Table 2.4-1 shows selected examples of odd- and even-order normalized elliptic functions, their pole and zero locations, and their quadratic denominator factors. A more complete table is given in App. B. Passive network elliptic realizations are given in App. A. For the even-order functions, cases *B* and *C* have been used. Additional realizations may be found in the references listed in the Bibliography.[2]

---

[2] The tables given in the references are usually catalogued using values of the reflection coefficient ρ, rather than of the passband ripple K_p. The values may be related by using $K_p = -10 \log [1 - (\rho/100)^2]$, where ρ is in percent. Some examples are given in the following table:

ρ (%)	1	2	5	10	15	20	25	50
K_p (dB)	0.00043	0.0017	0.011	0.044	0.098	0.18	0.28	1.25

In addition, the selection of the values of ω_s is made by the use of integer values of a modular angle θ, where $\omega_s = 1/\sin\theta$ ($\theta < 90°$) for case *A*. Unfortunately, the use of these parameters does not permit the specification of convenient numerical values of K_p and ω_s.

TABLE 2.4-1

Elliptic functions with poles at p_i and having the form $N(s) = H\prod_i \frac{s^2 + c_i}{s^2 + a_i s + b_i}$

Case	K_p, dB	n	ω_s	K_s, dB	c_i	p_i	a_i	b_i
					(a) Odd and case A even			
A	0.1	2	1.20	1.075	2.235990	$-0.236268 \pm j1.393844$	0.472537	1.998624
			2.00	7.418	7.464102	$-0.843443 \pm j1.581991$	1.686887	3.214092
Odd	0.1	3	1.20	6.691	1.699617	$-0.156766 \pm j1.170259$, -1.744102	0.313532	1.394082
			2.00	24.010	5.153209	$-0.381858 \pm j1.217905$, -1.116765	0.763717	1.628108
A	0.1	4	1.20	17.051	1.572430	$-0.108448 \pm j1.086869$	0.216897	1.193044
					6.224402	$-0.726853 \pm j0.798154$	1.453706	1.165365
			2.00	41.447	4.593261	$-0.670443 \pm j0.535639$	1.340886	0.736403
					24.227201	$-0.216254 \pm j1.116820$	0.432509	1.294053
A	1.0	2	1.20	6.150	2.235990	$-0.320565 \pm j1.064452$	0.641131	1.235820
			2.00	17.095	7.464102	$-0.499471 \pm j0.959482$	0.998942	1.170077
Odd	1.0	3	1.20	16.209	1.699617	$-0.136461 \pm j1.010059$, -0.701999	0.272923	1.038841
			2.00	34.454	5.153209	$-0.217034 \pm j0.981575$, -0.539958	0.434067	1.010594
A	1.0	4	1.20	27.432	1.572430	$-0.386971 \pm j0.560447$	0.773942	0.463847
					6.224402	$-0.075673 \pm j1.000256$	0.151346	1.006238
			2.00	51.906	4.593261	$-0.351273 \pm j0.442498$	0.702546	0.319197
					24.227201	$-0.121478 \pm j0.989176$	0.242957	0.993226
					(b) Case B (even)			
B	0.1	4	1.20	14.387	1.601406	$-0.118529 \pm j1.095964$	0.237059	1.215196
						$-0.746525 \pm j0.700487$	1.493659	1.045439
			2.00	38.697	4.716640	$-0.223115 \pm j1.118520$	0.446230	1.300868
						$-0.665767 \pm j0.521238$	1.331534	0.714935
B	1.0	4	1.20	24.700	1.601405	$-0.383555 \pm j0.523216$	0.767109	0.420870
						$-0.082727 \pm j1.000117$	0.165455	1.007078
			2.00	49.156	4.716540	$-0.348975 \pm j0.435853$	0.697949	0.311751
						$-0.124305 \pm j0.988442$	0.248611	0.992470
					(c) Case C even			
C	0.1	4	1.20	12.085	1.615455	$-0.128382 \pm j1.115527$	0.256764	1.260882
						$-0.953405 \pm j0.756606$	1.906811	1.481435
			2.00	36.023	4.733595	$-0.817435 \pm j0.520713$	1.634871	0.939343
						$-0.253437 \pm j1.142940$	0.506873	1.370542
C	1.0	4	1.20	22.293	1.615455	$-0.573215 \pm j0.472186$	1.146429	0.551534
						$-0.096950 \pm j1.003197$	0.193901	1.015804
			2.00	46.481	4.733595	$-0.499020 \pm j0.370743$	0.998040	0.386471
						$-0.145844 \pm j0.988871$	0.291689	0.999136

In using the tables it should be noted that a given elliptic network function and its realizations are characterized by four items of information: (1) the order n of the function; (2) the passband ripple constant K_p (see Fig. 2.4-1), expressed in decibels; (3) the stopband tolerance constant K_s similarly expressed; and (4) the stopband frequency ω_s. From data given in the table, we see that when any two of the three items 2 through 4 are specified, a network function of a given order is completely defined.

Example 2.4-2 Use of Table 2.4-1. It is desired to find an elliptic network function with the following specifications: (1) a passband of 0 to 1 rad/s with a ripple of 1.0 dB; (2) a stopband of $\omega \geq 2$ rad/s with a minimum attenuation of 34 dB; and (3) a maximum magnitude of unity in the passband. Using the 1.0-dB ripple entries of Table 2.4-1, we see that specifications 1 and 2 are satisfied by a third-order function. Using the tabulated poles and zeros, the function has the form

$$N(s) = \frac{H(s^2 + 5.153209)}{(s + 0.539958)(s^2 + 0.434067s + 1.010594)}$$

To satisfy specification 3, we evaluate $N(0) = H \times 5.153209/(0.539958 \times 1.010594) = 9.443675H = 1$, from which we obtain $H = 0.105891$. It should be noted that if a type A or B even-order function had been required to meet the specifications, then, to determine H, $N(0)$ would be set to 0.89125 (the magnitude equivalent of -1.0 dB) rather than to unity. This is because such even-order functions have their *minimum* passband magnitude at zero frequency, while odd-order and case C even-order functions have their *maximum* magnitude there.

Determination of Order

The determination of the order of an elliptic function follows the procedure outlined for the Butterworth function in Sec. 2.1 and the quantities shown in Fig. 2.4-1. The specifications are:

1. *Passband.* For $0 \leq \omega \leq \omega_p$ radians per second ($0 \leq f \leq f_p$ hertz), the ripple of the magnitude characteristic is K_p dB (the passband is normalized for $\omega_p = 1$ in Fig. 2.4-1).
2. *Stopband.* For $\omega \geq \omega_s$ radians per second ($f \geq f_s$ hertz), the equal-ripple attenuation varies between a minimum of K_s dB and infinity. This attenuation is measured from the greatest passband value of the characteristic.

To determine the order, we repeat the definitions given in (18) of Sec. 2.1:

$$\Omega = \frac{\omega_s}{\omega_p} = \frac{f_s}{f_p} \qquad M = \sqrt{\frac{10^{0.1K_s} - 1}{10^{0.1K_p} - 1}} \tag{9}$$

The elliptic order n_E is found by first determining the quantities[3]

$$C(M) = \frac{1}{16M^2}\left(1 + \frac{1}{2M^2}\right) \tag{10}$$

$$D(\Omega) = \frac{\sqrt{\Omega} - 1}{2(\sqrt{\Omega} + 1)} \tag{11}$$

The order n_E is given as

$$n_E = F_E(C)F_E(D) \tag{12}$$

where

$$F_E(x) = \frac{1}{\pi} \ln (x + 2x^5 + 15x^9) \tag{13}$$

The required order of the function is the next highest integer greater than n_E.

Example 2.4-3 Order of an elliptic function. An elliptic function must meet the following specifications (these are the same as were used for a Butterworth function in Example 2.1-6 and for a Chebyshev (equal-ripple) function in Example 2.2-3):

$$f = 1000 \text{ Hz} \qquad K = 3.0103 \text{ dB}$$
$$f = 1300 \text{ Hz} \qquad K = 22 \text{ dB}$$

From (9) we find $\Omega = 1.3$ and $M = 12.5495$. From (10) through (13) we find

$$C(M) = \frac{1}{16 \times (12.5495)^2}\left(1 + \frac{1}{2 \times (12.5495)^2}\right) = 0.0003981$$

$$D(\Omega) = \frac{\sqrt{1.3} - 1}{2(\sqrt{1.3} + 1)} = 0.03275$$

$$F_E(C) = \frac{1}{\pi} \ln (0.0003981) = -2.4920$$

$$F_E(D) = \frac{1}{\pi} \ln [0.03275 + 2(0.03275)^5] = -1.0883$$

$$n_E = (-2.4920)(-1.0883) = 2.7119$$

Note that the x^5 and x^9 terms are not needed in the computation of $F_E(C)$ and the x^9 term is not needed for $F_E(D)$. The specifications are met by a third-order elliptic function. Recall that 10th-order Butterworth and 5th-order Chebyshev functions were required for the same characteristics.

[3] Private communication to the author from Dr. George Szentermai. These relations have also appeared in P. M. Lin, "Single Curve for Determining the Order of an Elliptic Filter," *IEEE Trans. Circuits Syst.*, vol. 37, no. 9, September 1990, pp. 1181–1183.

Exercise 2.4-3. For each of the following sets of specifications, determine the order required for an elliptic function:

Passband		Stopband		Answers
ω_p (or f_p)	K_p, dB	ω_s (or f_s)	K_s, dB	n_E
1	3.0103	2	25	2.1210
1	3.0103	1.5	25	2.5786
1	3.0103	2	15	1.5373
1	3.0103	1.5	15	1.8689
1	1	2	25	2.4575
1	1	1.5	25	2.9877
1000	1	1100	25	4.4334

In this section we have introduced one of the most useful network functions for sharp cutoff filters, the elliptic one. Like the previously defined Butterworth, Chebyshev, and inverse-Chebyshev functions, the elliptic one has been defined for a low-pass characteristic. In the next section we shall see how all four of these characteristics can be used in high-pass and band-pass filters.

2.5 TRANSFORMATIONS OF THE COMPLEX-FREQUENCY VARIABLE

In the preceding sections of this chapter we have considered four methods for approximating magnitude characteristics. The techniques that were developed all apply to low-pass network functions. In this section we shall show how these approximations may be extended to other types of network functions. The types to be considered are the high-pass, band-pass, and band-elimination ones. The extension is done through the use of transformations made on the complex-frequency variable. We shall discuss the use of these transformations from three different viewpoints: (1) their effect on the magnitude characteristic, (2) their effect on the network function, and (3) their effect on the elements of a given network realization.

Low-Pass to High-Pass Transformation

The first transformation of the complex-frequency variable we shall describe is called the *normalized low-pass to high-pass transformation*. If we let $s = \sigma + j\omega$ be the original complex-frequency variable and $p = u + jv$ be the resulting transformed complex-frequency variable, then the transformation is defined as

$$s = \sigma + j\omega = \frac{1}{p} = \frac{1}{u + jv} = \frac{u}{u^2 + v^2} - j\frac{v}{u^2 + v^2} \tag{1}$$

In this relation, if we confine our range of interest to the sinusoidal steady-state case by

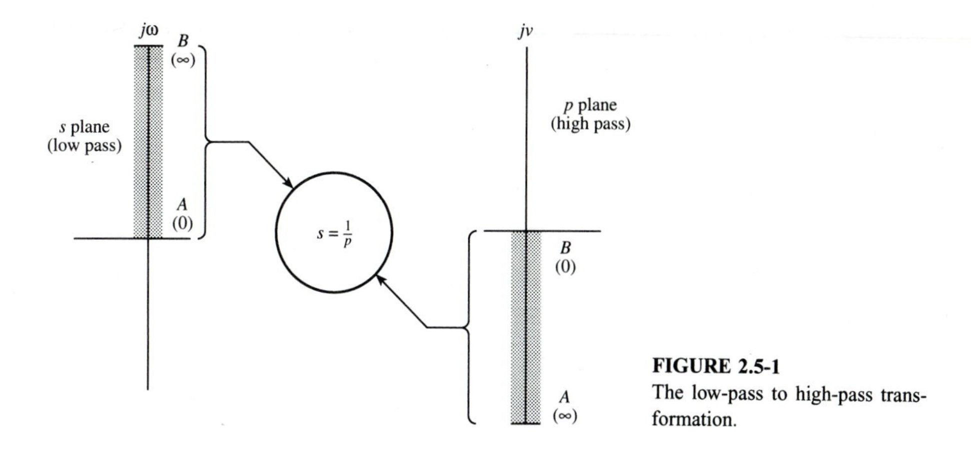

FIGURE 2.5-1
The low-pass to high-pass transformation.

letting $\sigma = 0$, then equating real and imaginary parts in (1), we obtain

$$u = 0 \qquad \omega = \frac{-1}{v} \tag{2}$$

The positive imaginary axis in the (original) s plane becomes the negative imaginary axis in the (transformed) p plane. In addition, as shown in Fig. 2.5-1, the points at the origin and infinity are interchanged as indicated by A and B in the figure. A similar transformation occurs between the negative imaginary axis in the s plane and the positive imaginary axis in the p plane. As a result of this interchange, a low-pass magnitude characteristic on the $j\omega$ axis is transformed to a high-pass one on the jv axis.[1] From (1) we also see that, under this transformation, corresponding points of the

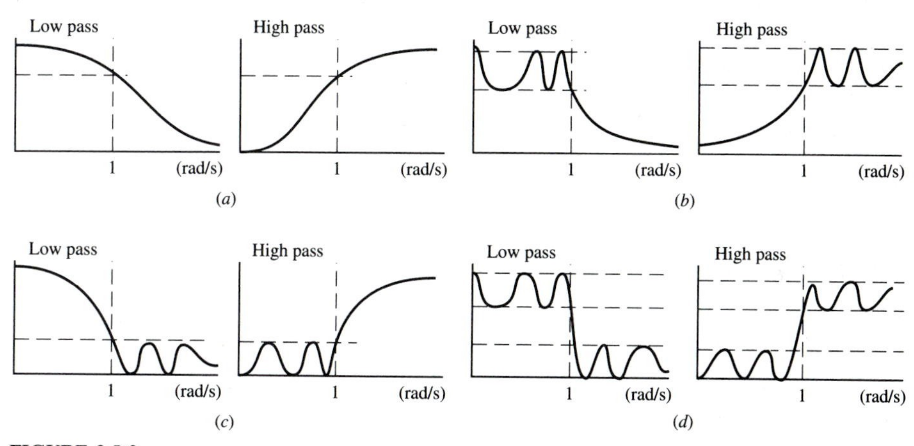

FIGURE 2.5-2
Examples of low-pass to high-pass transformations.

[1] Since magnitude characteristics are symmetrical about the origin, the reversal of sign has no effect.

magnitude characteristics on the $j\omega$ and jv axes are geometrically centered on the frequency 1 rad/s. Plots illustrating the application of the transformation to maximally flat, Chebyshev, inverse-Chebyshev, and elliptic low-pass magnitude characteristics are shown in Fig. 2.5-2.

Example 2.5-1 Low-pass and high-pass magnitude characteristics. As an example of the application of the normalized low-pass to high-pass transformation, consider a fourth-order low-pass Chebyshev function with a passband ripple of 3.0103 dB. Plots of the low-pass and high-pass magnitude are shown in Fig. 2.5-3. Note that the frequencies ω_i where the peaks and valleys occur in the low-pass plot and the frequencies v_i where corresponding peaks and valleys occur in the high-pass plot are reciprocal, that is, $\omega_i v_i = 1$.

In the preceding paragraph we have illustrated the effect of the transformation given in (1) on the magnitude characteristic. Now let us see what effect it has on the network function. We begin by considering a general low-pass function (with all its zeros at infinity) having the form

$$N_{LP}(s) = \frac{H}{a_0 + a_1 s + a_2 s^2 + \cdots + a_{n-1}s^{n-1} + s^n} \tag{3}$$

Applying the transformation and multiplying numerator and denominator by p^n, we obtain

$$N_{HP}(p) = N_{LP}\left(\frac{1}{p}\right) = \frac{Hp^n}{a_0 p^n + a_1 p^{n-1} + a_2 p^{n-2} + \cdots + a_{n-1}p + 1} \tag{4}$$

Obviously, the n zeros at infinity of $N_{LP}(s)$ have been transformed into n zeros at the origin in $N_{HP}(p)$.

Example 2.5-2 Low-pass and high-pass network functions. As an example of the effects of the low-pass to high-pass transformation on network functions, consider the two

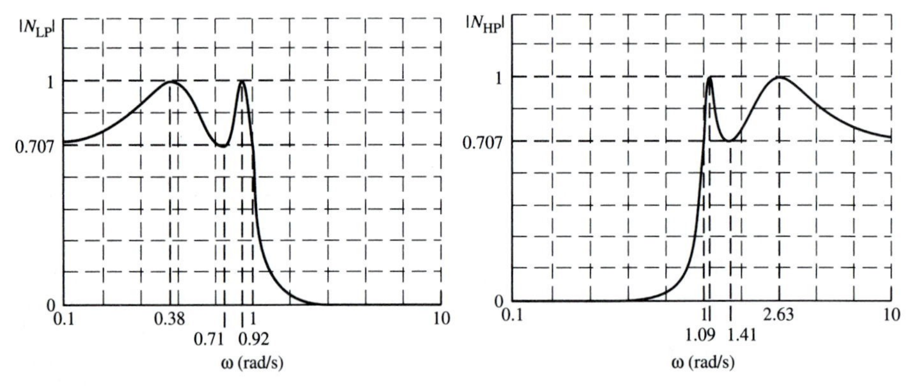

FIGURE 2.5-3
Low-pass and high-pass equal-ripple functions.

functions

$$N_{\mathrm{LP}}(s) = \frac{H}{s^3 + 2s^2 + 2s + 1} \qquad N_{\mathrm{HP}}(p) = \frac{Hp^3}{p^3 + 2p^2 + 2p + 1} \tag{5}$$

The function $N_{\mathrm{LP}}(s)$ is readily identified (see Table 2.1-3) as a third-order maximally flat magnitude low-pass function with a 3-dB down frequency of 1 rad/s. Since $N_{\mathrm{HP}}(p)$ is derived from $N_{\mathrm{LP}}(s)$ by the transformation $s = 1/p$, it is a third-order maximally flat magnitude high-pass function and its 3-dB down frequency is also 1 rad/s. For this function, the term *maximally flat magnitude* of course refers to its behavior at infinite frequency. The function $N_{\mathrm{HP}}(p)$ is easily modified to have its 3-dB down behavior at any desired frequency by applying a frequency denormalization. For example, for a 3-dB frequency of 1 kHz we choose $\Omega_n = 2\pi \times 10^3$ and obtain

$$N_{\mathrm{HP\text{-}Denorm}}(p) = \frac{Hp^3}{p^3 + 2\Omega_n p^2 + 2\Omega_n^2 p + \Omega_n^3}$$

From the derivation given above we see that, for normalized Butterworth low-pass functions with 3-dB down frequencies of 1 rad/s, due to the symmetry of the denominator polynomial coefficients, the pole locations for the normalized high-pass functions are the same as those of the low-pass ones. For other magnitude characteristics this is not the case, and as shown by (1), the high-pass poles are the reciprocal of the low-pass ones.

Exercise 2.5-2. Find the high-pass network function denominator polynomial (in the form $a_0 + a_1 p + \cdots + a_{n-1}p^{n-1} + p^n$) and its roots for each of the following cases:

Case	n	Type of low-pass characteristic	Starting frequency for high-pass passband, rad/s
(*a*)	4	Maximally flat	1
(*b*)	4	Maximally flat	2
(*c*)	3	1-dB equal ripple	1
(*d*)	3	1-dB equal ripple	2
(*e*)	4	1-dB equal ripple	1
(*f*)	4	1-dB equal ripple	2

Answers

Case	a_0	a_1	a_2	a_3	p_1	p_2
(*a*)	1.0000	2.6131	3.4142	2.6131	$-0.3827 \pm j0.9239$	-0.9239 ± 0.3827
(*b*)	16.0000	20.9050	13.6569	5.2263	$-0.7654 \pm j1.8478$	$-1.8478 \pm j0.7654$
(*c*)	2.0354	2.0117	2.5206	—	-2.0236	$-0.2485 \pm j0.9716$
(*d*)	16.2831	8.0466	5.0413	—	-4.0472	$-0.4971 \pm j1.9433$
(*e*)	3.6281	3.4569	5.2750	2.6943	$-1.2057 \pm j1.4579$	$-0.1414 \pm j0.9968$
(*f*)	58.0493	27.6550	21.0998	5.3886	$-2.4114 \pm j2.9158$	$-0.2829 \pm j1.9937$

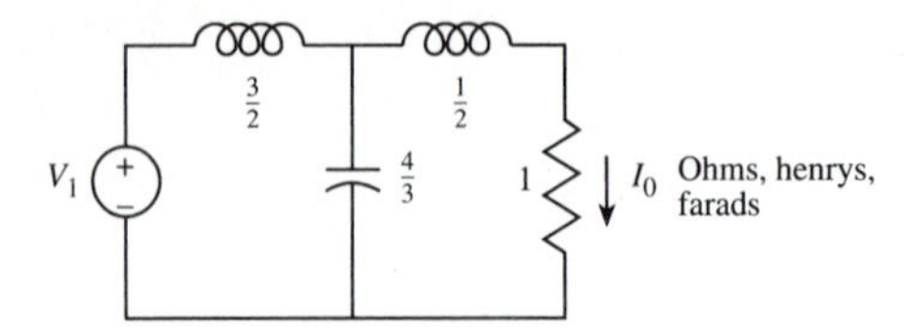

FIGURE 2.5-4
A third-order maximally flat magnitude low-pass filter.

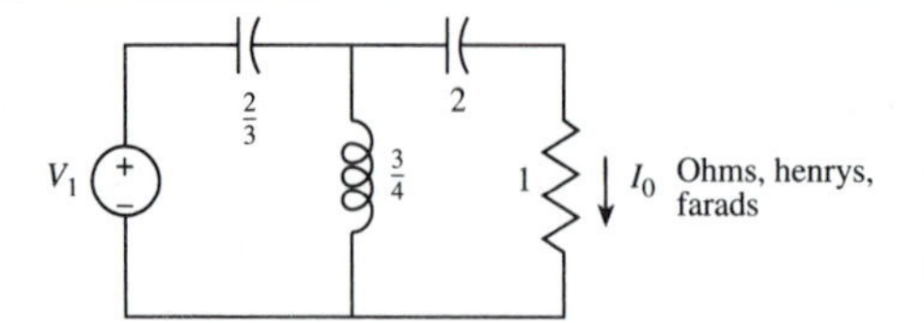

FIGURE 2.5-5
A third-order maximally flat magnitude high-pass filter obtained by transforming the filter of Fig. 2.5-4.

The transformation defined by (1) can also be applied directly to the elements of a network realization. Thus, an impedance $Z_{\text{LP}}(s) = Ks$, defining an inductor of K henrys, becomes, as a result of the transformation, an impedance $Z_{\text{HP}}(p) = K/p$, which defines a capacitor of $1/K$ farads. Similarly, an admittance $Y_{\text{LP}}(s) = Ks$, defining a capacitor of K farads, becomes an admittance $Y_{\text{HP}}(p) = K/p$, which defines an inductor of $1/K$ henrys.

Example 2.5-3 Low-pass and high-pass networks. As an example of applying the transformation of (1) directly to the elements of a network realization, consider the third-order maximally flat magnitude low-pass network realization shown in Fig. 2.5-4. This network is shown in App. A to have the transfer admittance given by $N_{\text{LP}}(s)$ in (5) (with $H = 1$). Applying the transformation $s = 1/p$ to the network elements, we obtain the realization shown in Fig. 2.5-5. Since applying the transformation to the network elements produces the same result as applying the transformation to the network function, we conclude that the transfer admittance of this network is given by $N_{\text{HP}}(p)$ in (5) (with $H = 1$).

Low-Pass to Band-Pass Transformation

The second transformation of the complex-frequency variable that we shall consider in this section is the *normalized low-pass to band-pass transformation*. Like the low-pass to high-pass transformation described above, this transformation may of course be applied to transform any type of magnitude characteristic; however, when it is applied to a low-pass characteristic, it produces a band-pass one. The transformation has the form

$$s = p + \frac{1}{p} = \frac{p^2 + 1}{p} \tag{6}$$

where s is the low-pass variable and p is the band-pass one. The resulting band-pass characteristic has a center frequency of unity and a bandwidth that is the same as that of the low-pass function. If we solve for p in (6), we obtain

$$p = \frac{s}{2} \pm \sqrt{\left(\frac{s}{2}\right)^2 - 1} \tag{7}$$

If we now confine our attention to the sinusoidal steady-state case by letting $s = \sigma + j\omega$ and set $\sigma = 0$, then for $p = u + jv$ we find from (7) that $u = 0$ and v is given by

$$v = \frac{\omega}{2} \pm \sqrt{\left(\frac{\omega}{2}\right)^2 + 1} \tag{8}$$

Thus, the imaginary axis in the s plane transforms to the imaginary axis in the p plane. The nature of the transformation may be further defined by noting that, from (7), the point $s = 0$ transforms into the two points $p = \pm j1$. Similarly, the point $s = \infty$ transforms into the two points $p = 0$ and $p = \infty$. Finally, using (8) we see that any arbitrary point on the positive imaginary axis of the s plane, defined as $s = j\omega_1$, transforms into two points jv_2 and $-jv_1$ on the p plane:

$$-v_1 = \frac{\omega_1}{2} - \sqrt{\left(\frac{\omega_1}{2}\right)^2 + 1} \qquad v_2 = \frac{\omega_1}{2} + \sqrt{\left(\frac{\omega_1}{2}\right)^2 + 1} \tag{9}$$

where v_1 and v_2 are both positive and $v_2 > v_1$. The point $-j\omega_1$ on the s plane is similarly transformed into the points jv_1 and $-jv_2$ on the p plane. Now let us consider the points v_1 and v_2 in more detail. From (9) we find that

$$v_1 v_2 = 1 \qquad v_2 - v_1 = \omega_1 \tag{10}$$

These results may be summarized as follows:

Summary 2.5-1 Properties of the low-pass to band-pass transformation. The normalized low-pass to band-pass transformation has the following effects on the magnitude characteristics:

1. Any frequency ω_1 designating a value of $|N_{LP}(j\omega_1)|$ for the low-pass magnitude is transformed by (9) into two frequencies v_1 and v_2 that determine the values $|N_{BP}(jv_1)|$ and $|N_{BP}(jv_2)|$ of the band-pass magnitude. The magnitudes have the same value, namely,

$$|N_{LP}(j\omega_1)| = |N_{BP}(jv_1)| = |N_{BP}(jv_2)|$$

2. The frequencies v_1 and v_2 satisfy the relation $v_1 v_2 = 1$, that is, their geometric mean is unity.
3. The frequencies v_1 and v_2 satisfy the relation $v_2 - v_1 = \omega_1$ (assuming $v_2 > v_1$), that is, the band-pass bandwidth is equal to the low-pass bandwidth.

Example 2.5-4 Low-pass and band-pass magnitude characteristics. As an example of the application of the normalized low-pass to band-pass transformation, consider a second-order Chebyshev function with a passband ripple of 3.0103 dB. Plots of the low-pass and band-pass magnitude are shown in Fig. 2.5-6. Note that for each frequency ω_i at which a characteristic low-pass behavior occurs, there are two band-pass frequencies v_{1i} and

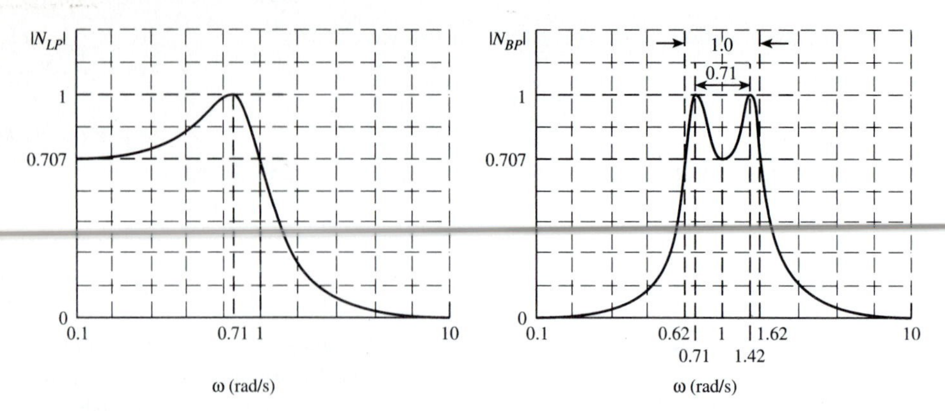

FIGURE 2.5-6
Low-pass and band-pass equal-ripple magnitude functions.

ν_{2i} at which the corresponding behavior occurs. For example, the low-pass passband peak generates two peaks in the band-pass passband characteristic. The low-pass bandwidth to the peak is the distance from the origin to the frequency $\omega_{\text{peak}} = 0.71$ rad/s. The band-passpassband peaks occur at $\nu_{1\text{-peak}} = 0.71$ and $\nu_{2\text{-peak}} = 1.412$ rad/s; thus $\nu_{2\text{-peak}} - \nu_{1\text{-peak}} = \omega_{\text{peak}}$. As another example, the low-pass passband bandwidth, the distance from the origin to the 3-dB cutoff frequency, is $\omega_{3\text{dB}} = 1$ rad/s. The band-pass passband edges occur at $\nu_{1,3\text{dB}} = 0.62$ and $\nu_{2,3\text{dB}} = 1.62$ rad/s; thus $\nu_{2,3\text{dB}} - \nu_{1,3\text{dB}} = \omega_{3\text{dB}}$. To produce different passband bandwidths, all that is necessary is to make an appropriate frequency denormalization of the low-pass function *before* applying the low-pass to band-pass transformation. As an example, in Fig. 2.5-7, the Chebyshev characteristic has been frequency denormalized so that the 3-dB frequency is 5 rad/s. The result, as shown in the figure, is a band-pass function with a 3-dB bandwidth of 5 rad/s. This would be called a *broadband* band-pass filter. If the frequency denormalization had been made in the opposite direction, a *narrow-band* band-pass filter would result.

Exercise 2.5-4. A low-pass network is defined with an ideal (brick wall) cutoff frequency of ω_s radians per second. For each of the cutoff frequency values given below, find the band-

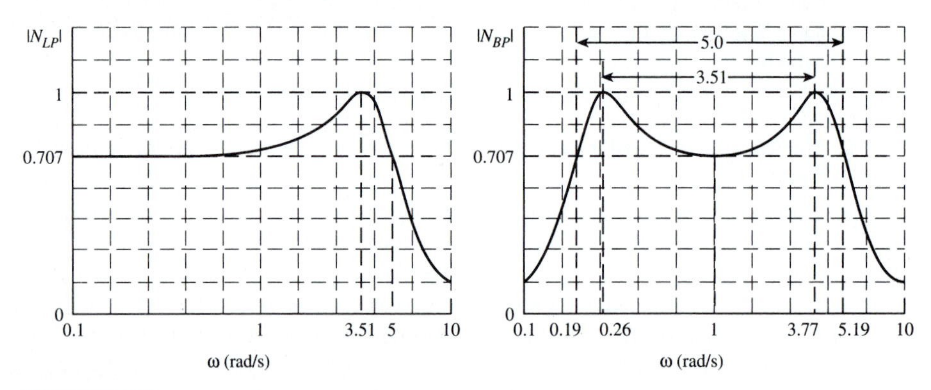

FIGURE 2.5-7
Low-pass and broadband band-pass equal-ripple magnitude functions.

pass cutoff frequencies v_1 and v_2 radians per second obtained from applying the normalized low-pass to band-pass transformation.

Low-pass cutoff frequency ω_s, rad/s	Answers	
	v_1	v_2
0.1	0.9512	1.0512
0.2	0.9050	1.1050
0.5	0.7807	1.2807
2.0	0.4142	2.4142
5.0	0.1926	5.1926

Now let us consider what happens when the transformation of (6) is applied to the general low-pass network function given in (3). After multiplying the numerator and denominator by p^n, we obtain

$$N_{\mathrm{BP}}(p) = \frac{Hp^n}{a_0 p^n + a_1 p^{n-1}(p^2+1) + a_2 p^{n-2}(p^2+1)^2 + \cdots + a_{n-1} p(p^2+1)^{n-1} + (p^2+1)^n} \tag{11}$$

The general resulting denominator polynomials for typical values of n are given in Table 2.5-1. Each resulting band-pass function has an nth-order zero at the origin and an nth-order zero at infinity.

Example 2.5-5 Low-pass and band-pass network functions. If the low-pass to band-pass transformation is applied to $N_{\mathrm{LP}}(s)$ given in (5), we obtain

$$N_{\mathrm{BP}}(p) = \frac{H}{[(p^2+1)/p]^3 + 2[(p^2+1)/p]^2 + 2[(p^2+1)/p] + 1}$$

$$= \frac{Hp^3}{p^6 + 2p^5 + 5p^4 + 5p^3 + 5p^2 + 2p + 1}$$

TABLE 2.5-1

Low-pass to band-pass transformation with normalized band-pass center frequency of 1 rad/s

$$N_{\mathrm{LP}}(s) = \frac{H}{a_0 + a_1 s + \cdots + a_{n-1} s^{n-1} + s^n}$$

$$N_{\mathrm{BP}}(p) = \frac{Hp^n}{1 + b_1 p + b_2 p^2 + \cdots + b_2 p^{2n-2} + b_1 p^{2n-1} + p^{2n}}$$

n	b_1	b_2	b_3	b_4	b_5
2	a_1	$a_0 + 2$			
3	a_2	$a_1 + 3$	$a_0 + 2a_2$		
4	a_3	$a_2 + 4$	$a_1 + 3a_3$	$a_0 + 2a_2 + 6$	
5	a_4	$a_3 + 5$	$a_2 + 4a_4$	$a_1 + 3a_3 + 10$	$a_0 + 2a_2 + 6a_4$

Note that, as indicated in Table 2.5-1, the coefficients are symmetrical in ascending and descending powers of p, and the zero-degree coefficient is unity. To find the pole locations, we let $s_1 = -1$ and apply (7). The computational steps are

$$\tfrac{1}{2}s_1 = -0.5 \qquad (\tfrac{1}{2}s_1)^2 = 0.25 \qquad (\tfrac{1}{2}s_1)^2 - 1 = -0.75$$

$$\sqrt{(\tfrac{1}{2}s_1)^2 - 1} = j0.8660$$

The band-pass poles are at $-0.5 \pm j0.8660$. Similarly, for the low-pass pole $s_2 = -0.5 + j0.8660$ we find

$$\tfrac{1}{2}s_2 = -0.25 + j0.4330 \qquad (\tfrac{1}{2}s_2)^2 = -0.125 - j0.2165 \qquad (\tfrac{1}{2}s_2)^2 - 1 = -1.125 - j0.2165$$

$$\sqrt{(\tfrac{1}{2}s_2)^2 - 1} = 0.1016 - j1.0655$$

The poles are at $-0.1484 - j0.6325$ (for the positive square root) and $-0.3516 + j1.4985$ (for the negative square root). The use of the low-pass pole $s_3 = -0.5 - j0.8660$ (the conjugate of s_2) gives the conjugate of these poles. We can now write the band-pass denominator in the factored form

$$N_{BP}(p) = \frac{Hp^3}{(p^2 + p + 1)(p^2 + 0.2968p + 0.4221)(p^2 + 0.7032p + 2.3691)}$$

Exercise 2.5-5. A set of low-pass maximally flat magnitude network functions are defined by their order n and their 3-dB frequency ω_p radians per second. Find the band-pass functions and pole locations that result from applying the low-pass to band-pass transformation. Use the notation for $N_{BP}(p)$ given in Table 2.5-1.

			Answers		
n	ω_p	b_1	b_2	b_3	**Poles**
2	1.0	1.4142	3.0000	—	$-0.2295 \pm j0.6541$, $-0.4776 \pm j1.3612$
2	0.5	0.7071	2.2500	—	$-0.1455 \pm j0.8237$, $-0.2080 \pm j1.1773$
3	0.5	1.0000	3.5000	2.1250	$-0.0984 \pm j0.7994$, $-0.1516 \pm j1.2324$, $-0.2500 \pm j0.9683$
3	0.1	0.2000	3.0200	0.4010	$-0.0151 \pm j0.6565$, $-0.0349 \pm j1.5225$, $-0.0500 \pm j0.9988$

The transformation defined by (6) can also be applied directly to the elements of a given network realization. Thus an inductor of K henrys, with an impedance $Z_{LP}(s) = Ks$, becomes $Z_{BP}(p) = Kp + K/p$, a *series connection* of an inductor of K henrys and a capacitor of $1/K$ farads; while a capacitor of K farads with an admittance $Y_{LP}(s) = Ks$ becomes $Y_{BP}(p) = Kp + K/p$, a *parallel connection* of a capacitor of K farads and an inductor of $1/K$ henrys.

Example 2.5-6 Low-pass and band-pass networks. As an example of applying the transformation of (6) directly to a low-pass network, the network realization for a third-order maximally flat magnitude low-pass transfer admittance shown in Fig. 2.5-4 may be directly transformed to realize a sixth-order maximally flat magnitude transfer admittance

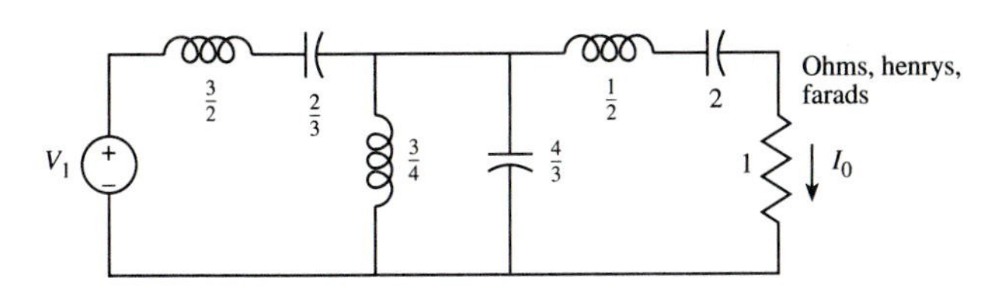

FIGURE 2.5-8
A sixth-order maximally flat magnitude band-pass filter obtained by transforming the network of Fig. 2.5-4.

with a bandwidth of 1 rad/s and 3-dB down frequencies of 0.618 and 1.618 rad/s. The resulting network is shown in Fig. 2.5-8. The transfer admittance of this network is given in Example 2.5-5 (with $H = 1$).

The normalized low-pass to high-pass and low-pass to band-pass transformations defined in (1) and (6), coupled with appropriate frequency normalizations, may be applied in various sequences to obtain any desired combination of center frequency and bandwidth. In addition, if a low-pass to band-pass transformation is applied to a high-pass network, a *band-elimination* characteristic results. An example of this follows.

Example 2.5-7 Band-elimination filter. A sixth-order band-elimination filter realizing a transfer admittance is to have a resistance termination of 1000 Ω, a bandwidth of 1 kHz, and a (geometric) center frequency of 5 kHz. The specifications require a monotonic magnitude characteristic and an attenuation of 15 dB at the band edges. We first normalize the bandwidth and center-frequency specifications to 0.2 and 1 rad/s, respectively. These normalized specifications can be met by applying the low-pass to band-pass transformation to a maximally flat high-pass filter with 15 dB of attenuation at 0.2 rad/s. The prototype low-pass filter would thus require 15 dB of attenuation at 5 rad/s. From (21) of Sec. 2.1 we find that a normalized third-order low-pass Butterworth filter has 15 dB of attenuation at 1.7688 rad/s. The network shown in Fig. 2.5-4 has this property. If we frequency normalize the filter by $5/1.7688 = 2.8268$, the 15-dB frequency will be moved to 5 rad/s. The result is shown in Fig. 2.5-9(*a*). Applying the low-pass to high-pass transformation, we obtain the network in Fig. 2.5-9(*b*). This has 15 dB of attenuation at 0.2 rad/s. Applying the low-pass to band-pass transformation, we obtain the filter of Fig. 2.5-9(*c*). This is a band-elimination filter with 15-dB frequencies 0.2 rad/s apart and a center frequency of 1 rad/s. Finally, applying a frequency denormalization of $2\pi \times 5000$ and an impedance denormalization of 1000, we obtain the filter shown in Fig. 2.5-9(*d*), which meets the specifications.

A summary of the effects on network elements of the various transformations introduced in this section is given in Table 2.5-2. For convenience, various denormalization factors have been included in the table.

Narrow-Band Approximation

If the bandwidth BW of a band-pass function satisfies the relation $\text{BW} \le 0.1\omega_0$ where ω_0 is the center frequency, the result is defined as a *narrow-band* band-pass function. In this case, as an alternate procedure to the use of (6) or (7), we may develop a simplified

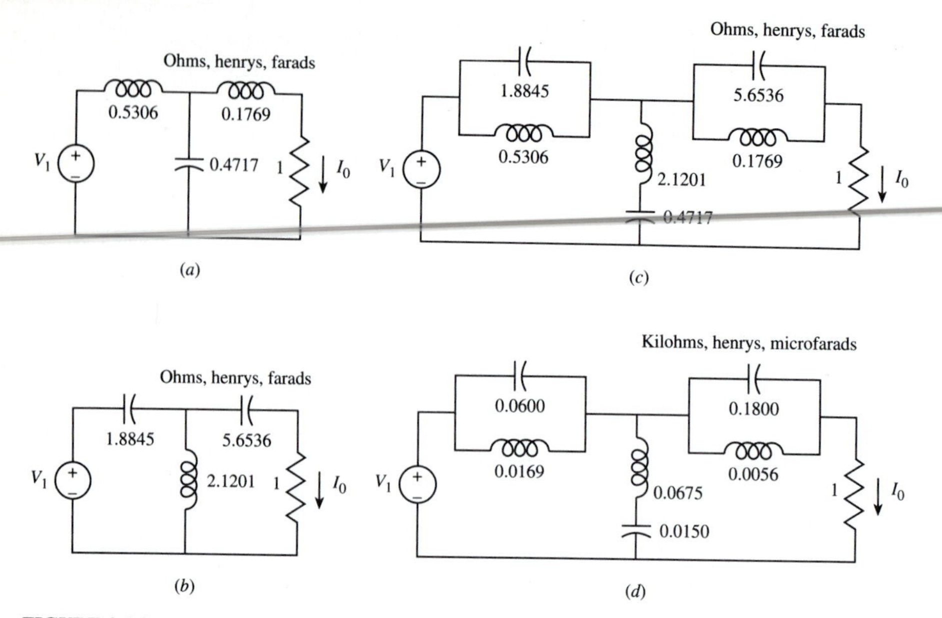

FIGURE 2.5-9
The band-elimination filter realized in Example 2.5-7.

method for directly determining the pole positions resulting from the low-pass to band-pass transformation. This allows us to produce a band-pass function with its denominator polynomial in factored form. The procedure is referred to as the *narrow-band approximation.* To see how it operates, let us assume that the center frequency of the band-pass function is to be 1 rad/s. We now define an intermediate frequency variable p' by the relation

$$p = p' + j1 \tag{12}$$

where p is the complex-frequency variable of the band-pass function. This transformation puts the origin of the p' plane at the point $j1$ in the p plane. Substituting this transformation in (6) and using the binomial expansion, we obtain

$$\begin{aligned} s &= (p' + j1) + \frac{1}{p' + j1} = (p' + j1) + \frac{-j}{1 + p'/j} \\ &= (p' + j1) - j\left(1 - \frac{p'}{j} + \cdots\right) \approx 2p' \end{aligned} \tag{13}$$

where the last relation given in the right member of (13) is a first-order approximation valid for $p' \approx 0$, that is, valid in the vicinity of the center-frequency location $j1$ in the p plane. This relation together with (12) defines the narrow-band approximation, which may be directly applied to pole and zero locations in the (low-pass) s plane to determine

TABLE 2.5-2
Changes of network elements under frequency transformations

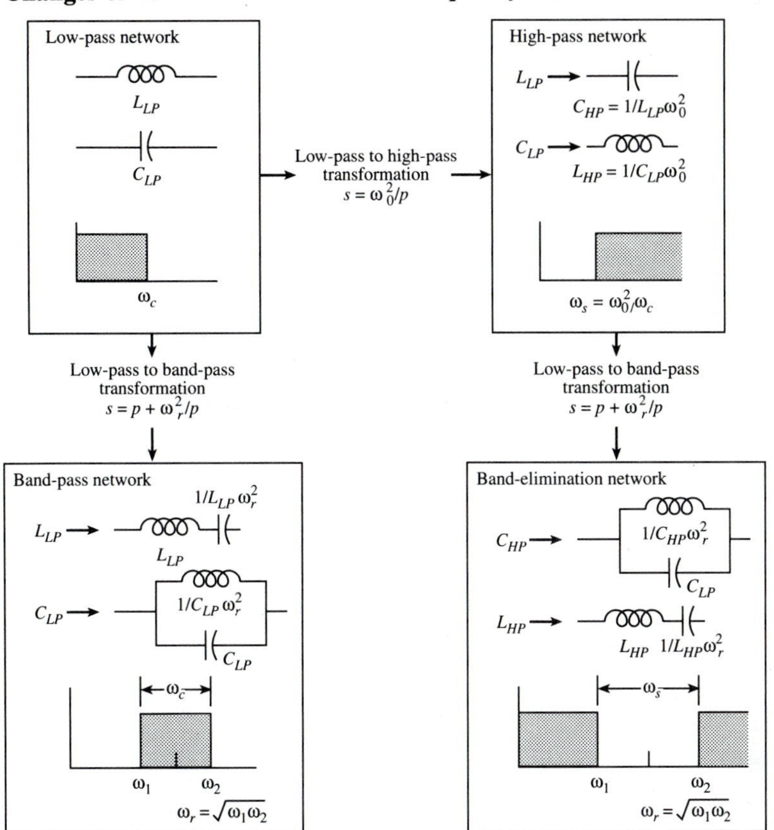

locations in the (band-pass) p plane. Specifically, comparing the relation of the p-plane poles to the point $p = j1$, with the relation of the s-plane poles to the origin, we see that the angles are identical but that the p-plane poles are only half as far away from the referenced point. Note that the narrow-band approximation only produces the upper left half-plane pole locations. The conjugate locations and the $\frac{1}{2}n$ zeros at the origin must be added to complete the network function.

Example 2.5-8 Narrow-band approximation. Consider the use of the narrow-band approximation to determine the pole locations for a four-pole maximally flat magnitude band-pass function with a 3-dB bandwidth of 0.05 rad/s and a center frequency of 1 rad/s. The s-plane pole locations for a two-pole low-pass maximally flat magnitude function with a 3-dB bandwidth of 0.05 rad/s are $-0.05/\sqrt{2} \pm j0.05/\sqrt{2}$, as shown in Fig. 2.5-10(a). The p-plane pole locations determined by (12) and (13) are $-0.025/\sqrt{2} \pm j(1 \pm 0.025/\sqrt{2})$, as shown in Fig. 2.5-10(b). Adding the obvious requirements of a second-order zero at the

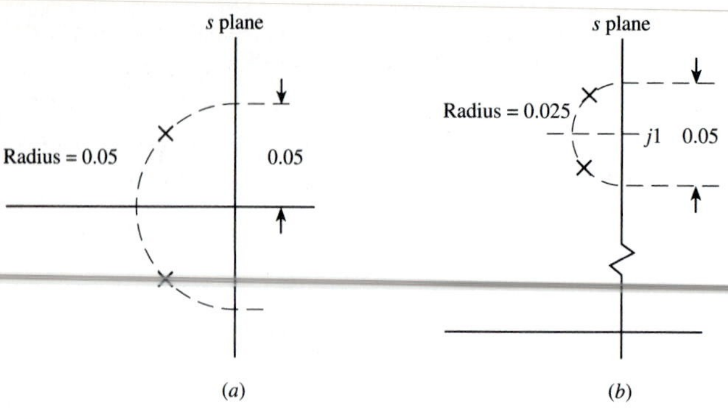

FIGURE 2.5-10
An example of the use of the narrow-band approximation.

origin and the conjugate poles, the resulting network function $N_{NB}(p)$ is

$$N_{NB}(p) = \frac{Hp^2}{\left[p + \frac{0.025}{\sqrt{2}} + j\left(1 + \frac{0.025}{\sqrt{2}}\right)\right]\left[p + \frac{0.025}{\sqrt{2}} - j\left(1 + \frac{0.025}{\sqrt{2}}\right)\right] \times \left[p + \frac{0.025}{\sqrt{2}} + j\left(1 - \frac{0.025}{\sqrt{2}}\right)\right]\left[p + \frac{0.025}{\sqrt{2}} - j\left(1 - \frac{0.025}{\sqrt{2}}\right)\right]}$$

$$= \frac{Hp^2}{p^4 + 0.0707p^3 + 2.0025p^2 + 0.0707p + 1.0000}$$

The actual network function $N_{BP}(p)$ obtained by directly applying (6) is identical with the unfactored form above within the given number of significant figures.

Exercise 2.5-8. A set of low-pass maximally flat magnitude network functions are defined by their order n and their 3-dB frequency ω_p radians per second. Use the narrow-band approximation to find the band-pass pole locations resulting from using the normalized low-pass to band-pass transformation.

n	ω_p	Answers (pole locations)
2	0.1	$-0.03535 \pm j(1 \pm 0.03535)$
3	0.1	$-0.05 \pm j1,\ -0.025 \pm j(1 \pm 0.04330)$
3	0.05	$-0.025 \pm j1,\ -0.0125 \pm j(1 \pm 0.02165)$
3	0.01	$-0.005 \pm j1,\ -0.0025 \pm j(1 \pm 0.004330)$

2.6 PHASE APPROXIMATION

In the preceding sections of this chapter we have discussed various methods for approximating a magnitude function. In many filtering specifications, the magnitude characteristic is of dominant importance. For example, in voice or audio-frequency applications, phase is of minor concern due to the relative insensitivity of the ear to changes in phase. In other applications, for example, the transmission of video or digital signals, phase characteristics become the dominant factor. In this section we consider the approximation of a phase function. Specifically, we desire to find the pole and zero locations for a network function that has some specified phase characteristic.

The Function $A(\omega)$

To begin our study of phase approximation, let us consider a general rational network function that has been put in the form

$$N(s) = \frac{C(s)}{B(s)} = \frac{m_1(s) + n_1(s)}{m_2(s) + n_2(s)} \tag{1}$$

where $m_1(s)$ and $n_1(s)$ are the even and odd parts, respectively, of the numerator polynomial $C(s)$ and $m_2(s)$ and $n_2(s)$ are the even and odd parts of the denominator polynomial $B(s)$. If we now consider $N(s)$ under sinusoidal steady-state conditions, i.e., by letting $s = j\omega$, then $m_1(j\omega)$ and $m_2(j\omega)$ being even are real, while $n_1(j\omega)$ and $n_2(j\omega)$ being odd are imaginary. Thus, the function $N(j\omega)$ may be written

$$N(j\omega) = \frac{\text{Re } C(j\omega) + j \text{ Im } C(j\omega)}{\text{Re } B(j\omega) + j \text{ Im } B(j\omega)} = \frac{m_1(j\omega) + n_1(j\omega)}{m_2(j\omega) + n_2(j\omega)} \tag{2}$$

Rationalizing this by multiplying the numerator and denominator by the conjugate of the denominator, $m_2(j\omega) - n_2(j\omega)$, we obtain

$$N(j\omega) = \text{Re } N(j\omega) + j \text{ Im } N(j\omega)$$

$$= \frac{m_1 m_2 - n_1 n_2}{m_2^2 - n_2^2} + \frac{m_2 n_1 - m_1 n_2}{m_2^2 - n_2^2} \tag{3}$$

where, for convenience, we have deleted the functional notation on the quantities m_i and n_i. The phase or argument of $N(j\omega)$ is now defined as

$$\arg N(j\omega) = \tan^{-1} \frac{\text{Im } N(j\omega)}{\text{Re } N(j\omega)} = \tan^{-1} \frac{1}{j} \frac{m_2 n_1 - m_1 n_2}{m_1 m_2 - n_1 n_2} \tag{4}$$

It is now convenient to define a function $A(\omega)$ as

$$A(\omega) = \tan\,[\arg N(j\omega)] = \frac{1}{j} \frac{m_2 n_1 - m_1 n_2}{m_1 m_2 - n_1 n_2} \tag{5}$$

From (5) we conclude that a necessary condition for the function $A(\omega)$ is that it be an odd rational function, that is, the ratio of an odd polynomial to an even one.

Example 2.6-1 Determination of the function $A(\omega)$. As an example, consider the network function

$$N(s) = \frac{H}{s^2 + \sqrt{2}s + 1} = \frac{H}{(s^2 + 1) + \sqrt{2}s}$$

From (2) we find

$$m_1(j\omega) = H \qquad n_1(j\omega) = 0 \qquad m_2(j\omega) = 1 - \omega^2 \qquad n_2(j\omega) = j\sqrt{2}\omega$$

From (5) we obtain

$$A(\omega) = \frac{1}{j}\frac{(1 - \omega^2)(0) - (H)(j\sqrt{2}\omega)}{(H)(1 - \omega^2) - (0)(j\sqrt{2}\omega)} = \frac{-\sqrt{2}\omega}{1 - \omega^2}$$

Finding $N(s)$ from $A(\omega)$

The requirement that $A(\omega)$ be an odd rational function is not only necessary, it is sufficient for the determination of a rational network function $N(s)$ from $A(\omega)$. To show the sufficiency, we first form the function $1 + jA(\omega)$. From (5) we obtain

$$1 + jA(\omega) = \frac{m_1 m_2 - n_1 n_2 + m_2 n_1 - m_1 n_2}{m_1 m_2 - n_1 n_2} = \frac{(m_1 + n_1)(m_2 - n_2)}{m_1 m_2 - n_1 n_2} \tag{6}$$

From the form of the right member of the above relation we conclude that the numerator of the function $1 + jA(s/j)$ must contain, as factors, the zeros and the right-half-plane reflection of the poles of the network function $N(s)$, which is related to the function $A(\omega)$ by (5). Any desired assignment of these factors that is consistent with stability requirements thus satisfies the sufficiency condition. An example follows.

Example 2.6-2 Determination of a network function from its argument. As an example of the procedure given above consider the function

$$A(\omega) = \frac{-\omega^3 + 11\omega}{5\omega^2 - 15}$$

Thus we see that

$$1 + jA(\omega)\Big|_{\omega=s/j} = \frac{-j\omega^3 + 5\omega^2 + j11\omega - 15}{5\omega^2 - 15}\Bigg|_{\omega=s/j}$$

$$= \frac{s^3 - 5s^2 + 11s - 15}{-5s^2 - 15}$$

The numerator of the right member of the above is readily factored as $(s - 3)(s^2 - 2s + 5)$. Choosing $s - 3$ as $m_1 + n_1$ and $s^2 - 2s + 5$ as $m_2 - n_2$, we find the resulting network function $N(s)$ to be

$$N(s) = \frac{s - 3}{s^2 + 2s + 5}$$

For this network function, tan [arg $N(j\omega)$] is given as the function $A(\omega)$. Alternately, we could choose the factor $(s - 3)$ as $m_2 - n_2$ and $s^2 - 2s + 5$ as $m_1 + n_1$ and obtain

$$N(s) = -\frac{s^2 - 2s + 5}{s + 3}$$

as a second network function with the same phase characteristic. Finally, we could choose $(s-3)(s^2-2s+5)$ as $m_2 - n_2$ and 1 as $m_1 + n_1$ and obtain

$$N(s) = -\frac{1}{(s+3)(s^2+2s+5)}$$

as a third network function with the same phase characteristic. For all three of the network functions given above, tan [arg $N(j\omega)$] is given as the function $A(\omega)$.

Exercise 2.6-2. For each of the following odd rational functions $A(\omega)$, find all the related network functions $N(s)$ (stable and unstable):

$A(\omega)$	Answers [$N(s)$]
ω	$s+1, \dfrac{1}{-s+1}$
$\dfrac{\omega}{1-\omega^2}$	$s^2+s+1, \dfrac{1}{s^2-s+1}$
$\dfrac{-\omega}{1-\omega^2}$	$s^2-s+1, \dfrac{1}{s^2+s+1}$
$\dfrac{\omega^3-2\omega}{1-2\omega^2}$	$(s^2-s+1)(-s+1), \dfrac{1}{(s+1)(s^2+s+1)},$
	$\dfrac{-s+1}{s^2+s+1}, \dfrac{s^2-s+1}{s+1}$

Summary 2.6-1 Properties of the argument of a network function. The necessary and sufficient properties for the phase of a network function $N(s)$ are that the function $A(\omega)$ defined in (5) be a ratio of odd over even polynomials in ω.

Ideal Transmission

When we discussed the approximation of magnitude characteristics in the preceding sections, our goal was usually to keep this quantity constant within some specified tolerance in the passband. The goal for phase approximation, however, is usually quite different. We can define this latter goal for many cases by first investigating the concept of *ideal transmission*. For a network to provide ideal transmission of some arbitrary input excitation signal $e(t)$, we would like the output response signal $r(t)$ to have the same information content, i.e., the same waveform. In such a case there are only two operations the network may be permitted to perform on $e(t)$, namely, a magnitude scaling and a time shifting (a delay). Thus we may define ideal transmission by the relation

$$r(t) = Ke(t-t_0) \qquad t_0 > 0 \tag{7}$$

In other words, the output waveform will be identical to the input waveform except that its magnitude is multiplied by K and it is delayed by t_0 seconds. Assuming $e(t)$ is zero for $t < 0$, the Laplace transformation of (7) gives $R(s) = KE(s)e^{-t_0 s}$ where $R(s)$ and $E(s)$ are the Laplace transforms of $r(t)$ and $e(t)$. Letting $s = j\omega$, we now obtain

$$\frac{\mathcal{R}}{\mathcal{E}} = N(j\omega) = Ke^{-j\omega t_0} \tag{8}$$

where $\mathcal{R}$ and $\mathcal{E}$ are output and input phasors and $N(j\omega)$ is the network function. From (8) we see that ideal transmission requires that $|N(j\omega)| = K$ (that is, the magnitude is a constant independent of frequency) and $\arg N(j\omega) = -\omega t_0$ (that is, *the phase is linearly proportional to frequency*).

Let us now consider how we may approximate the linear phase that, as shown above, produces ideal transmission of a given band of frequencies. For the low-pass case, one way of obtaining the desired linearity is to make the first derivative of the phase function be nonzero but to set as many as possible of the higher order derivatives to zero at $\omega = 0$. This approach is similar to that used to obtain maximally flat magnitude performance in Sec. 2.1. Assuming that the general low-pass network function $N(s)$ has the form

$$N(s) = \frac{1}{a_0 + a_1 s + a_2 s^2 + a_3 s^3 + a_4 s^4 + \cdots} \tag{9}$$

the corresponding phase function is

$$\arg N(j\omega) = \tan^{-1} \frac{-a_1\omega + a_3\omega^3 - \cdots}{a_0 - a_2\omega^2 + a_4\omega^4 - \cdots} \tag{10}$$

To expand this, we may use the series

$$\tan^{-1} x = x - \tfrac{1}{3}x^3 + \tfrac{1}{5}x^5 - \cdots \tag{11}$$

An example of the procedure follows:

Example 2.6-3 Determination of a linear phase function. Consider the second-order low-pass function

$$N(s) = \frac{1}{s^2 + a_1 s + a_0}$$

The argument function may be put in the form

$$\arg N(j\omega) = -\tan^{-1} \frac{1}{a_0} \frac{a_1\omega}{1 - \omega^2/a_0}$$

Using (11) this may be written

$$\arg N(j\omega) = \frac{-1}{a_0} \frac{a_1\omega}{1 - \omega^2/a_0} + \frac{1}{3a_0^3}\left(\frac{a_1\omega}{1 - \omega^2/a_0}\right)^3 - \cdots$$

If we now apply the binomial expansion $(1 + x)^n \approx 1 + nx + \cdots$ to the terms $(1 - \omega^2/a_0)^{-1}$

and $(1 - \omega^2/a_0)^{-3}$ and simplify the result, we obtain

$$\arg N(j\omega) = -\frac{a_1}{a_0}\omega + \left(-\frac{a_1}{a_0^2} + \frac{a_1^3}{3a_0^3}\right)\omega^3 - \cdots$$

Defining $a_1/a_0 = t_0$ and setting the coefficient of ω^3 to zero in the above will provide an approximation to the linear phase $\arg N(j\omega) = -\omega t_0$. Doing this we obtain the following relations:

$$a_0 = \frac{3}{t_0^2} \qquad a_1 = t_0 a_0$$

For example, for $t_0 = 1$ we require $a_0 = 3$ and $a_1 = 3$. Thus, the second-order low-pass network function having a linear phase with a slope of -1 (at direct current) is

$$N(s) = \frac{1}{s^2 + 3s + 3}$$

Linear Phase (Thomson) Functions

If we extend the procedure given in Example 2.6-3 to higher order low-pass functions, we obtain a series of polynomials that are related to Bessel polynomials. The network functions that use these polynomials for their denominators are called *Thomson functions.*[1] For such functions, the coefficients of the denominator of (9) corresponding to the approximation of a linear phase with a slope of -1 (a delay of 1 s at direct current) may be found from the relation

$$a_k = \frac{(2n-k)!}{2^{n-k}k!(n-k)!} \qquad k = 0, 1, \ldots, n-1 \tag{12}$$

where n is the degree of the denominator. The highest degree coefficient is unity. A listing of polynomial coefficients, roots, and quadratic factors is given in Table 2.6-1. The recursion formula

$$B_n(s) = (2n-1)B_{n-1}(s) + s^2 B_{n-2}(s) \tag{13}$$

may also be used to derive the denominator polynomials starting with $B_1(s) = s + 1$ and $B_2(s) = s^2 + 3s + 3$.

Delay Functions

The treatment of ideal transmission given above can be extended by considering what happens when an impulse is applied to an arbitrary network characterized by a network function $N(j\omega)$. Such an excitation is convenient since its peak value is readily identified in the output waveform. The response $h(t)$ to the impulse excitation is found by taking

[1] W. E. Thomson, "Delay Networks Having Maximally Flat Frequency Characteristics," *Proc. IEE*, part 3, vol. 96, November 1949, pp. 487–490.

TABLE 2.6-1*a*

The denominator coefficients of linear phase (Thomson) low-pass functions of the form $a_0 + a_1 s + a_2 s^2 + \cdots + a_{n-1} s^{n-1} + s^n$ with normalized delay (at dc) of 1 s

n	a_0	a_1	a_2	a_3	a_4	a_5	a_6	a_7	a_8	a_9
2	3	3								
3	15	15	6							
4	105	105	45	10						
5	945	945	420	105	15					
6	10,395	10,395	4,725	1,260	210	21				
7	135,135	135,135	62,370	17,325	3,150	378	28			
8	2,027,025	2,027,025	945,945	270,270	51,975	6,930	630	36		
9	34,459,425	34,459,425	16,216,200	4,729,725	945,945	135,135	13,860	990	45	
10	654,729,075	654,729,075	310,134,825	91,891,800	18,918,900	2,837,835	315,315	25,740	1,485	55

TABLE 2.6-1*b*

Pole locations and quadratic factors $(a_0 + a_1 s + s^2)$ of linear phase (Thomson) low-pass functions with normalized delay (at dc) of 1 s

n	Poles	a_0	a_1	Q	ω_n
2	$-1.50000 \pm j0.86603$	3.00000	3.00000	0.57735	1.73205
3	$-1.83891 \pm j1.75438$	6.45943	3.67781	0.69105	2.54154
	-2.32219				
4	$-2.10379 \pm j2.65742$	11.48780	4.20758	0.80554	3.38937
	$-2.89621 \pm j0.86723$	9.14013	5.79242	0.52193	3.02326
5	$-2.32467 \pm j3.57102$	18.15632	4.64935	0.91648	4.26102
	$-3.35196 \pm j1.74266$	14.27248	6.70391	0.56354	3.77789
	-3.64674				
6	$-2.51593 \pm j4.49267$	26.51403	5.03186	1.02331	5.14918
	$-3.73571 \pm j2.62627$	20.85282	7.47142	0.61119	4.56649
	$-4.24836 \pm j0.86751$	18.80113	8.49672	0.51032	4.33603
7	$-2.68568 \pm j5.42069$	36.59679	5.37135	1.12626	6.04953
	$-4.07014 \pm j3.51717$	28.93655	8.14028	0.66082	5.37927
	$-4.75829 \pm j1.73929$	25.66644	9.51658	0.53236	5.06621
	-4.97179				
8	$-2.83898 \pm j6.35391$	48.43202	5.67797	1.22567	6.95931
	$-4.36829 \pm j4.41444$	38.56925	8.73658	0.71085	6.21041
	$-5.20484 \pm j2.61618$	33.93474	10.40968	0.55961	5.82535
	$-5.58789 \pm j0.86761$	31.97723	11.17577	0.50599	5.65484
9	$-2.97926 \pm j7.29146$	62.04144	5.95852	1.32191	7.87664
	$-4.63844 \pm j5.31727$	49.78850	9.27688	0.76061	7.05610
	$-5.60442 \pm j3.49816$	43.64665	11.20884	0.58941	6.60656
	$-6.12937 \pm j1.73785$	40.58927	12.25874	0.51971	6.37097
	-6.29702				
10	$-3.10892 \pm j8.23270$	77.44270	6.21783	1.41531	8.80015
	$-4.88622 \pm j6.22499$	62.62559	9.77244	0.80979	7.91363
	$-5.96753 \pm j4.38495$	54.83916	11.93506	0.62047	7.40535
	$-6.92204 \pm j0.86767$	48.66755	13.84409	0.50391	6.97621
	$-6.61529 \pm j2.61157$	50.58236	13.23058	0.53755	7.11213

the inverse Fourier transformation. Thus we obtain

$$h(t) = \frac{1}{2\pi}\int_{-\infty}^{\infty} N(j\omega)e^{j\omega t}\, d\omega = \frac{1}{\pi}\int_{0}^{\infty} |N(j\omega)| \cos\,[\omega t + \arg N(j\omega)]\, d\omega \tag{14}$$

In this expression, the principal contribution to the integral occurs when the argument of the cosine function in the integrand is constant. Thus, the peak value of $h(t)$ occurs when

$$\frac{d}{d\omega}[\omega t + \arg N(j\omega)] = 0 \tag{15}$$

Solving this expression we may define a *delay function* $D(\omega)$[2] for an arbitrary network as

$$D(\omega) = -\frac{d}{d\omega}\arg N(j\omega) \tag{16}$$

Obviously, for the ideal transmission case discussed above, in which $\arg N(j\omega) = -\omega t_0$, $D(\omega) = t_0$, and a delay of t_0 seconds occurs for all frequency components of the input signal. Considering the above we see that the maximally linear (at the origin) phase characteristic of Thomson filters produces a delay characteristic that is maximally flat (at the origin). Thus these filters are frequently referred to as *maximally flat delay* (MFD) filters. For ideal transmission, the magnitude, phase, and delay characteristics for a network function $N(s)$ are shown by the solid lines in Fig. 2.6-1. When low-pass characteristics are used to limit the frequency range, then the magnitude characteristic is modified as shown by the dashed line in Fig. 2.6-1(*a*). Maximum phase is determined by the order of the network function. As a result, the ideal phase and delay characteristics are modified as shown by the dashed lines in Figs. 2.6-1(*b*) and (*c*).

The characteristics of delay functions are frequently displayed in plot form. Such plots are readily constructed by separately evaluating the quadratic factors of the numerator and denominator polynomials. To see this, consider a network function of

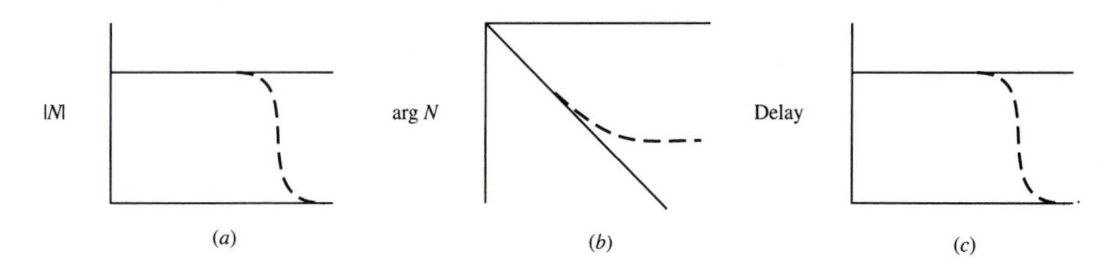

FIGURE 2.6-1
Magnitude, phase, and delay curves for ideal transmission.

---

[2] This is also referred to as the *group delay function*.

the form

$$N(s) = \frac{Q_{n1}(s)Q_{n2}(s)\cdots}{Q_{d1}(s)Q_{d2}(s)\cdots} \tag{17}$$

where Q_{ni} are the numerator quadratic factors and Q_{di} are the denominator ones. For this we find

$$\arg N(j\omega) = \arg Q_{n1}(j\omega) + \arg Q_{n2}(j\omega) + \cdots - \arg Q_{d1}(j\omega) - \arg Q_{d2}(j\omega) - \cdots \tag{18}$$

If we now define individual delays for each of the quadratic terms as $D_{ni}(\omega) = -(d/d\omega)\arg Q_{ni}(j\omega)$ and $D_{di}(\omega) = -(d/d\omega)\arg Q_{di}(j\omega)$, then the total delay $D(\omega)$ for $N(s)$ is given as

$$D(\omega) = D_{n1}(\omega) + D_{n2}(\omega) + \cdots - D_{d1}(\omega) - D_{d2}(\omega) - \cdots \tag{19}$$

Let each of the quadratic factors have the form $Q(s) = s^2 + a_1 s + a_0$. Then

$$\arg Q(j\omega) = \arg(a_0 - \omega^2 + ja_1\omega) = \tan^{-1}\frac{a_1\omega}{a_0 - \omega^2} \tag{20}$$

For each of these quadratic factors the delay $D_q(\omega)$ is found by using $(d/d\omega)\tan^{-1}x = [1/(1+x^2)](dx/d\omega)$. From this we obtain

$$D_q(\omega) = -\frac{d}{d\omega}\tan^{-1}\frac{a_1\omega}{a_0 - \omega^2} = -\frac{a_1\omega^2 + a_1 a_0}{\omega^4 + \omega^2(a_1^2 - 2a_0) + a_0^2} \tag{21}$$

Any first-order numerator or denominator terms can be similarly evaluated. Let the first-order term have the form $F(s) = s + a$. The corresponding delay term $D_f(\omega)$ is found as

$$D_f(\omega) = -\frac{d}{d\omega}\tan^{-1}\frac{\omega}{a} = -\frac{a}{\omega^2 + a^2} \tag{22}$$

Terms of this type are readily included in (19).

Example 2.6-4 Plot of a delay function. The third-order Thomson function has the form

$$N_{3T}(s) = \frac{H}{s^3 + 6s^2 + 15s + 15} = \frac{H}{(s^2 + 3.67781s + 6.45943)(s + 2.32219)}$$

From (19), (21), and (22) we find

$$D_{3T}(\omega) = \frac{3.67781\omega^2 + 23.75656}{\omega^4 + 0.60743\omega^2 + 41.72424} + \frac{2.32219}{\omega^2 + 5.39257}$$

A plot of this function is shown (marked $n = 3$) in Fig. 2.6-2.

Plots of the delay functions for various Thomson, Butterworth, and Chebyshev functions are shown in Figs. 2.6-2 to 2.6-5. Note that different ordinate and abscissa scales have been chosen to emphasize the characteristics of the different plots.

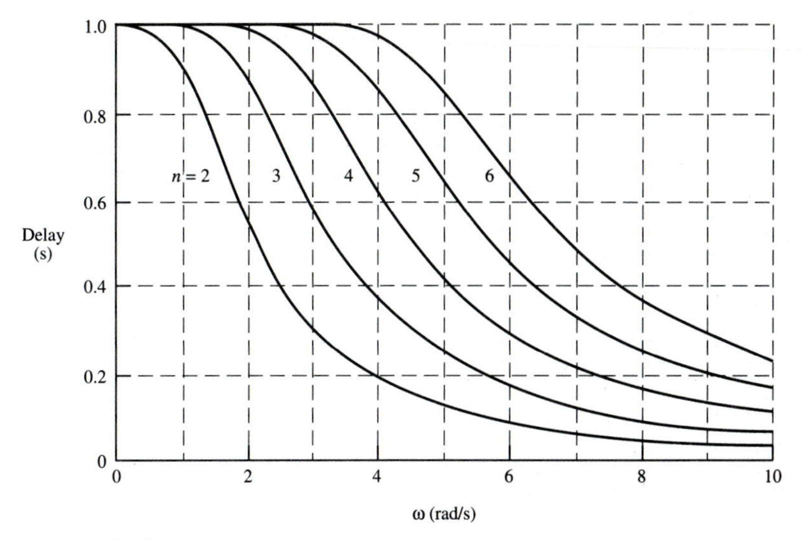

FIGURE 2.6-2
Delay of Thomson functions.

Determination of Order

The actual characteristics provided by the Thomson functions tabulated in Table 2.6-1 differ from the ideal in both delay and magnitude, as shown in Fig. 2.6-6. The deviation decreases as the order increases. The deviation in delay at any frequency is readily found using the expressions for delay given in (21) and (22). The deviation from the magnitude is found from the functions given in Table 2.6-1.

Example 2.6-5 Delay and magnitude errors in a Thomson function. As an example, let us determine the delay and magnitude errors in a third-order Thomson function at

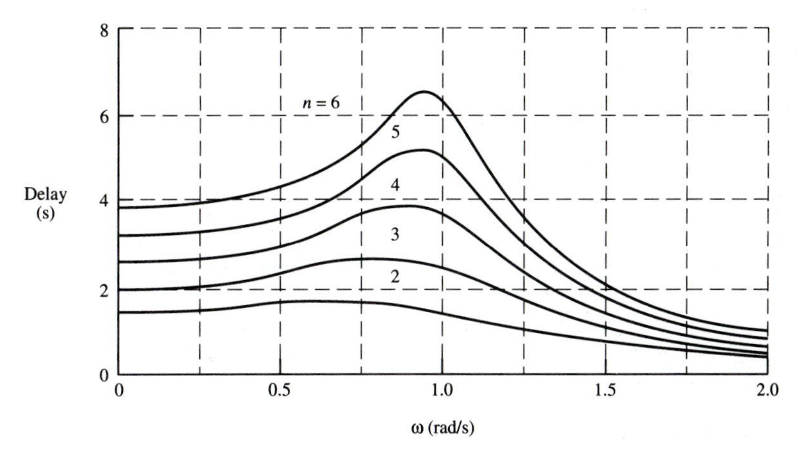

FIGURE 2.6-3
Delay of Butterworth functions.

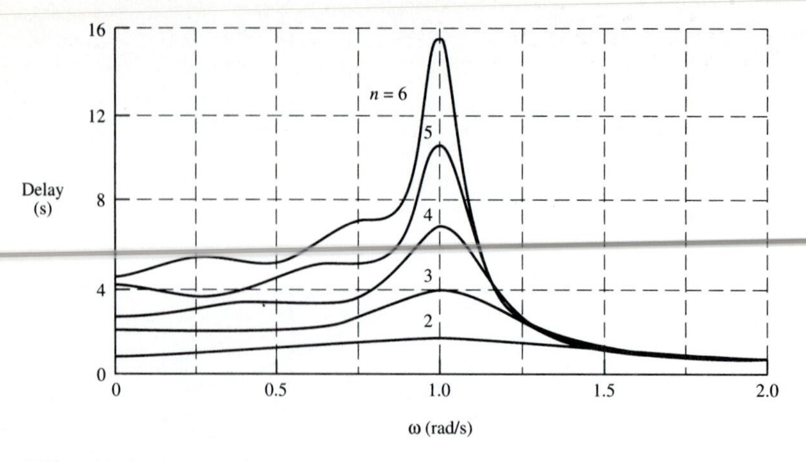

FIGURE 2.6-4
Delay of Chebyshev functions with 0.5-dB passband ripple.

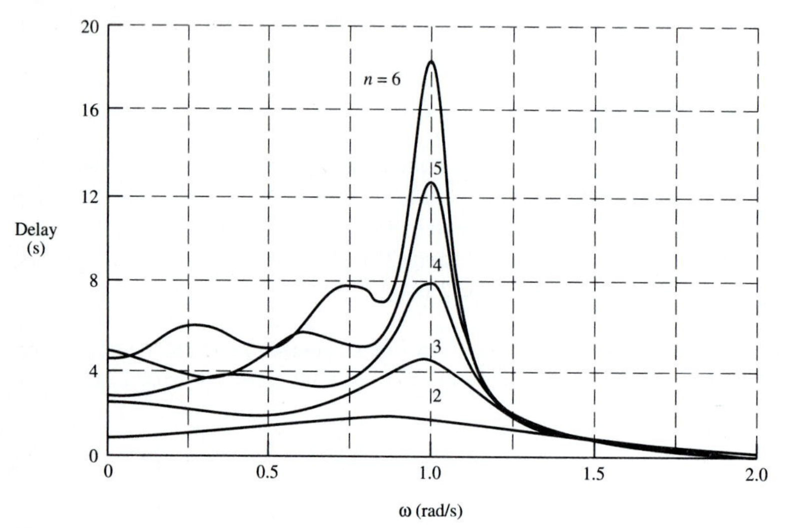

FIGURE 2.6-5
Delay of Chebyshev functions with 1-dB passband ripple.

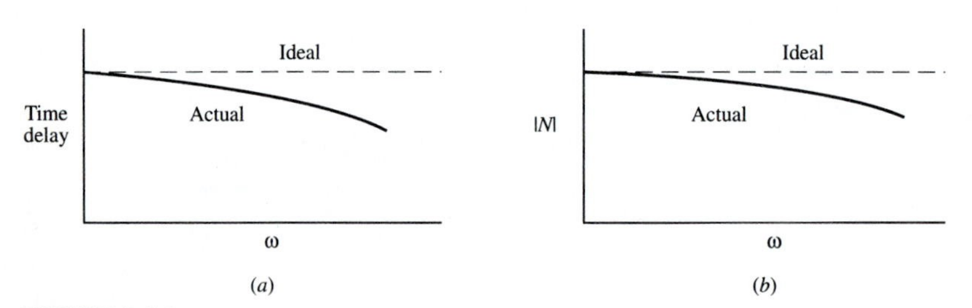

FIGURE 2.6-6
Ideal and nonideal characteristics for low-pass functions for ideal transmission.

$\omega = 1.5$ rad/s. The delay error is found by evaluating the expression for $D_{3T}(\omega)$ given in Example 2.6-4. We obtain $D_{3T}(1.5) = 0.96905$ s. The percentage of error is

$$\text{Error(\% Delay)} = \frac{1.0 - 0.96905}{0.96905} \times 100 = 3.2\%$$

The magnitude error is found by evaluating the expression for $N_{3T}(s)$ given in Example 2.6-4. We obtain $|N_{3T}(j1.5)| = H/19.18373$. The error in decibels is found as

$$\text{Error(dB} - \text{magnitude)} = 20 \log_{10} \frac{H/15}{H/19.18373} = 2.14 \text{ dB}$$

Figures showing the delay and magnitude errors as a function of frequency for various orders of Thomson functions are given in Figs. 2.6-7 and 2.6-8. Both of these figures use ωT, where T is the ideal delay, as abscissas. Since this product is dimensionless, it is not affected by any frequency normalization. In applying the figures to determine the order required for a given filter, if both delay and magnitude tolerances are to be met, the figure giving the higher value will determine the required order. The values of network elements for various orders of Thomson filters may be found in App. A.

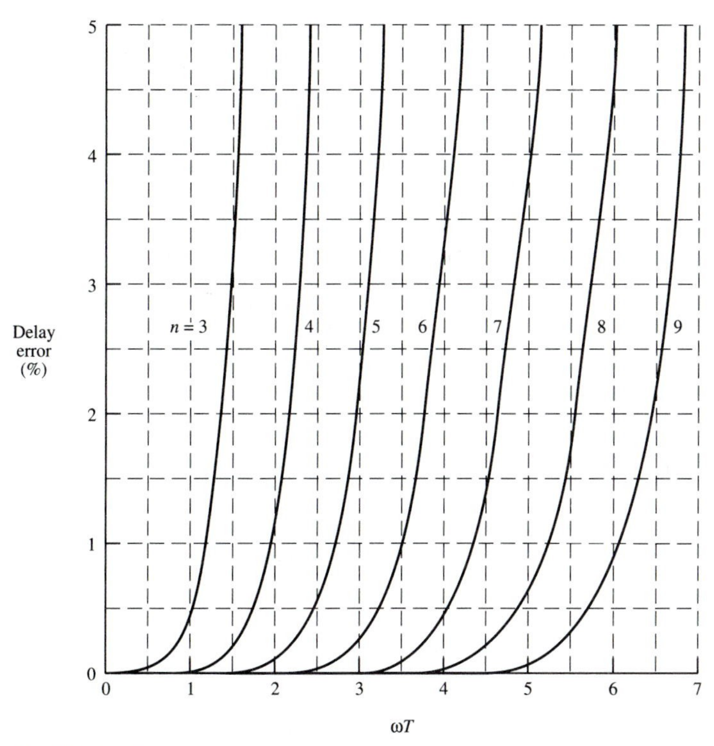

FIGURE 2.6-7
Delay error of the functions given in Table 2.6-1.

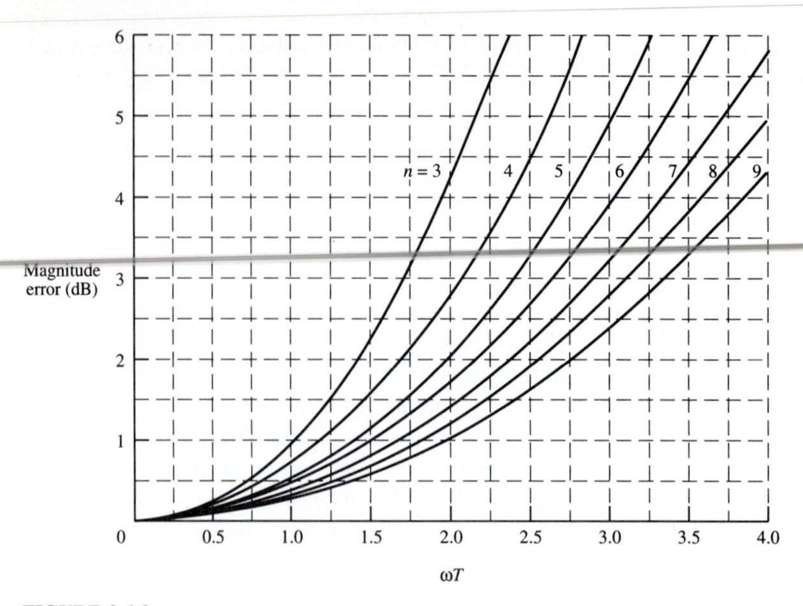

FIGURE 2.6-8
Magnitude error of the functions given in Table 2.6-1.

Example 2.6-6 Order of a Thomson function. A low-pass function with a normalized delay of 1 s (at direct current) is required to have a maximum delay error 0.5 percent at 2 rad/s and a maximum magnitude error of 1.5 dB at 2 rad/s. From Fig. 2.6-7 we see that a fifth-order function will meet the delay error requirement. From Fig. 2.6-8, however, we see that a seventh-order function is required to satisfy the magnitude specification. Thus the requirements call for a seventh-order function.

Delay Equalizers

All-pole low-pass network functions (ones with all their zeros at infinity) may be used as a prototype to derive another type of function that has similar phase and delay characteristics but is not low pass. To see this, consider a network function having the form

$$N(s) = \frac{H}{m(s) + n(s)} \tag{23}$$

where $m(s)$ is the even part of the denominator polynomial and $n(s)$ is the odd part. The phase characteristic is defined as

$$\arg N(j\omega) = -\tan^{-1} \frac{n(j\omega)/j}{m(j\omega)} \tag{24}$$

Now consider what happens if we form the function

$$N_{\text{AP}}(s) = H \frac{m(s) - n(s)}{m(s) + n(s)} \tag{25}$$

For this, we find $|N(j\omega)| = H$, that is, the magnitude is constant for all frequencies. Such a function is called an *all-pass function*. The phase characteristic of this function is

$$\arg N_{AP}(j\omega) = -2 \tan^{-1} \frac{n(j\omega)/j}{m(j\omega)} \tag{26}$$

The phase characteristics (and the delay function) are identical with that of the original low-pass function of (23) except that they have been scaled by a factor of 2.

Example 2.6-7 An all-pass maximally flat delay function. The second-order Thomson function given in Table 2.6-1 has the form

$$N_{2T}(s) = \frac{H}{s^2 + 3s + 3}$$

It has been shown to have a maximally flat delay normalized to 1 s (at direct current). From (23) and (25) we see that a second-order all-pass function

$$N_{2AP}(s) = H\frac{s^2 - 3s + 3}{s^2 + 3s + 3}$$

will also have a maximally flat delay characteristic. The normalization of this function is 2 s (at direct current).

All-pass functions provide a simple means of correcting the delay characteristic of low-pass functions to achieve the constant delay required for ideal transmission. This is accomplished by using all-pass networks in cascade with a low-pass network so that the overall network function is the product of the low-pass function and the all-pass ones. The all-pass functions are chosen so as to provide the required correction for the low-pass delay. In such an application the all-pass functions are referred to as a *delay equalizer.*

The form of the delay characteristics used as corrections are easily visualized by considering an all-pass function of the form

$$N_{AP}(s) = \frac{s^2 - a_1 s + a_0}{s^2 + a_1 s + a_0} \tag{27}$$

Plots of $D_{AP}(\omega)$ for various values of a_1, assuming a normalized value of $a_0 = 1$, are shown in Fig. 2.6-9. For a given application these charts provide visual assistance in selecting the parameters of the delay equalizer for a given low-pass characteristic. A more accurate determination will usually require the use of computer-based optimization methods. The general approach is illustrated in the following example.

Example 2.6-8 Delay equalizer. A filter application requires a normalized (1-rad/s passband) low-pass characteristic that is satisfied by a fourth-order Chebyshev function with a 1-dB passband ripple. Its delay characteristic $D_C(\omega)$ is shown in Fig. 2.6-10. To provide better transmission of pulses (a flatter delay), we may add two all-pass functions. The first will provide a peak to fill in the valley of $D_C(\omega)$ near 0.7 rad/s. Before frequency denormalization, this will appear something like the curves for $a_1 = 1$ and $a_1 = \frac{2}{3}$ in Fig.

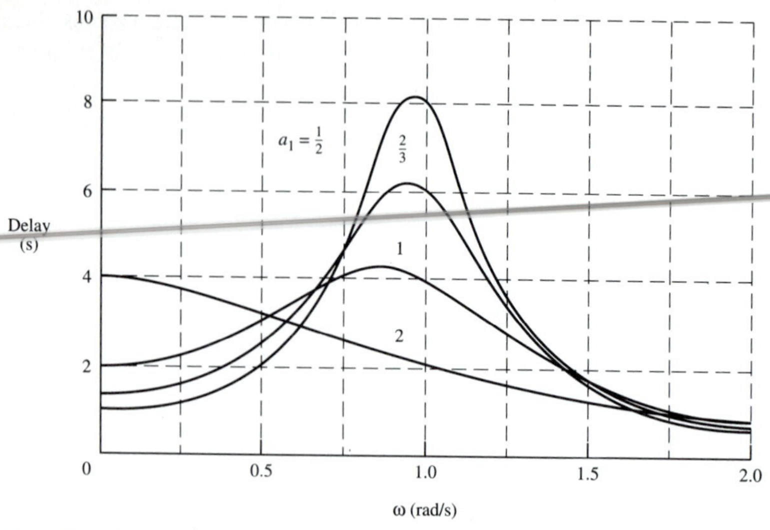

FIGURE 2.6-9
Delay of all-pass functions of (27) with $a_0 = 1$.

2.6-9. Let this be $N_{\text{AP1}}(s)$. The second all-pass function will be used to add delay in the low-frequency region. Before frequency normalization it will appear something like the curve for $a_1 = 2$. Let this be $N_{\text{AP2}}(s)$. After optimizing, we obtain

$$N_{\text{AP1}}(s) = \frac{s^2 - 0.558s + 0.605}{s^2 + 0.558s + 0.605}$$

$$N_{\text{AP2}}(s) = \frac{s^2 - 0.709s + 0.178}{s^2 + 0.709s + 0.178}$$

The corresponding delay functions $D_{\text{AP1}}(\omega)$ and $D_{\text{AP2}}(\omega)$ are shown in Fig. 2.6-10. The total delay produced by the product of the low-pass function and the two all-pass functions is

$$D_{\text{total}}(\omega) = D_C(\omega) + D_{\text{AP1}}(\omega) + D_{\text{AP2}}(\omega)$$

As indicated in the figure, this provides a considerably flatter delay response than the original Chebyshev function even though the low-pass magnitude characteristic is unchanged.

2.7 TIME-DOMAIN CONSIDERATIONS

In the preceding sections of this chapter we have considered some of the *frequency domain* properties of network functions. In this section we present a brief discussion of some of their *time-domain* properties. Unlike the frequency-domain situation in which completely general expressions relating the ratio of response and excitation may be developed, in the time-domain situation the response must usually be separately determined for each form of excitation. For simplicity of presentation, we shall restrict our attention to the case that has the most practical interest, namely, the step response

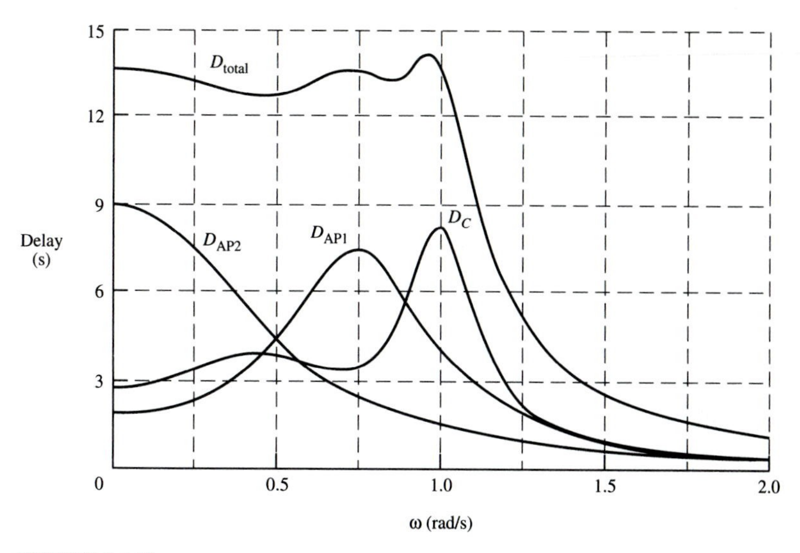

FIGURE 2.6-10
Delay equalizer example.

of low-pass all-pole functions. This case is of considerable importance in studying pulse and other digital transmission systems.

Step Response

We now consider the situation where the excitation function $e(t)$ is a unit step occurring at $t = 0$. This may be written as

$$e(t) = u(t)\begin{cases} = 0 & t < 0 \\ = 1 & t > 0 \end{cases} \tag{1}$$

where $u(t)$ is the unit step function. It is shown in Fig. 2.7-1. The response $r(t)$ of a filter function to such an excitation is called its *step response*. For an ideal network function, this response would be determined by (7) of Sec. 2.6 as

$$r(t) = u(t - t_0)\begin{cases} = 0 & t < t_0 \\ = 1 & t > t_0 \end{cases} \tag{2}$$

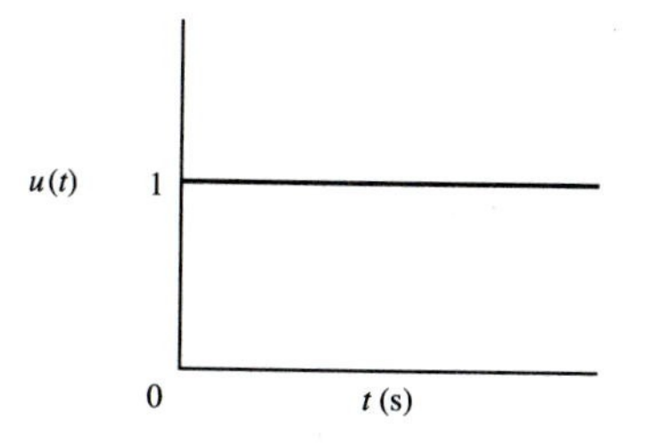

FIGURE 2.7-1
A unit step.

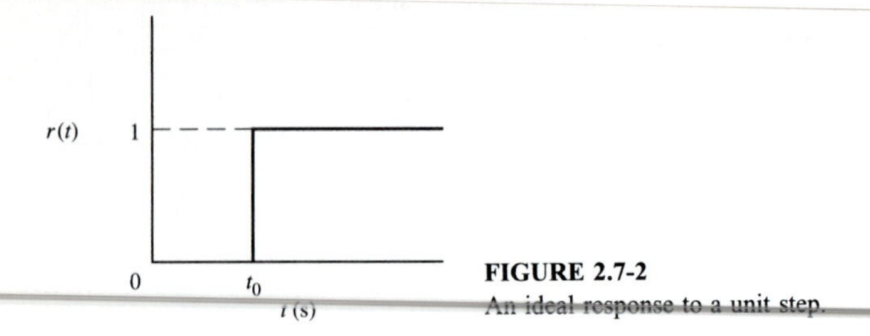

FIGURE 2.7-2
An ideal response to a unit step.

This is shown in Fig. 2.7-2. Actually, for the nonideal low-pass function, the step response will usually have a form more like the one shown in Fig. 2.7-3; that is, it will demonstrate an oscillatory overshoot about its final value. This is referred to as *ringing*. Also in Fig. 2.7-3, several figures of merit commonly used to evaluate a step response are defined. These are the delay time t_d, the time required for the response to reach 50 percent of its final or steady-state value r_{final}; $t_{10\%}$, the time for the response to reach 10 percent of its final value; and $t_{90\%}$, the time for it to first reach 90 percent of its final value. In terms of these latter two quantities, we define the rise time t_r as

$$t_r = t_{90\%} - t_{10\%} \tag{3}$$

Other step-response quantities of interest illustrated in Fig. 2.7-3 are: t_s, the settling time, meaning the time after which the response remains within some specified range, usually ± 2 percent of its final value; and the overshoot, usually defined in terms of a

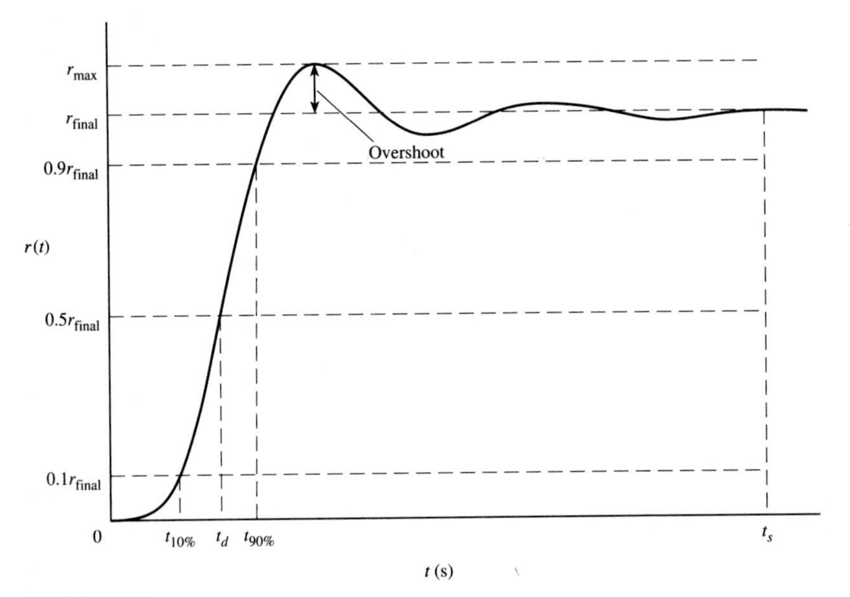

FIGURE 2.7-3
An actual response to a unit step.

peak-overshoot ratio (POR) given (in percentage) as

$$\text{POR} = \frac{r_{\max} - r_{\text{final}}}{r_{\text{final}}} \times 100 \tag{4}$$

In general, good time-domain performance means small values of t_r, t_s, and POR.

Elimination of Ringing

A time-domain step response in which no overshoot (or ringing) occurs is desirable in many filtering applications, especially those in which pulse information is to be transmitted. Thomson functions are very desirable for such applications since they typically have a POR of less than 1 percent. A response that is completely free of ringing can be shown to be obtainable from a network function that has a normalized gaussian magnitude characteristic and a linear phase characteristic. These would have the form

$$|N(j\omega)| = e^{[-(\ln 2)/2]\omega^2} \qquad \arg N(j\omega) = -\tfrac{1}{2}n\pi\omega \tag{5}$$

The constant (ln 2)/2 in the exponent for the magnitude is chosen so that the 3-dB frequency is 1 rad/s. A plot of the magnitude function is shown in Fig. 2.7-4. The resulting step response can be shown to have the characteristics (the *slope rise time* is the inverse of the response function slope at 50 percent of its final value)

$$\text{POR} = 0 \qquad \tau_{\text{Slope Rise Time}} = \sqrt{2\pi \ln 2} = 2.087 \tag{6}$$

The characteristics defined by (5) are not directly realizable by a rational network function. An approximation to them, however, can be made by using a Taylor series expansion for the magnitude function. We obtain

$$|N(j\omega)|^2 = e^{-\omega^2} = \frac{1}{e^{\omega^2}} = \frac{1}{1 + \omega^2 + \omega^4/2! + \omega^6/3! + \cdots} \tag{7}$$

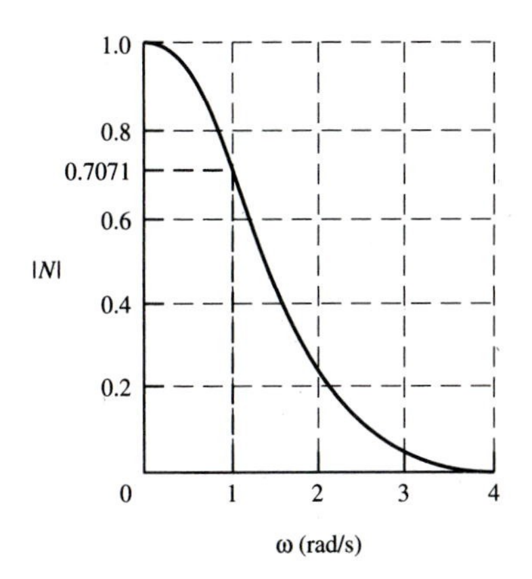

FIGURE 2.7-4
Gaussian magnitude response.

The resulting denominator polynomial can be truncated to obtain the desired order of gaussian function.

Example 2.7-1 Network function with gaussian response. From (7) we see that the approximation to the second-order gaussian response is

$$|N(j\omega)|^2 = \frac{1}{1 + \omega^2 + \omega^4/2}$$

The techniques discussed in Sec. 2.1 may be applied to obtain

$$N(s)N(-s) = \frac{2}{s^4 - 2s^2 + 2}$$

The left-half-plane poles at $s = -1.0986 \pm j0.45509$ define the second-order gaussian function

$$N(s) = \frac{\sqrt{2}}{s^2 + 2.19737s + \sqrt{2}}$$

To compare this function with a Thomson one, we frequency normalize it so that its zero- and first-degree coefficients are the same, thus normalizing the delay (at direct current) to 1.0 s. To do this, we define the normalized function

$$N_{fn}(s) = \frac{\sqrt{2}\Omega_n^2}{s^2 + 2.19737\Omega_n s + \sqrt{2}\Omega_n^2}$$

and set $2.19737\Omega_n = \sqrt{2}\Omega_n^2$. The result is $\Omega_n = 1.55377$. The resulting normalized gaussian function is

$$N_{1\,\text{s delay}}(s) = \frac{3.41420}{s^2 + 3.41420s + 3.41420}$$

Comparing this with the Thomson second-order poles at $s = -1.5 \pm j0.86603 = 1.73204e^{\pm j150°}$, we see that the gaussian function has poles that are closer to the real axis, namely, at $s = -1.70711 \pm j0.70711 = 1.84775e^{\pm j157.5°}$.

Relation between Time- and Frequency-Domain Parameters

Some of the step-response parameters defined above may be approximately related to the frequency-domain parameters used in this chapter. As an example of this, for low-pass network functions in which the POR is small (less than 5 percent), the rise time t_r and the -3-dB bandwidth ω_c are approximately related as

$$t_r\omega_c \approx 2.2 \tag{8}$$

As an illustration of this relation, consider the network shown in Fig. 2.7-5. Its voltage transfer function is

$$\frac{V_2(s)}{V_1(s)} = \frac{1}{s+1} \tag{9}$$

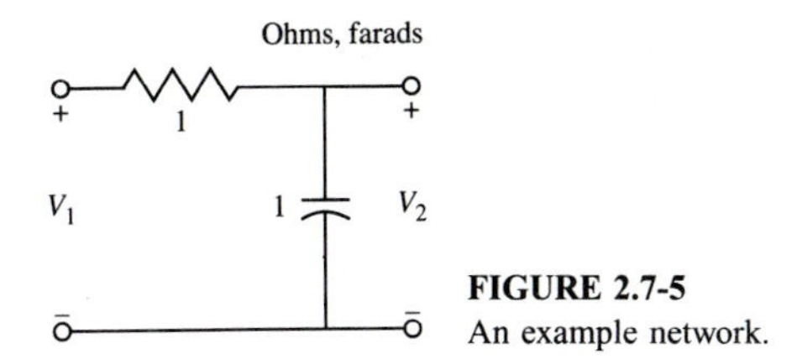

FIGURE 2.7-5
An example network.

The bandwidth ω_c is readily shown to be 1 rad/s. The step response may be determined as

$$v_2(t) = (1 - e^{-t})u(t) \tag{10}$$

From this relation we readily find $t_{10\%} = 0.1054$ s and $t_{90\%} = 2.3026$ s. Thus $t_r = 2.1972$ s and $t_r \times \omega_c = 2.1972$, in good agreement with the approximate value of 2.2 predicted in (8). Another example of the use of this relation is for the value of rise time for the gaussian function given in (6).

Example Step Responses

Some examples of step responses for various orders of the Butterworth and Chebyshev (1-dB ripple) functions introduced earlier in this chapter are shown in Figs. 2.7-6 and 2.7-7. From these we may observe another general correlation between the time and frequency domains: overshoot and settling time for Butterworth and Chebyshev filters increase as the cutoff of the network function's magnitude response is made sharper. As

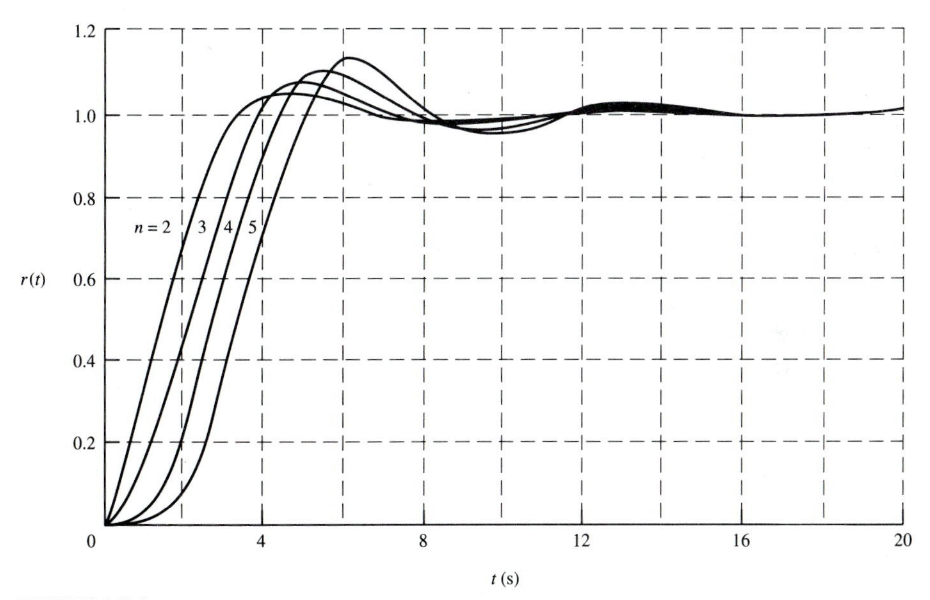

FIGURE 2.7-6
Step responses for Butterworth functions.

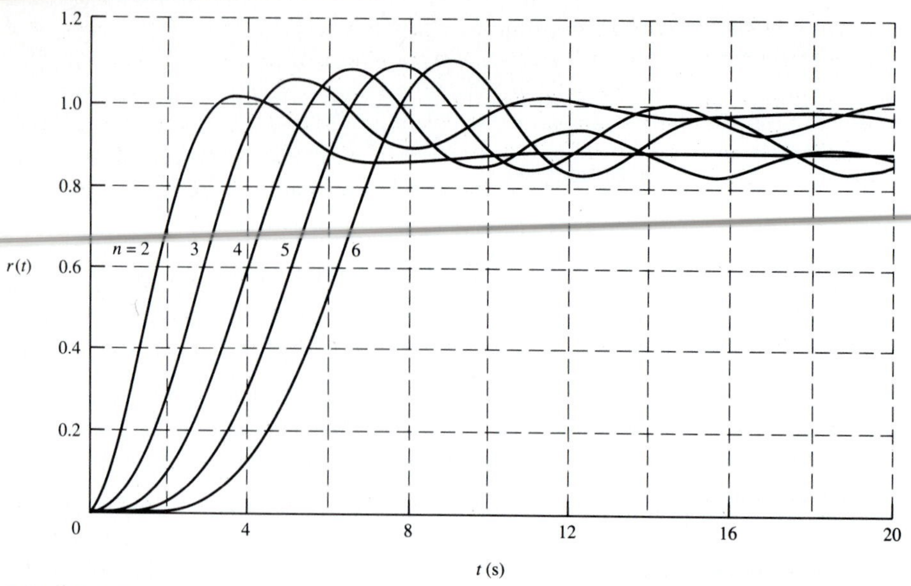

FIGURE 2.7-7
Step responses for 1-dB-ripple Chebyshev functions.

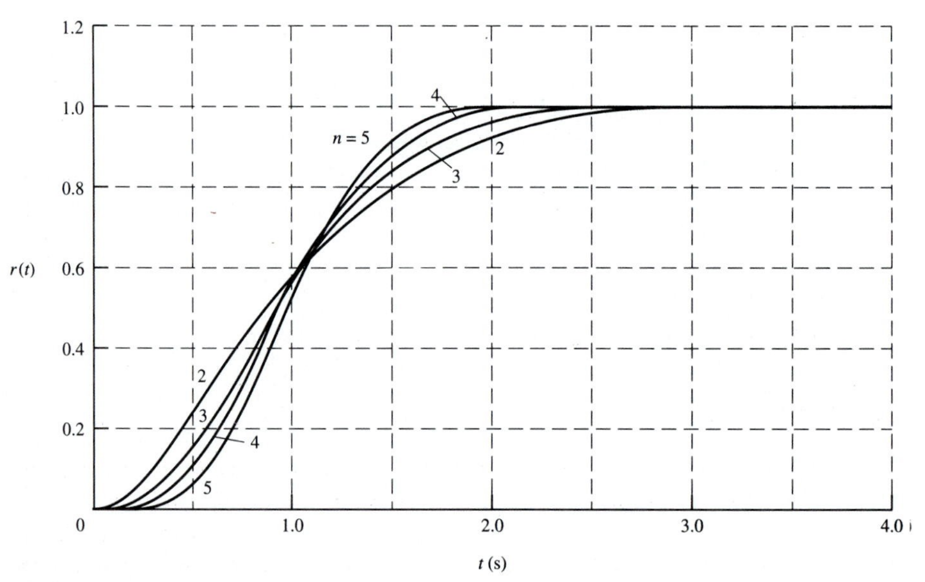

FIGURE 2.7-8
Step responses for Thomson functions.

a result, higher order functions show more ringing in their step response than lower order ones do, and Chebyshev functions have more than Butterworth ones. Note in Fig. 2.7-7 that odd-order functions have a different final value than do even-order ones. Another interesting property of these types of functions is that (for the constant bandwidth normalization chosen) their delay times increase with increasing order, while their rise times are very nearly constant. The properties of the Bessel (or Thomson) functions discussed in Sec. 2.6 are quite different. Their step responses are illustrated in Fig. 2.7-8. In this figure, we see some dramatic results of the linear phase properties that these functions have, and the constant delay time normalization chosen for them. First, there is almost no overshoot. Second, the rise time decreases with increasing order. Obviously the Bessel functions have far better time-domain performance than do the Butterworth, which in turn are better than the Chebyshev.

2.8 APPROXIMATION USING COMPUTERS

Algorithms for all of the approximations introduced in this section can be readily implemented on a PC (personal computer). Many shareware and commercial PC programs are available, and they provide a wide range of capabilities. In this section we describe a useful and full-featured program developed by Professor David J. Willis of Oregon Institute of Technology called Filter Master. The program provides the generation of the following approximations:

Butterworth
Chebyshev
Inverse Chebyshev
Elliptic
Thomson

The program also permits the input of a user-specified function. The filter functions either supplied directly or found by applying transformations of the type given in Sec. 2.5 are:

Low pass
High pass
Band pass
Band elimination

The desired order of the approximation may be specified by the user. Alternately, passband and stopband specifications may be entered, in which case the order is determined automatically. Output displays of pole and zero positions and quadratic factors and Q are given. Numerical output over a range of frequencies giving the values of magnitude (decimal and decibel), phase (degrees and radians), and delay is provided. Plots are displayed for magnitude, phase, and delay versus frequency. Active filter implementations of the approximations are also made available. Some examples of the use of the program follow.

Example 2.8-1 A fifth-order Chebyshev approximation. As an example of the use of Filter Master, it is desired to generate magnitude, phase, and delay plots for a fifth-order low-pass Chebyshev network function with 1-dB ripple in the normalized passband from 0 to 1 rad/s. The pole locations generated by the program are the same as those shown in Table 2.2-3*c*. Plots of the magnitude, phase, and delay are shown in Figs. 2.8-1(*a*), (*b*), and (*c*), respectively.

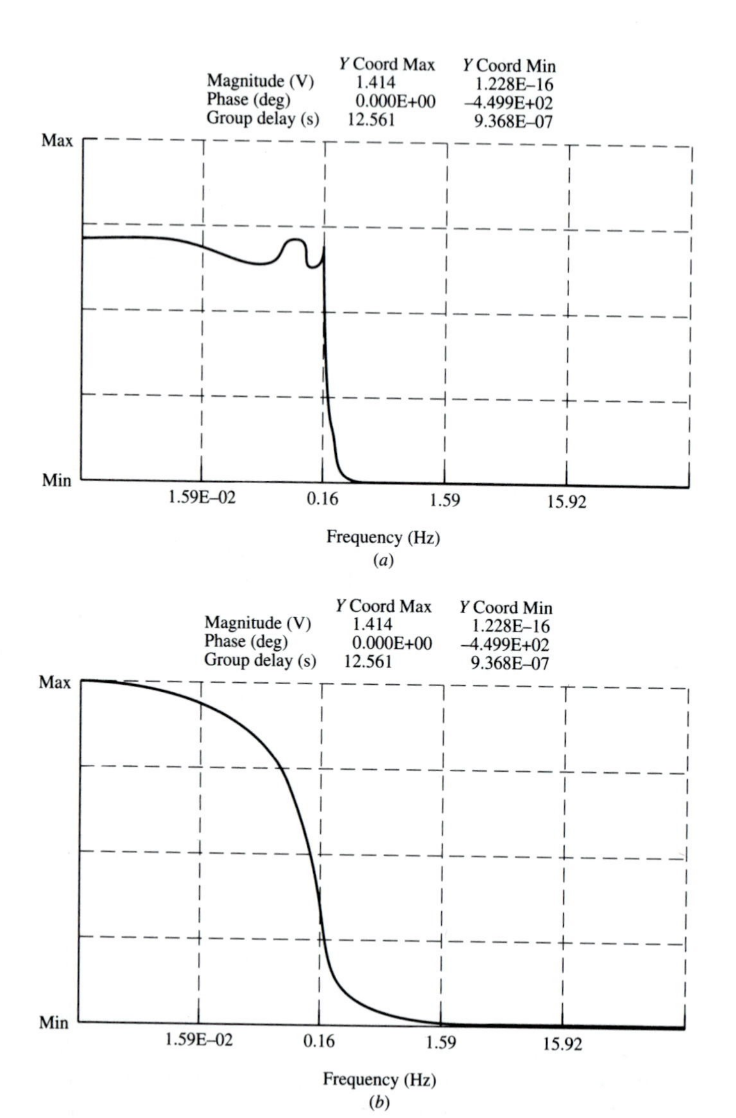

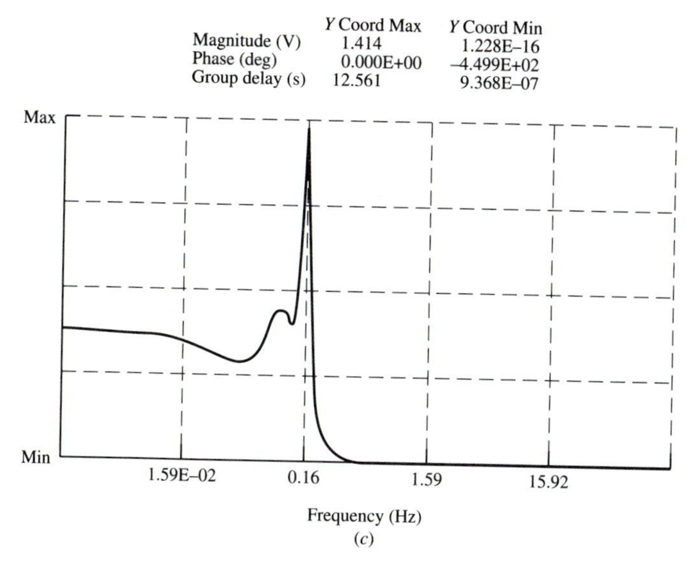

FIGURE 2.8-1
Fifth-order Chebyshev approximation.

Example 2.8-2 A tenth-order band-pass function. As an example of the transformation capabilities provided by Filter Master, a 5th-order low-pass Butterworth function with a bandwidth of 1 rad/s was generated and transformed using the normalized low-pass to band-pass transformation given in (6) of Sec. 2.5. The low-pass poles are the same as those given in Table 2.1-3*b*. The result is a 10th-order band-pass function with a center frequency of 1 rad/s and a bandwidth of 1 rad/s. The pole and zero locations found by the program are:

Poles	Zeros
$-0.500000 \pm j0.866025$	$0.000000 + j0.000000$
$-0.281748 \pm j0.674516$	$0.000000 + j0.000000$
$-0.087624 \pm j0.622983$	$0.000000 + j0.000000$
$-0.221393 \pm j1.574040$	$0.000000 + j0.000000$
$-0.527269 \pm j1.262302$	$0.000000 + j0.000000$

The magnitude characteristic of the band-pass function is shown in Fig. 2.8-2.

More information concerning Filter Master and its availability may be obtained from Professor David J. Willis, Electronics Engineering Technology, Oregon Institute of Technology, 3201 Campus Drive, Klamath Falls, Oregon 97601-8801.

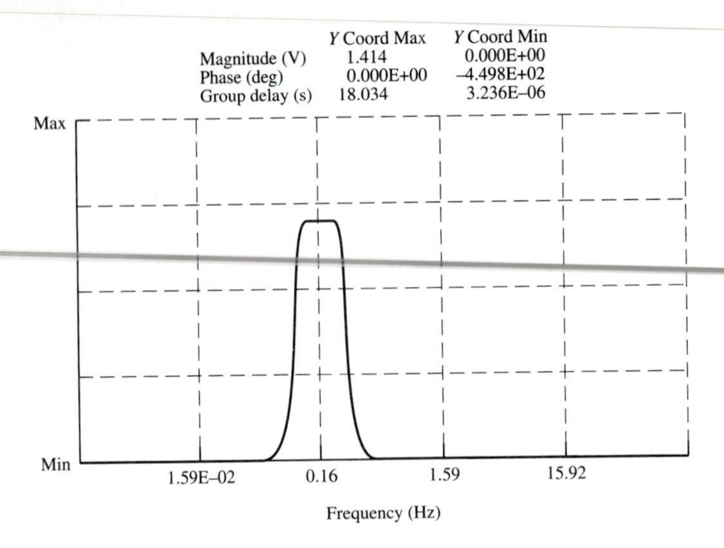

FIGURE 2.8-2
Tenth-order band-pass function.

PROBLEMS

Note: Tables of passive filter realizations are given in App. A. These may be used for any of the problems that require a filter to be designed. Additional information on using the tables is given in Sec. 4.9.

Section 2.1

1. (*a*) Find the corresponding values in dB of $|N(j\omega)|$ for the following decimal values: 0.8, 0.7, 0.6, 0.3.
   (*b*) Find the corresponding decimal values of $|N(j\omega)|$ for the following dB values: -0.2, -0.7, -3, -15.
2. For each of the following network functions $N_i(s)$ $(i = a, b, c)$, find the magnitude-squared function $|N_i(j\omega)|^2$ and show that it is a ratio of even polynomials:

   (*a*) $N_a(s)\dfrac{H}{s^2 + as + b}$ (*b*) $N_b(s) = \dfrac{H(s + c)}{s^2 + as + b}$ (*c*) $N_c(s) = \dfrac{H}{s^3 + 2s^2 + 2s + 1}$
3. Show that for each of the following network functions $N_i(s)$ $(i = a, b)$, the quantities $|N_i(j\omega)|$ are constants, independent of frequency. Such functions are called *all-pass functions*.

   (*a*) $N_a(s) = \dfrac{s - a}{s + a}$ (*b*) $N_b(s) = H\dfrac{s^2 - as + b}{s^2 + as + b}$
4. Find the quadratic factors of each of the following polynomials $P_i(s)$ $(i = a, b)$:
   (*a*) $P_a(s) = s^4 + 1$ (*b*) $P_b(s) = s^4 - 1$
5. Find the quadratic factors of each of the following polynomials $P_i(s)$ $(i = a, b)$:
   (*a*) $P_a(s) = s^6 + 1$ (*b*) $P_b(s) = s^6 - 1$

6. Find $N_i(s)$ $(i = a, b, c)$ for any of the following for which it exists:

(*a*) $|N_a(j\omega)|^2 = \dfrac{1-\omega^2}{\omega^4 - 4\omega^2 + 8}$ (*b*) $|N_b(j\omega)|^2 = \dfrac{1+\omega^2}{\omega(\omega^4+1)}$

(*c*) $|N_c(j\omega)|^2 = \dfrac{\omega^4 - 2\omega^2 + 1}{(\omega^2+1)(\omega^4 - 3\omega^2/2 + 25/16)}$

7. A network is required to have

$$\left|\frac{V_2(s)}{V_1(s)}\right|^2_{s=j\omega} = |N(j\omega)|^2 = \frac{4(\omega^2+1)}{(-\omega^2+1)^2}$$

In addition, the excitation $v_1(t) = \frac{1}{2}e^t$ (assuming no initial conditions) will produce the response $v_2(t) = -\sin t$. Find $N(s)$.

8. Determine the value of the constant a in the following network function $N(s)$ so that $|N(j\omega)|$ satisfies a maximally flat magnitude criterion:

$$N(s) = \frac{s+1}{s^2 + as + 1}$$

9. It is desired to make the following network function $N(s)$ satisfy a maximally flat magnitude criteria. Find the necessary relation that must exist between the coefficients a and b.

$$N(s) = \frac{s+b}{s^2 + as + 1}$$

10. Repeat Prob. 9 for the following network function $N(s)$:

$$N(s) = \frac{s^2 + bs + 2}{s^2 + as + 1}$$

11. Prove that the function $|N(j\omega)|$ defined by (12) is monotonic.

12. Use (15) to find the pole locations and quadratic factors of a Butterworth function for $n = 6$. Use (17) to find the denominator polynomial for the same order.

13. Find the order required for a normalized Butterworth ($\omega_p = 1$ rad/s, $K_p = -3.0103$ dB) function with a stopband attenuation K_s of at least 20 dB at each of the following values of stopband frequency ω_s: (*a*) 1.6 rad/s, (*b*) 1.8 rad/s, (*c*) 2.0 rad/s.

14. Find the order required for a normalized Butterworth ($\omega_p = 1$ rad/s, $K_p = -3.0103$ dB) function for each of the following stopband attenuations K_s, all specified at $\omega_s = 1.8$ rad/s: (*a*) 15 dB, (*b*) 20 dB, (*c*) 25 dB.

15. (*a*) For the specifications given in Example 2.1-6, find the maximum value to which K_s may be changed without raising the order of the filter.
(*b*) Find the maximum value of K_s that will permit a ninth-order filter to be used.

16. (*a*) For the specifications given in Example 2.1-6, find the minimum value to which f_s may be changed without raising the order of the filter.
(*b*) Find the minimum value of f_s that will permit a ninth-order filter to be used.

17. (*a*) Find the required order for a maximally flat magnitude low-pass network function that is down 2 dB at 2 rad/s and down at least 25 dB at 6 rad/s.
(*b*) Find a filter realization for the specifications given in part (*a*) for the case where the network function is the voltage transfer ratio of a single-resistance-terminated (1-Ω) lossless ladder network.
(*c*) Repeat part (*b*) for a double-resistance-terminated (1-Ω) filter.

18. (*a*) Find a low-pass maximally flat magnitude network function whose magnitude characteristic is 3.0103 dB down (from its dc value) at 1 kHz and a minimum of 20 dB down at all frequencies greater than 2.5 kHz.

(*b*) Find a filter realization that meets the specifications given in part (*a*) for the case where the network function is the voltage transfer ratio of a single-resistance-terminated (1-kΩ) lossless ladder network.

(*c*) Repeat part (*b*) for a double-resistance-terminated (1-kΩ) filter.

19. A network function has a magnitude relation approximated as

$$|N(j\omega)| = \frac{1}{\sqrt{1+\varepsilon^2[f(\omega)]^n}}$$

where $f(\omega)$ satisfies the following relations: $f(0) = 0$ and $f(1) = 1$. Find an expression for the value of n required to obtain an attenuation K_s dB at ω_s radians per second. The expression should contain only the variables K_s, ω_s, and ε. The expression should use $\log_{10}$, not the logarithm to any other base.

20. (*a*) Find and plot the pole locations on the s plane for a Butterworth function with $n = 2$ and $\omega_p = 1$ for the following values of K_p: -3.0103, 2, and 1 dB.

(*b*) Repeat part (*a*) for $n = 3$.

21. (*a*) Find a third-order maximally flat magnitude network function that is 1 dB down (from its zero-frequency value) at 1 rad/s. Also find its pole locations.

(*b*) Find a filter realization that meets the specifications of part (*a*) for the case where the network function is the voltage transfer ratio of a single-resistance-terminated (1-Ω) lossless ladder network.

(*c*) Repeat part (*b*) for a double-resistance-terminated (1-Ω) filter.

22. Repeat Prob. 21 for the case where the function is 1 dB down at 1 kHz.

23. A low-pass filter is constructed as a noninteracting cascade of four identical second-order maximally flat magnitude filter stages. Each of the stages has $K_p = 3.0103$ dB and $\omega_p =$ 1 rad/s. Find the frequency at which the entire cascade is down 3.0103 dB (from its dc value).

24. Find a second-order maximally flat magnitude network function that has the property that a noninteracting cascade of four such functions will have a 3-dB down frequency of 1 rad/s.

Section 2.2

1. (*a*) Find the values of a third-order Chebyshev polynomial at values of ω of 0.6 and 1.3 rad/s using (3). Confirm these values using (4).

(*b*) Repeat part (*a*) for a fourth-order polynomial.

2. (*a*) Find an expression for the derivative of the $|N(j\omega)|^2$ function defined by (1) with $\varepsilon = 1$ that is valid for the frequency range $\omega \geq 1$ rad/s. Use this expression to show that the function is monotonic in this region.

(*b*) Use the expression for the derivative of $|N(j\omega)|^2$ for a Butterworth function derived in Prob. 11 of Sec. 2.1 and compare the result with that obtained in part (*a*) for the case $n = 6$, $\omega = 1.2$ rad/s, and (for the Chebyshev function) $\varepsilon = 1$.

3. Determine the Chebyshev polynomial $C_7(\omega)$ using the relations of (3).

4. Starting with the trigonometric expression given in (4*a*), derive the last expression given in (3).

5. (*a*) Determine a value for the multiplicative constant H in Example 2.2-2 such that $|N(j0)| = 1$.

(*b*) Determine a value for the multiplicative constant H in Example 2.2-2 such that

$$\max |N(j\omega)|_{0\le\omega\le 1} = 1$$

6. Find the pole locations and the network function for a fourth-order low-pass Chebyshev characteristic with a passband $0 \le \omega \le 2$ rad/s with a 1-dB ripple. Determine the value of the multiplicative numerator constant such that the maximum passband magnitude of the function is unity.

7. (*a*) Find a second-order low-pass network function that has a 2-dB equal-ripple magnitude characteristic in a 0-to-2-rad/s passband. Determine the value of the numerator constant such that the maximum passband magnitude of the function is 1.0.
(*b*) At a frequency of 4 rad/s, how many dB will the magnitude characteristic be down from its zero-frequency value?

8. (*a*) Find a third-order low-pass network function that has a 3-dB equal-ripple magnitude characteristic in a 0-to-1-rad/s passband. Determine the value of the multiplicative constant such that the maximum passband value is 1.0.
(*b*) At a frequency of 2 rad/s, how many dB will the magnitude characteristic be down from its zero-frequency value?

9. For a third-order normalized ($\omega_p = 1$ rad/s) Chebyshev low-pass function, the denominator $D(s)$ is factored in the form

$$D(s) = (s^2 + as + b)(s + c)$$

Find any relations that exist among the coefficients a, b, and c.

10. Determine the order of an equal-ripple magnitude function that has a 0.5-dB ripple in the passband 0 to 1 kHz and a 3-dB down frequency that is no greater than 1.1 kHz.

11. (*a*) Determine the voltage transfer function of a low-pass filter having a 1-dB equal-ripple characteristic in the passband 0 to 20 krad/s and a minimum of 28 dB attenuation at all frequencies over 50 krad/s. Determine the value of the multiplicative constant so that the dc transmission is 0.5.
(*b*) Find a double-resistance-terminated (1-kΩ resistor) filter realization for the function found in part (*a*). Does this network also realize the multiplicative constant determined in part (*a*)?

12. Find the order required for a normalized Chebyshev ($\omega_p = 1$ rad/s) function with 1-dB passband ripple and a stopband attenuation K_s of at least 28 dB for each of the following values of stopband frequency ω_s: (*a*) 1.4 rad/s, (*b*) 1.6 rad/s, (*c*) 1.8 rad/s.

13. Find the order required for a normalized Chebyshev ($\omega_p = 1$ rad/s) function with 1-dB passband ripple and a stopband frequency $\omega_s = 1.6$ rad/s for each of the following values of stopband attenuation K_s: (*a*) 20 dB, (*b*) 28 dB, (*c*) 36 dB.

14. (*a*) For the specifications given in Example 2.2-3, find the maximum value to which K_s may be changed without raising the order of the filter.
(*b*) Find the maximum value of K_s that will permit a fourth-order filter to be used.

15. (*a*) For the specifications given in Example 2.2-3, find the minimum value to which f_s may be changed without raising the value of the filter.
(*b*) Find the minimum value of f_s that will permit a fourth-order filter to be used.

16. (*a*) Find and plot the pole locations on the s plane for a Chebyshev function with $n = 2$ and $\omega_p = 1$ for the following values of K_p: 0.5, 1.0, and 2.0 dB.
(*b*) Repeat part (*a*) for $n = 3$.

17. Find a second-order equal-ripple network function that has the property that a noninteracting cascade of four such functions will have a passband of 0 to 1 rad/s with a 1-dB ripple.

Section 2.3

1. Find a second-order network function $N(s)$ that has the property that $|N(j\omega)|$ has a monotonic behavior bounded by values of 1 and 0.7071 in the frequency range $0 \le \omega \le 1$ rad/s and an equal-ripple behavior bounded by values of 0.7071 and 0 in the frequency range $1 \le \omega \le \infty$ rad/s.
2. Find the order of an inverse-Chebyshev function with $\omega_p = 0.8$ rad/s, $K_p = 1$ dB, $\omega_s = 1$ rad/s, and $K_s = 30$ dB.
3. For the specifications given in Example 2.3-1, find the actual attenuation at $\omega = 0.5$ rad/s by determining the value of ε^2 and using (3).
4. (*a*) For the specifications given in Example 2.3-1, find the maximum value to which ω_p may be changed without raising the order of the filter.
   (*b*) Find the maximum value of ω_p that will permit a second-order filter to be used.
5. (*a*) For the specifications given in Example 2.3-1, find the minimum value to which K_p may be changed without raising the value of the filter.
   (*b*) Find the minimum value of K_p that will permit a second-order filter to be used.
6. Derive the expressions given in (14) and (15).
7. (*a*) Find a third-order inverse-Chebyshev function with a 20-dB-attenuation stopband of $1 \le \omega \le \infty$ rad/s. Determine the multiplicative constant such that $|N(0)| = 1$.
   (*b*) Find a single-resistance-terminated (1-Ω resistor) filter realization for the function found in part (*a*). Does this network also realize the multiplicative constant determined in part (*a*)?
8. Repeat Prob. 7 for a fourth-order function.
9. (*a*) Find the frequency $\omega_{3\text{dB}}$ at which a second-order inverse-Chebyshev function with 20 dB attenuation in a stopband $1 \le \omega \le \infty$ rad/s has 3.0103 dB attenuation from its dc value.
   (*b*) Repeat part (*a*) for a function with 30 dB stopband attenuation.
   (*c*) Repeat part (*a*) for a function with 40 dB stopband attenuation.
10. (*a*) Find a third-order inverse-Chebyshev function with a 30-dB-attenuation stopband of $10 \le \omega \le \infty$ kHz. Determine the multiplicative constant such that $|N(0)| = 1$.
    (*b*) Find a single-resistance-terminated (1-kΩ resistor) filter realization for the function defined in part (*a*). Does this network also realize the multiplicative constant determined in part (*a*)?
11. Find the zero locations of an 11th-order inverse-Chebyshev function.
12. (*a*) Make a plot on the s plane showing how the pole positions of a third-order inverse-Chebyshev function change as K_s is given the value 20, 30, and 40 dB.
    (*b*) Repeat part (*a*) for a fourth-order function.
13. Find a second-order inverse-Chebyshev function that has 25 dB attenuation in a stopband $1 \le \omega \le \infty$ rad/s and a maximum magnitude of unity. Give the denominator in polynomial form and also give the complex locations of the poles.
14. Find a third-order inverse-Chebyshev function that has 25 dB attenuation in a stopband $1 \le \omega \le \infty$ rad/s and a maximum magnitude of unity. Give the denominator in polynomial form and also give the complex locations of the poles.
15. (*a*) Find an inverse-Chebyshev network function meeting the following specifications: (1) peak passband magnitude is unity; (2) passband of 0 to 1 rad/s with a maximum of 3.0103 dB attenuation at 1 rad/s; (3) a minimum of 24 dB attenuation at all frequencies greater than or equal to 2 rad/s.
    (*b*) What is the attenuation of the inverse-Chebyshev function at 1 rad/s?
    (*c*) Find a Chebyshev network function that meets the specifications given in part (*a*).
    (*d*) What is the attenuation of the Chebyshev function at 2 rad/s?

Section 2.4

1. (*a*) Find an elliptic network function with a 1-dB-ripple passband $0 \leq \omega \leq 1$ rad/s, a 45-dB-attenuation stopband $\omega \geq 2$ rad/s, and a maximum passband magnitude of unity. Use case *C* (if possible) if an even-order function is required.
   (*b*) Find a double-resistance-terminated (1-Ω input resistor) filter realization for the function found in part (*a*). Find the value of the multiplicative constant for the network function determined in part (*a*) that is actually realized by the filter.
2. (*a*) Find an elliptic network function with a 1-dB-ripple passband $0 \leq \omega \leq 1$ rad/s, a 48-dB-attenuation stopband $\omega \geq 2$ rad/s, and a maximum passband magnitude of unity. Use case *B* (if possible) if an even-order function is required.
   (*b*) Find a double-resistance-terminated (1-Ω input resistor) filter realization for the function found in part (*a*). Find the value of the multiplicative constant for the network function determined in part (*a*) that is actually realized by the filter.
3. Find an elliptic network function with a 1-dB-ripple passband $0 \leq \omega \leq 1$ rad/s, a 50-dB-attenuation stopband $\omega \geq 2$ rad/s, and a maximum passband magnitude of unity. Use case *A* (if possible) if an even-order function is required.
4. (*a*) Find the required order for a maximally flat magnitude function that has a 0-to-1-rad/s passband with a maximum attenuation of 1 dB and also has 38 dB attenuation at 1.5 rad/s.
   (*b*) Repeat part (*a*) for an equal-ripple function.
   (*c*) Repeat part (*a*) for an elliptic function.
   (*d*) If even-order case *A* functions are excluded, what order of elliptic function is required.
5. (*a*) Find the required order for a maximally flat magnitude function that has a 0-to-1-rad/s passband with a maximum attenuation of 1 dB and also has 29 dB attenuation at 1.1 rad/s.
   (*b*) Repeat part (*a*) for an equal-ripple function.
   (*c*) Repeat part (*a*) for an elliptic function.
6. (*a*) For the specifications given in Example 2.4-3, find the minimum value to which f_s may be changed without raising the order of the filter.
   (*b*) Find the minimum value of f_s that will permit a second-order filter to be used.
7. (*a*) For the specifications given in Example 2.4-3, find the minimum value to which K_p may be changed without raising the order of the filter.
   (*b*) Find the minimum value of K_p that will permit a second-order filter to be used.
8. (*a*) For the specifications given in Example 2.4-3, find the maximum value to which K_s may be changed without raising the order of the filter.
   (*b*) Find the maximum value of K_s that will permit a second-order filter to be used.
9. Find a low-pass elliptic filter realization that has a maximum of 1 dB attenuation in a 0-to-1-kHz passband and a minimum of 34 dB attenuation at all frequencies greater than 2 kHz. The filter should have equal 1000-Ω terminating resistors at the input and output.

Section 2.5

1. (*a*) Find the order of a high-pass filter that is to have a 10-to-∞-kHz passband with a maximum ripple of 1 dB and a stopband attenuation of 29 dB at all frequencies less than 6.667 kHz. The magnitude characteristic should be of the Butterworth type.
   (*b*) Repeat part (*a*) for an equal-ripple passband characteristic.
   (*c*) Repeat part (*a*) for an elliptic characteristic.
2. Make a sketch of the magnitude characteristic for the three high-pass network functions defined in Prob. 1. For the elliptic case show the frequencies at which the stopband nulls occur.

3. (*a*) Find the normalized (passband 1 to ∞ rad/s) high-pass network function and the pole locations that will produce a third-order 0.5-dB-ripple passband magnitude characteristic. Find a multiplicative constant that will produce a maximum passband magnitude of unity.
   (*b*) Find a double-resistance-terminated (1-Ω terminating resistor) filter realization for a voltage transfer function meeting the specifications given in part (*a*). Find the value of the multiplicative constant for the network function determined in part (*a*) that is actually realized by the filter.
4. Find a fourth-order maximally flat magnitude band-pass filter with a −3-dB-bandwidth of 100 Hz and a (geometric) center frequency of 1 kHz. The filter should have equal 1-kΩ resistive terminations.
5. Find a realization for a filter with a voltage transfer function having the characteristics defined in Prob. 1(*c*). The resistive terminations should have a value of 1 kΩ.
6. (*a*) A $\frac{1}{2}$-dB-ripple Chebyshev second-order low-pass function is transformed to provide a band-pass function with a (normalized) center frequency of 1 rad/s and a bandwidth of 1 rad/s. Find the network function and the pole positions.
   (*b*) Find a realization of a filter with a voltage transfer function that meets the specifications given in part (*a*).
7. (*a*) Use the low-pass to band-pass transformation to find the network function for a fourth-order maximally flat magnitude band-elimination filter with a (normalized) notch center frequency of 1 rad/s and a bandwidth (defined by −3 dB attenuation from maximum passband values at zero and infinity) of 1 rad/s. Express the result as a ratio of polynomials.
   (*b*) Show the error that would occur if the narrow-band approximation was used to meet these specifications by determining the network function using this approximation and comparing the values of the coefficients with those found in part (*a*).
   (*c*) Find a realization of a filter whose voltage transfer function meets the specifications given in part (*a*).
8. The band-elimination network shown in Fig. P2.5-8 may be shown to have a maximally flat magnitude characteristic in the passband of its voltage transfer function. The stopband has a center frequency of 1 rad/s and a bandwidth (−3 dB attenuation from maximum passband values) of $\frac{1}{3}$ rad/s. Find appropriate frequency transformations that may be made on this filter so as to derive a prototype maximally flat magnitude second-order low-pass network with a bandwidth of 1 rad/s.

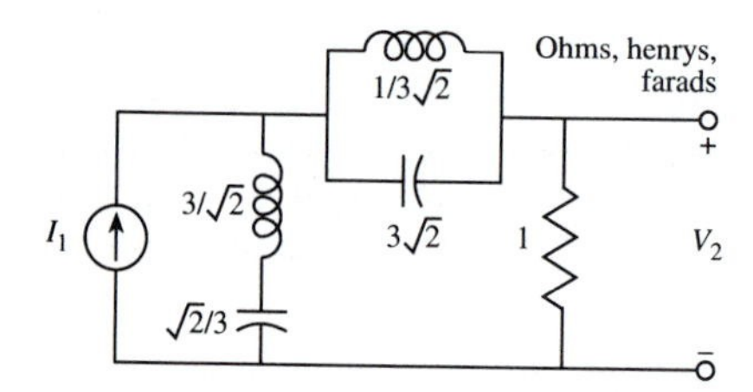

FIGURE P2.5-8

9. Find a fourth-order network function that has a magnitude varying within 1 dB in a passband from 0.6180 to 1.6180 rad/s and has nulls at 0.3269 and 3.059 rad/s.
10. The network shown in Fig. P2.5-10 has a maximally flat magnitude low-pass characteristic with a −3-dB frequency of 1 rad/s for its transfer admittance $Y_{21}(s) = V_2(s)/I_1(s)$. It is desired to use frequency transformations to obtain a band-elimination network with a (geometric)

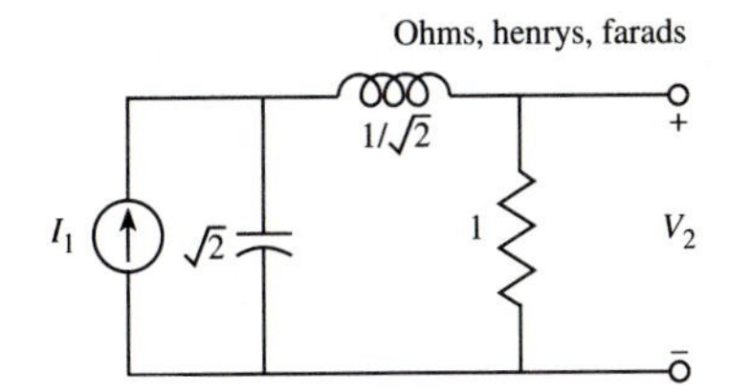

FIGURE P2.5-10

center frequency of 1 rad/s and a bandwidth of $\frac{1}{3}$ rad/s defined in terms of the −3-dB-attenuation frequencies (from maximum passband transmission). Two different procedures are to be followed: (*a*) First transform the given network to a high-pass one (−3 dB at 1 rad/s), then frequency normalize the high-pass network, and finally apply the low-pass to band-pass transformation; (*b*) first frequency normalize the low-pass network, then apply the low-pass to band-pass transformation, and then apply the low-pass to high-pass transformation. If the resulting networks are not the same, explain the reason for any difference.

11. (*a*) Use the narrow-band approximation to find the pole locations for a band-pass network function that has a sixth-order equal-ripple characteristic with a 1-dB ripple in the passband. The passband is to be 0.1 rad/s with a center frequency of 1 rad/s.
(*b*) Find a filter with a voltage transfer function that matches the specifications given in part (*a*).

12. (*a*) Use the narrow-band approximation to find the pole locations of a fourth-order band-pass network function with a center frequency of 1 rad/s, a bandwidth of 0.08 rad/s, and a maximally flat magnitude characteristic.
(*b*) From the results of part (*a*) find the network function expressed as a ratio of polynomials.
(*c*) Use the low-pass to band-pass transformation to find the network function directly and compare the results with those obtained from part (*b*).

13. Find a fourth-order network function that has a magnitude characteristic that varies within 1 dB in a 0.05-rad/s passband with a (geometric) center frequency of 1 rad/s. The function should have a minimum of 11 dB attenuation at all frequencies less than 0.9625 rad/s and greater than 1.0375 rad/s.

14. Find a band-pass double-terminated (1-Ω resistor) filter realization that has *two* passbands, one with center frequency 0.95 rad/s, the other with center frequency 1.05 rad/s. Each passband is to have a bandwidth of 0.005 rad/s and a maximally flat magnitude characteristic. *Note*: In making the various transformations required by this problem, assume that the arithmetic mean can be used rather than having to calculate the geometric mean. This will greatly simplify the calculations required.

15. A low-pass network function is assumed to have the (idealized) magnitude characteristic shown in Fig. P2.5-15. If the function is transformed by applying the low-pass to band-pass transformation *two times*, draw the resulting magnitude characteristic. On the drawing include numerical values for all the cutoff frequencies.

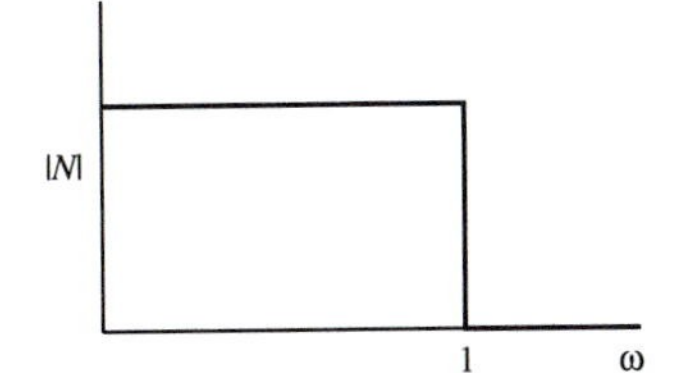

FIGURE P2.5-15

16. A band-pass network function with the idealized magnitude characteristic shown in Fig. P2.5-16 is transformed by applying the low-pass to band-pass transformation *two times*. Draw the resulting magnitude characteristic. On your drawing give numerical values for all the cutoff frequencies.

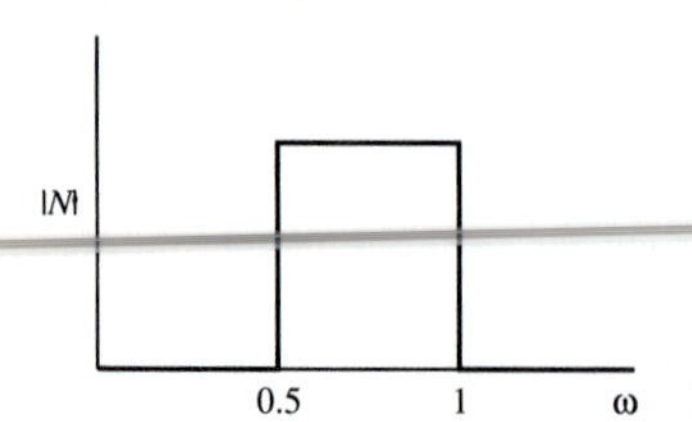

FIGURE P2.5-16

17. Identify the characteristics of the *low-pass* approximation used as the prototype for each of the *band-pass* pole-zero plots shown in Fig. P2.5-17. The shaded areas may contain three, four, or five poles. Your identification should list an approximation characteristic (Butterworth, Chebyshev, inverse Chebyshev, or elliptic) and an order. For even-order elliptic you should also indicate the case(s) that apply.

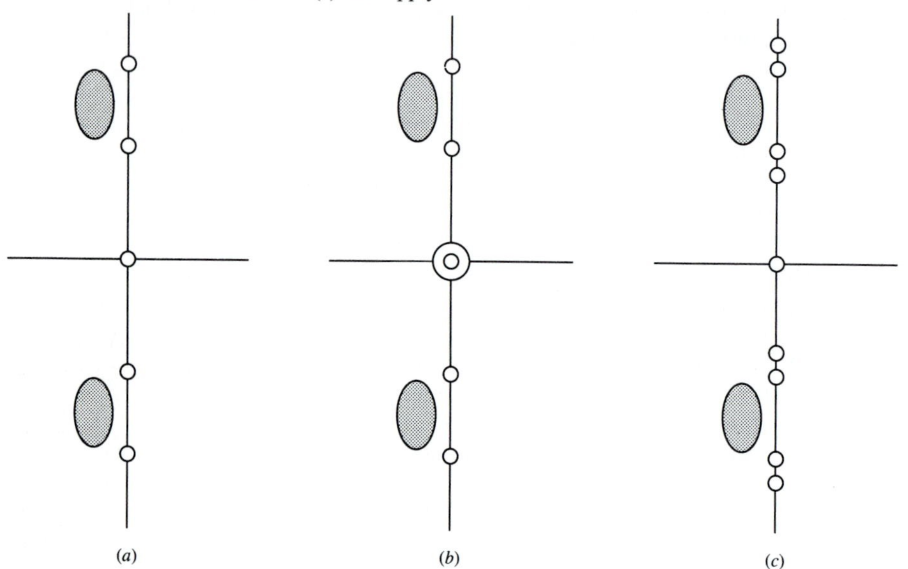

FIGURE P2.5-17

Section 2.6

1. Determine whether the function

$$A(\omega) = \tan\,[\arg N(j\omega)] = \frac{\omega^5 - 2\omega^3 - 2\omega}{-2\omega^4 - \omega^2 + 2}$$

is the phase function of any of the network functions $N_i(s)$ $(i = a, b, c, d)$ given below:

(*a*) $N_a(s) = \dfrac{s+1}{(s^2+s+1)(s^2+2s+2)}$ (*b*) $N_b(s) = \dfrac{s^2-s+1}{(s^2+2s+2)(s+1)}$

(*c*) $N_c(s) = \dfrac{s-1}{(s^2+s+1)(s^2+2s+2)}$ (*d*) $N_d(s) = \dfrac{(s+1)(s^2-2s+2)}{s^2+s+1}$

2. (*a*) Find $A(\omega)$ for a normalized third-order Butterworth function.
   (*b*) Repeat part (*a*) for a fourth-order function.
3. (*a*) For the function

$$N(s) = \frac{H}{s^2 + 3s + 3}$$

   find the value of arg $N(j\omega)$ for values of ω of 0.1, 0.2, and 0.5.
   (*b*) Repeat the computation using the expansion

$$\tan^{-1} x \approx x - \tfrac{1}{3}x^3$$

   (*c*) Repeat part (*b*) but include the fifth-degree term in the expansion
   (*d*) Repeat part (*c*) but include the seventh-degree term in the expansion.
4. (*a*) Find $N(s)$ as an all-pole function for the following expression for $A(\omega)$:

$$A(\omega) = \frac{-15\omega + \omega^3}{15 - 6\omega^2}$$

   (*b*) Find a lower order $N(s)$ that will have the same phase characteristics.
5. (*a*) Find $N(s)$ as an all-pole function for the following expression for $A(\omega)$:

$$A(\omega) = \frac{-105\omega + 10\omega^3}{105 - 45\omega^2 + \omega^4}$$

   (*b*) Find a lower order $N(s)$ that will have the same phase characteristics.
6. Find $N(s)$ for the following expression for $A(\omega)$:

$$A(\omega) = \frac{-4\omega + \omega^3}{10.8 - 6.7\omega^2 + \omega^4}$$

7. Apply the procedure given in Example 2.6-3 to determine the denominator coefficients of a third-order all-pole network function that has a maximally linear phase characteristic with a slope of -1. Verify the results using Table 2.6-1.
8. (*a*) In the following network function $N(s)$, find the value of the coefficient a so that a maximally linear phase characteristic results:

$$N(s) = \frac{H(s + a)}{s^2 + 2s + 1}$$

   (*b*) Find the value of the delay (at direct current) produced by the function.
9. Find the values of the coefficients a, b, and c for the third-order network function $N(s)$ given below such that arg $N(j\omega)$ has the specified series expansion:

$$N(s) = \frac{H}{s^3 + as^2 + bs + c} \qquad \arg N(j\omega) = -\omega - \tfrac{1}{3}\omega^3 + 0\omega^5 + \cdots$$

10. Use the expression given in (12) to determine the denominator coefficients of the network function for a fourth-order normalized (1 s delay at direct current) Thomson function. Verify the results using Table 2.6-1.
11. (*a*) Use the expression given in (13) to determine the denominator coefficients of a fifth-order normalized (1 s delay at direct current) Thomson function. Verify the results using Table 2.6-1.

(b) Find a double-resistance-terminated (1-Ω terminating resistor) filter realization for a voltage transfer function meeting the specifications given in part (a). Find the value of the multiplicative constant for the network function determined in part (a) that is actually realized by the filter.

12. Find a double-resistance-terminated (1-kΩ terminating resistor) filter realizing the voltage transfer function of a third-order Thomson approximation with a delay time of 1 ms.

13. (a) Find the frequency at which a normalized (1 s delay at direct current) second-order Thomson function is 3.0103 dB down from its dc (zero-frequency) magnitude.

(b) Repeat part (a) for a third-order function.

14. (a) Determine the expression for $D(\omega)$ for a second-order Butterworth function. Evaluate the expression at $\omega = 1$ rad/s and compare your result with Fig. 2.6-3.

(b) Repeat part (a) for a third-order function.

15. (a) Determine the required order for a maximally flat delay network function with a 1-ms delay error no greater than 2.5 percent and a loss no greater than 3 dB for all frequencies up to 1500 rad/s.

(b) Realize the function as the transfer admittance of a lossless ladder network with a single 1-kΩ terminating resistance.

16. (a) Determine the magnitude error of a third-order Thomson function at $\omega = 2$ rad/s and compare the value with Fig. 2.6-8.

(b) Repeat part (a) for the delay error. Use Fig. 2.6-7.

17. Determine and sketch the delay functions for each of the following network functions $N_i(s)$ $(i = a, b, c)$:

$$N_a(s) = \frac{1}{s^2 + s + 1} \qquad N_b(s) = \frac{s^2 - s + 1}{s^2 + s + 1} \qquad N_c(s) = \frac{s^2 + 1}{s^2 + s + 1}$$

18. For the following network function, determine, as qualitatively as possible, any properties possessed by (a) $|N(j\omega)|$, (b) $\arg N(j\omega)$, and (c) $D(\omega)$:

$$N(s) = \frac{s^2 - 9s + 27}{s^2 + 9s + 27}$$

19. For the following network function, determine, as qualitatively as possible, any properties possessed by (a) $|N(j\omega)|$, (b) $\arg N(j\omega)$, and (c) $D(\omega)$:

$$N(s) = \frac{-s^3 + 3s^2 - \frac{15}{4}s + \frac{15}{8}}{s^3 + 3s^2 + \frac{15}{4}s + \frac{15}{8}}$$

Section 2.7

1. Find a third-order network function whose magnitude-squared characteristic approximates a gaussian response. Frequency normalize the function so that the delay (at direct current) is 1.0 s.

2. Repeat Prob. 1 for a fourth-order function.

3. For the network function

$$N(s) = \frac{2s + 3}{(s + 1)(s + 2)}$$

find the rise time t_r and the 3-dB bandwidth ω_c and compare their product with the value given in (8).

4. For a second-order Butterworth function, find the rise time t_r and the 3-dB bandwidth ω_c and compare their product with the value given in (8).

5. Repeat Prob. 4 for a second-order Thomson function.

6. Repeat Prob. 4 for the second-order gaussian function defined in Example 2.7-1.

CHAPTER 3

SENSITIVITY

One of the problems that continually face a network designer is the evaluation of his or her design, especially in comparison with other possible realizations that meet the same specifications. To do this, the designer must be concerned with the sensitivity of the filter. By *sensitivity*, we mean a measure of the change in some performance characteristic of the network (or the network function) resulting from some change in the nominal value of one or more of the elements of the network. Even if a network realization is attractive from theoretical considerations, high sensitivities may make it useless in practice. Thus, in the design of filters we are interested both in choosing realizations that have low sensitivities and in minimizing the sensitivities of realizations we desire to use. We shall see examples of both processes in the chapters that follow.

3.1 RELATIVE SENSITIVITY RELATIONS

The symbol S is used to denote sensitivity. In addition, a superscript character is used to indicate the performance characteristic being evaluated and a subscript character to indicate the network element that is causing the change. If we let y be the performance characteristic and x be the element, our symbol looks like

$$S_x^y \begin{array}{l} \leftarrow \text{characteristic} \\ \leftarrow \text{element} \end{array} \tag{1}$$

Definition of Sensitivity

To define sensitivity, consider the way in which a characteristic $y(x)$ can depend on an element x. If the nominal value of x is x_0, then variations in $y(x)$ produced by changes in x can be expressed by the Taylor series

$$y(x) = y(x_0) + \frac{\partial y}{\partial x}\bigg|_{x=x_0} dx + \frac{1}{2}\frac{\partial^2 y}{\partial x^2}\bigg|_{x=x_0} (dx)^2 + \cdots \tag{2}$$

For "small" variations in x we may ignore all the higher order derivative terms in (2), retaining only the first-order one. The result may be written

$$\Delta y(x_0) = y(x) - y(x_0) = \frac{\partial y}{\partial x}\bigg|_{x=x_0} dx \tag{3}$$

where we have defined $\Delta y(x_0)$ as the change in y resulting from the variation in x. Since we are interested in relative (rather than absolute) changes in y and x, we may add normalizing terms to (3) to obtain

$$\frac{\Delta y(x_0)}{y(x_0)} = \left[\frac{\partial y}{\partial x}\frac{x}{y(x)}\right]\bigg|_{x=x_0} \frac{dx}{x_0} \tag{4}$$

The relation between $\Delta y(x_0)/y(x_0)$ and dx/x_0 is the sensitivity function. We may now formally define

$$S_x^y = \frac{\partial y}{\partial x}\frac{x}{y} = \frac{\partial y/y}{\partial x/x} = \frac{\partial(\ln y)}{\partial(\ln x)} \tag{5}$$

Because of the normalization of both y and x, this is referred to as *relative sensitivity.* Many choices are common for the performance characteristic y. For example, it may be the actual network function (with variable s or ω). Other choices are pole (and zero) locations, coefficients of the numerator and denominator polynomials, properties such as the Q of the network function, and so forth. A treatment of the various types of sensitivity will be given in later sections of this chapter.

Properties of Relative Sensitivity

Any relative sensitivity function defined as shown in (5) will be characterized by a set of algebraic properties. Some of the most important ones are derived in the following paragraphs. The numbers of the properties refer to the entries in Table 3.1-1.

Properties 1 and 2. *The sensitivity of any characteristic multiplied by a constant is the same as the original sensitivity.* To see this, we write

$$S_x^{ky} = \frac{\partial(ky)}{\partial x}\frac{x}{ky} = \frac{\partial(ky)}{\partial y}\frac{\partial y}{\partial x}\frac{x}{ky} = \frac{\partial y}{\partial x}\frac{x}{y} = S_x^y$$

where k is not a function of x. Similarly, $S_{kx}^y = S_x^y$.

TABLE 3.1-1
Properties of the relative sensitivity function of (5)

Property no.	Relation
1	$S_x^{ky} = S_{kx}^y = S_x^y$
2	$S_x^x = S_x^{kx} = S_{kx}^{kx} = 1$
3	$S_{1/x}^y = S_x^{1/y} = -S_x^y$
4	$S_x^{y_1 y_2} = S_x^{y_1} + S_x^{y_2}$
5	$S_x^{\prod_{i=1}^n y_i} = \sum_{i=1}^n S_x^{y_i}$
6	$S_x^{y^n} = nS_x^y$
7	$S_x^{x^n} = S_x^{kx^n} = n$
8	$S_{x^n}^y = \frac{1}{n} S_x^y$
9	$S_{x^n}^x = S_{kx^n}^x = \frac{1}{n}$
10	$S_x^{y_1/y_2} = S_x^{y_1} - S_x^{y_2}$
11	$S_{x_1}^y = S_{x_2}^y S_{x_1}^{x_2}$
12†	$S_x^y = S_x^{\lvert y \rvert} + j \arg y \, S_x^{\arg y}$
13†	$S_x^{\arg y} = \frac{1}{\arg y} \operatorname{Im} S_x^y$
14†	$S_x^{\lvert y \rvert} = \operatorname{Re} S_x^y$
15	$S_x^{y+z} = \frac{1}{y+z} (yS_x^y + zS_x^z)$
16	$S_x^{\sum_{i=1}^n y_i} = \frac{\sum_{i=1}^n y_i S_x^{y_i}}{\sum_{i=1}^n y_i}$
17	$S_x^{\ln y} = \frac{1}{\ln y} S_x^y$

† In this relation y is a complex quantity and x is a real quantity.

Property 3. *The sensitivity to a reciprocal quantity is the negative of the sensitivity to the original quantity.* To see this, we may write

$$S_{1/x}^y = \frac{\partial y}{\partial(1/x)} \frac{1/x}{y} = \frac{\partial y}{\partial x} \frac{\partial x}{\partial(1/x)} \frac{1/x}{y} = -\frac{\partial y}{\partial x} \frac{x}{y} = -S_x^y$$

Similarly, $S_x^{1/y} = -S_x^y$.

Properties 4 through 9. *The sensitivity of a product of characteristics is the sum of the sensitivities of the individual characteristics.* To see this, we write

$$S_x^{y_1 y_2} = \frac{\partial(y_1 y_2)}{\partial x}\frac{x}{y_1 y_2} = \frac{y_2\,\partial y_1}{\partial x}\frac{x}{y_1 y_2} + \frac{y_1\,\partial y_2}{\partial x}\frac{x}{y_1 y_2}$$

$$= \frac{\partial y_1}{\partial x}\frac{x}{y_1} + \frac{\partial y_2}{\partial x}\frac{x}{y_2} = S_x^{y_1} + S_x^{y_2}$$

Extending this, we see that the sensitivity of a characteristic to the nth power is n times the sensitivity to the first power, that is, $S_x^{y^n} = nS_x^y$. Similarly, $S_{x^n}^y = S_x^y/n$.

Properties 10 through 11. *The sensitivity of a ratio of characteristics is the difference of the individual sensitivities.* Here we use properties 4 and 3 to obtain

$$S_x^{y_1/y_2} = S_x^{y_1} + S_x^{1/y_2} = S_x^{y_1} - S_x^{y_2}$$

Properties 12 through 14. *The sensitivity of a complex characteristic is also complex. The real part is the magnitude sensitivity, and the imaginary part is the phase sensitivity multiplied by the phase.*[1] To see this, let Y be the magnitude of y and ϕ be the phase. Thus $y = Ye^{j\phi}$. Using (5) and property 4 we obtain

$$S_x^{Ye^{j\phi}} = S_x^Y + S_x^{e^{j\phi}} = S_x^Y + \frac{\partial(\ln e^{j\phi})}{\partial x/x}$$

$$= S_x^Y + j\frac{\partial \phi}{\partial x/x} = S_x^Y + j\phi S_x^\phi$$

Table 3.1-1 contains a summary of the relations derived above as well as several other useful ones.

Application of Sensitivity

Frequently we will use sensitivity information to predict changes in some network characteristic y as a result of some incremental change in a network element x. In this case (5) is written in the form

$$\frac{\Delta y}{y} = S_x^y \frac{\Delta x}{x} \tag{6}$$

We see that if the sensitivity function is known, the normalized incremental change in the characteristic y can be found as the product of the sensitivity function and the normalized incremental change in the element x. In general, the result given by (6) can only be used for small incremental changes for which the differential quantities of (5) are validly approximated by the incremental quantities of (6). An example of this will be given in the next section.

---

[1] This result assumes that x is real.

Example 3.1-1 Sensitivity of an amplifier. A noninverting voltage-controlled voltage source (VCVS) amplifier is realized by an operational amplifier and two resistors R_1 and R_2 as shown in Fig. 3.1-1(a). For this circuit the amplifier gain K is found to be $K = (R_1 + R_2)/R_1$. The values $R_1 = 10\ \text{k}\Omega$ and $R_2 = 90\ \text{k}\Omega$ are chosen to produce a gain of 10. From (5) we find

$$S_{R_1}^K = \frac{-R_2}{R_1 + R_2} = -0.9 \qquad S_{R_2}^K = \frac{R_2}{R_1 + R_2} = 0.9$$

From (6) we can find the results of a 1 percent increase in R_1, namely,

$$\frac{\Delta K}{K} = S_{R_1}^K \frac{\Delta R_1}{R_1} = (-0.9)(0.01) = -0.009$$

A -0.9 percent change in the gain K results. Similarly, a 1 percent increase in R_2 will produce a (positive) 0.9 percent change in K.

Exercise 3.1-1. For the inverting VCVS amplifier shown in Fig. 3.1-1(b), the amplifier gain is $K = -R_2/R_1$. Use (6) to find the percentage change in K produced by 1 percent increases in R_1 and R_2.

Answers For a 1 percent increase in R_1, a 1 percent decrease in $\Delta K/K$ occurs ($|K|$ becomes smaller). For a 1 percent increase in R_2, a 1 percent increase in $\Delta K/K$ occurs ($|K|$ becomes larger).

3.2 FUNCTION SENSITIVITY

In this section we introduce the first of the sensitivities having the form given in (5) of Sec. 3.1 that we shall consider.

Definition of Function Sensitivity

Function sensitivity is defined by choosing the characteristic y of (5) of Sec. 3.1 as the network function $N(s)$. The element x is usually chosen as some passive or active element in the circuit realization of the function. The sensitivity is called *function*

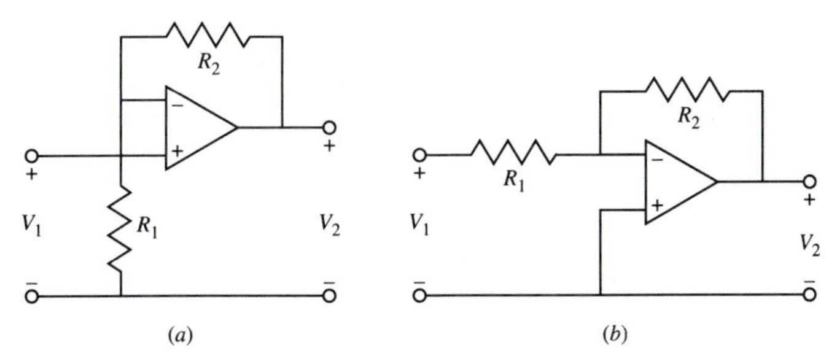

FIGURE 3.1-1
VCVS realizations: (a) noninverting; (b) inverting.

sensitivity. It is defined as[2]

$$S_x^{N(s)} = \frac{\partial N(s)}{\partial x}\frac{x}{N(s)} \tag{1}$$

If the network function is written as a ratio of polynomials $A(s)$ and $B(s)$

$$N(s) = \frac{A(s)}{B(s)} \tag{2}$$

then we may derive the following convenient form of (1):

$$S_x^{N(s)} = x\left[\frac{A'(s)}{A(s)} - \frac{B'(s)}{B(s)}\right] \tag{3}$$

where

$$A'(s) = \frac{\partial A(s)}{\partial x} \qquad B'(s) = \frac{\partial B(s)}{\partial x}$$

Under conditions of sinusoidal steady state, applying property 12 of Table 3.1-1 we find that

$$S_x^{N(j\omega)} = S_x^{|N(j\omega)|} + j\frac{\partial \arg N(j\omega)}{\partial x/x} \tag{4}$$

Summary 3.2-1 Function sensitivity for $N(j\omega)$. The sensitivity $S_x^{N(j\omega)}$ defined in (4) for the network function $N(j\omega)$ has the following properties:

1. The real part of $S_x^{N(j\omega)}$ is equal to $S_x^{|N(j\omega)|}$, the sensitivity of the magnitude of the network function.
2. The imaginary part of $S_x^{N(j\omega)}$ is equal to the change in $\arg N(j\omega)$ with respect to the change in the normalized element value. This is proportional to $S_x^{\arg N(j\omega)}$.

Example 3.2-1 Function sensitivity of a passive *RLC* network. As an example of function sensitivity consider the series *RLC* network shown in Fig. 3.2-1. For this

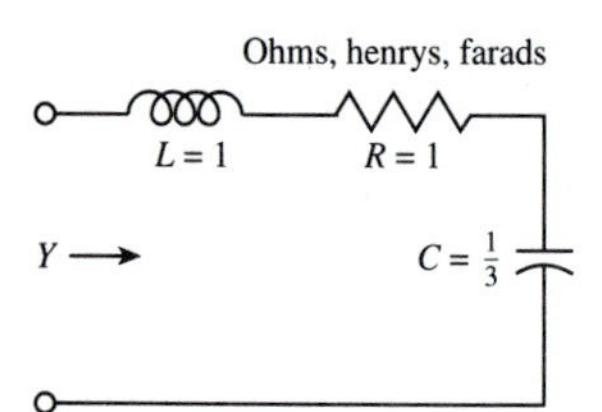

FIGURE 3.2-1
The *RLC* network used in Example 3.2-1.

[2] This is also sometimes referred to as *classical* or *Bode sensitivity.* It was originally presented (as the reciprocal of the relation given here) in H. W. Bode, *Network Analysis and Feedback Amplifier Design*, D. Van Nostrand Company, Inc., Princeton, N.J., 1945, p. 52.

network

$$Y(s) = \frac{s(1/L)}{s^2 + s(R/L) + 1/LC} \tag{5}$$

Applying (3), we find

$$\begin{aligned} S_R^{Y(s)} &= \frac{-sR/L}{s^2 + s(R/L) + 1/LC} \\ S_L^{Y(s)} &= \frac{-s^2}{s^2 + s(R/L) + 1/LC} \\ S_C^{Y(s)} &= \frac{1/LC}{s^2 + s(R/L) + 1/LC} \end{aligned} \tag{6}$$

Further examining the sensitivity to R by inserting the nominal element values shown in Fig. 3.2-1 and letting $s = j\omega$, we obtain

$$S_R^{Y(j\omega)} = \frac{-\omega^2}{(3 - \omega^2)^2 + \omega^2} + j\frac{-\omega(3 - \omega^2)}{(3 - \omega^2)^2 + \omega^2} \tag{7}$$

Plots of the magnitude and phase of the network function for values of R of 1 and 1.1 Ω and of the sensitivity (for $R = 1$ Ω) are shown in Fig. 3.2-2.

Differential versus Incremental Sensitivity

As pointed out in the preceding section, it is important to realize that the relations defined by sensitivity functions are only exact for differential quantities. They provide a linear relation that considers only the first-order effects of the element on the characteristic. When incremental quantities are used, since higher order effects are ignored, the values calculated for the sensitivity functions may not be exact.

Example 3.2-2 Effect of incremental changes. As an example of the difference between differential and incremental quantities, consider the network used in Example 3.2-1. If we evaluate the sensitivity to the element R (for a nominal value of $R = 1$) given in (7) at $\omega = 2$ rad/s, we obtain $S_R^{Y(j2)} = -0.8 + j0.4$. If we actually let R change value by 10 percent, then, for convenience writing $Y(j2)$ as $Y(j2, R)$, we may compute

$$\frac{\Delta|Y|}{\Delta R}\frac{R}{|Y|} = \frac{|Y(j2, 1.1)| - |Y(j2, 1)|}{1.1 - 1}\frac{1}{|Y(j2, 1)|} = -0.7471$$

$$\frac{\Delta \arg Y}{\Delta R/R} = \frac{\arg Y(j2, 1.1) - \arg Y(j2, 1)}{(1.1 - 1)/(1)} = 0.3702$$

Thus the value of the sensitivity function that relates the actual changes in the magnitude and phase of $Y(j2)$ to the change in R is $-0.747 + j0.370$. As the percentage change in R is made smaller, of course, this value will approach the value $-0.8 + j0.4$ more closely.

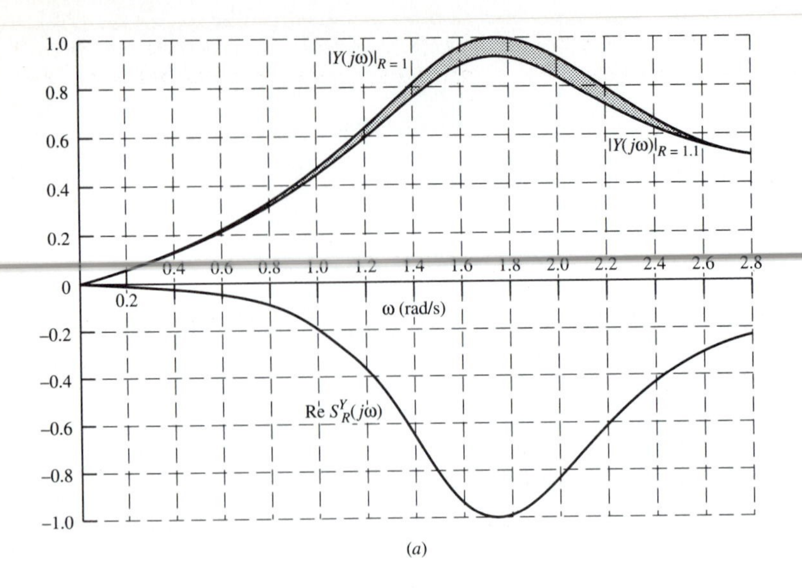

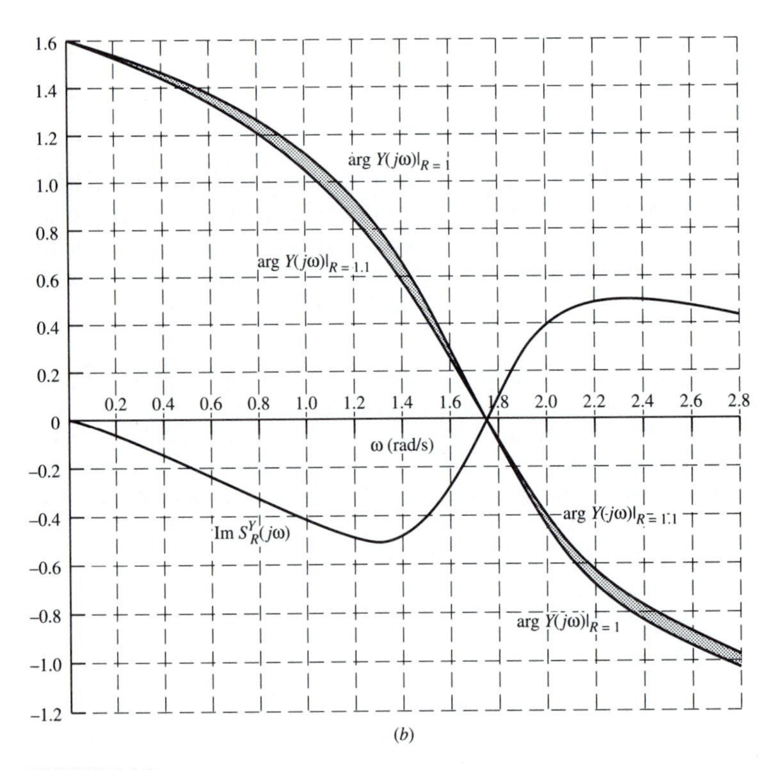

FIGURE 3.2-2
Function sensitivities for the network of Fig. 3.2-1

Exercise 3.2-2. Compute the actual sensitivity $S_R^{Y(j2)}$ as was done in the preceding example but consider the case where R changes by 1 percent, that is, goes from a value of 1 to a value of 1.01.

Answer. $S_R^{Y(j2)} = -0.7944 + j0.3968$.

Multiparameter Sensitivity

The development of function sensitivity given in this section is readily extended to determine the effects on a network function of several parameters simultaneously. The extension is the direct result of the linear nature of the sensitivity functions. To see this, we may write (1) in the form

$$\frac{dN(s)}{N(s)} = d[\ln N(s)] = \sum_{i=1}^{n} S_{x_i}^{N(s)} \frac{dx_i}{x_i} \tag{8}$$

where n is the number of elements being considered. As an extension of this result, using (4), we note that

$$\frac{d|N(j\omega)|}{N(j\omega)} = \sum_{i=1}^{n} \text{Re } S_x^{N(j\omega)} \frac{dx_i}{x_i} \tag{9}$$

$$d \arg N(j\omega) = \sum_{i=1}^{n} \text{Im } S_{x_i}^{N(j\omega)} \frac{dx_i}{x_i} \tag{10}$$

Thus (8) compactly evaluates changes in both magnitude and phase.

Example 3.2-3 Multiparameter sensitivity of a passive *RLC* network. As an example of multiparameter sensitivity, consider the passive *RLC* network used in Example 3.2-1. For this we find at resonance ($\omega = \sqrt{3}$), $S_R^{Y(j\sqrt{3})} = -1$, $S_L^{Y(j\sqrt{3})} = -j\sqrt{3}$, and $S_C^{Y(j\sqrt{3})} = -j\sqrt{3}$; thus

$$\left.\frac{dY(s)}{Y(s)}\right|_{s=j\sqrt{3}} = \frac{d|Y(j\sqrt{3})|}{|Y(j\sqrt{3})|} + j\,d[\arg Y(j\sqrt{3})]$$

$$= (-1)\frac{dR}{R} + (-j\sqrt{3})\frac{dL}{L} + (-j\sqrt{3})\frac{dC}{C} \tag{11}$$

Some other observations concerning this circuit may be made from (11); namely, at resonance, the magnitude is unaffected by changes in the values of L or C, and the magnitude of the sensitivity to R is unity. Similarly, changes in R will not affect the phase (at resonance).

Exercise 3.2-3. Determine the multiparameter sensitivity for the passive *RLC* network used in Example 3.2-1 at $\omega = 1$ rad/s.

Answer

$$\left.\frac{dY(s)}{Y(s)}\right|_{s=j1} = (-0.2)\frac{dR}{R} + (0.4)\frac{dL}{L} + (1.2)\frac{dC}{C}$$

$$+ j\left[(-0.4)\frac{dR}{R} + (-0.2)\frac{dL}{L} + (-0.6)\frac{dC}{C}\right]$$

Sensitivity of Lossless Ladder Filters

The use of function sensitivity provides a convenient means of illustrating an important property of the commonly used double-resistance-terminated lossless ladder filters catalogued in App. A. For such filters, at any passband frequency of maximum gain, it may be shown that the source delivers maximum available power to the load. Thus at any such frequency a change, either plus or minus, in the value of any L or C component can only cause the gain to decrease. As a result, when only first-order effects are considered at these frequencies, the magnitude sensitivity, i.e., the real part of the function sensitivity, must be zero.[2]

Example 3.2-4 Sensitivity of a fourth-order band-pass filter. As an example, consider the fourth-order broadband band-pass filter shown in Fig. 3.2-3(*a*). We will let its voltage transfer function be specified as $N(s) = V_2(s)/V_1(s)$. As shown by its magnitude characteristic in Fig. 3.2-3(*b*), it has a center frequency of 1 rad/s, a 3-dB down bandwidth of 1 rad/s, and a Butterworth (maximally flat magnitude) characteristic. Thus its frequency of maximum gain is at the center frequency, 1 rad/s. Plots of $S^{|N(j\omega)|}$ for the various elements are given in Fig. 3.2-4. The zero-sensitivity property at the center frequency is readily seen in these plots. It should be noted, however, that as the passband of such a filter is narrowed, the magnitude sensitivities become larger. As an example of this, consider a filter similar

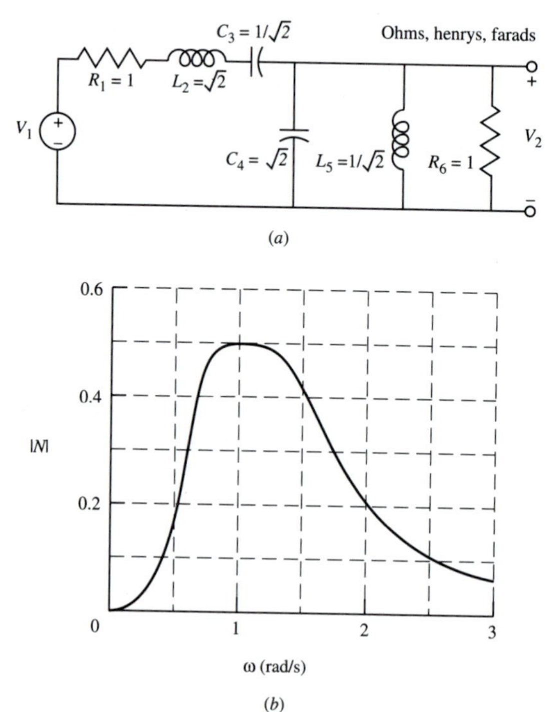

FIGURE 3.2-3
A fourth-order broadband band-pass filter and its magnitude characteristic.

---

[2] H. J. Orchard, "Inductorless Filters," *Electron. Lett.*, vol. 2, no. 6, June 1966, pp. 224–225.

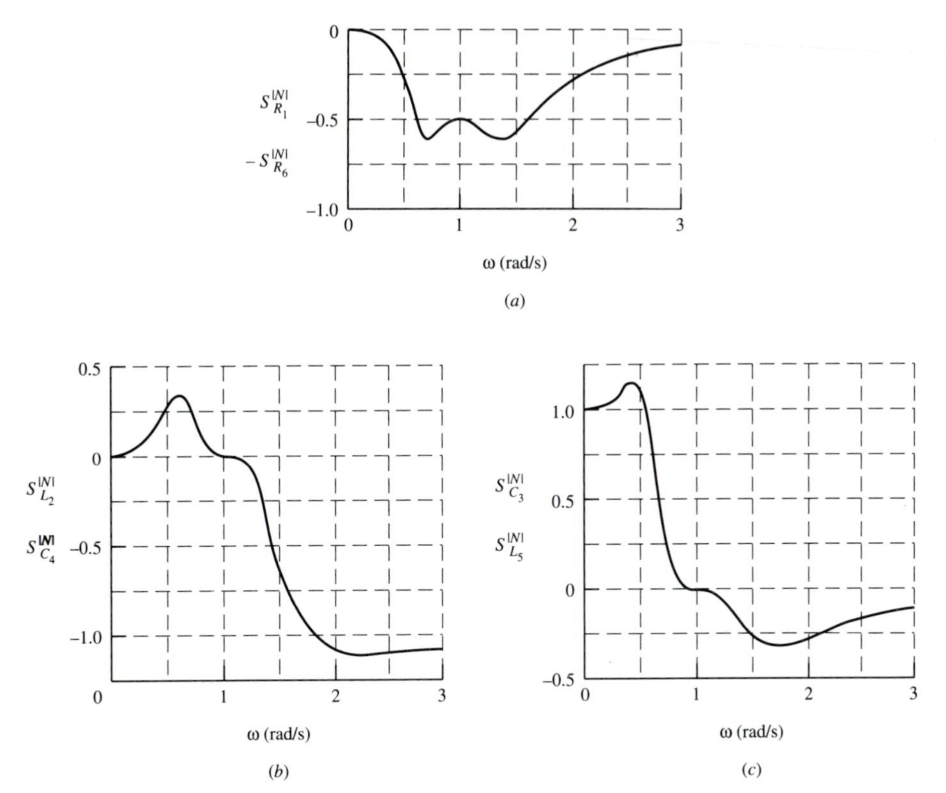

FIGURE 3.2-4
Plots of $S^{|N(j\omega)|}$ for the network of Fig. 3.2-3.

to the one described above but with element values $L_2 = C_4 = 10\sqrt{2}$ and $C_3 = L_5 = \frac{1}{10}\sqrt{2}$. The resulting magnitude characteristic is shown in Fig. 3.2-5. The center frequency is still 1 rad/s, but the filter is now narrow band with a bandwidth of 0.1 rad/s. Plots of $S^{|N(j\omega)|}$ for the various elements are shown in Fig. 3.2-6. In comparing these plots with the ones given in Fig. 3.2-4, note that different ordinate scales have been used. From the plots we note that despite the increased magnitude of the sensitivities at the edges of the passband, the sensitivity at the resonant frequency remains zero.

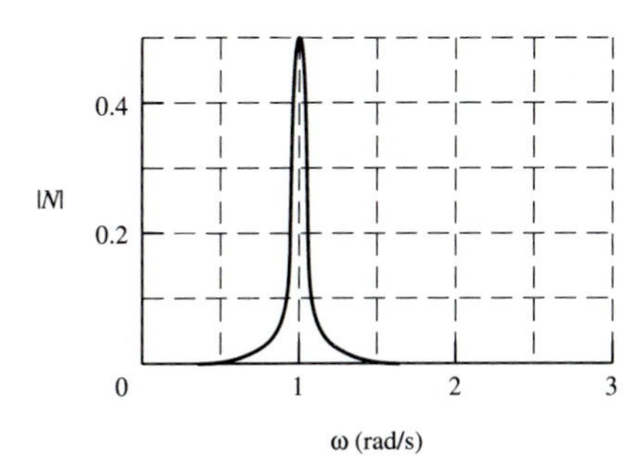

FIGURE 3.2-5
The magnitude characteristic of a narrow-band band-pass filter.

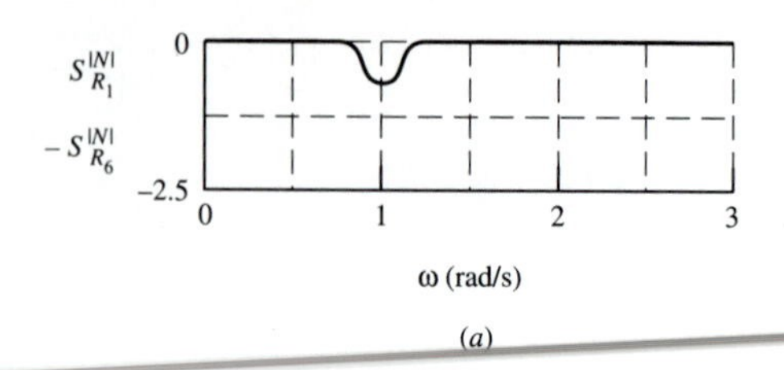

(a)

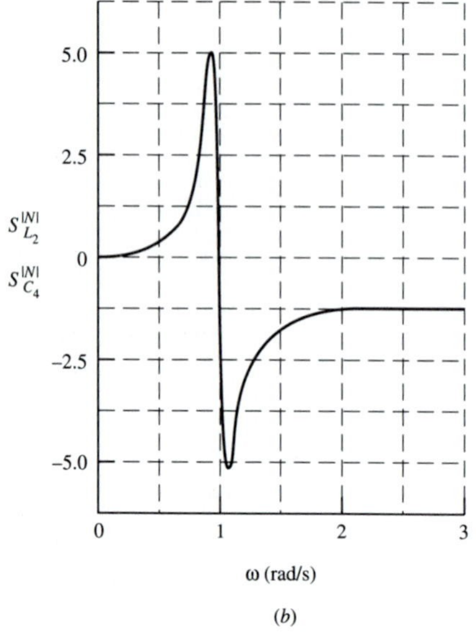

(b)

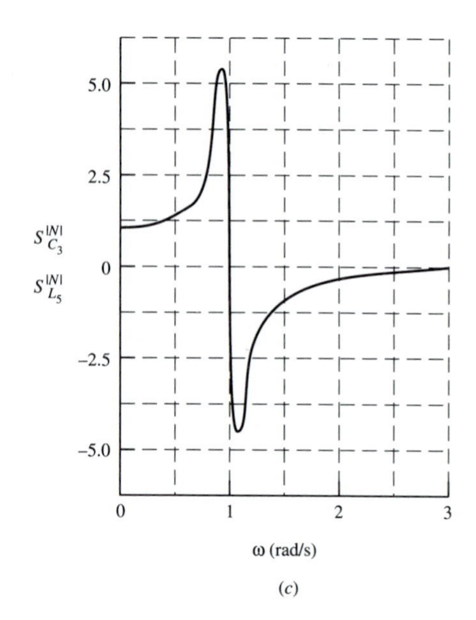

(c)

FIGURE 3.2-6
Plots of $S^{|N(j\omega)|}$ for the narrow-band band-pass filter.

3.3 COEFFICIENT SENSITIVITY

In this section we introduce another type of sensitivity that has the relative form defined in (5) of Sec. 3.1. In general, a network function $N(s)$ for any active or passive lumped network is a ratio of polynomials having the form

$$N(s) = \frac{A(s)}{B(s)} = \frac{a_0 + a_1 s + a_2 s^2 + \cdots + a_m s^m}{b_0 + b_1 s + b_2 s^2 + \cdots + b_n s^n} \tag{1}$$

in which the coefficients a_i and b_i are real and are functions of the network elements x_i. Thus, for any arbitrary network element x we may define sensitivities that are relative in form, called *coefficient sensitivities*, as follows:

$$S_x^{a_i} = \frac{\partial a_i}{\partial x}\frac{x}{a_i} \qquad S_x^{b_i} = \frac{\partial b_i}{\partial x}\frac{x}{b_i} \tag{2}$$

Bilinear Dependence

One of the reasons for the importance of coefficient sensitivity is the manner in which a network function $N(s)$ is dependent on any element x. The relationship is called a *bilinear dependence*. Thus, $N(s)$ of (1) may also be written in the form

$$N(s) = \frac{A(s)}{B(s)} = \frac{E(s) + xF(s)}{C(s) + xD(s)} \tag{3}$$

where $C(s)$, $D(s)$, $E(s)$, and $F(s)$ are polynomials with real coefficients that are not functions of the network element x. This is true whether x is chosen to be the value of a passive resistor or capacitor or the gain of some amplifier or controlled source.

Example 3.3-1 Bilinear dependence of a network function. As an example of bilinear dependence, consider the *RLC* network used in Example 3.2-1. Using the indicated nominal element values, we may write $Y(s)$ as a bilinear function of R as

$$Y(s) = \frac{s}{(s^2 + 3) + Rs} \tag{4}$$

For this case, from (3) we have $E(s) = s$, $F(s) = 0$, $C(s) = s^2 + 3$, and $D(s) = s$. Similarly, for the other network elements we may write

$$Y(s) = \frac{s/L}{s^2 + (1/L)(s + 3)} \qquad Y(s) = \frac{s}{(s^2 + s) + 1/C} \tag{5}$$

Note that in these cases we have used a reciprocal valued element. From property 3 of Table 3.1-1, the coefficient sensitivity for the nonreciprocal valued element is simply the negative of that found for the reciprocal valued one. The expressions given in (5) can also be written in the form

$$Y(s) = \frac{s}{(s + 3) + Ls^2} \qquad Y(s) = \frac{Cs}{1 + C(s^2 + s)} \tag{6}$$

For these relations, the values of the coefficient sensitivities will be different than those obtained from (5).

Form of Coefficient Sensitivities

Because of bilinear dependence, there are three cases that specify the way in which a coefficient a_i (or b_i) may depend on a positive-valued network element x.

CASE 1. $a_i = kx$ where k may be positive or negative:

$$\mathcal{S}_x^{a_i} = 1 \qquad \text{(magnitude of sensitivity} = 1)$$

This is a "good" sensitivity case in that the percentage change of the parameter is the same as that of the element. As an example of this case, consider the network shown in Fig. 3.2-1. The admittance is

$$Y(s) = \frac{s/L}{s^2 + sR/L + 1/LC} = \frac{s}{s^2 + s + 3} = \frac{a_1}{b_2 s^2 + b_1 s + b_0}$$

TABLE 3.3-1
Coefficient sensitivities for network of Fig. 3.2-1

	Coefficient			
Element	a_1	b_0	b_1	b_2
R	0	0	1	0
L	−1	−1	−1	0
C	0	−1	0	0

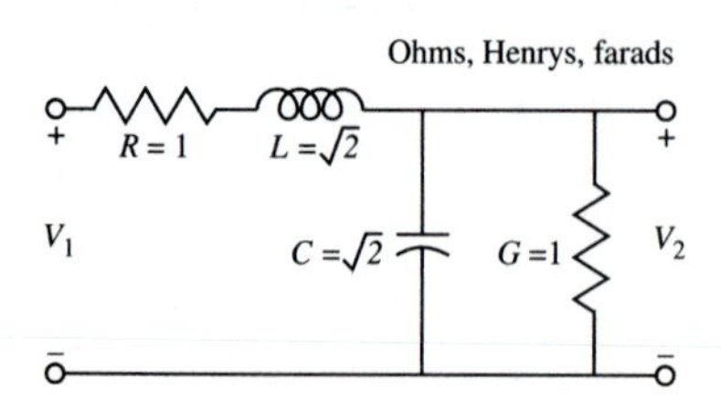

FIGURE 3.3-1
A network with coefficient sensitivities ≤ 1.

The coefficient sensitivities are listed in Table 3.3-1. Note that we have used property 3 of Table 3.1-1 for the cases where x appears as a reciprocal quantity.

CASE 2. $a_i = k_0 + k_1 x$ where k_0 and k_1 have the same polarity:

$$S_x^{a_i} = \frac{k_1 x}{k_0 + k_1 x} \qquad \text{(magnitude of sensitivity} < 1)$$

This is a "good" sensitivity case, in which the percentage change of the parameter is less than the percentage change of the element. As an example of this case, consider the network shown in Fig. 3.3-1. The voltage transfer function is

$$\frac{V_2(s)}{V_1(s)} = \frac{1/LC}{s^2 + s(R/L + G/C) + (RG + 1)/LC}$$
$$= \frac{0.5}{s^2 + \sqrt{2}s + 1} = \frac{a_0}{b_2 s^2 + b_1 s + b_0}$$

The coefficient sensitivities are given in Table 3.3-2.

TABLE 3.3-2
Coefficient sensitivities for network of Fig. 3.3-1

	Coefficient			
Element	a_0	b_0	b_1	b_2
R	0	0.5	0.5	0
L	−1	−1	−0.707	0
C	−1	−1	−0.707	0
G	0	0.5	0.5	0

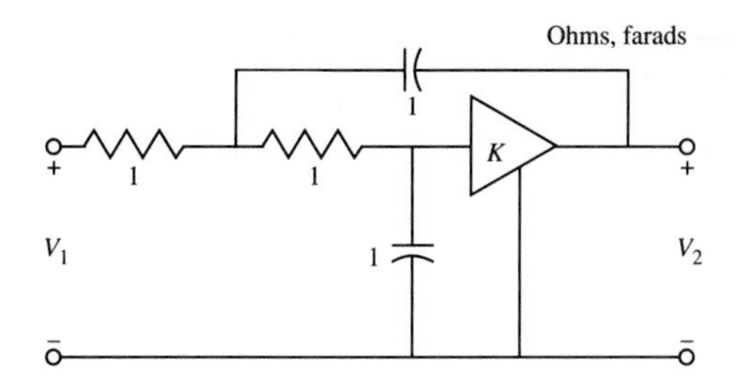

FIGURE 3.3-2
A network with sensitivity magnitude > 1.

CASE 3. $a_i = k_0 + k_1x$ where k_0 and k_1 have opposite polarity and $|a_i|$ is less than either $|k_0|$ or $|k_1x|$:

$$S_x^{a_i} = \frac{k_1x}{k_0 + k_1x} \qquad \text{(magnitude of sensitivity > 1)}$$

This is the "bad" sensitivity case; the large value results from the fact that $|k_0 + k_1x|$ can be much smaller than $|k_1x|$. As an example of this case, consider the network shown in Fig. 3.3-2. The voltage transfer function for $K = 2.5$ is

$$\frac{V_2(s)}{V_1(s)} = \frac{K}{s^2 + (3 - K)s + 1} = \frac{2.5}{s^2 + 0.5s + 1} = \frac{a_0}{b_2s^2 + b_1s + b_0}$$

The coefficient sensitivities are $S_K^{a_0} = 1$ and $S_K^{b_1} = -K/(3 - K) = -5$.

A summary of the three types of coefficient sensitivity is given in Table 3.3-3.

Relation between Function Sensitivity and Coefficient Sensitivities

The coefficient sensitivities defined in this section are readily related to the function sensitivity introduced in the preceding section by inserting (1) and (2) into (3) of Sec. 3.2. Thus we obtain

$$S_x^{N(s)} = \frac{\sum_{i=0}^{m} S_x^{a_i} a_i s^i}{A(s)} - \frac{\sum_{i=0}^{n} S_x^{b_i} b_i s^i}{B(s)} \tag{7}$$

Example 3.3-2 Deriving function sensitivity from coefficient sensitivities. The function sensitivity $S_L^{Y(s)}$ for the network shown in Fig. 3.2-1 can be derived from the coefficient

TABLE 3.3-3
Cases for coefficient sensitivities

Case	Form of a_i	Restrictions	$S_x^{a_i}$	$\lvert S_x^{a_i}\rvert$
1	kx		1	1
2	$k_0 + k_1x$	k_0 and k_1x of same sign	$k_1x/(k_0 + k_1x)$	<1
3	$k_0 + k_1x$	k_0 and k_1x of opposite sign	$k_1x/(k_0 + k_1x)$	>1

sensitivities for L given in Table 3.3-1. From (5) of Sec. 3.2, using the values $R = 1$ and $C = \frac{1}{3}$, we obtain

$$Y(s) = \frac{s(1/L)}{s^2 + s(1/L) + (3/L)} = \frac{a_1 s}{b_2 s^2 + b_1 s + b_0}$$

If we apply (7) above, we obtain

$$S_L^{Y(s)} = \frac{S_L^{a_1} a_1 s}{a_1 s} - \frac{S_L^{b_0} b_0 + S_L^{b_1} b_1 s}{b_2 s^2 + b_1 s + b_0}$$

For the value $L = 1$ this becomes

$$S_L^{Y(s)} = \frac{(-1)(1)s}{(1)s} - \frac{(-1)(3) + (-1)(1)s}{(1)s^2 + (1)s + (3)} = \frac{-s^2}{s^2 + s + 3}$$

This agrees with the expression given in (6) of Sec. 3.2.

Exercise 3.3-2. Repeat the derivation given in Example 3.3-2 for the network shown in Fig. 3.2-1 to find the sensitivities $S_R^{Y(s)}$ and $S_C^{Y(s)}$.

3.4 UNNORMALIZED ROOT SENSITIVITY

One of the most meaningful criteria that is used in determining the manner in which a network's properties vary as some element changes is a determination of how the zeros and poles of the network function, i.e., how the roots of the numerator and denominator polynomials, change as the value of the element changes. In this section we consider these effects by studying root sensitivity.

Unnormalized Sensitivity

The use of relative sensitivity as defined in (5) of Sec. 3.1 is not appropriate when considering roots because normalization of the characteristic y would usually involve division by a complex number. Instead we define an *unnormalized* (semirelative) sensitivity by the relation

$$US_x^r = \frac{\partial r}{\partial x/x} \tag{1}$$

where r is the root location and x is the network element. The usefulness of this definition is shown in the following example.

Example 3.4-1 Unnormalized versus normalized root sensitivity. As an example of the use of unnormalized root sensitivity, consider the polynomial

$$B(s) = s^2 + 2s + x$$

This has a root at $s_1 = -1 + j\sqrt{x - 1}$ (and a conjugate one at $s_2 = -1 - j\sqrt{x - 1}$). The unnormalized sensitivity is

$$US_x^{s_1} = \frac{\partial(-1 + j\sqrt{x-1})}{\partial x/x} = \frac{jx}{2\sqrt{x-1}}$$

If we assume a nominal value $x = 2$, for which $s_1 = -1 + j1$, the results of a 1 percent change in x are found as

$$\Delta s_1 = US_x^{s_1}\Bigg|_{x=2} \frac{\Delta x}{x} = (j1)(0.01) = j0.01$$

We conclude that there is a corresponding 0.01 change in the imaginary part of the root location. The new root (as predicted by linear sensitivity theory) is at $-1 + j1.01$. This is verified by solving $B(s) = s^2 + 2s + 2.02 = 0$ for which we find $s = -1 + j1.00995$. Note that the use of relative sensitivity would give

$$S_x^{s_1} = \frac{j/\sqrt{x-1}}{-1 + j\sqrt{x-1}}\Bigg|_{x=2} = \frac{1-j}{2}$$

As a result of the division by the root, such a relative sensitivity does not give meaningful information on the actual change of the root location.

Exercise 3.4-1. Find the change predicted by unnormalized sensitivity using (1) and the actual change in the upper half-plane root for each of the following cases:

			Answers	
Polynomial	**Nominal value of x**	**Percentage change in x**	**Change in root predicted by *US* of (1)**	**Actual change in root**
$s^2 + 2s + x$	2	10	$j0.1$	$j0.0955$
$s^2 + 2s + x$	5	1	$j0.0125$	$j0.01246$
$s^2 + xs + 2$	2	1	$-0.01 - j0.01$	$-0.01 - j0.0101$
$s^2 + xs + 2$	2	10	$-0.1 - j0.1$	$-0.1 - j0.1112$

Determination of Sensitivity

We may now define the *unnormalized* (semirelative) *root sensitivities*

$$US_x^{p_i} = \frac{\partial p_i}{\partial x/x} \qquad US_x^{z_i} = \frac{\partial z_i}{\partial x/x} \tag{2}$$

where p_i and z_i are the poles and zeros of the network function. Since both of these sensitivities refer to the roots of polynomials, we need to examine only the pole sensitivity in detail. The treatment of the zero sensitivity is identical.

Let $B(s)$ be the denominator polynomial of a network function $N(s)$. Because of the bilinear dependence discussed in Sec. 3.3, the dependence of this polynomial on any parameter x may be put in the form

$$B(s) = C(s) + xD(s) \tag{3}$$

where $C(s)$ and $D(s)$ are polynomials with real coefficients that are not functions of x. Evaluating (3) at any pole p_i of $N(s)$, we obtain

$$B(p_i) = C(p_i) + xD(p_i) = 0 \tag{4}$$

To determine the effect of a variation of x on p_i in this equation, we may replace x by $x + \Delta x$ and p_i by $p_i + \Delta p_i$. Using a series expansion terminated after the first-order term, we see that $C(s + \Delta s) = C(s) + \Delta s\, C'(s)$, where $C'(s) = dC(s)/ds$. Similarly, $D(s + \Delta s) = D(s) + \Delta s\, D'(s)$. Inserting these results in (4), we obtain

$$C(p_i) + \Delta p_i\, C'(p_i) + (x + \Delta x)[D(p_i) + \Delta p_i\, D'(p_i)] = 0 \tag{5}$$

Retaining only first-order incremental terms, this may be put in the form

$$\frac{\Delta p_i}{\Delta x} = \frac{-D(p_i)}{B'(p_i)} \tag{6}$$

In the limit as Δx approaches zero, we have

$$US_x^{p_i} = \frac{\partial p_i}{\partial x/x} = \frac{-xD(p_i)}{B'(p_i)} \tag{7}$$

as the defining relation for our unnormalized pole sensitivity. An example follows.

Example 3.4-2 Unnormalized pole sensitivities of an active *RC* filter. As an example of unnormalized pole sensitivity, consider the voltage transfer function for the active *RC* network shown in Fig. 3.3-2. This is

$$\frac{V_2(s)}{V_1(s)} = \frac{K}{s^2 + (3 - K)s + 1}$$

The poles are

$$p_1 = \tfrac{1}{2}(K - 3) + j\sqrt{1 - \tfrac{1}{4}(3 - K)^2} = p_2^*$$

From (2) or (7) we find

$$US_K^{p_1} = \frac{K}{2} + j\frac{K(3 - K)/4}{\sqrt{1 - (3 - K)^2/4}} = (US_K^{p_2})^*$$

where, since the poles are complex conjugates, the sensitivities are also complex conjugates. As a further illustration, if the nominal value of K is 2, then $p_1 = p_2^* = -0.5 + j0.866$ and $US_K^{p_1} = US_K^{p_2^*} = 1 + j0.577$. From these sensitivities we see that a 10 percent change in K (K going to a value of 2.2) produces $\Delta p_l = 0.1 + j0.0577$; that is, the pole moves to the right and upward, the change in the real part being almost twice as great as the change in the imaginary part. The resulting value of p_1 as found from sensitivity calculations is $-0.4 + j0.924$. The exact value of p_1 is readily found to be $-0.4 + j0.9165$. A smaller perturbation of K would, of course, produce a closer correlation between the two results.

Exercise 3.4-2. For the active *RC* filter used in Example 3.4-2, find the upper half-plane pole positions predicted by unnormalized sensitivity and the actual pole positions resulting from changes in K for each of the cases shown below:

		Answers	
Nominal value of *K*	**Percentage change of *K***	**New pole position predicted by use of unnormalized sensitivity**	**Actual new pole position**
2.0	1	$-0.4900 + j0.8718$	$-0.4900 + j0.8717$
2.5	1	$-0.2375 + j0.9715$	$-0.2375 + j0.9714$
2.5	10	$-0.1250 + j1.0005$	$-0.1250 + j0.9922$
2.8	1	$-0.0860 + j0.9964$	$-0.0860 + j0.9963$
2.8	5	$-0.0300 + j1.0020$	$-0.0300 + j0.9995$

The unnormalized root sensitivity introduced above effectively defines the root locus of a given pole position as a function of the value of the specific element. For second-order network functions, these loci will always be circles or straight lines. Some typical loci for various elements of the networks discussed in Examples 3.2-1 and 3.3-1 are given in Fig. 3.4-1 (only the upper half of the complex-frequency plane is shown). These loci may be determined by matching the real and imaginary parts of the pole locations to the equation for the locus of a circle. For example, for the pole locations

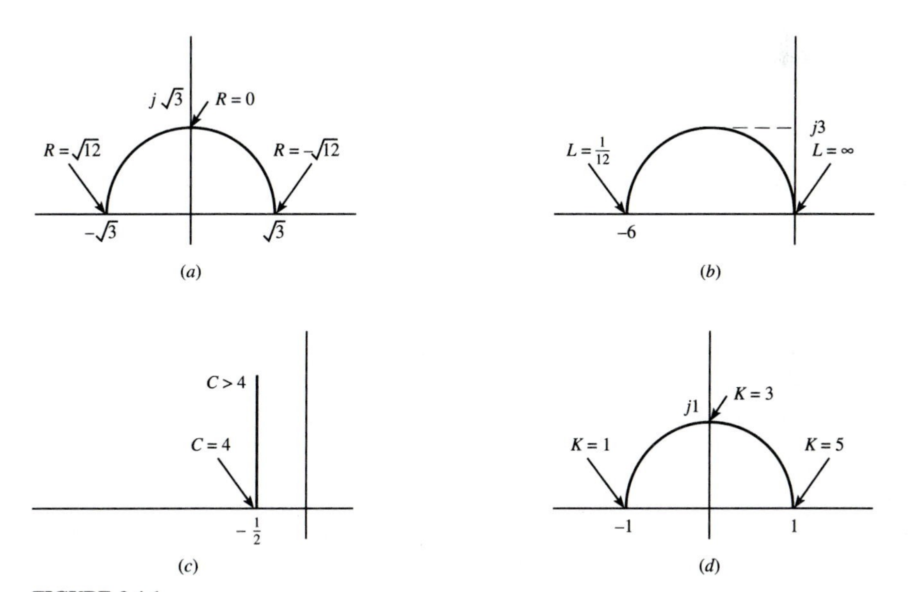

FIGURE 3.4-1
The loci of the poles of network functions produced by varying the network parameters: (*a*) R in Fig. 3.2-1; (*b*) L in Fig. 3.2-1; (*c*) C in Fig. 3.2-1; (*d*) K in Fig. 3.3-2.

given in Example 3.4-2, using $\sigma = \mathrm{Re}(p_1)$ and $\omega = \mathrm{Im}(p_1)$, we obtain

$$\sigma^2 + \omega^2 = \left(\frac{K-3}{2}\right)^2 + \left[1 - \left(\frac{K-3}{2}\right)^2\right] = 1$$

As shown in Fig. 3.4-1(*d*), this defines a circle centered at the origin of the s plane, with a radius of unity.

Relation between Unnormalized Root Sensitivity and Coefficient Sensitivities

The unnormalized root sensitivity is readily related to the coefficient sensitivity introduced in Sec. 3.3. To see this, in (3) we define

$$C(s) = c_0 + c_1 s + c_2 s^2 + \cdots \qquad D(s) = d_0 + d_1 s + d_2 s^2 + \cdots \tag{8}$$

Thus the polynomial $B(s)$ of (3) may be written

$$\begin{aligned} B(s) &= b_0 + b_1 s + b_2 s^2 + \cdots \\ &= (c_0 + x\,d_0) + (c_1 + x\,d_1)s + (c_2 + x\,d_2)s^2 + \cdots \end{aligned} \tag{9}$$

An alternate form for $D(s)$ is thus readily seen to be

$$\begin{aligned} D(s) &= \frac{\partial b_0}{\partial x} + \frac{\partial b_1}{\partial x}s + \frac{\partial b_2}{\partial x}s^2 + \cdots \\ &= \frac{b_0}{x}S_x^{b_0} + \frac{b_1}{x}S_x^{b_1}s + \frac{b_2}{x}S_x^{b_2}s^2 + \cdots \end{aligned} \tag{10}$$

The polynomial $B'(s)$ is also readily found to be

$$B'(s) = b_1 + 2b_2 s + 3b_3 s^2 + \cdots \tag{11}$$

Inserting (10) and (11) in (9), we obtain

$$US_x^{p_i} = \left.\frac{-\sum_{i=0}^{n} b_i s^i S_x^{b_i}}{\sum_{i=0}^{n-1} (i+1) b_{i+1} s^i}\right|_{s=p_i} \tag{12}$$

Example 3.4-3 Unnormalized pole sensitivity for a passive network. For the passive network shown in Fig. 3.2-1 and the indicated element values, we may use (12) to find the unnormalized pole sensitivity to R from the coefficient sensitivities given in Table 3.3-1. We define

$$B(s) = b_0 + b_1 s + b_2 s^2 = 3 + Rs + s^2 = 3 + s + s^2$$

for which $p_1 = -0.5 + j1.6583$. From (12) we obtain

$$US_R^{p_1} = \left.\frac{-b_1 s S_R^{b_1}}{b_1 + 2b_2 s}\right|_{s=p_1} = \left.\frac{-(1)(s)(1)}{1+2s}\right|_{s=p_1} = \frac{0.5 - j1.6583}{j3.3166} = -0.5 - j0.1508$$

The relative changes in the real and imaginary parts of p_1 are in good agreement with the locus shown in Fig. 3.4-1(*a*).

Sensitivity for Nonsimple Roots

The sensitivity definition given in (7) applies only to simple roots of $B(s)$, since if p_i is a multiple-order root, $B'(p_i) = 0$, and thus the sensitivity is infinite. This does *not*, of course, mean that an infinite change of the pole position results from a change in the element x. What it does mean can be seen by considering a network function $N(s)$ with a denominator polynomial of the form of (3) and replacing x by $x+\Delta x$. The new roots are then found by solving the equation

$$C(s) + (x + \Delta x)D(s) = 0 \tag{13}$$

To do this, we first define a function

$$G(s) = \frac{xD(s)}{B(s)} \tag{14}$$

In terms of this function, (13) may be written in the form

$$1 + \frac{\Delta x}{x} G(s) = 0 \tag{15}$$

The poles of $G(s)$ are obviously the same as those of $N(s)$. Thus, assuming simple poles, the partial-fraction expansion of $G(s)$ will have the form

$$G(s) = \sum_{i=1}^{n} \frac{K_i^{(p)}}{s - p_i} + K_0^{(p)} \tag{16}$$

where the superscript (p) notation is used to indicate that the residues refer to poles and where the term $K_0^{(p)}$ only occurs if the order of $D(s)$ is the same as the order of $B(s)$. For values of s in the vicinity of the pole at p_i, the ith term in the expansion dominates. Thus from (16)

$$\lim_{s \to p_i} \left[1 + \frac{\Delta x}{x} G(s) \right] = 1 + \frac{\Delta x}{x} \frac{K_i^{(p)}}{s - p_i} = 1 + \frac{\Delta x}{x} \frac{K_i^{(p)}}{\Delta p_i} = 0 \tag{17}$$

where we have defined $s - p_i$ as Δp_i. From the second last member of the above, as Δx approaches zero, we find

$$US_x^{p_i} = \frac{\partial p_i}{\partial x/x} = -K_i^{(p)} \tag{18}$$

That is, the residues of $G(s)$ are the negative of the pole sensitivities. Now consider the case where a kth-order pole is located at p_1. The second last member of (17) now becomes

$$1 + \frac{\Delta x}{x} \left[\frac{K_{11}^{(p)}}{\Delta p_1} + \frac{K_{12}^{(p)}}{(\Delta p_1)^2} + \cdots + \frac{K_{1k}^{(p)}}{(\Delta p_1)^k} \right] = 0 \tag{19}$$

This may be written in the form

$$(\Delta p_1)^k + \frac{\Delta x}{x} [K_{11}^{(p)}(\Delta p_1)^{k-1} + K_{12}^{(p)}(\Delta p_1)^{k-2} + \cdots + K_{1k}^{(p)}] = 0 \tag{20}$$

The solution to this kth-order equation produces k values of Δp_1; that is, the kth-order root splits into k simple roots. For small values of Δp_1, we find that

$$\Delta p_1 = \left(\frac{-\Delta x}{x} K_{1k}^{(p)} \right)^{1/k} \tag{21}$$

Thus, the new simple roots are (for small changes of x) equiangularly spaced in a circle around p_1. An example follows.

Example 3.4-4 An active *RC* filter with a second-order pole. As an example of a multiorder root, consider the network function given in Example 3.4-2. For a nominal value of $K = 1$, using the notation of (3), $B(s) = s^2 + 2s + 1$, $C(s) = s^2 + 3s + 1$, and $D(s) = -s$. Thus there is a second-order pole $p_1 = -1$. From (16)

$$G(s) = \frac{-s}{(s+1)^2} = \frac{-1}{s+1} + \frac{1}{(s+1)^2}$$

Thus $K_{12}^{(p)} = 1$ and $\Delta p_1 = \sqrt{-\Delta K}$. For $\Delta K = 0.01$ we obtain roots at $-1 \pm j0.1$, while for $\Delta K = -0.01$, the roots move to -0.9 and -1.1. These approximate pole locations are illustrated in Fig. 3.4-2.

A development similar to that given above holds for the zeros of a network function $N(s) = A(s)/B(s)$ with a numerator polynomial $A(s) = E(s) + xF(s)$. If we define (for the simple zero case)

$$H(s) = \frac{xF(s)}{A(s)} = \sum_{i=1}^{m} \frac{K_i^{(z)}}{s - z_i} + K_0^{(z)} \tag{22}$$

where the superscript (z) notation is used to indicate that the residues refer to zeros, then the zero sensitivities are obtained as

$$US_x^{z_i} = \frac{\partial z_i}{\partial x/x} = -K_i^{(z)} \tag{23}$$

Similarly, for a kth-order zero at z_1 we obtain (for small Δx)

$$\Delta z_1 = \left(-\frac{\Delta x}{x} K_{1k}^{(z)} \right)^{1/k} \tag{24}$$

where $K_{1k}^{(z)}$ is the coefficient of the highest order term in the expansion for z_1 in $H(s)$.

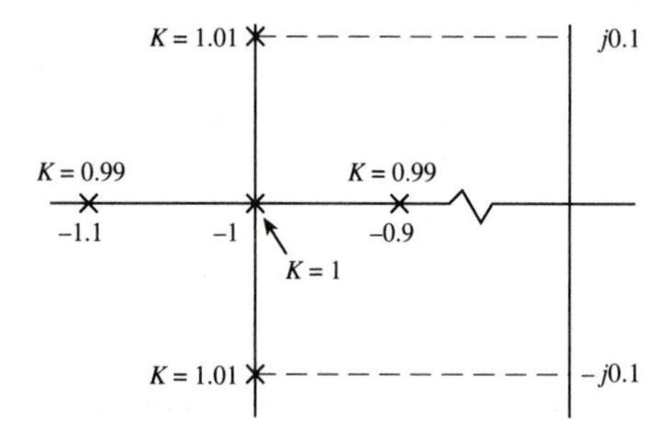

FIGURE 3.4-2
Changes in the second-order pole of Example 3.4-4.

Relation between Function Sensitivity and Unnormalized Root Sensitivities

The functions $G(s)$ and $H(s)$ defined in (14) and (22) may be used to relate the unnormalized root sensitivities discussed in this section to the function sensitivity defined in Sec. 3.2. Using the general bilinear form for a network function

$$N(s) = \frac{A(s)}{B(s)} = \frac{E(s) + xF(s)}{C(s) + xD(s)} \tag{25}$$

and using (3) of Sec. 3.2, we obtain

$$S_x^{N(s)} = \frac{xF(s)}{A(s)} - \frac{xD(s)}{B(s)} = H(s) - G(s) \tag{26}$$

Substituting from the above, we obtain (for simple roots)

$$S_x^{N(s)} = \sum_{i=1}^{n} \frac{US_x^{p_i}}{s - p_i} - \sum_{i=1}^{m} \frac{US_x^{z_i}}{s - z_i} - K_0^{(p)} + K_0^{(z)} \tag{27}$$

Thus we see that the classical sensitivity may be expressed as a weighted sum of the unnormalized root sensitivities.

3.5 Q AND ω_n SENSITIVITIES

Second-Order Case

In considering the performance of a network under conditions of sinusoidal steady state, the function sensitivity defined in (4) of Sec. 3.2 is frequently too cumbersome to be of much use, since it specifies the performance of the network over the entire frequency range. This is especially true in the second-order bandpass case, in which it is more advisable to use criteria that emphasize the resonant, i.e., frequency-selective, nature of such functions. Such criteria are primarily the resonant frequency ω_n, at which the peak of the magnitude response occurs, and the relative sharpness or quality factor Q of that peak. This latter quantity is defined as

$$Q = \frac{\omega_n}{\text{BW}} \tag{1}$$

where BW is the bandwidth defined as the difference between the frequencies at which the network function is 3 dB down from its peak magnitude at ω_n. These quantities are illustrated in Fig. 3.5-1, and they may be used to define a general second-order bandpass function

$$N(s) = \frac{A(s)}{B(s)} = \frac{Hs}{s^2 + b_1 s + b_0} = \frac{Hs}{s^2 + s(\omega_n/Q) + \omega_n^2} \tag{2}$$

where HQ/ω_n is the gain at resonance ($s = j\omega_n$). The quantities in the above expression

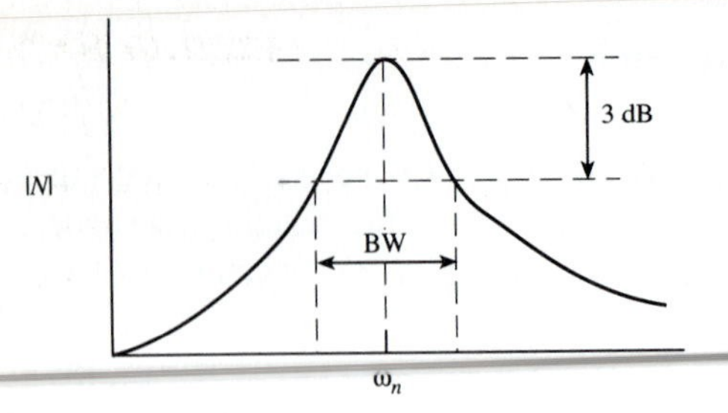

FIGURE 3.5-1
A second-order band-pass function.

are related to the pole positions p_0 and p_0^* and to each other by the relations

$$p_0 = \sigma_0 + j\omega_0 = \frac{-b_1}{2} + j\sqrt{b_0 - \left(\frac{b_1}{2}\right)^2} = \frac{-\omega_n}{2Q} + j\frac{\omega_n}{2Q}\sqrt{4Q^2 - 1}$$
$$Q = \frac{|p_0|}{2|\sigma_0|} = \frac{\omega_n}{2|\sigma_0|} = \frac{\sqrt{b_0}}{b_1} \qquad \omega_n = \sqrt{b_0} \tag{3}$$

In terms of these quantities we may define Q and ω_n sensitivities as

$$S_x^Q = \frac{\partial Q/Q}{\partial x/x} \qquad S_x^{\omega_n} = \frac{\partial \omega_n/\omega_n}{\partial x/x} \tag{4}$$

Obviously these are relative in form. For the second-order case the above sensitivities are readily evaluated by relating them to the coefficient sensitivities using the expressions of (3) and those given in Table 3.1-1. Thus we obtain

$$S_x^Q = \tfrac{1}{2}S_x^{b_0} - S_x^{b_1} \qquad S_x^{\omega_n} = \tfrac{1}{2}S_x^{b_0} \tag{5}$$

Also, for the second-order bandpass case, the real part of the function sensitivity discussed in Sec. 3.2, which gives the sensitivity of the magnitude of the network function, can be related to the Q and ω_n sensitivities at resonance ($\omega = \omega_n$), for which it can be shown

$$S_x^{|N(j\omega_n)|} = \mathrm{Re}\, S_x^{N(j\omega_n)} = S_x^H + S_x^Q - S_x^{\omega_n} \tag{6}$$

Example 3.5-1 Q and ω_n sensitivities of example networks. As an example of the determination of Q and ω_n sensitivities, using Table 3.3-1 for the coefficient sensitivities of the network of Fig. 3.2-1, and the relations of (5), we find

$$Q = \frac{1}{R}\sqrt{\frac{L}{C}} \qquad S_R^Q = -1 \qquad S_L^Q = \tfrac{1}{2} \qquad S_C^Q = -\tfrac{1}{2}$$
$$\omega_n = \frac{1}{\sqrt{LC}} \qquad S_R^{\omega_n} = 0 \qquad S_L^{\omega_n} = -\tfrac{1}{2} \qquad S_C^{\omega_n} = -\tfrac{1}{2}$$

Similarly, from Example 3.4-2 for the network shown in Fig. 3.3-2

$$Q = \frac{1}{3-K} \qquad \omega_n = 1 \qquad S_K^Q = \frac{K}{3-K} \qquad S_K^{\omega_n} = 0$$

Note that in this latter example, as K increases ($K < 3$), Q and the Q sensitivity also increase, and that as $K \to 3$, they both go to infinity.

Third-Order Case

The techniques outlined above are readily extended to the third-order case. In this situation the denominator polynomial will have the form

$$D(s) = s^3 + d_2 s^2 + d_1 s + d_0 \tag{7}$$

This may be factored into a first-order term and a quadratic term. Thus, using the notation of (2), we may write

$$D(s) = (s+g)\left(s^2 + \frac{\omega_n}{Q}s + \omega_n^2\right) \tag{8}$$

For the third-order case we may define three sensitivity functions, namely, S_x^g, S_x^Q, and $S_x^{\omega_n}$. These are most conveniently calculated from the coefficient sensitivities $S_x^{d_i}$. The relations are found by first putting (8) in the form

$$D(s) = s^3 + s^2\left(g + \frac{\omega_n}{Q}\right) + s\left(\omega_n^2 + \frac{g\omega_n}{Q}\right) + g\omega_n^2 \tag{9}$$

Equating corresponding coefficients of (7) and (9), we obtain the relations

$$d_0 = g\omega_n^2 \qquad d_1 = \omega_n^2 + \frac{g\omega_n}{Q} \qquad d_2 = g + \frac{\omega_n}{Q} \tag{10}$$

Taking the partial derivatives of these expressions, we obtain

$$\begin{bmatrix} 2 & 0 & 1 \\ \left[2\omega_n^2 + \dfrac{g\omega_n}{Q}\right]\dfrac{1}{d_1} & -\dfrac{g\omega_n}{Qd_1} & \dfrac{g\omega_n}{Qd_1} \\ \dfrac{\omega_n}{Qd_2} & -\dfrac{\omega_n}{Qd_2} & \dfrac{g}{d_2} \end{bmatrix} \begin{bmatrix} S_x^{\omega_n} \\ S_x^Q \\ S_x^g \end{bmatrix} = \begin{bmatrix} S_x^{d_0} \\ S_x^{d_1} \\ S_x^{d_2} \end{bmatrix} \tag{11}$$

Solving the set of simultaneous equations, we find that[1]

$$\begin{bmatrix} S_x^{\omega_n} \\ S_x^Q \\ S_x^g \end{bmatrix} = \frac{1}{\Delta} \begin{bmatrix} (Qg - \omega_n)d_0 & Q\omega_n^2 d_1 & -Q\omega_n^2 g d_2 \\ (2\omega_n Q^2 + gQ - \omega_n)d_0 & (\omega_n^2 Q - 2\omega_n g Q^2)d_1 & (Qg - 2\omega_n Q^2)\omega_n^2 d_2 \\ 2\omega_n^2 Q d_0/g & -2\omega_n^2 Q d_1 & 2\omega_n^2 Q g d_2 \end{bmatrix} \begin{bmatrix} S_x^{d_0} \\ S_x^{d_1} \\ S_x^{d_2} \end{bmatrix} \tag{12}$$

where

$$\Delta = 2\omega_n^2[Q(\omega_n^2 + g^2) - \omega_n g] \tag{13}$$

As an example of the use of this relation, consider the normalized third-order

---

[1] M. A. Soderstrand and S. K. Mitra, "Sensitivity Analysis of Third-Order Filters," *Intern. J. Electron.*, vol. 30, no. 3, 1971, pp. 265–272.

Butterworth function. In this case, from Table 2.1-3, we have

$$D(s) = s^3 + 2s^2 + 2s + 1 = (s+1)(s^2+s+1) \tag{14}$$

Thus, $d_0 = 1$, $d_1 = 2$, $d_2 = 2$, $g = 1$, $\omega_n = 1$, and $Q = 1$. Substituting these values in (12), we obtain

$$\begin{bmatrix} S_x^{\omega_n} \\ S_x^Q \\ S_x^g \end{bmatrix} = \begin{bmatrix} 0 & 1 & -1 \\ 1 & -1 & -1 \\ 1 & -2 & 2 \end{bmatrix} \begin{bmatrix} S_x^{d_0} \\ S_x^{d_1} \\ S_x^{d_2} \end{bmatrix} \tag{15}$$

For example, the Q sensitivity in this case is given as

$$S_x^Q = S_x^{d_0} - S_x^{d_1} - S_x^{d_2} \tag{16}$$

Example 3.5-2 Sensitivities of a third-order active *RC* filter. A third-order low-pass active *RC* network is shown in Fig. 3.5-2. The voltage transfer function for this circuit may be shown to be

$$\frac{V_2(s)}{V_1(s)} = \frac{Kd_0}{s^3 + d_2 s^2 + d_1 s + d_0}$$

where
$$d_0 = G_1 G_2 G_3 S_1 S_2 S_3$$
$$d_1 = G_2 G_3 S_2 S_3 + G_2 G_3 S_1 S_2 + (G_2 G_3 S_1 S_3 + G_1 G_3 S_1 S_3)(1-K) + G_1 G_3 S_1 S_2 + G_1 G_2 S_1 S_2$$
$$d_2 = G_1 S_1 + G_3 S_2 + G_2 S_2 + G_2 S_1 + G_3 S_3 (1-K)$$

and where, for convenience in taking the partial derivatives, we have used conductance $G_i = 1/R_i$ and susceptance $S_i = 1/C_i$. A solution for the Butterworth case of (14) is given by $K = 2$, $R_1 = 1.565$, $R_2 = 1.469$, $R_3 = 0.435$, and unity for all capacitors. The coefficient sensitivities to the gain K are readily found to be

$$S_K^{d_0} = 0 \qquad S_K^{d_1} = -\frac{K(G_2 G_3 S_1 S_3 + G_1 G_3 S_1 S_3)}{d_1} = -3.0338$$

$$S_K^{d_2} = -\frac{K G_3 S_3}{d_2} = -2.2989$$

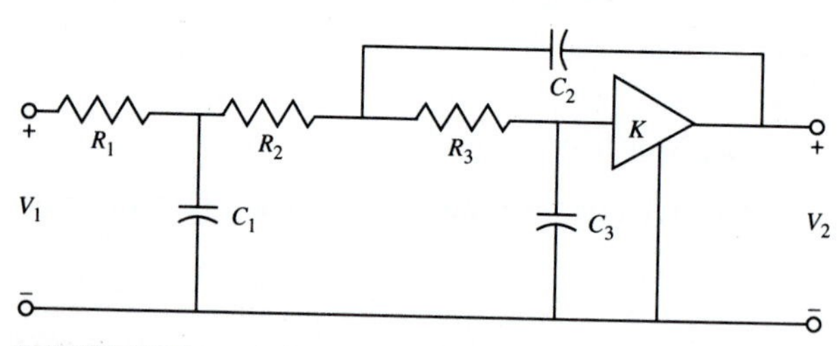

FIGURE 3.5-2
The third-order low-pass network used in Example 3.5-2.

From (16) we see that

$$S_K^Q = 0 + 3.0338 + 2.2988 = 5.3327$$

Cancellation of Denominator Factors

One other application of the third-order sensitivity expressions derived above occurs for the case where a network realizes a second-order network function but uses three independent reactive elements to do this. The resulting network function usually has a denominator polynomial of third order, but it also has a first-order factor in its numerator. In such a case a typical low-pass network function has the form

$$N(s) = \frac{A(s)}{D(s)} = \frac{H(s+g)}{s^3 + d_2 s^2 + d_1 s + d_0} \tag{17}$$

where the coefficients H, g, and d_i are functions of the element values. The design procedure for such a network will normally specify element values such that the numerator factor $s + g$ will also appear in the denominator; thus cancellation occurs, and the network function is effectively of second order. In studying the sensitivity of such a function, it is important to note that, in general, the numerator and denominator factors are *different* functions of the network elements. As a result, cancellation of the factors occurs only for nominal element values, not for perturbed ones. Because of this, the Q and ω_n sensitivities must be determined by using the third-order denominator expression. An example follows.

Example 3.5-3 A second-order active *RC* filter with three reactive elements. A bandpass circuit with three independent capacitors is shown in Fig. 3.5-3. If we choose $K = -(4Q - 1)$, the resulting voltage transfer function is

$$\frac{V_2(s)}{V_1(s)} = \frac{-(4Q-1)s(s+1)/2Q}{(s+1)[s^2 + s(1/Q) + 1]} = \frac{-(4Q-1)s/2Q}{s^2 + s(1/Q) + 1}$$

Comparing this expression with that given in (8), we see that $\omega_n = g = 1$, and thus we find that $d_0 = 1$ and $d_1 = d_2 = 1 + 1/Q$. The coefficient sensitivities with respect to the gain K are found to be $S_K^{d_0} = 0$, $S_K^{d_1} = S_K^{d_2} = -(4Q+1)/4Q(Q+1)$. Substituting these values in (12), we obtain $S_K^Q = 1 - 1/4Q$. Thus, the sensitivity is very low, although the gain of $-(4Q - 1)$ required is high.

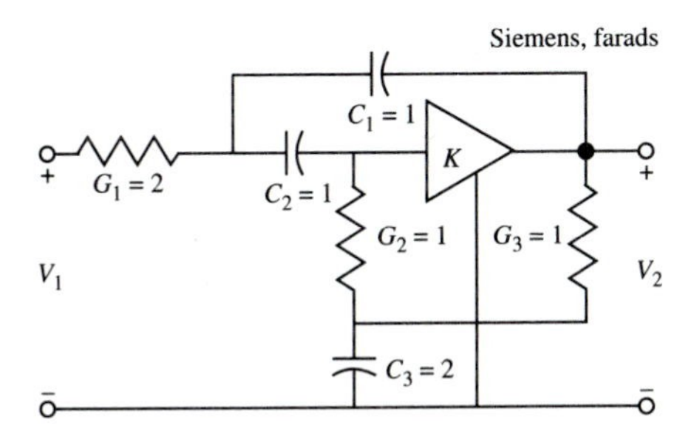

FIGURE 3.5-3
The second-order band-pass network used in Example 3.5-3.

3.6 PARASITIC SENSITIVITY

Parasitic Elements

One of the most interesting and practical applications of sensitivity is in the determination of the effects on various network characteristics of a particular element type called a parasitic. We define a *parasitic element* as one whose nominal (ideal) value is zero. Examples of parasitic elements are:

The resistance used in series with an ideal inductance to model core losses, winding resistance, and other dissipative quantities

The conductance used in parallel with an ideal capacitor to model leakage effects

The resistance used in series with an ideal voltage source to model an amplifier's internal resistance

The conductance used in parallel with an ideal current source to model the source's internal conductance

The reciprocal of the gain of an ideal operational amplifier

Sensitivity for Parasitic Elements

The relative sensitivity $S_x^y = (\partial y/\partial x)(x/y)$ defined in (5) of Sec. 3.1 cannot be used for parasitic elements. The reason for this is that the nominal zero value of the element results in the sensitivity also being zero. Instead, we may define a *parasitic sensitivity*

$$PS_v^y = \frac{\partial y}{\partial v}\frac{1}{y} \tag{1}$$

where v is a parasitic element and y is some characteristic. Note that parasitic sensitivity is a semirelative sensitivity in that the characteristic y is normalized but the element v is not. Another example of a semirelative sensitivity, the unnormalized root sensitivity, was discussed in Sec. 3.4. The application of parasitic sensitivity will typically be to find the normalized change in y as a result of a change in v. Thus, (1) will be used in the form

$$\frac{\Delta y}{y} = PS_v^y\,\Delta v \tag{2}$$

The steps used in the determination of parasitic sensitivity are described in the following summary.

Summary 3.6-1 Determination of parasitic sensitivity. To find a parasitic sensitivity of the form given in (1), the following procedure may be followed:

1. Determine an expression for the characteristic y that includes the parasitic element v as a literal quantity.

2. Take the derivative $\partial y/\partial v$ and form the function $(\partial y/\partial v)(1/y)$.
3. In the expression $(\partial y/\partial v)(1/y)$ set $v = 0$.

The result is the parasitic sensitivity PS_v^y. Just as is the case for relative sensitivity, parasitic sensitivity is a first-order sensitivity, valid only for small variations of v.

Example 3.6-1 Parasitic sensitivity of a simple network. As an example of the use of parasitic sensitivity, consider the low-pass RC network shown in Fig. 3.6-1 in which the capacitor C has a parasitic conductance G in shunt with it. If we ignore G, the voltage transfer function for the network is

$$\frac{V_2(s)}{V_1(s)} = \frac{1/RC}{s + 1/RC}$$

which has a pole at $-1/RC$. It is desired to find the change in the pole location caused by the parasitic conductance going from 0 to 0.01. The pole is real so we need not use an unnormalized form for the characteristic. Following the steps given above, we first find the network function with G included,

$$\frac{V_2(s)}{V_1(s)} = \frac{1/RC}{s + (RG + 1)/RC}$$

which has a pole p at $-(RG + 1)/RC$. Next we take the partial derivative of the pole and multiply the result by $1/p$:

$$\frac{\partial p}{\partial G}\frac{1}{p} = \frac{-R}{RC}\left(\frac{-RC}{RG + 1}\right) = \frac{R}{RG + 1}$$

If we set $G = 0$ in this last expression, we obtain

$$PS_G^p = \frac{\partial p}{\partial G}\frac{1}{p}\bigg|_{G=0} = R$$

From (2) we now find

$$\frac{\Delta p}{p} = PS_G^p\,\Delta G = R\,\Delta G = R(0.01)$$

As a numerical example, if $R = 2$ and $C = 1$, $p = -\frac{1}{2}$ and $\Delta p/p = 0.02$ or $\Delta p = -0.01$. We see that parasitic sensitivity predicts that changing G from 0 to 0.01 (changing the shunt resistance from infinity to 100 Ω) will move the pole from -0.5 to -0.51. This result is easily verified by direct computation.

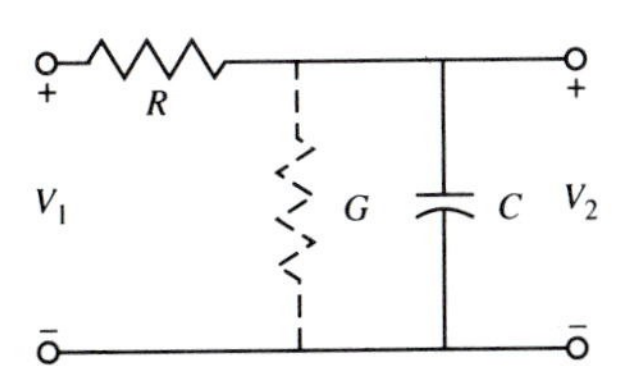

FIGURE 3.6-1
Network with a parasitic element.

Operational Amplifier Parasitic Sensitivity

One of the most important uses of parasitic sensitivity is to determine the effect of the finite dc gain of an operational amplifier, a quantity assumed to be infinite in most analyses. This can be done by treating the reciprocal of the gain as a parasitic element with a nominal value of zero. To see how this is done, consider the operational amplifier model shown in Fig. 3.6-2. For the nonideal case we may write

$$V_0 = A(V_+ - V_-) \tag{3}$$

where A is the dc gain, a large number ideally infinite. To treat this as a parasitic sensitivity, we define $B = 1/A$ as a parasitic quantity and write (3) in the form

$$V_0 = \frac{1}{B}(V_+ - V_-) \tag{4}$$

where B is a very small number, ideally zero. For a given circuit, the parasitic sensitivity can be determined using the steps outlined in Summary 3.6-1. An example follows.

Example 3.6-2 Parasitic sensitivity in a damped integrator. A circuit using an operational amplifier is shown in Fig. 3.6-3. It consists of the conductances G_1 and G_2 and the capacitor C. If the operational amplifier is ideal, the voltage transfer function is

$$\frac{V_2(s)}{V_1(s)} = \frac{-G_1/C}{s + G_2/C}$$

which has a pole at $-G_2/C$. The circuit is called a *damped integrator*. Note that for $G_2 = 0$ (an open circuit), this becomes the voltage transfer function of an ideal integrator. Now let us consider the effect of using a nonideal operational amplifier as defined by (4) with $V_+ = 0$. The new network function is

$$\frac{V_2(s)}{V_1(s)} = \frac{-G_1}{B(G_1 + G_2 + sC) + (sC + G_2)}$$

which has a pole at

$$p = -\frac{B(G_1 + G_2) + G_2}{BC + C}$$

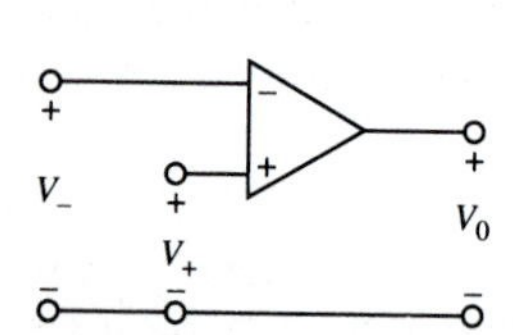

FIGURE 3.6-2
Model of an operational amplifier.

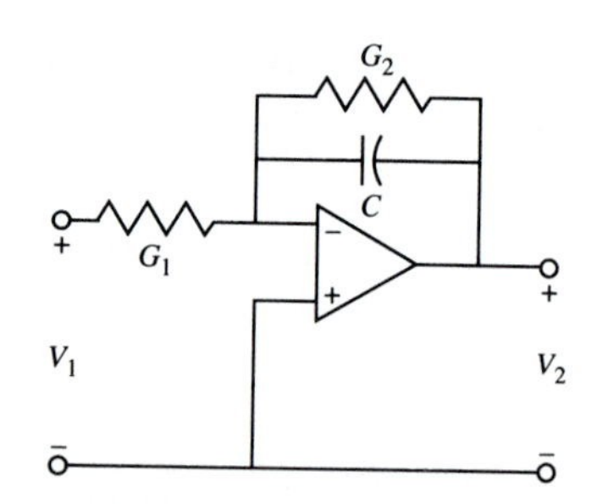

FIGURE 3.6-3
Damped integrator.

If we take the derivative of p with respect to B, we obtain

$$\frac{\partial p}{\partial B} = -\frac{(G_1 + G_2)(BC + C) - C[B(G_1 + G_2) + G_2]}{(BC + C)^2}$$

Multiplying this by $1/p$ and setting $B = 0$, we find

$$PS_B^p = \frac{\partial p}{\partial B}\frac{1}{p}\bigg|_{B=0} = \frac{G_1}{G_2}$$

As an example of the use of this result, consider an application in which $G_1 = 10$, $G_2 = 0.1$ ($R_1 = 0.1\ \Omega$ and $R_2 = 10\ \Omega$), $C = 1$ F, and the operational amplifier has a gain $A = 10^4$ ($B = 10^{-4}$). Ideally, for $B = 0$, the pole is located at -0.1. Applying the parasitic sensitivity, we find

$$\frac{\Delta p}{p} = PS_B^p\ \Delta B = \frac{G_1}{G_2}\Delta B = \frac{10}{0.1}10^{-4} = 0.01$$

The change in pole location is easily found by multiplying this result by p. We obtain $\Delta p = -0.001$. Thus the pole location becomes -0.101 as the result of the nonideality of the operational amplifier. The same procedure may be applied to determine the effects of the operational amplifier nonideality on any other characteristic, for example, the voltage transfer function.

Multiparameter Sensitivity

We may now form a general multiparameter expression that gives the normalized change in some characteristic y as the combined effects of the relative and parasitic sensitivities. We define

$$y = y(x_1, \ldots, x_i, \ldots, x_n, v_1, \ldots, v_j, \ldots, v_m) \tag{5}$$

where the quantities x_i are regular (*non*-parasitic) elements and the v_j are parasitic ones. If only first-order effects are considered, we may write

$$\Delta y = \sum_{i=1}^{n} \frac{\partial y}{\partial x_i}\Delta x_i + \sum_{j=1}^{m} \frac{\partial y}{\partial v_j}\Delta v_j. \tag{6}$$

If we divide both sides of this equation by y, we obtain

$$\frac{\Delta y}{y} = \sum_{i=1}^{n} S_{x_i}^y \frac{\Delta x_i}{x_i} + \sum_{j=1}^{m} PS_{v_j}^y\ \Delta v_j \tag{7}$$

The normalized change in y is seen to be the sum of the effects of the normalized changes in the nonparasitic elements x_i and the nonnormalized changes in the parasitic elements v_j.

Example 3.6-3. The damped integrator shown in Fig. 3.6-3 is assumed to have nominal element values $G_1 = 10$, $G_2 = 0.1$, $C = 1$, and operational amplifier gain $A = \infty$. For these values, the pole p is located at -0.1. To find the change in the pole location caused by 1 percent changes in the conductances G_1 and G_2 and an actual operational amplifier gain of 10^4, we may use the multiparameter sensitivity defined in (7) to write

$$\frac{\Delta p}{p} = S^p_{G_1}\frac{\Delta G_1}{G_1} + S^p_{G_2}\frac{\Delta G_2}{G_2} + PS^p_B\,\Delta B$$

The results of Example 3.6-2 may be applied to obtain

$$p = \frac{-G_2}{C} = -0.1 \qquad S^p_{G_1} = 0 \qquad S^p_{G_2} = 1 \qquad PS^p_B = \frac{G_1}{G_2} = 100$$

For these results we obtain

$$\frac{\Delta p}{p} = (0 \times 0.01) + (1 \times 0.01) + (100 \times 10^{-4}) = 0.02$$

Multiplying this result by the value of p, we obtain $\Delta p = -0.002$. The pole location becomes -0.102 as a result of the specified changes in the parameters.

PROBLEMS

Section 3.1

1. (*a*) Derive property 7 in Table 3.1-1.
   (*b*) Repeat for property 13.
2. (*a*) Derive property 15 in Table 3.1-1.
   (*b*) Repeat for property 16.
   (*c*) Repeat for property 17.
3. (*a*) A performance characteristic y is related to an element x by the coefficients a and b as follows:

   $$y = ax + b$$

   For the values $a = 2.5$ and $b = 4.9$ and for a nominal value of $x = 2$, find S^y_x.
   (*b*) Repeat part (*a*) for the coefficient values $a = 2.5$ and $b = -4.9$.
4. (*a*) For the performance characteristic defined in Prob. 3(*a*), use the sensitivity function to find $\Delta y/y$ for a 1 percent change in x.
   (*b*) For the performance characteristic defined in Prob. 3(*b*), use the sensitivity function to find $\Delta y/y$ for a 1 percent change in x.

Section 3.2

1. (*a*) Draw plots of $S^{|Y(j\omega)|}_L$, Im $S^{Y(j\omega)}_L$, and $S^{\arg Y(j\omega)}_L$ for Example 3.2-1.
   (*b*) Draw plots of $S^{|Y(j\omega)|}_C$, Im $S^{Y(j\omega)}_C$, and $S^{\arg Y(j\omega)}_C$ for Example 3.2-1.
2. (*a*) Compute the sensitivity $S^{|Y(j2)|}_L$ for Example 3.2-1.
   (*b*) Find the actual ratio between the percentage changes in $|Y(j2)|$ and L for a 10 percent change in L and compare it with the result obtained in part (*a*).
   (*c*) Repeat part (*b*) for a 1 percent change in L.
3. (*a*) Compute the sensitivity $S^{\arg Y(j2)}_L$ for Example 3.2-1.
   (*b*) Find the actual ratio between the percentage changes in arg $Y(j2)$ and L for a 10 percent change in L and compare it with the result obtained in part (*a*).
   (*c*) Repeat part (*b*) for a 1 percent change in L.

4. (*a*) Compute the sensitivity $S_C^{|Y(j2)|}$ for Example 3.2-1.
(*b*) Find the actual ratio between the percentage changes in $|Y(j2)|$ and C for a 10 percent change in C and compare it with the result obtained in part (*a*).
(*c*) Repeat part (*b*) for a 1 percent change in C.

5. (*a*) Compute the sensitivity $S_C^{\arg Y(j2)}$ for Example 3.2-1.
(*b*) Find the actual ratio between the percentage changes in arg $Y(j2)$ and C for a 10 percent change in C and compare it with the result obtained in part (*a*).
(*c*) Repeat part (*b*) for a 1 percent change in C.

6. (*a*) Find the sensitivity $S_x^{N(s)}$ for the network shown in Fig. P3.2-6, where $N(s)$ is the open-circuit voltage transfer function $V_2(s)/V_1(s)$ and x is the inductance L.
(*b*) Repeat for the capacitance C.
(*c*) Repeat for the resistance R.
(*d*) Repeat for the conductance G.

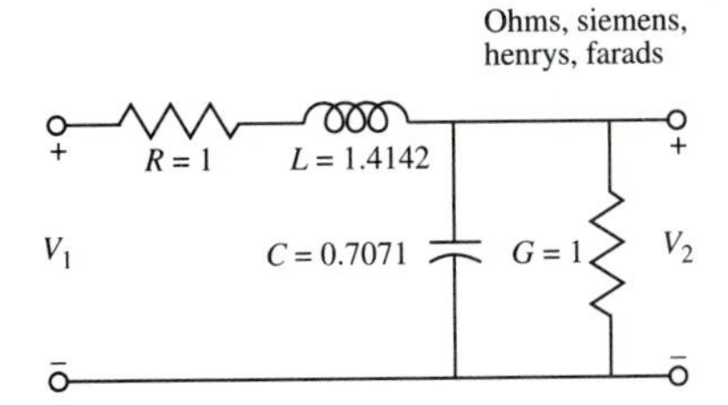

FIGURE P3.2-6

7. Find an expression for the multiparameter sensitivity for $\Delta|N(j1)|/|N(j1)|$ for Prob. 6.

8. A filter has the following network function:

$$N(s) = \frac{1}{s^3(R_1\Gamma_2R_3C_4) + s^2(R_1\Gamma_2) + s(R_1C_4 + R_3C_4) + 1}$$

where $R_1 = \frac{3}{2}\ \Omega$, $\Gamma_2 = \frac{4}{3}\ \text{H}^{-1}$, $R_3 = \frac{1}{2}\ \Omega$, and $C_4 = 1$ F. For these numerical values, find expressions for $E(\omega)$, $F(\omega)$, and $G(\omega)$, where

$$S_{\Gamma_2}^{N(j\omega)} = \frac{E(\omega)}{G(\omega)} + j\frac{F(\omega)}{G(\omega)}$$

9. (*a*) For the network shown in Fig. P3.2-9, find expressions for the sensitivities

$$S_L^{|Z(j\omega)|} \qquad \text{and} \qquad S_L^{\arg Z(j\omega)}$$

(*b*) Draw plots of $|Z(j\omega)|$, arg $Z(j\omega)$, $S_L^{|Z(j\omega)|}$, and $S_L^{\arg Z(j\omega)}$.

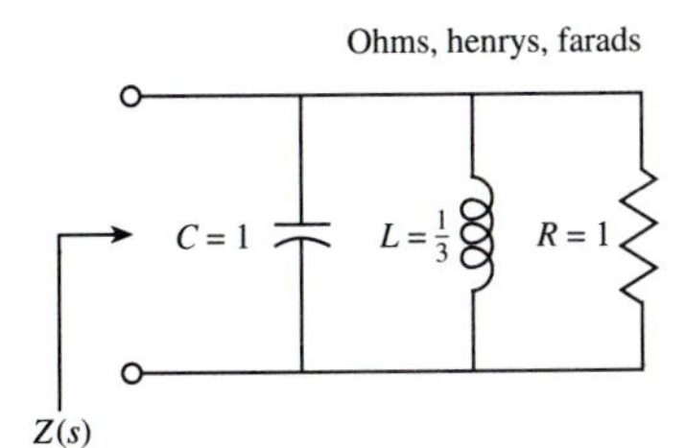

FIGURE P3.2-9

10. (*a*) For the broadband band-pass network described in Example 3.2-4, it is desired to have a maximum deviation of ±1 percent in the magnitude of the network function at the lower 3-dB down frequency of 0.618 rad/s. Find the maximum tolerance required for each of the elements R_1, L_2, and C_3. *Hint*: Use the function sensitivity plots given in Fig. 3.2-4.

(*b*) Repeat part (*a*) for a maximum deviation of ±1 percent in the magnitude at the resonant frequency of 1 rad/s. Comment on whether the answers obtained for L_2 and C_3 are factual.

Section 3.3

1. (*a*) Find the coefficient sensitivities for the network function $N(s)$ defined in Prob. 6 of Sec. 3.2 with respect to the inductance L.

(*b*) Repeat for the capacitance C.

(*c*) Repeat for the resistance R.

(*d*) Repeat for the conductance G.

2. (*a*) Use the results of Prob. 1 to derive the function sensitivity for $N(s)$ obtained in Prob. 6 of Sec. 3.2 with respect to the inductance L.

(*b*) Repeat for the capacitance C.

(*c*) Repeat for the resistance R.

(*d*) Repeat for the conductance G.

3. Use the coefficient sensitivities with respect to the gain K for the active filter shown in Fig. 3.3-2 to find the function sensitivity. Verify the result by directly determining the function sensitivity.

4. For the network function that follows, determine the coefficient sensitivities:

$$N(s) = \frac{a_0}{s^2 + b_1 s + b_0} = \frac{K/(RC)^2}{s^2 + s(3 - K)/RC + (1/RC)^2}$$

5. The following two network functions realize the same numerator and denominator polynomials, but the coefficients a_1 and b_1 are different functions of some parameter x:

$$N_1(s) = \frac{1}{s^2 + a_1 s + a_0} \qquad N_2(s) = \frac{1}{s^2 + b_1 s + b_0}$$

In addition, the following restrictions apply:

$$|S_x^{a_1}| > |S_x^{b_1}| \qquad S_x^{a_1} > 0 \qquad S_x^{b_1} > 0 \qquad S_x^{a_0} = S_x^{b_0} = 0$$

What, if anything, can be concluded about the relative values of each of the following pairs of quantities [give the appropriate (in)equality: <, >, or =].

(*a*) $S_x^{|N_1(j\omega)|}$ versus $S_x^{|N_2(j\omega)|}$

(*b*) $|S_x^{|N_1(j\omega)|}|$ versus $|S_x^{|N_2(j\omega)|}|$

(*c*) $S_x^{\arg N_1(j\omega)}$ versus $S_x^{\arg N_2(j\omega)}$

(*d*) $|S_x^{\arg N_1(j\omega)}|$ versus $|S_x^{\arg N_2(j\omega)}|$

6. We define the following network function:

$$N(s) = \frac{a_0}{s^2 + b_1 s + b_0} = \frac{2}{s^2 + s + 1}$$

The coefficient sensitivities with respect to a specific parameter x are found to be (any coefficient sensitivity not listed is zero)

$$S_x^{a_0} = 1 \qquad S_x^{b_1} = \frac{-x}{3 - x}$$

(*a*) Find the function sensitivity $S_x^{|N(j\omega)|}$.
(*b*) Find the maximum magnitude of $S_x^{|N(j\omega)|}$.

Section 3.4

1. Verify the relation given for $US_K^{p_1}$ in Example 3.4-2.

2. (*a*) Find the unnormalized pole sensitivity for the network of Example 3.2-1 for the resistance R.
(*b*) Repeat for the inductance L.
(*c*) Repeat for the capacitance C.

3. (*a*) Prove that the loci of the poles of the network function of Example 3.2-1 with respect to variations in the resistance R make up a circle.
(*b*) Repeat for the inductance L.
(*c*) Show that for variations in the capacitance C, the loci of the poles are straight lines.

4. (*a*) Find the loci of the poles of the network function for the network shown in Fig. P3.2-6 with respect to variations in the resistance R.
(*b*) Repeat for variations in Γ ($= 1/L$).

5. (*a*) Find the unnormalized pole sensitivity for the network function $N(s)$ defined in Prob. 6 of Sec. 3.2 with respect to the inductance L.
(*b*) Repeat for the capacitance C.
(*c*) Repeat for the resistance R.
(*d*) Repeat for the conductance G.

6. (*a*) Use the coefficient sensitivities given in Table 3.3-1 to find the unnormalized pole sensitivities for the network of Example 3.2-1 for the inductance L. Compare the results with those obtained in Prob. 2 and with the locus shown in Fig. 3.4-1(*b*).
(*b*) Repeat part (*a*) for the capacitance C. Compare the results with the locus shown in Fig. 3.4-1(*c*).

7. (*a*) Use the coefficient sensitivities determined in Prob. 1 of Sec. 3.3 to find the unnormalized pole sensitivities for the network shown in Fig. P3.2-6 for the inductance L. Compare the results with those obtained in Prob. 5.
(*b*) Repeat for the capacitance C.
(*c*) Repeat for the resistance R.
(*d*) Repeat for the conductance G.

8. Use the coefficient sensitivities for the gain K of the VCVS used in the filter shown in Fig. 3.3-2 to find the unnormalized pole sensitivity. Compare the result with that given in Example 3.4-2.

9. (*a*) For the network shown in Fig. 3.2-1, assume the element values $L = 1$, $C = 1$, and $R = 2$. Use the techniques for nonsimple poles to determine the change in pole positions for a +1 percent change in R.
(*b*) Repeat for a -1 percent change in R.

10. Repeat Prob. 9 using L as the parameter that is changed.

11. Repeat Prob. 9 using C as the parameter that is changed.

12. (*a*) In a given third-order filter realization, the denominator of the network function has the form

$$D(s) = s^3 + 3s^2 + 3s + K|_{K=1} = (s+1)^3$$

Find the pole locations resulting from a +1 percent change in the parameter K from its nominal value of unity.

(*b*) Repeat for a -1 percent change in K.

13. (*a*) In a given third-order filter realization, the denominator of the network function has the form

$$D(s) = s^3 + 3s^2 + Ks + 1|_{K=3} = (s+1)^3$$

Find the pole locations resulting from a +1 percent change in the parameter K from its nominal value of 3.

(*b*) Repeat for a -1 percent change in K.

14. (*a*) Starting with the unnormalized pole sensitivity found in Prob. 2, derive the function sensitivity for the resistance R given in Example 3.2-1.

(*b*) Repeat for the inductance L.

(*c*) Repeat for the capacitance C.

Section 3.5

1. The denominator of a network function has the form given below. Find the Q sensitivity to K in terms of the coefficients a, b, and c and in terms of Q. The resulting expression should *not* be a function of K:

$$s^2 + s(a - bK) + c$$

2. The denominator of a network function has the form given below. Find the Q sensitivity to $|K|$ in terms of the coefficients a, b, and c and in terms of Q. The resulting expression should *not* be a function of K:

$$S^2 + as + (b + |K|c)$$

3. A second-order network function has the following expressions determining Q and ω_n in terms of a VCVS gain K and the network elements R_1, R_3, C_2, and C_4. Find expressions for the Q and ω_n sensitivities:

$$\frac{1}{Q} = \sqrt{\frac{R_3C_4}{R_1C_2}} + \sqrt{\frac{R_1C_4}{R_3C_2}} + (1-K)\sqrt{\frac{R_1C_2}{R_3C_4}} \qquad \omega_n = \frac{1}{\sqrt{R_1R_3C_2C_4}}$$

4. A second-order network function has the following expressions determining Q and ω_n in terms of the network elements R_1, R_2, R_3, C_5, and C_6. Find expressions for the Q and ω_n sensitivities:

$$\frac{1}{Q} = \sqrt{\frac{C_6}{C_5}}\left(\frac{\sqrt{R_2R_3}}{R_1} + \sqrt{\frac{R_3}{R_2}} + \sqrt{\frac{R_2}{R_3}}\right) \qquad \omega_n = \frac{1}{\sqrt{R_2R_3C_5C_6}}$$

5. (*a*) Find the Q and ω_n sensitivities for the network defined in Prob. 6, Sec. 3.2, with respect to the inductance L.

(*b*) Repeat for the capacitance C.

(*c*) Repeat for the resistance R.

(*d*) Repeat for the conductance G.

6. For the following band-pass frequency-normalized network function, find the maximum magnitude of $S_Q^{|N(j\omega)|}$:

$$N(s) = \frac{Hs}{s^2 + s(1/Q) + 1}$$

7. The low-pass network shown in Fig. 3.5-2 with a network function as defined in Example 3.5-2 is used to realize a third-order Chebyshev approximation with a 0.5-dB ripple. The required element values are $K = 2$, $R_1 = 1.876\ \Omega$, $R_2 = 2.778\ \Omega$, $R_3 = 0.2681\ \Omega$, and unity for all the capacitors. Find the Q and ω_n sensitivities with respect to K.

8. For the network used in Example 3.5-2, assume that all the resistors and capacitors have unity value and $K = 2$. Find the value of the Q sensitivity to K.

Section 3.6

Note: The analysis given in App. C will be of assistance for some of the problems in this section.

1. (*a*) The low-pass network shown in Fig. P3.6-1 includes an inductor that has a parasitic resistor R_p in series with it. Find an expression for the parasitic sensitivity $PS_{R_p}^{p}$, where p is the pole location.

(*b*) For the case where $L = 1$ H and $R = 10^4\ \Omega$, use the results of part (*a*) to find $\Delta p/p$ for a value $R_p = 0.01\ \Omega$.

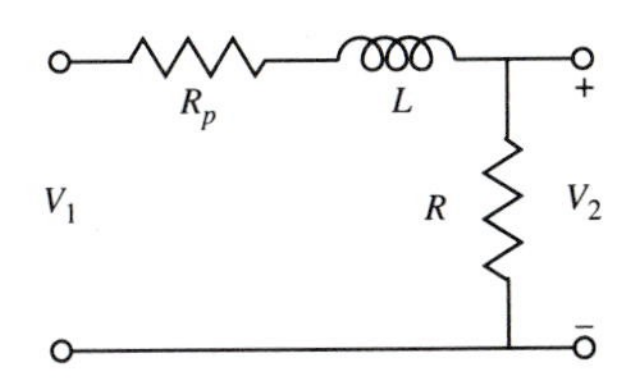

FIGURE P3.6-1

2. (*a*) A noninverting VCVS is realized with an operational amplifier and two resistors R_1 and R_2 as shown in Fig. P3.6-2. The VCVS gain is defined as $K = V_2/V_1$. The operational amplifier open-loop gain is defined as $A = 1/B$ where B is a parasitic with an ideal value of zero. Find the parasitic sensitivity PS_B^K in terms of R_1 and R_2.

(*b*) For the resistor values $R_1 = 1\ \text{k}\Omega$ and $R_2 = 100\ \text{k}\Omega$, use the results of part (*a*) to determine $\Delta K/K$ if $A = 10^4$ rather than infinity.

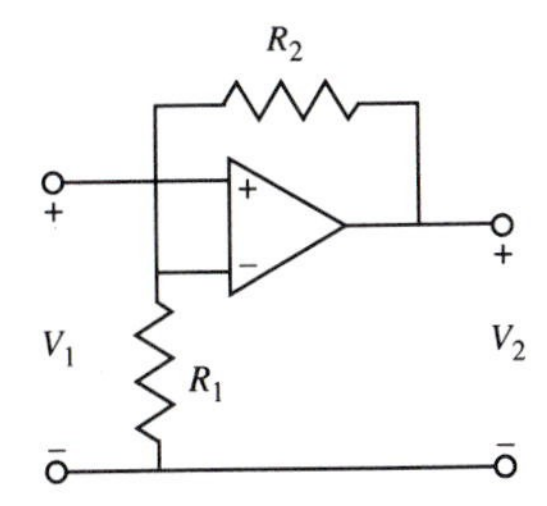

FIGURE P3.6-2

3. (*a*) A noninverting VCVS is realized with an operational amplifier and two resistors R_1 and R_2 as shown in Fig. P3.6-2. The VCVS gain is defined as $K = V_2/V_1$. The parasitic output resistance of the operational amplifier is defined as R_0 with an ideal value of zero. Find the parasitic sensitivity $PS_{R_0}^K$ in terms of R_1 and R_2.

(*b*) For the resistor values $R_1 = 1\ \text{k}\Omega$ and $R_2 = 100\ \text{k}\Omega$, use the results of part (*a*) to determine $\Delta K/K$ if $R_0 = 0.01\ \Omega$ rather than zero.

4. (*a*) A noninverting VCVS is realized with an operational amplifier and two resistors R_1 and R_2 as shown in Fig. P3.6-2. The VCVS gain is defined as $K = V_2/V_1$. The parasitic input conductance between the terminals of the operational amplifier is defined as G_i with an ideal value of zero. Find the parasitic sensitivity $PS_{G_i}^{K}$ in terms of R_1 and R_2.

(*b*) For the resistor values $R_1 = 1\ \text{k}\Omega$ and $R_2 = 100\ \text{k}\Omega$, use the results of part (*a*) to determine $\Delta K/K$ for if $R_i = 1/G_i = 100\ \Omega$ rather than infinity.

5. (*a*) For the filter shown in Fig. P3.6-5, let A be the (positive) open-loop gain of the operational amplifier and let $N(s) = V_2(s)/V_1(s)$ be the voltage transfer function. A parasitic B is defined as $B = -1/A$. Find the parasitic sensitivity $PS_{B}^{N(s)}$.

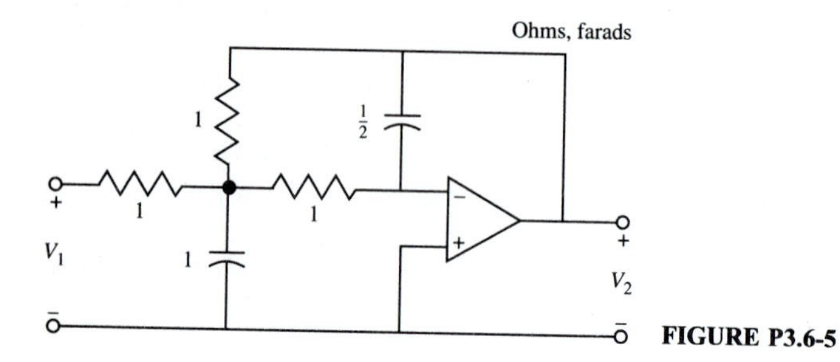

FIGURE P3.6-5

(*b*) If the operational amplifier (positive) open-loop gain A at a sinusoidal frequency of $\sqrt{2}$ rad/s is 10^4 rather than the ideal value of infinity, use the results of part (*a*) to find the percentage change in $|N(j\sqrt{2})|$ caused by this deviation from the ideal.

CHAPTER 4

PASSIVE NETWORK SYNTHESIS

In this chapter we introduce a first topic in filter design, the synthesis of passive filters. We shall see that this subject is both interesting and useful. It is interesting because of the clarity and sophistication of the mathematical processes involved. It is useful because many of the results are directly applicable to the various types of active filters that will be covered in the chapters that follow. It should be emphasized, however, that the technical literature covering passive network synthesis spans several decades and is vast in scope. As a result, our treatment given here must, of necessity, be quite selective. We will only present the most fundamental techniques, and we will minimize proofs so that we may emphasize properties and results that have the greatest application.

4.1 ENERGY FUNCTIONS FOR *LC* NETWORKS

In this section we develop some general properties of the driving-point functions of *LC* networks.

Incidence Matrix

In this section we analyze a network consisting of two-terminal elements to derive some relations between the voltage and current variables. As an example, consider the network shown in Fig. 4.1-1. Let v_k and i_k ($k = 1, 2, 3, 4, 5$) be the branch voltages and branch

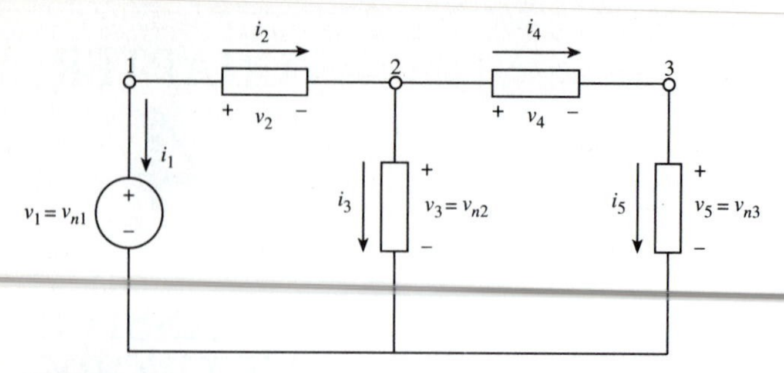

FIGURE 4.1-1
Network for incidence matrix determination.

currents and v_{nk} ($k = 1, 2, 3$) be the node voltages. The Kirchhoff current law equations for the network are

$$\begin{aligned} i_1 + i_2 \qquad\qquad &= 0 \\ -i_2 + i_3 + i_4 \qquad &= 0 \\ -i_4 + i_5 &= 0 \end{aligned} \tag{1}$$

In matrix form this becomes

$$\begin{bmatrix} 1 & 1 & 0 & 0 & 0 \\ 0 & -1 & 1 & 1 & 0 \\ 0 & 0 & 0 & -1 & 1 \end{bmatrix} \begin{bmatrix} i_1 \\ i_2 \\ i_3 \\ i_4 \\ i_5 \end{bmatrix} = \begin{bmatrix} 0 \\ 0 \\ 0 \end{bmatrix} \tag{2}$$

In the general case the equations may be written in matrix form as

$$\mathbf{Ai}(t) = \mathbf{0} \tag{3}$$

where **A** is defined as the incidence matrix, $\mathbf{i}(t)$ is the branch current vector, and **0** is the null vector. The elements a_{kj} of **A** are defined as

$$a_{kj} = \begin{cases} +1 & \text{if reference direction of } i_j \text{ is away from node } k \\ -1 & \text{if reference direction of } i_j \text{ is toward node } k \\ 0 & \text{if } i_j \text{ does not flow into or out of node } k \end{cases}$$

The incidence matrix can also be used to relate the branch voltages to the node voltages. For the example of Fig. 4.1-1

$$\begin{bmatrix} v_1 \\ v_2 \\ v_3 \\ v_4 \\ v_5 \end{bmatrix} = \begin{bmatrix} 1 & 0 & 0 \\ 1 & -1 & 0 \\ 0 & 1 & 0 \\ 0 & 1 & -1 \\ 0 & 0 & 1 \end{bmatrix} \begin{bmatrix} v_{n1} \\ v_{n2} \\ v_{n3} \end{bmatrix} \tag{4}$$

Note that the rectangular matrix in this equation is simply the transpose of the incidence matrix. Thus, for the general case, this equation has the form

$$\mathbf{v}(t) = \mathbf{A}^{\mathrm{t}}\mathbf{v}_n(t) \tag{5}$$

where $\mathbf{v}(t)$ is the branch voltage vector and $\mathbf{v}_n(t)$ the node voltage vector. From (3) and (5) we may derive

$$\mathbf{v}^{\mathrm{t}}(t)\mathbf{i}(t) = [\mathbf{A}^{\mathrm{t}}\mathbf{v}_n(t)]^{\mathrm{t}}\mathbf{i}(t) = \mathbf{v}_n^{\mathrm{t}}(t)\mathbf{A}\mathbf{i}(t) = \mathbf{v}_n^{\mathrm{t}}(t)\mathbf{0} = 0 \tag{6}$$

The left member of (6), however, is the product of a row and a column matrix. Thus this expression may be written

$$\mathbf{v}^{\mathrm{t}}(t)\mathbf{i}(t) = \sum_{k=1}^{b} v_k(t)i_k(t) = 0 \tag{7}$$

where b is the number of branches. This, we note, is a well-known result; namely, that the sum of the branch powers is zero at every instant of time.

Application to Laplace Transform

The preceding development can be extended to obtain a result that is important to our study of driving-point functions. Applying the Laplace transform to (3), we obtain

$$\mathbf{A}\mathbf{I}(s) = \mathbf{0} \tag{8}$$

where $\mathbf{I}(s)$ is the transformed branch current vector whose elements are the transformed branch currents $I_k(s)$. Taking the complex conjugate of both sides of this, we obtain

$$\mathbf{A}\mathbf{I}^*(s) = \mathbf{0} \tag{9}$$

similarly, the Laplace transform of (5) is

$$\mathbf{V}(s) = \mathbf{A}^{\mathrm{t}}\mathbf{V}_n(s) \tag{10}$$

where $\mathbf{V}(s)$ and $\mathbf{V}_n(s)$ are, respectively, the transformed branch voltage and node voltage vectors. From (9) and (10) we may make a development similar to that given in (6):

$$\mathbf{V}^{\mathrm{t}}(s)\mathbf{I}^*(s) = [\mathbf{A}^{\mathrm{t}}\mathbf{V}_n(s)]^{\mathrm{t}}\mathbf{I}^*(s) = \mathbf{V}_n^{\mathrm{t}}(s)\mathbf{A}\mathbf{I}^*(s) = \mathbf{V}_n^{\mathrm{t}}(s)\mathbf{0} = 0 \tag{11}$$

Following the development of (7), we may write

$$\mathbf{V}^{\mathrm{t}}(s)\mathbf{I}^*(s) = \sum_{k=1}^{b} V_k(s)I_k^*(s) = 0 \tag{12}$$

Let us consider, without loss of generality, that the node pair for branch 1 defines the input point for our general *RLC* network as shown in Fig. 4.1-2(*a*). From (12)

$$-V_1(s)I_1^*(s) = \sum_{k=2}^{b} V_k(s)I_k^*(s) \tag{13}$$

Dividing both sides of this equation by $|I_1(s)|^2 = I_1(s)I_1^*(s)$, we find

$$\frac{V_1}{-I_1} = \frac{1}{|I_1(s)|^2}\sum_{k=2}^{b} V_k(s)I_k^*(s) \tag{14}$$

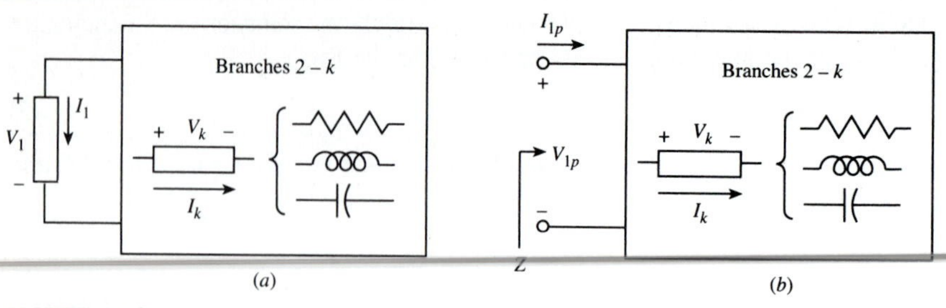

FIGURE 4.1-2
A general one-port *RLC* network.

Using conventional port reference directions (the positive current reference is into the network) as shown in Fig. 4.1-2(*b*), the input impedance $Z(s)$ of the general network consisting of branches 2 through b is defined as the negative of the ratio of the branch 1 voltage and current. Thus,

$$Z(s) = \frac{V_{1p}(s)}{I_{1p}(s)} = \frac{V_1(s)}{-I_1(s)} = \frac{1}{|I_1(s)|^2} \sum_{k=2}^{b} V_k(s) I_k^*(s) \tag{15}$$

where $V_{1p}(s)$ and $I_{1p}(s)$ are the port voltage and current variables.

Energy Functions

Now consider the expressions relating the variables in the different branches of the network. For the cases where the branch is an inductor or capacitor, respectively, we have

$$V_k(s) = sL_k I_k(s) \qquad V_k(s) = \frac{1}{sC_k} I_k(s) \tag{16}$$

For these cases the product $V_k(s)I_k^*(s)$ becomes

$$V_k(s)I_k^*(s) = sL_k |I_k(s)|^2 \qquad V_k(s)I_k^*(s) = \frac{1}{sC_k} |I_k(s)|^2 \tag{17}$$

If we use $\sum_L$ and $\sum_C$ to indicate sums only over the branches containing inductors and capacitors, respectively, the general expression for the input impedance $Z(s)$ of (15) may be written

$$Z(s) = \frac{1}{|I_1(s)|^2} \left[s \sum_L L_k |I_k(s)|^2 + \frac{1}{s} \sum_C \frac{1}{C_k} |I_k(s)|^2 \right] \tag{18}$$

Each of the two summation terms inside the brackets is a real nonnegative quantity. To emphasize this, we define two quantities called *energy functions*[1]:

$$T_0(s) = \sum_L L_K |I_k(s)|^2 \geq 0 \qquad V_0(s) = \sum_C \frac{1}{C_k} |I_k(s)|^2 \geq 0 \tag{19}$$

[1] The term *energy functions* comes from the fact that the expressions of (19) are related to the sinusoidal steady-state expressions for the energy stored in the magnetic fields of inductors and the electric fields of capacitors.

Thus, the general expression for the *driving-point impedance* of an arbitrary *LC* network may always be put in the form

$$Z(s) = \frac{V_{1p}(s)}{I_{1p}(s)} = \frac{1}{|I_1(s)|^2}\left[sT_0(s) + \frac{1}{s}V_0(s)\right] \tag{20}$$

A development similar to that given above shows that

$$Y(s) = \frac{I_{1p}(s)}{V_{1p}(s)} = \frac{1}{|V_1(s)|^2}\left[s^*T_0(s) + \frac{1}{s^*}V_0(s)\right] \tag{21}$$

The details are left as an exercise for the reader. This expression defines the *driving-point admittance* $Y(s)$ of a passive *LC* network in terms of the energy functions.

> **Example 4.1-1 Energy functions.** As a simple example to illustrate the notation used above, consider the network shown in Fig. 4.1-3. For this circuit,
>
> $$V_0(s) = 2|I_2(s)|^2 \qquad T_0(s) = 3|I_3(s)|^2$$
>
> By inspection we readily find
>
> $$-I_1(s) = I_{1p}(s) = I_2(s) = I_3(s) = \frac{sV_1}{2+3s^2}$$
>
> Thus, from (20),
>
> $$Z(s) = \left|\frac{2+3s^2}{sV_1}\right|^2\left[\frac{2}{s}\left|\frac{sV_1}{2+3s^2}\right|^2 + 3s\left|\frac{sV_1}{2+3s^2}\right|^2\right] = \frac{2+3s^2}{s}$$
>
> A more complex example is given in the problems.

The expression for $Z(s)$ given in (20), in general, is of little use in analyzing a network. Such an operation is far more easily accomplished using conventional mesh or node analysis techniques. The importance of (20) is that it allows us to develop general properties that apply to driving-point functions for arbitrary *LC* networks. This is done in the following section.

4.2 PROPERTIES OF *LC* DRIVING-POINT FUNCTIONS

Pole and Zero Locations

In the preceding section we showed how the driving-point impedance of an arbitrary *LC* network could be expressed in terms of a set of two energy functions $T_0(s)$ and $V_0(s)$,

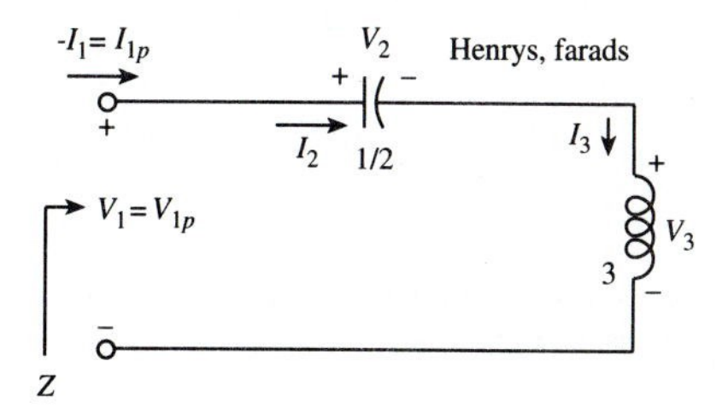

FIGURE 4.1-3
An example *LC* network.

which are always real and nonnegative. Now let us see how to apply those results in determining some of the properties of driving-point functions for LC networks, i.e., networks containing no dissipative (resistive) elements. In this case, from (20) in Sec. 4.1, the general driving-point impedance $Z_{LC}(s)$ may be expressed as

$$Z_{LC}(s) = \frac{1}{|I_1(s)|^2}\left[sT_0(s) + \frac{1}{s}V_0(s)\right] \tag{1}$$

Setting the right member of this expression equal to zero, we find that the zeros of $Z_{LC}(s)$ are located at those values of s that satisfy the relation

$$s = \pm j\sqrt{\frac{V_0(s)}{T_0(s)}} \tag{2}$$

In the right member of this relation, even though we cannot evaluate $T_0(s)$ and $V_0(s)$ [since they include expressions for the branch currents $I_k(s)$], we know that they are always real and nonnegative. Therefore we see that *the zeros of* $Z_{LC}(s)$ *must lie on the* $j\omega$ *axis* (including the origin and infinity). A similar development may be made starting with (21) of Sec. 4.1 to show that a general LC driving-point admittance function $Y_{LC}(s)$ will also have its zeros on the $j\omega$ axis. Since $Z_{LC}(s) = 1/Y_{LC}(s)$, we conclude that *both the poles and the zeros of LC driving-point immittances lie on the* $j\omega$ *axis.*[1] On the *finite* $j\omega$ axis (excluding the origin and infinity) they must, of course, occur in conjugate pairs.

Reactance Function

Now let us consider (1) in more detail. Our goal is to find information concerning $Z(j\omega)$. If we let $s = \sigma + j\omega$, then, letting $U(\sigma, \omega)$ and $W(\sigma, \omega)$ be, respectively, the real and imaginary parts of $Z_{LC}(s)$ and assuming, without loss of generality, that $|I_1(s)|$ is normalized to unity, we may write

$$\begin{aligned} Z_{LC}(\sigma + j\omega) &= U(\sigma, \omega) + jW(\sigma, \omega) \\ &= \left[\sigma T_0 + \frac{\sigma}{\sigma^2 + \omega^2}V_0\right] + j\left[\omega T_0 - \frac{\omega}{\sigma^2 + \omega^2}V_0\right] \end{aligned} \tag{3}$$

where, for convenience, we have omitted the functional notation on T_0 and V_0. From (3), we see that on the $j\omega$ axis ($\sigma = 0$), $Z_{LC}(s)$ is purely imaginary, i.e., we may write

$$Z_{LC}(j\omega) = jX(\omega) \tag{4}$$

that is, *when* $Z_{LC}(s)$ *is evaluated for* $s = j\omega$, *the result is a purely imaginary quantity.* This property defines a *reactance function* and $X(\omega)$ is its imaginary part. To obtain still more information about the $j\omega$ axis behavior of $Z_{LC}(s)$, we now investigate the quantity $\partial W(\sigma, \omega)/\partial\omega$ under the condition $\sigma = 0$. From the Cauchy–Riemann conditions we

[1] Immittance is a general term used to include both *impedance* and *admittance*.

know that[2]

$$\frac{\partial U(\sigma, \omega)}{\partial \sigma} = \frac{\partial W(\sigma, \omega)}{\partial \omega} \tag{5}$$

If we now examine the quantities $U(\sigma, \omega)$ and $W(\sigma, \omega)$ in (3), it is apparent that, anticipating setting σ to zero, it is easier to evaluate the left member of (5) than the right one. Performing the indicated differentiation, we obtain

$$\frac{\partial U(\sigma, \omega)}{\partial \sigma} = T_0 + \sigma \frac{\partial T_0}{\partial \sigma} + V_0 \left[\frac{1}{\sigma^2 + \omega^2} - \frac{2\sigma^2}{(\sigma^2 + \omega^2)^2} \right] + \frac{\sigma}{(\sigma^2 + \omega^2)} \frac{\partial V_0}{\partial \sigma} \tag{6}$$

Evaluating this expression for $\sigma = 0$ and using (5), we obtain the following result for the derivative of $X(\omega)$:

$$\frac{dX(\omega)}{d\omega} = \left. \frac{\partial W(\sigma, \omega)}{\partial \omega} \right|_{\sigma=0} = T_0 + \frac{V_0}{\omega^2} \tag{7}$$

We conclude that *the slope of $X(\omega)$, the imaginary part of the reactance function, is always positive*. A similar development is easily made to define $B(\omega)$ as the imaginary part of the admittance $Y_{LC}(j\omega)$. Thus

$$Y_{LC}(j\omega) = jB(\omega) \tag{8}$$

It is easily shown that the derivative $dB(\omega)/d\omega$ is also always positive.

Alternation of Poles and Zeros

The properties of $Z_{LC}(j\omega)$ developed above allow us to construct plots of $X(\omega)$ by noting that since the poles and zeros of $Z_{LC}(s)$ occur on the $j\omega$ axis, they also define the values of ω at which $X(\omega)$ is infinite or zero. Some typical plots, drawn for positive values of ω, are shown in Fig. 4.2-1. From these plots, we see that the positive-slope property requires that *the poles and zeros be simple and that they alternate along the $j\omega$ axis*. Thus, the corresponding pole-zero diagrams appear as shown in Fig. 4.2-2. In addition, as shown in the plots, *the zero- and infinite-frequency behavior of $Z_{LC}(s)$ must be that of a pole or a zero*. To see why this is so, consider a function $F(s)$ with poles and zeros on the $j\omega$ axis not including the origin. It may be written in the form

$$F(s) = H \frac{\prod_{i=1}^{m} (s^2 + \omega_i^2)}{\prod_{j=1}^{n} (s^2 + \omega_j^2)} \tag{9}$$

where H is an arbitrary positive multiplicative constant. Evaluating this function at the origin, we see that $F(0)$ is real. Driving-point functions for LC networks, however, must

[2] The Cauchy–Riemann conditions include the relation given in (5). They may be found in any standard textbook on complex-variable theory. For example, see R. V. Churchill, *Complex Variables and Applications*, McGraw-Hill Book Co., New York, 1984.

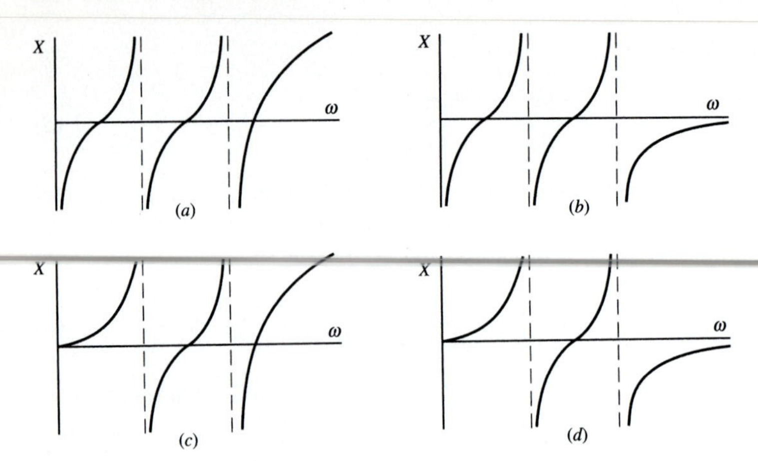

FIGURE 4.2-1
Typical plots of $X(\omega)$.

be imaginary at all points on the $j\omega$ axis, and the origin is certainly on the $j\omega$ axis. Thus, a term of the type s or $1/s$ must also be present if the $F(s)$ given in (9) is to satisfy this requirement. Such terms obviously represent a zero or a pole at the origin. Similarly, for $F(\infty)$ to be imaginary, the degrees of its numerator and denominator polynomials must differ by an odd number. In addition, since poles and zeros can only be simple, these degrees must differ by exactly unity, i.e., there must be a pole (if the numerator is of higher degree than the denominator) or a zero (if the denominator is of higher degree than the numerator) at infinity.

Example 4.2-1 Functions not *LC* realizable. As a result of the properties developed in the preceding paragraphs, we see that it is *not* possible for any of the following functions

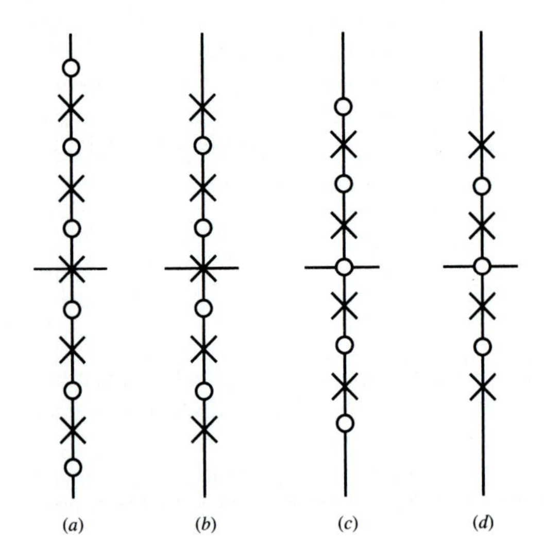

FIGURE 4.2-2
Pole-zero plots for the functions $X(\omega)$ shown in Fig. 4.2-1.

to be the driving-point immittance of a passive *LC* network:

Function	Reason
$F_a(s) = \dfrac{(s^2+1)(s^2+2)}{s(s^2+3)}$	The poles and zeros do not alternate.
$F_b(s) = \dfrac{s(s^2+2)(s^2+4)}{(s+2)(s^2+3)}$	There is a pole on the negative real axis.
$F_c(s) = \dfrac{(s^2+1)(s^2+3)}{(s^2+2)}$	There is a constant value at the origin; the pole at infinity is of second order.
$F_d(s) = \dfrac{(s^2+2)(s^2+4)}{(s^2+1)(s^2+3)}$	There is a constant value at both the origin and infinity.

As a consequence of the pole and zero requirements discussed above, we may write four general types of expressions for *LC* driving-point immittance functions, illustrating the various possible zero- and infinite-frequency behaviors. These are given in Table 4.2-1. An examination of these expressions shows that if the numerator is an even polynomial, i.e., a polynomial containing only even powers of s, then the denominator must be an odd polynomial. Similarly, if the numerator is odd, the denominator must be even. A network function with this property is referred to as an *odd rational function*. Note that the numerator and denominator polynomials differ in degrees by exactly 1.

TABLE 4.2-1
Four types of *LC* driving-point functions

Case	Behavior at origin	Behavior at infinity	General form of network function	Condition
1	Pole	Pole	$F(s) = H\dfrac{(s^2+\omega_1^2)(s^2+\omega_3^2)\cdots(s^2+\omega_m^2)}{s(s^2+\omega_2^2)(s^2+\omega_4^2)\cdots(s^2+\omega_n^2)}$	$m = n+1$
2	Pole	Zero	$F(s) = H\dfrac{(s^2+\omega_1^2)(s^2+\omega_3^2)\cdots(s^2+\omega_m^2)}{s(s^2+\omega_2^2)(s^2+\omega_4^2)\cdots(s^2+\omega_n^2)}$	$m = n-1$
3	Zero	Pole	$F(s) = H\dfrac{s(s^2+\omega_2^2)(s^2+\omega_4^2)\cdots(s^2+\omega_m^2)}{(s^2+\omega_1^2)(s^2+\omega_3^2)\cdots(s^2+\omega_n^2)}$	$m = n+1$
4	Zero	Zero	$F(s) = H\dfrac{s(s^2+\omega_2^2)(s^2+\omega_4^2)\cdots(s^2+\omega_m^2)}{(s^2+\omega_1^2)(s^2+\omega_3^2)\cdots(s^2+\omega_n^2)}$	$m = n-1$

Note: $0 < \omega_1 < \omega_2 < \omega_3 < \cdots$, $H > 0$.

Example 4.2-2 Examples of *LC* functions. Some simple examples of the different types of *LC* functions defined in Table 4.2-1 follow:

$$\text{Case 1:}\quad \frac{s^2+1}{s} \qquad \text{Case 2:}\quad \frac{s^2+1}{s(s^2+2)}$$

$$\text{Case 3:}\quad \frac{s(s^2+2)}{s^2+1} \qquad \text{Case 4:}\quad \frac{s}{s^2+1}$$

Exercise 4.2-2. For each of the following functions, identify the case as given in Table 4.2-1.

$$(a)\ \frac{s(s^2+2)}{(s^2+1)(s^2+3)} \qquad (b)\ \frac{(s^2+1)(s^2+3)}{s(s^2+2)}$$

$$(c)\ \frac{s(s^2+2)(s^2+4)}{(s^2+1)(s^2+3)} \qquad (d)\ \frac{(s^2+1)(s^2+3)}{s(s^2+2)(s^2+4)}$$

Answers. (*a*) Case 4. (*b*) Case 1. (*c*) Case 3. (*d*) Case 2.

Partial-Fraction Expansion

Now let us consider the partial-fraction expression of an *LC* driving-point impedance function $Z_{LC}(s)$. In general this will have the form

$$Z_{LC}(s) = \frac{k_0}{s} + k_\infty s + \sum_i \left[\frac{c_i}{s - j\omega_i} + \frac{c_i^*}{s + j\omega_i}\right] \tag{10}$$

where k_0, k_∞, and c_i are the residues of the poles at the origin, at infinity, and on the $j\omega$ axis, respectively. If we examine $Z_{LC}(s)$ in the limit as s takes on values in the vicinity of a pole at $s = j\omega_i$, then the term $c_i/(s - j\omega_i)$ in the above expression becomes very much larger than all the other terms (since its denominator becomes very small). Thus, we may write[3]

$$\lim_{s \to j\omega_i} Z_{LC}(s) \cong \frac{c_i}{s - j\omega_i} \tag{11}$$

If we now also constrain s to be on the $j\omega$ axis, we may write

$$\lim_{\omega \to \omega_i} Z_{LC}(j\omega) \cong \frac{c_i}{j\omega - j\omega_i} = j\frac{c_i}{\omega_i - \omega} \tag{12}$$

The quantity $c_i/(\omega_i - \omega)$ is simply the limiting value of $X(\omega_i)$, the imaginary part of the reactance function. Since this value must always be real, we see that c_i is required to be real and thus that $c_i^* = c_i$. In addition, from the plots of $X(\omega)$ shown in Fig. 4.2-1, we see that when $\omega < \omega_i$, the quantity $c_i/(\omega_i - \omega)$ approaches a large positive value. This can

---

[3] Formally, this step may be justified by writing a Laurent expansion of $Z_{LC}(s)$ around the point $s = j\omega_i$. See Churchill, op. cit.

only be true if c_i is positive. These results are readily extended to the case of poles at the origin and at infinity. Thus we conclude that as a result of the alternation properties for poles and zeros, *the residues of the poles of an LC driving-point impedance function are real and positive.* The same result holds for an *LC* driving-point admittance function, since its partial-fraction expansion is identical with the one given in (10).

Example 4.2-3 Residues at $j\omega$-axis poles. If a function is *not* realizable as the driving-point immittance of an *LC* network, then in most cases, the residues at any $j\omega$-axis poles will not be real. As an example of this, consider

$$F(s) = \frac{s+1}{s^2+1} = \frac{c_1}{s-j1} + \frac{c_1^*}{s+j1}$$

Using conventional partial-fraction determination techniques, the residues are found to be

$$c_1 = (s-j1)F(s)\big|_{s=j1} = \left.\frac{s+1}{s+j}\right|_{s=j1} = \tfrac{1}{2} - j\tfrac{1}{2}$$

$$c_1^* = (s+j1)F(s)\big|_{s=-j1} = \left.\frac{s+1}{s-j1}\right|_{s=-j1} = \tfrac{1}{2} + j\tfrac{1}{2}$$

In this section we have derived a set of *necessary* properties for driving-point immittance functions. They are collected in Summary 4.2-1. In the next section we shall apply the properties of *LC* driving-point functions presented in this chapter to develop actual design procedures for such functions and thus to show the sufficiency of the conditions.

Summary 4.2-1 Necessary conditions for *LC* driving-point functions

1. The poles are simple and on the $j\omega$ axis.
2. The zeros are simple and on the $j\omega$ axis.
3. The poles and zeros alternate.
4. There is a pole or a zero at the origin.
5. There is a pole or a zero at infinity.
6. The residues of the poles are real and positive.
7. The functions are reactance functions whose value along the $j\omega$ axis is purely imaginary, i.e., $Z_{LC}(j\omega) = jX(\omega)$ and $Y_{LC}(j\omega) = jB(\omega)$.
8. $dX(\omega)/d\omega$ and $dB(\omega)/d\omega$ are always positive.
9. The functions are odd rational functions.

4.3 SYNTHESIS OF *LC* DRIVING-POINT FUNCTIONS

Series Foster Form

In the preceding section we derived a set of *necessary* properties for *LC* driving-point immittances. In this section we shall develop synthesis procedures for such immittances.

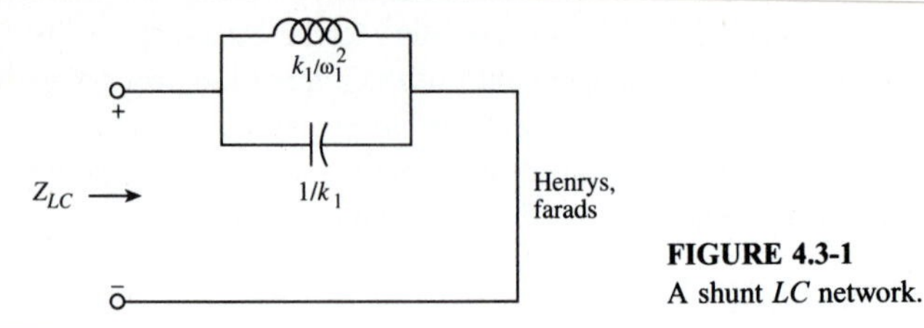

FIGURE 4.3-1
A shunt LC network.

In the process we shall show the sufficiency of various subsets of the necessary conditions given in Summary 4.2-1. The first such procedure that we shall discuss is based on a *partial-fraction expansion*. As a preliminary step, let us first consider the partial-function expansion of a function $Z_{LC}(s)$ consisting only of a pair of complex-conjugate poles at $s = \pm j\omega_1$. It has the form

$$Z_{LC}(s) = \frac{c_1}{s - j\omega_1} + \frac{c_1}{s + j\omega_1} = \frac{c_1(s + j\omega_1) + c_1(s - j\omega_1)}{s^2 + \omega_1^2}$$
$$= \frac{2c_1 s}{s^2 + \omega_1^2} = \frac{k_1 s}{s^2 + \omega_1^2} \tag{1}$$

where c_1 is the residue at each of the complex-conjugate poles and where we have defined $k_1 = 2c_1$. In Sec. 4.2, c_1 was shown to be real and positive; thus, obviously, k_1 will be real and positive. A term such as that given in the right member of (1) may be directly synthesized as a parallel (shunt) LC network as shown in Fig. 4.3-1.

Now let us consider a general function $Z_{LC}(s)$. From (10) of Sec. 4.2, its partial-fraction expansion will have the form

$$Z_{LC}(s) = k_\infty s + \frac{k_0}{s} + \sum_i \frac{k_i s}{s^2 + \omega_i^2} \tag{2}$$

where we have used (1) to combine the terms representing the complex-conjugate poles. Each of the terms in the right member of (2) is directly realizable as an inductor, a capacitor, or a parallel LC network. Thus, we may realize the general function $Z_{LC}(s)$ as a series connection of subnetworks, each realizing one of the terms of (2). The form of the resulting network is shown in Fig. 4.3-2. This configuration is called a *series Foster form*. Of course, if there is no pole at infinity ($k_\infty = 0$), the single inductor will be absent. Similarly, if there is no pole at the origin ($k_0 = 0$), the single capacitor will be missing. And so on. The various residues are found as follows[1]:

$$k_\infty = \lim_{s \to \infty} \frac{1}{s} Z_{LC}(s) \tag{3a}$$

$$k_0 = sZ_{LC}(s)\big|_{s=0} \tag{3b}$$

$$k_i = \frac{s^2 + \omega_i^2}{s} Z_{LC}(s)\Bigg|_{s^2 = -\omega_i^2} \tag{3c}$$

[1] The various quantities k_i are not all, strictly speaking, residues; however, it is convenient to refer to them as such.

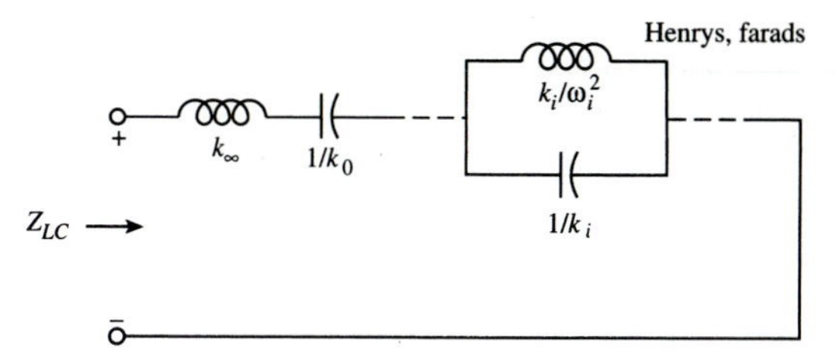

FIGURE 4.3-2
Series Foster form.

Example 4.3-1 The series Foster form. As an example of this synthesis procedure, consider the function

$$Z(s) = 4\frac{(s^2+1)(s^2+9)}{s(s^2+4)} = k_\infty s + \frac{k_0}{s} + \frac{k_1 s}{s^2+4}$$

The residues are found as follows:

$$k_\infty = \lim_{s\to\infty} \frac{1}{s} Z(s) = \lim_{s\to\infty} 4\frac{(s^2+1)(s^2+9)}{s^2(s^2+4)} = 4$$

$$k_0 = sZ(s)\big|_{s=0} = 4\frac{(s^2+1)(s^2+9)}{s^2+4}\bigg|_{s=0} = 9$$

$$k_1 = \frac{s^2+4}{s}Z(s)\bigg|_{s^2=-4} = 4\frac{(s^2+1)(s^2+9)}{s^2}\bigg|_{s^2=-4} = 15$$

Thus, the partial-fraction expansion has the form

$$Z(s) = 4s + \frac{9}{s} + \frac{15s}{s^2+4}$$

The network realization is shown in Fig. 4.3-3.

Exercise 4.3-1. Find a realization of the following LC driving-point impedances:

$$Z_a(s) = \frac{s^4+3s^2+1}{s(s^2+1)} \qquad Z_b(s) = 5\frac{s^4+3s^2+1}{s(s^2+1)}$$

Answers. The term $Z_a(s)$ has the form shown in Fig. 4.3-3 with all elements having unity value; $Z_b(s)$ has the form shown in Fig. 4.3-3 with inductors of 5 H and capacitors of $\frac{1}{5}$ F.

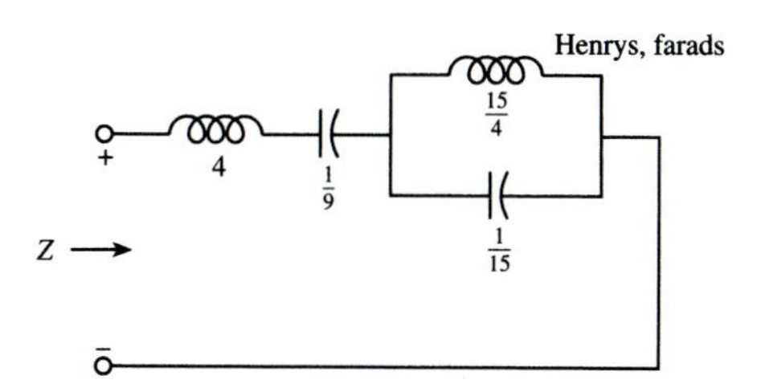

FIGURE 4.3-3
Network for Example 4.3-1.

Shunt Foster Form

A synthesis procedure similar to the one given above is obtained by starting with a driving-point admittance function $Y_{LC}(s)$. After combining the terms representing pairs of complex-conjugate poles, the partial-fraction expansion for such a function will have the form

$$Y_{LC}(s) = k_{\infty}s + \frac{k_0}{s} + \sum_i \frac{k_i s}{s^2 + \omega_i^2} \tag{4}$$

Each of the terms in this expansion is directly realizable as a capacitor, an inductor, or a series LC network. Thus, the general function $Y_{LC}(s)$ may be realized as a parallel connection of subnetworks, each of which realizes one of the terms of (4). The form of the resulting network is shown in Fig. 4.3-4. It is called a *shunt* or *parallel Foster form*. The expressions for finding the various residues are identical to those given in (3) [with $Y_{LC}(s)$ substituted for $Z_{LC}(s)$].

Example 4.3-2 The shunt Foster form. As an example of this synthesis procedure, consider again the function given in Example 4.3-1. Inverting the function, we obtain

$$Y(s) = \frac{1}{Z(s)} = \frac{s(s^2+4)}{4(s^2+1)(s^2+9)} = \frac{k_1 s}{s^2+1} + \frac{k_2 s}{s^2+9}$$

The residues are found as follows:

$$k_1 = \frac{s^2+1}{s} Y(s)\bigg|_{s^2=-1} = \frac{s^2+4}{4(s^2+9)}\bigg|_{s^2=-1} = \frac{3}{32}$$

$$k_2 = \frac{s^2+9}{s} Y(s)\bigg|_{s^2=-9} = \frac{s^2+4}{4(s^2+1)}\bigg|_{s^2=-9} = \frac{5}{32}$$

Thus, the partial-fraction expansion is

$$Y(s) = \frac{3s/32}{s^2+1} + \frac{5s/32}{s^2+9}$$

The resulting network realization is shown in Fig. 4.3-5.

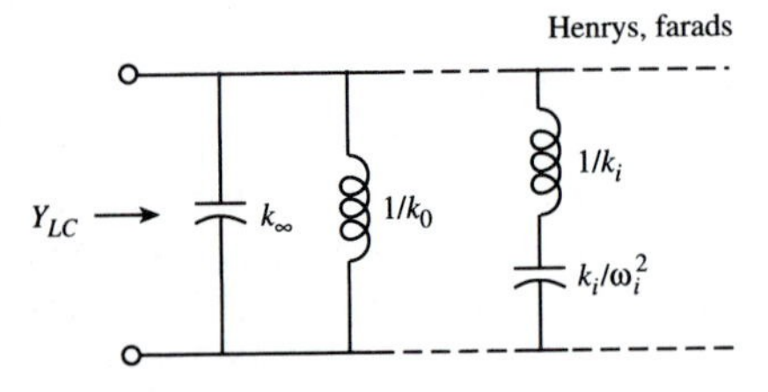

FIGURE 4.3-4
Shunt Foster form.

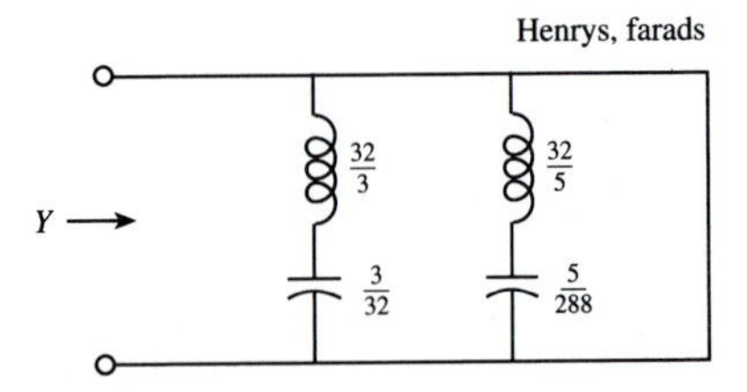

FIGURE 4.3-5
Network for Example 4.3-2.

Exercise 4.3-2. For each of the functions given below find the (*LC*-realizable) partial-fraction expansion. Assume that the functions are driving-point admittances and find a parallel Foster form realization.

Functions	Answers (see Fig. 4.3-4)
$F_1(s) = \dfrac{s^4 + 4s^2 + 3}{s^3 + 2s}$	$F_1(s) = s + \dfrac{1.5}{s} + \dfrac{0.5s}{s^2 + 2}$
	Shunt capacitor 1 F, shunt inductor $\frac{2}{3}$ H, series *LC* circuit 2 H and $\frac{1}{4}$ F
$F_2(s) = \dfrac{s^3 + 2s}{s^4 + 4s^2 + 3}$	$F_2(s) = \dfrac{0.5s}{s^2 + 1} + \dfrac{0.5s}{s^2 + 3}$
	First series *LC* circuit 2 H and $\frac{1}{2}$ F, second series *LC* circuit 2 H and 0.16667 F
$F_3(s) = \dfrac{s^4 + 4s^2 + 3}{s^5 + 6s^3 + 8s}$	$F_3(s) = \dfrac{0.375}{s} + \dfrac{0.25s}{s^2 + 2} + \dfrac{0.375s}{s^2 + 4}$
	Shunt inductor $\frac{8}{3}$ H, second series *LC* circuit 4 H and $\frac{1}{8}$ F, second series *LC* circuit $\frac{8}{3}$ H and 0.09375 F
$F_4(s) = \dfrac{s^5 + 6s^3 + 8s}{s^4 + 4s^2 + 3}$	$F_4(s) = s + \dfrac{1.5s}{s^2 + 1} + \dfrac{0.5s}{s^2 + 3}$
	Shunt capacitor 1 F, first series *LC* circuit $\frac{2}{3}$ H and $\frac{3}{2}$ F, second series *LC* circuit 2 H and 0.16667 F

Considering both of the synthesis procedures described above, we conclude that for a rational function to be realizable as the driving-point immittance of an *LC* network, it is sufficient that the poles be simple, be on the $j\omega$ axis (including the origin and infinity), and have positive real residues. Thus, properties 1 and 6 of Summary 4.2-1 are necessary and sufficient. Equivalently, if (2) is rationalized, it is sufficient to require that the poles and zeros be simple and alternate on the $j\omega$ axis and that there be a pole or a zero at the origin and infinity.[2] Thus, properties 1 through 5 of Summary 4.2-1 are also necessary and sufficient.

Removal-of-Poles-at-Infinity Cauer Form

The Foster form realizations based on partial-fraction expansions presented in the preceding paragraphs are only one of several types of synthesis methods that may be used to realize *LC* driving-point functions. Such realizations are called *canonic* since they use the minimum possible number of elements. A second canonic realization method is based on the use of *continued fractions*. The resulting networks are called *Cauer forms*. The continued fractions are generated by the removal of poles from functions that are *LC* realizable. The basis of the technique follows from considering a rational function $F(s)$ that is *LC* realizable and thus may have poles at infinity, the origin and complex-conjugate poles on the $j\omega$ axis. Removing any of the poles of $F(s)$ leaves a function that is also *LC* realizable.

[2] In this case it is also necessary to require that the multiplicative constant for the function be positive.

Example 4.3-3 Removal of poles. Consider the LC-realizable function

$$F(s) = s + \frac{2}{s} + \frac{3s}{s^2+1} = \frac{(s^2+0.3542)(s^2+5.6458)}{s(s^2+1)}$$

If we define $F_a(s)$, $F_b(s)$, and $F_c(s)$ as the functions produced by removing the pole at infinity, the origin, and the pole pair at $\pm j1$, respectively, we obtain

$$F_a(s) = F(s) - s = \frac{2}{s} + \frac{3s}{s^2+1} = \frac{5(s^2+0.4)}{s(s^2+1)}$$

$$F_b(s) = F(s) - \frac{2}{s} = s + \frac{3s}{s^2+1} = \frac{s(s^2+4)}{s^2+1}$$

$$F_c(s) = F(s) - \frac{3s}{s^2+1} = s + \frac{2}{s} = \frac{s^2+2}{s}$$

Note that the functions $F_a(s)$, $F_b(s)$, and $F_c(s)$ are all LC realizable.

To see how this approach is used, consider an LC driving-point impedance function $Z_{LC}(s)$ that has a pole at infinity. Its partial-fraction expansion will have the form

$$Z_{LC}(s) = k_\infty^{(Z_{LC})} s + \frac{k_0}{s} + \sum_i \frac{k_i s}{s^2+\omega_i^2} \tag{5}$$

Let us now define a driving-point function $Z_1(s)$ as the original function $Z_{LC}(s)$ from which we have removed the pole at infinity. Thus,

$$Z_1(s) = Z_{LC}(s) - k_\infty^{(Z_{LC})} s = \frac{k_0}{s} + \sum_i \frac{k_i s}{s^2+\omega_i^2} \tag{6}$$

where $k_\infty^{(Z_{LC})}$ is the residue of $Z_{LC}(s)$ at the pole at $s = \infty$. The term $Z_1(s)$ satisfies properties 1 and 6 of Summary 4.2-1. Therefore it is also realizable as an LC driving-point impedance function. Such functions must have either a pole or a zero at infinity, and since $Z_1(s)$ obviously has no pole there, it must have a zero there. Therefore, its reciprocal $Y_1(s) = 1/Z_1(s)$ must have a pole at infinity. Thus, we may write $Y_1(s)$ in the form

$$Y_1(s) = k_\infty^{(Y_1)} s + Y_2(s) \tag{7}$$

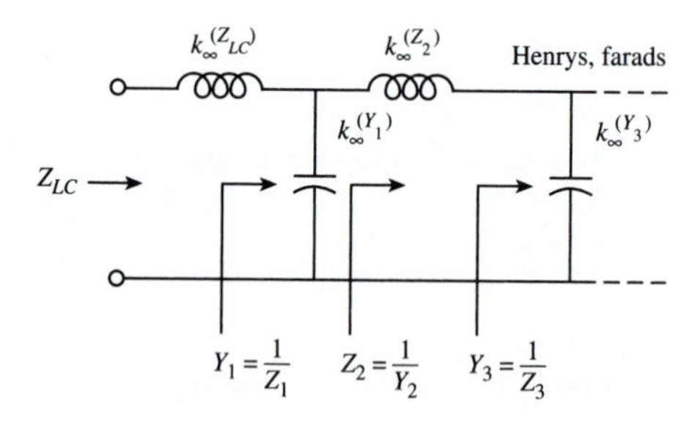

FIGURE 4.3-6
Removal-of-poles-at-infinity Cauer form.

where $Y_2(s)$ is an LC-realizable driving-point admittance with a zero at infinity. Thus, $Z_2(s) = 1/Y_2(s)$ will have a pole at infinity and may be written

$$Z_2(s) = k_\infty^{(Z_2)} s + Z_3(s) \tag{8}$$

The process may be continued until the entire function $Z_{LC}(s)$ has been realized. The form of the resulting network is shown in Fig. 4.3-6. Note that if the original function $Z_{LC}(s)$ does not have a pole at infinity, the first inductor shown in the figure will be missing, i.e, the removal of poles at infinity will start with the admittance function. The procedure generates a continued fraction having the form

$$Z_{LC}(s) = k_\infty^{(Z_{LC})} s + \cfrac{1}{k_\infty^{(Y_1)} s + \cfrac{1}{k_\infty^{(Z_2)} s + \cfrac{1}{k_\infty^{(Y_3)} s + \cdots}}} \tag{9}$$

Example 4.3-4 The removal-of-poles-at-infinity Cauer form. As an example of the procedure, consider the function used in Examples 4.3-1 and 4.3-2:

$$Z(s) = 4\frac{(s^2+1)(s^2+9)}{s(s^2+4)} = \frac{4s^4 + 40s^2 + 36}{s^3 + 4s}$$
$$= 4s + \frac{24s^2 + 36}{s^3 + 4s} = 4s + Z_1(s)$$

where the last form was obtained by dividing the denominator of the original function into the numerator, which gives the remainder $24s^2 + 36$. The next step is to divide $s^3 + 4s$ by $24s^2 + 36$. Thus we obtain

$$Y_1(s) = \frac{1}{Z_1(s)} = \frac{s^3 + 4s}{24s^2 + 36} = \frac{s}{24} + \frac{5s/2}{24s^2 + 36} = \frac{s}{24} + Y_2(s)$$

The final step is

$$Z_2(s) = \frac{1}{Y_2(s)} = \frac{24s^2 + 36}{5s/2} = \frac{48s}{5} + \frac{36}{5s/2} = \frac{48s}{5} + Z_3(s)$$

The network realization is shown in Fig. 4.3-7. The actual division steps are readily carried out using a tabular form as follows:

$$\begin{array}{l}
s^3 + 4s \,\rfloor\, 4s^4 + 40s^2 + 36 \,\lfloor\, 4s \\
\qquad\qquad 4s^4 + 16s^2 \\
\qquad\qquad\qquad 24s^2 + 36 \,\rfloor\, s^3 + 4s \,\lfloor\, s/24 \\
\qquad\qquad\qquad\qquad\qquad\quad s^3 + 3s/2 \\
\qquad\qquad\qquad\qquad\qquad\quad 5s/2 \,\rfloor\, 24s^2 + 36 \,\lfloor\, 48s/5 \\
\qquad\qquad\qquad\qquad\qquad\qquad\qquad 24s^2 \\
\qquad\qquad\qquad\qquad\qquad\qquad\qquad 36 \,\rfloor\, 5s/2 \,\lfloor\, 5s/72 \\
\qquad\qquad\qquad\qquad\qquad\qquad\qquad\qquad 5s/2 \\
\qquad\qquad\qquad\qquad\qquad\qquad\qquad\qquad 0
\end{array}$$

Divisor becomes dividend

Remainder becomes dividend

Remainder becomes dividend

Exercise 4.3-4. Find a realization of the following LC driving-point impedances:

$$Z_a(s) = \frac{s^4 + 3s^2 + 1}{s(s^2 + 2)} \qquad Z_b(s) = 3\frac{s^4 + 3s^2 + 1}{s(s^2 + 2)}$$

Answer. The term $Z_a(s)$ will have the form shown in Fig. 4.3-7, but all the elements will have unity value; $Z_b(s)$ will have the form shown in Fig. 4.3-7, but the inductors will have value 3 H and the capacitors will have value $\frac{1}{3}$ F.

Removal-of-Poles-at-the-Origin Cauer Form

The Cauer form network realization developed above is based on the removal of poles at infinity from LC driving-point immittances. Another Cauer form may be defined based on the removal of poles *at the origin*. In this case the continued fraction will have the form

$$Z_{LC}(s) = a/s + \cfrac{1}{b/s + \cfrac{1}{c/s + \cdots}} \tag{10}$$

where $a, b, c, \ldots$ are positive real constants. The resulting network configuration will have the form shown in Fig. 4.3-8. If the function $Z_{LC}(s)$ does not have a pole at the origin, then the first capacitor will not be present, i.e., the synthesis process will start with the admittance function. A tabular format may again be used for ease of computation of the element values. In this case, the polynomials should first be written in ascending order.

Example 4.3-5 The removal-of-poles-at-the-origin Cauer form. For the network function given in Examples 4.3-1, 4.3-2, and 4.3-4, the first two steps are

$$Z(s) = \frac{36 + 40s^2 + 4s^4}{4s + s^3} = \frac{9}{s} + \frac{31s^2 + 4s^4}{4s + s^3} = \frac{9}{s} + Z_1(s)$$

$$Y_1(s) = \frac{1}{Z_1(s)} = \frac{4s + s^3}{31s^2 + 4s^4} = \frac{4}{31s} + \frac{15s^3/31}{31s^2 + 4s^4} = \frac{4}{31s} + Y_2(s)$$

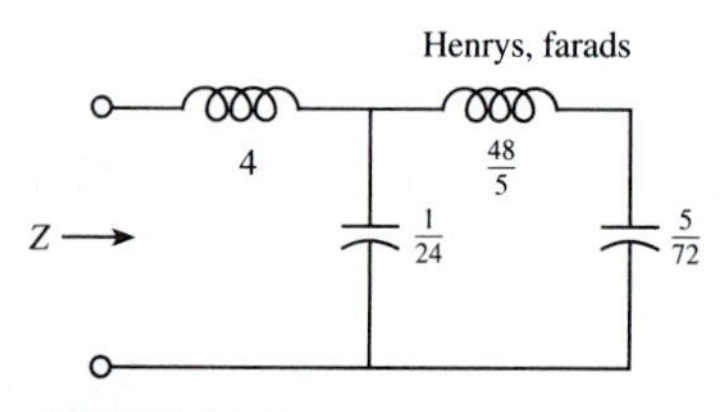

FIGURE 4.3-7
Network for Example 4.3-4.

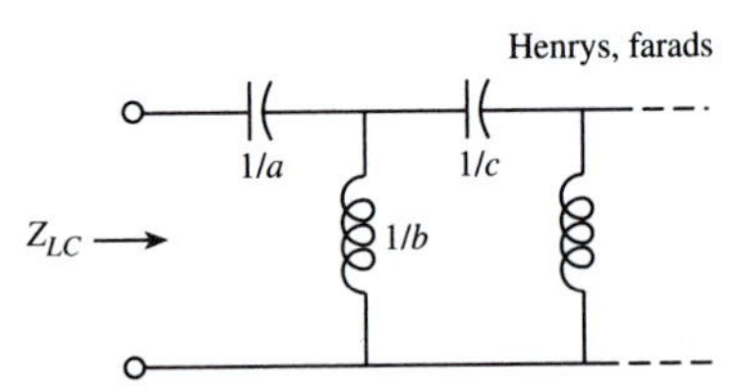

FIGURE 4.3-8
Removal-of-poles-at-the-origin Cauer form.

These and the remaining steps are shown in the following tabular division:

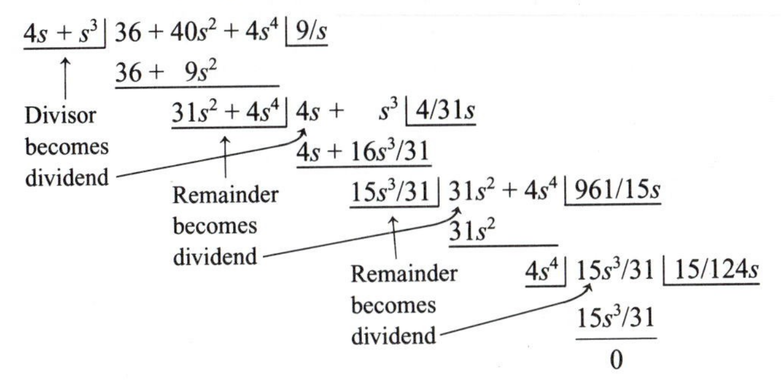

The network realization is shown in Fig. 4.3-9.

Exercise 4.3-5. Find a realization of the following *LC* driving-point impedances:

$$Z_a(s) = \frac{s^4 + 3s^2 + 1}{2s^3 + s} \qquad Z_b(s) = 4\,\frac{s^4 + 3s^2 + 1}{2s^3 + s}$$

Answer. The term $Z_a(s)$ will have the form shown in Fig. 4.3-9, but all the elements will have unity value; $Z_b(s)$ will have the form shown in Fig. 4.3-9, but the inductors will have a value of 4 H and the capacitors will have the value $\frac{1}{4}$ F.

From the developments given in this section, it should be apparent that synthesis is not a unique process. As an example of this fact, we have here presented four different network configurations each of which realizes the same network function. Thus, from any measurement of voltage or current that we may make at the input terminal pairs, these circuits are indistinguishable. Other characteristics of these circuits, however, may be quite different. For example, the circuit shown in Fig. 4.3-3 requires only 7.75 total henrys of inductance, while the one shown in Fig. 4.3-5 requires over 17 total henrys. Obviously, if the cost of inductors is a major factor, the first network is preferable. As another example, the designer may be constrained to only using capacitors that have one terminal grounded. In this case, only the networks of Figs. 4.3-5 and 4.3-7 are satisfactory. Many other examples of constraints on the design procedure may be cited. The important point here is that, in actual design practice, it is not enough to merely *realize* a given network function. The designer must also look for the *best possible such realization*, either by evaluating many different realizations or by looking for ways of modifying the configuration selected.

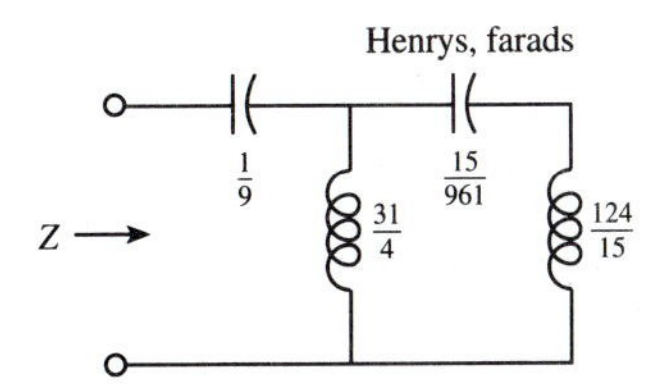

FIGURE 4.3-9
Network for Example 4.3-5.

4.4 POLES OF TRANSFER FUNCTIONS

In the preceding sections we discussed the properties of passive *LC* networks and developed necessary and sufficient conditions for their use in the synthesis of *driving-point* functions. In this chapter we continue our study of passive networks by considering their use in the realization of *transfer* functions. Such functions represent the most commonly encountered filtering situation, in which the frequency content of some input signal is to be altered (or filtered) by being passed (or transferred) through a network in order to produce an output signal with a modified frequency content.

Relation to Poles of Driving-Point Functions

Let us consider how the poles of driving-point and transfer functions are related. For a transfer function to be defined for a given network, the network must have both an input and an output port (or terminal pair) at which voltage and current variables are specified. We will use the usual two-port network convention in which the variables are designated as $V_1(s)$ and $I_1(s)$ at the input port and $V_2(s)$ and $I_2(s)$ at the output port. Thus, the general network configuration for defining transfer functions is as shown in Fig. 4.4-1. In most practical situations, there is a common ground for both of the terminal pairs. In such a case the configuration is also referred to as a *three-terminal network.*[1] To relate the various voltage and current variables defined in the figure, a set of four network functions, called *network parameters*, are required. The most useful parameters in synthesis applications are the z and y parameters defined by the relations[2]

$$\mathbf{V}(s) = \begin{bmatrix} V_1(s) \\ V_2(s) \end{bmatrix} = \begin{bmatrix} z_{11}(s) & z_{12}(s) \\ z_{21}(s) & z_{22}(s) \end{bmatrix} \begin{bmatrix} I_1(s) \\ I_2(s) \end{bmatrix} = \mathbf{Z}(s)\mathbf{I}(s) \tag{1}$$

$$\mathbf{I}(s) = \begin{bmatrix} I_1(s) \\ I_2(s) \end{bmatrix} = \begin{bmatrix} y_{11}(s) & y_{12}(s) \\ y_{21}(s) & y_{22}(s) \end{bmatrix} \begin{bmatrix} V_1(s) \\ V_2(s) \end{bmatrix} = \mathbf{Y}(s)\mathbf{V}(s) \tag{2}$$

where, for passive networks, the square matrices $\mathbf{Z}(s)$ and $\mathbf{Y}(s)$ are symmetric, i.e., $z_{12}(s) = z_{21}(s)$ and $y_{12}(s) = y_{21}(s)$.

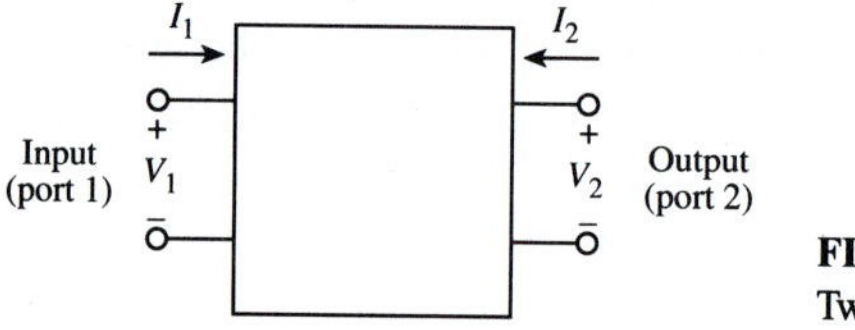

FIGURE 4.4-1
Two-port network.

[1] Such a network is also referred to in the literature as an *unbalanced* network.

[2] Detailed descriptions of these parameters and methods for their determination may be found in most of the standard books on circuit theory. For example, see L. P. Huelsman, *Basic Circuit Theory*, 3rd ed., Prentice-Hall, Inc., Englewood Cliffs, N.J., 1991.

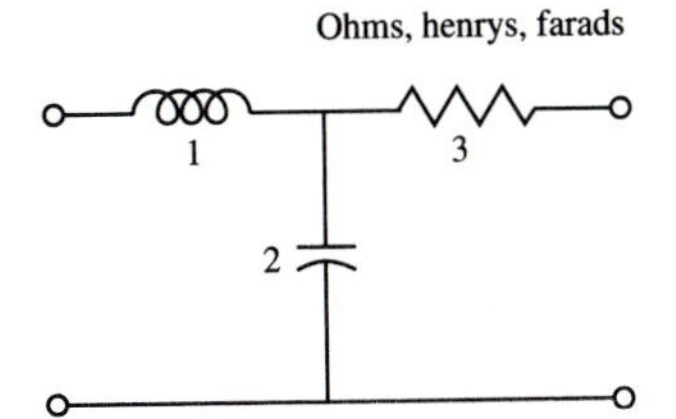

FIGURE 4.4-2
Network for z parameters in Example 4.4-1.

The parameters $z_{11}(s)$, $z_{22}(s)$, $y_{11}(s)$, and $y_{22}(s)$ are *driving-point functions*. The remaining parameters are *transfer functions* since they relate an excitation at one port to a response at the other port.

Example 4.4-1 z and y parameters. As an example of driving-point and transfer functions in a two-port network, consider the circuit shown in Fig. 4.4-2. The z-parameter matrix is

$$\mathbf{Z}(s) = \begin{bmatrix} s + 1/2s & 1/2s \\ 1/2s & 3 + 1/2s \end{bmatrix} \tag{3}$$

As a second example, the network shown in Fig. 4.4-3 has the y-parameter matrix (note that the value of the resistor is given in siemens rather than ohms)

$$\mathbf{Y}(s) = \begin{bmatrix} 2s + 1/s & -2s \\ -2s & 2s + 3 \end{bmatrix} \tag{4}$$

Now let us consider some of the properties of the network parameters defined above. For passive networks comprised of lumped elements, each of the parameters will be a *rational function*, i.e., a ratio of polynomials in the complex-frequency variable s. In the following section of this chapter we shall see that for transfer functions the zeros[3] may be located anywhere in the complex-frequency plane. We shall also find, however, that the topological form of the network places some additional restrictions on the locations of these zeros. The poles of transfer functions are the natural frequencies of the network, just as are the poles of the driving-point functions. The locations of these is determined by the type of elements in the network. For LC elements (see Sec. 4.2) the poles are simple and on the $j\omega$ axis. For RC and RL networks the poles are simple and on the negative real axis. For RLC networks the poles may be anywhere in the left-half plane and need not be simple.

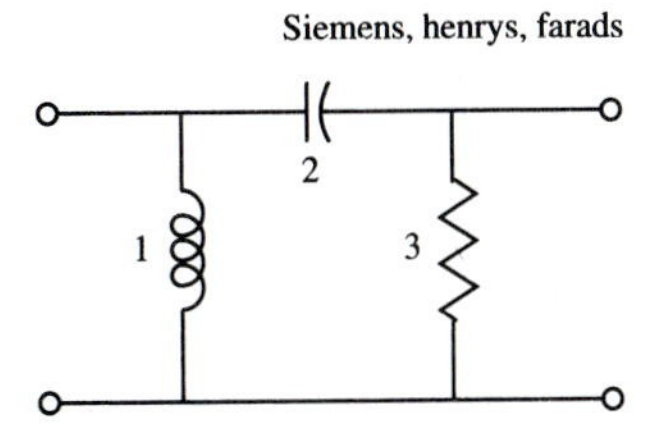

FIGURE 4.4-3
Network for y parameters in Example 4.4-1.

[3] The zeros of transfer functions are also referred to as *transmission zeros*.

Private Poles

In general, the transfer parameters for a given network will have the same poles as the driving-point parameters. There are, however, some important exceptions. The first occurs in the situation shown in Fig. 4.4-4 in which a current source provides the excitation and the first element of the network is a series-connected impedance $Z_1(s)$. Obviously, the current $I_0(s)$ is the same as the current $I_1(s)$. Thus, although the element $Z_1(s)$ will affect the driving-point impedance, $z_{11}(s)$, it will not affect the transfer impedance $z_{21}(s)$. A similar situation occurs at the output port of a network. Assume, as shown in Fig. 4.4-4, that the last element in the network is a series-connected impedance $Z_2(s)$. For any transfer function in which the output is open circuited, the current in this impedance is zero. Thus, although it will affect the driving-point impedance $z_{22}(s)$, it has no effect on the transfer impedance $z_{21}(s)$. As a result, *the driving-point parameters may have poles not present in the transfer parameters.* Such poles are referred to as *private poles.* As an example of this, consider the z parameters given in (3) for the network shown in Fig. 4.4-2. The parameter $z_{11}(s)$ has poles at infinity and at the origin, while the transfer parameters only have poles at the origin. Thus the pole at infinity is a private pole of $z_{11}(s)$.

The second situation in which private poles appear is when a voltage source is used to provide the excitation and the first element of the network is a parallel-connected admittance $Y_1(s)$, as shown in Fig. 4.4-5. In this case, since the voltage $V_0(s)$ is the same as the voltage $V_1(s)$, the element $Y_1(s)$ has no effect on the transfer admittance $y_{21}(s)$. A similar situation occurs at the output port of a network. Assume, as shown in Fig. 4.4-5, that the last element is a shunt-connected admittance $Y_2(s)$. For any transfer function in which the output is short circuited, the voltage across this admittance will be zero. Thus, although it will affect the driving-point admittance $y_{22}(s)$, it will have no effect on the transfer admittance $y_{21}(s)$. As an example of this, consider the y parameters given in (4) for the network shown in Fig. 4.4-3. The parameter $y_{11}(s)$ has a private pole at the origin that is not present in the transfer parameters.

Residue Condition for $j\omega$-Axis Poles

At this point we might ask whether it is possible for the converse of the situation discussed above to occur, i.e., for a transfer immittance such as $z_{12}(s)$ to have a pole not

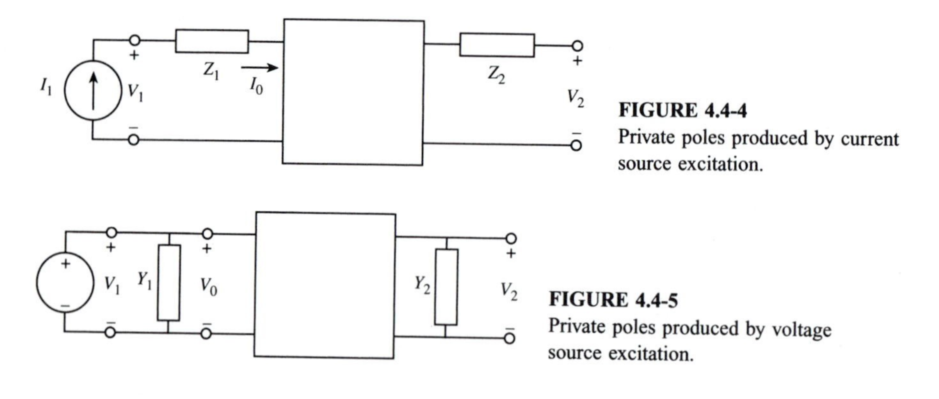

FIGURE 4.4-4
Private poles produced by current source excitation.

FIGURE 4.4-5
Private poles produced by voltage source excitation.

shared by $z_{11}(s)$ and $z_{22}(s)$. To answer this question, let us start by assuming that the network has natural frequencies on the $j\omega$ axis at $s = \pm j\omega_i$. We also assume that all the functions $z_{ij}(s)$ have poles at these locations. At values of s in the vicinity of the pole at $s = j\omega_i$

$$\lim_{s \to j\omega_i} z_{ij}(s) = \frac{k_{ij}}{s - j\omega_i} \tag{5}$$

where k_{ij} is the residue of $z_{ij}(s)$ at the pole at $s = j\omega_i$. It may be shown that the residues k_{ij} must satisfy the relation (we assume that $k_{12} = k_{21}$ as is always the case for passive networks)

$$k_{11}k_{22} - k_{12}^2 \geq 0 \tag{6}$$

This is called the *residue condition*. It provides the answer to our original question since if k_{12} is nonzero, satisfaction of the inequality of (6) requires that k_{11} and k_{22} also be nonzero. We conclude that *if $z_{12}(s)$ has a $j\omega$-axis pole, the same pole must be present in both $z_{11}(s)$ and $z_{22}(s)$*. An analysis similar to that given above may be made to show that an identical restriction exists for the y parameters, i.e., $j\omega$-axis poles of these parameters must also satisfy a residue condition similar to (6). It should be noted that since poles at the origin and at infinity are on the $j\omega$ axis, the residue condition must be satisfied for these poles in both z and y parameters.

Example 4.4-2 Residue condition. As examples of the residue condition, consider the networks shown in Fig. 4.4-6. All the elements are assumed to have unity value. For the network shown in Fig. 4.4-6(a) we find

$$z_{11}(s) = \frac{1}{s} + \frac{s}{s^2+1} \qquad z_{12}(s) = \frac{s}{s^2+1} \qquad z_{22}(s) = 1 + \frac{s}{s^2+1}$$

For the pole at $s = j1$ the residues are $k_{11} = k_{12} = k_{22} = \frac{1}{2}$ and the residue condition is satisfied

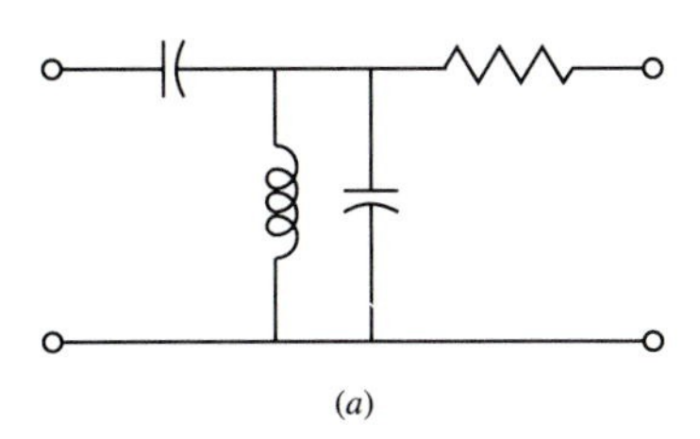

(a)

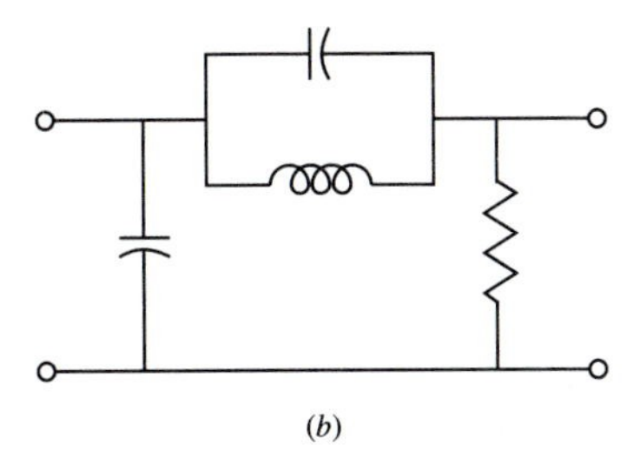

(b)

FIGURE 4.4-6
Examples of the residue condition.

with an equality. For the network shown in Fig. 4.4-6(*b*) we find

$$y_{11}(s) = 2s + \frac{1}{s} \qquad y_{12}(s) = -s - \frac{1}{s} \qquad y_{22}(s) = s + 1 + \frac{1}{s}$$

For the pole at $s = \infty$ the residues are $k_{11} = 2$, $k_{12} = -1$, and $k_{22} = 1$ and the residue condition is satisfied with an inequality.

Poles in $z_{12}(s)$ Not Present in $z_{11}(s)$ and $z_{22}(s)$

Now let us continue our examination of the same problem, i.e., whether $z_{12}(s)$ can have a pole that is not present in $z_{11}(s)$ or $z_{22}(s)$. This time, however, we will consider non-$j\omega$-axis poles. In this case we find the opposite conclusion, i.e., neither $z_{11}(s)$ nor $z_{22}(s)$ is required to have a pole present in $z_{12}(s)$. An example of such a situation is shown in Fig. 4.4-7 for which the z parameters are $z_{11}(s) = z_{22}(s) = 1$, $z_{12}(s) = s/(s+1)$. Obviously, $z_{12}(s)$ has a pole at $s = -1$ that is not present in $z_{11}(s)$ or $z_{22}(s)$. In this case the impedance of the series *RC* branch and the shunt *RL* branch are said to be *complementary impedances* since their sum is a constant.

From the discussions given above we conclude that, in general, the driving-point and transfer functions of a given network have the same poles. The exceptions are summarized as follows.

Summary 4.4-1 Cases relating the poles of driving-point and transfer parameters. For a given network, the poles of driving-point functions and transfer functions will be the same except for the following restrictions:

1. If the first network element connected to a port is a series one, it can produce private poles in the driving-point function that are not present in the transfer function. This will occur when the port is excited by a current source or when it is open circuited.
2. If the first network element connected to a port is a shunt one, it can produce private poles in the driving-point function that are not present in the transfer function. This will occur when the port is excited by a voltage source or when it is short circuited.
3. If a transfer parameter has a pole on the $j\omega$ axis (including the origin and infinity), both driving-point parameters will also have that pole (the residue condition).
4. If a transfer parameter has a non-$j\omega$-axis pole, this pole need not be present in the driving-point functions.

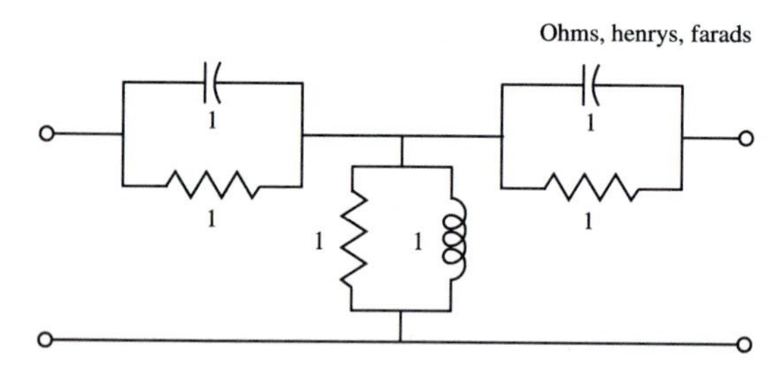

FIGURE 4.4-7
Network in which $z_{12}(s)$ has a pole not present in $z_{11}(s)$ or $z_{22}(s)$.

4.5 OTHER PROPERTIES OF TRANSFER FUNCTIONS

In the preceding section we used z and y parameters to determine some of the properties of transfer functions. Specifically, we showed how the poles of driving-point and transfer functions are related. In this section we continue our study of the properties of transfer functions.

Fialkow Condition

To begin our treatment, let us first consider a network consisting of a *wye configuration* (also called a T configuration) of passive admittances $Y_a(s)$, $Y_b(s)$, and $Y_c(s)$ as shown in Fig. 4.5-1(a). If we apply a *wye-delta transformation* to this network, we convert it to a *delta configuration* (also called a π configuration) of admittances $Y_1(s)$, $Y_2(s)$, and $Y_3(s)$, as shown in Fig. 4.5-1(b), and we obtain

$$Y_1(s) = \frac{Y_b(s)Y_c(s)}{Y_d(s)} \tag{1a}$$

$$Y_2(s) = \frac{Y_a(s)Y_c(s)}{Y_d(s)} \tag{1b}$$

$$Y_3(s) = \frac{Y_a(s)Y_b(s)}{Y_d(s)} \tag{1c}$$

where
$$Y_d(s) = Y_a(s) + Y_b(s) + Y_c(s) \tag{1d}$$

Since the admittances of the wye are passive, all the coefficients of their numerator and denominator polynomials will be nonnegative. The operations in (1) used in finding $Y_1(s)$, $Y_2(s)$, and $Y_3(s)$, however, involve only multiplications and additions. Thus, the coefficients of these admittances must also be nonnegative (although the admittances will not necessarily be realizable).

Now let us consider the inverse of the transformation described above, namely, a *delta-wye transformation*. In this case, using the admittances defined in Fig. 4.5-1, we find

$$Y_a(s) = \frac{Y_0(s)}{Y_1(s)} \qquad Y_b(s) = \frac{Y_0(s)}{Y_2(s)} \qquad Y_c(s) = \frac{Y_0(s)}{Y_3(s)} \tag{2a}$$

where
$$Y_0(s) = Y_1(s)Y_2(s) + Y_1(s)Y_3(s) + Y_2(s)Y_3(s) \tag{2b}$$

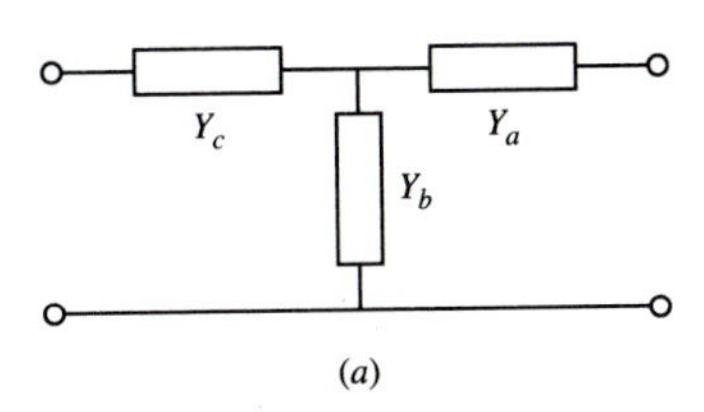

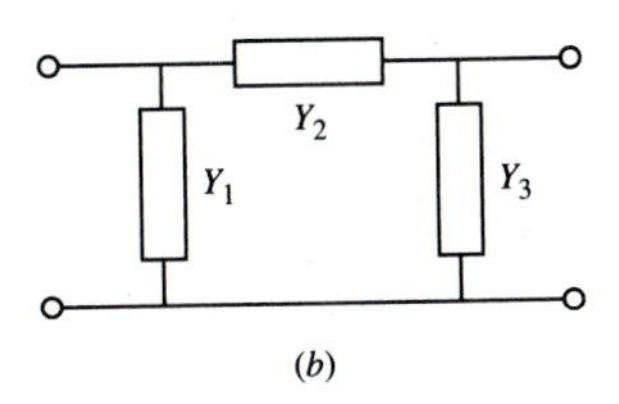

FIGURE 4.5-1
Simple network configurations: (a) wye; (b) delta.

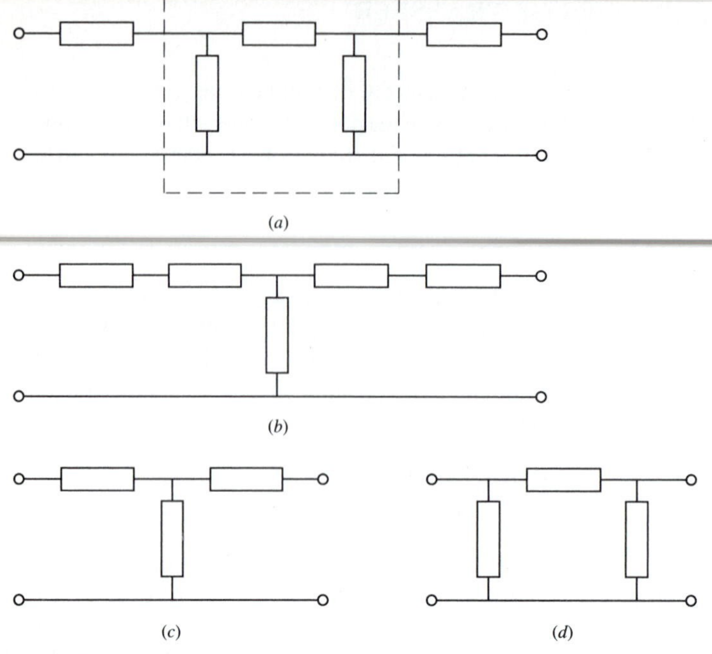

FIGURE 4.5-2
Reduction of a ladder network.

Again we see that the equations used in the transformation must retain the positive-coefficient property for the admittances $Y_a(s)$, $Y_b(s)$, and $Y_c(s)$.

The two transformations described above can be used to reduce any *ladder network* (a network consisting of series and shunt branches) to a simple delta network in which the admittances have only nonnegative coefficients.[1] For example, consider the ladder network shown in Fig. 4.5-2(*a*). If we apply the delta-wye transformation to the elements enclosed in the dashed lines, we obtain the result shown in Fig. 4.5-2(*b*). The series impedances can be combined to produce the network shown in Fig. 4.5-2(*c*). The result of applying the wye-delta transformation now gives the simple delta network shown in Fig. 4.5-2(*d*).

Now let us determine the y parameters of a delta network that results from the reduction process described above. Using the nomenclature of Fig. 4.5-1(*b*), we find that

$$y_{11}(s) = Y_1(s) + Y_2(s) \tag{3a}$$

$$y_{12}(s) = -Y_2(s) \tag{3b}$$

$$y_{22}(s) = Y_2(s) + Y_3(s) \tag{3c}$$

[1] Although our primary interest here is in ladder networks, the reduction technique can be applied to more general network configurations by using generalizations of the wye-delta and delta-wye transformations called *star-mesh* and *mesh-star* ones, respectively.

Solving these relations, we obtain

$$Y_1(s) = y_{11}(s) + y_{12}(s) \tag{4a}$$
$$Y_2(s) = -y_{12}(s) \tag{4b}$$
$$Y_3(s) = y_{22}(s) + y_{12}(s) \tag{4c}$$

We conclude that the numerator and denominator polynomials of the quantities $-y_{12}(s)$, $y_{11}(s) + y_{12}(s)$, and $y_{22}(s) + y_{12}(s)$ for an arbitrary passive network can have only nonnegative coefficients. Now assume that the y parameters have the form

$$-y_{12}(s) = \frac{a_0 + a_1 s + \cdots + a_n s^n}{D(s)} \tag{5a}$$

$$y_{11}(s) = \frac{b_0 + b_1 s + \cdots + b_n s^n}{D(s)} \tag{5b}$$

$$y_{22}(s) = \frac{c_0 + c_1 s + \cdots + c_n s^n}{D(s)} \tag{5c}$$

From (4*b*), the coefficients a_i of the numerator of $-y_{12}(s)$ must all be positive (or zero). From (5) we find that

$$y_{11}(s) + y_{12}(s) = \frac{(b_0 - a_0) + (b_1 - a_1)s + \cdots + (b_n - a_n)s^n}{D(s)} \tag{6a}$$

$$y_{22}(s) + y_{12}(s) = \frac{(c_0 - a_0) + (c_1 - a_1)s + \cdots + (c_n - a_n)s^n}{D(s)} \tag{6b}$$

Again applying the results of (4), we see that none of the quantities $(b_i - a_i)$ or $(c_i - a_i)$ can be negative. Thus we require that

$$b_i \geq a_i \qquad c_i \geq a_i \qquad i = 0, 1, 2, \ldots, n \tag{7}$$

We conclude that for a passive network the numerator coefficients of the transfer admittance $-y_{12}(s)$ can only be positive (or zero) and that they can be no greater than the corresponding numerator coefficients of $y_{11}(s)$ and $y_{22}(s)$, whichever are less. A similar proof shows that the numerator coefficients of $z_{12}(s)$ [and $z_{21}(s)$] can only be positive (or zero) and can be no greater than the corresponding numerator coefficients of $z_{11}(s)$ and $z_{22}(s)$, whichever is less. This result is called the *Fialkow condition*.[2] Since none of the coefficients can be negative, it follows from Descartes' rule of signs that zeros on the positive real axis are excluded.[3]

---

[2] For this result to hold, the denominators of all the functions $y_{ij}(s)$ [or $z_{ij}(s)$] must be the same, and any canceled common factors must be restored.

[3] For balanced networks the coefficients a_i of (5*a*) may be negative; thus the restrictions corresponding to (7) are $b_i \geq |a_i|$ and $c_i \geq |a_i|$. In this case zeros on the positive real axis are permitted.

Example 4.5-1 The Fialkow condition. As an example of the Fialkow condition, consider the network shown in Fig. 4.5-3. The z parameters are given by

$$\mathbf{Z}(s) = \begin{bmatrix} \frac{1}{s} + 2s + 4 & 2s + 4 \\ 2s + 4 & 2s + 9 \end{bmatrix} = \begin{bmatrix} \frac{2s^2 + 4s + 1}{s} & \frac{2s^2 + 4s}{s} \\ \frac{2s^2 + 4s}{s} & \frac{2s^2 + 9s}{s} \end{bmatrix}$$

where, in the right member of the equation, we have put all the z parameters over a common denominator. Note that the numerator coefficients of $z_{11}(s)$ and $z_{22}(s)$ are all greater than or equal to the corresponding coefficients of $z_{12}(s)$ [and $z_{21}(s)$].

Dimensionless Transfer Functions

In general, it is customary to consider input excitation as being applied at the left (number 1) port of a two-port network and the output taken at the right (number 2) port. This convention corresponds with the definition of the transfer parameters $z_{21}(s)$ and $y_{21}(s)$. For passive (reciprocal) networks, these are, of course, equal to $z_{12}(s)$ and $y_{12}(s)$. In addition there are two other basic transfer functions that are frequently encountered. The first is the dimensionless voltage transfer function relating an output voltage $V_2(s)$ to an input voltage $V_1(s)$. The output voltage is measured under the condition that no current flows from the output terminals, i.e., that these terminals are open circuited. Thus this transfer function may be called an *open-circuit voltage transfer function*. We will designate it as $V_{OC}(s)$. It may be directly related to the z and y parameters as follows:

$$V_{OC}(s) = \frac{z_{21}(s)}{z_{11}(s)} = \frac{-y_{21}(s)}{y_{22}(s)} \tag{8}$$

To determine where the poles of such a transfer function may occur, let us define

$$z_{11}(s) = \frac{N_{11}(s)}{P(s)D(s)} \qquad z_{21}(s) = \frac{N_{21}(s)}{D(s)} \tag{9}$$

where $P(s)$ is a polynomial containing the private poles (if any) of $z_{11}(s)$, $D(s)$ is a polynomial containing the other natural frequencies, and $N_{11}(s)$ and $N_{21}(s)$ are the numerator polynomials of $z_{11}(s)$ and $z_{21}(s)$, respectively. Inserting these relations in (8),

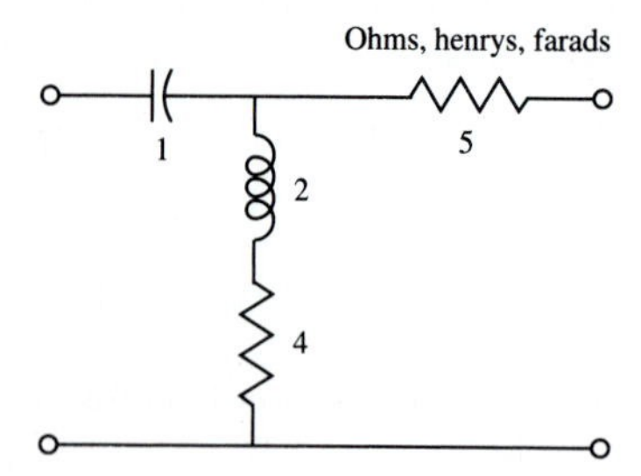

FIGURE 4.5-3
Example network for Fialkow condition.

we obtain[4]

$$V_{OC}(s) = \frac{P(s)N_{21}(s)}{N_{11}(s)} \tag{10}$$

Thus we see that *the poles of $V_{OC}(s)$ are the zeros of a driving-point parameter.* As such they have the properties developed for the driving-point functions of various classes of networks. The zeros of $V_{OC}(s)$ come from two places. First, they contain the zeros of a transfer parameter, which, as has been pointed out, may occur anywhere in the complex-frequency plane. In addition, the zeros of $V_{OC}(s)$ include the private poles of the driving-point parameter. Finally, the coefficients of the polynomials $P(s)N_{21}(s)$ and $N_{11}(s)$ are related by the Fialkow condition of (7).

A second dimensionless transfer function is the current transfer function. It is the ratio of an output current $I_2(s)$ to an input current $I_1(s)$. The output current is measured under the condition that no voltage is present at the output terminals, i.e., that these terminals are short circuited. Thus this transfer function is sometimes called a *short-circuit current transfer function.* We will designate it as $I_{SC}(s)$. It may be expressed as

$$I_{SC}(s) = \frac{-z_{21}(s)}{z_{22}(s)} = \frac{y_{21}(s)}{y_{11}(s)} \tag{11}$$

The conclusions developed for the location of the poles and zeros of $V_{OC}(s)$ and the restrictions imposed by the Fialkow condition also apply to $I_{SC}(s)$.

Real-Part Conditions

An important property of network functions concerns the behavior of their real part when they are evaluated on the $j\omega$ axis. For a driving-point function, a negative value of the real part at any frequency results in energy being supplied from the network to the driving source. This is not possible for a passive network. As a result, for any passive network a necessary condition is that $\operatorname{Re} Z(j\omega) \geq 0$ and $\operatorname{Re} Y(j\omega) \geq 0$ for any driving-point immittance. Energy conditions also place a restriction on the real part of the transfer immittances of a passive network. The z parameters of such networks must satisfy

$$\operatorname{Re} z_{11}(j\omega) \operatorname{Re} z_{22}(j\omega) - [\operatorname{Re} z_{21}(j\omega)]^2 \geq 0 \tag{12}$$

Similarly, for the y parameters

$$\operatorname{Re} y_{11}(j\omega) \operatorname{Re} y_{22}(j\omega) - [\operatorname{Re} y_{21}(j\omega)]^2 \geq 0 \tag{13}$$

These conditions are called the *real-part conditions.*

---

[4] Here, for simplicity, we ignore the possible poles of $z_{21}(s)$, which are not present in $z_{11}(s)$ due to complementary impedances.

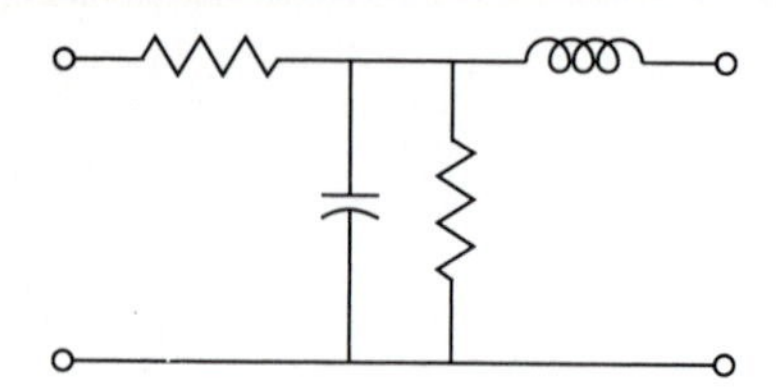

FIGURE 4.5-4
Example network for real-part condition.

Example 4.5-2 Real-part conditions. For the network shown in Fig. 4.5-4, assuming all the elements have unity value, we find

$$\operatorname{Re} z_{11}(j\omega) = \operatorname{Re}\left[\frac{s+2}{s+1}\right]_{s=j\omega} = \frac{2+\omega^2}{1+\omega^2}$$

$$\operatorname{Re} z_{21}(j\omega) = \operatorname{Re}\left[\frac{1}{s+1}\right]_{s=j\omega} = \frac{1}{1+\omega^2}$$

$$\operatorname{Re} z_{22}(j\omega) = \operatorname{Re}\left[s+\frac{1}{s+1}\right]_{s=j\omega} = \frac{1}{1+\omega^2}$$

When these values are used in (12), the inequality is readily verified.

As an application of these conditions, consider the possible behavior of $V_{OC}(s)$ at the origin, i.e., at $s = 0$. From (8), for it to have a pole, it is required that $z_{11}(s)$ be zero and $z_{21}(s)$ be nonzero, i.e., to have a pole or to be a constant. A pole of $z_{21}(s)$, however, would violate the residue condition of (6) of Sec. 4.4, while a constant would violate the real-part condition of (12). Thus we conclude that if $z_{11}(0)$ is zero, $z_{21}(0)$ must also be zero and thus that $V_{OC}(s)$ cannot have a pole at the origin. A similar argument shows that a pole at infinity is prohibited. These locations for poles are also not possible in the transfer function $I_{SC}(s)$.

Ladder Networks

In this and the preceding section we have introduced some general characteristics of various transfer functions. Some additional properties occur as a result of the topology, that is, the configuration of the elements in the circuit used to realize the transfer function. A configuration that is frequently used for realizations is the *ladder network*. It consists of alternating series and shunt branches as shown in Fig. 4.5-5(*a*). To illustrate the properties of this configuration, let us show only the first two branches explicitly and let the rest of the ladder have parameters $z'_{ij}(s)$, as shown in Fig. 4.5-5(*b*). For the overall network we find that

$$z_{21}(s) = \frac{V_2(s)}{I_1(s)} = \frac{z'_{21}(s)Z_1(s)}{z'_{11}(s) + Z_2(s) + Z_1(s)} \tag{14}$$

From an examination of this relation we conclude that the zeros of $z_{21}(s)$ can occur at any value of s where (1) $z'_{21}(s)$ is zero, (2) $Z_1(s)$ is zero [and $z'_{11}(s) + Z_2(s)$ is *not* zero], or (3) $Z_2(s)$ is infinite. It is easily shown that subsequent series and shunt elements in

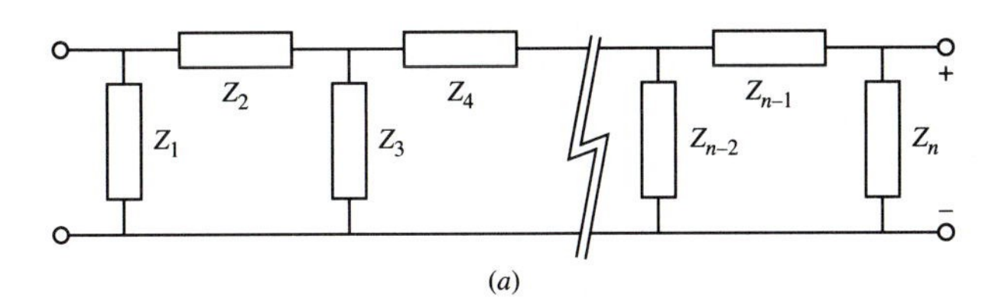

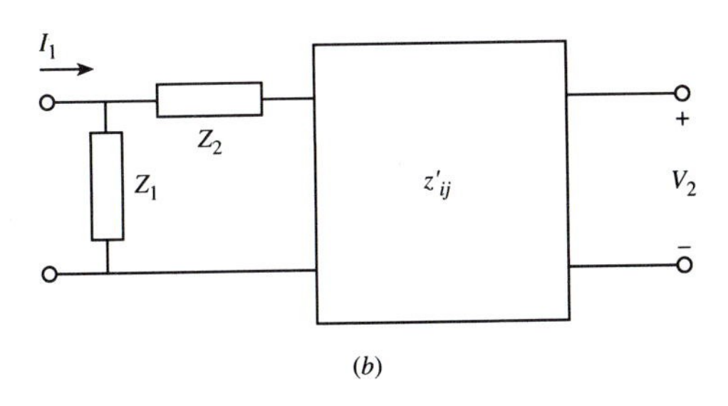

FIGURE 4.5-5
Ladder network.

the ladder will have similar effects to those of $Z_1(s)$ and $Z_2(s)$. Thus, in general, *the poles of series impedances and the zeros of shunt impedances (the poles of shunt admittances) produce transmission zeros in a ladder network.* Since such poles and zeros are those of driving-point functions, however, they are restricted to the left-half plane and the $j\omega$ axis. Thus *the transmission zeros of ladder networks can only lie in this region.* The transmission zeros can, however, be multiple.

Example 4.5-3 Transmission zeros of a ladder network. As an illustration of these conclusions, consider the network shown in Fig. 4.5-6. The impedance of the series-connected resistor and capacitor forming the shunt branch is $(s+2)/s$. Thus the transmission parameter $z_{21}(s)$ will have a zero at $s=-2$. This zero will be present in any transfer function for which the output is open circuited. Thus, it will also be present in $V_{OC}(s)$. The zero will *not* be present in any transfer function for which the output, and thus the branch, is short circuited such as $y_{21}(s)$ or $I_{SC}(s)$. The series branch consisting of the parallel-connected inductor and capacitor has an impedance $s/(s^2+1)$ and thus will produce transmission zeros at $s=\pm j1$ in any transfer function defined using a voltage excitation. This means that these zeros will be present in $y_{21}(s)$ and $V_{OC}(s)$. They will *not* be present, as shown in the discussion of private poles in Sec. 4.4, in any transfer function defined using a current excitation. Therefore, they will not be present in $z_{21}(s)$ or $I_{SC}(s)$. To further

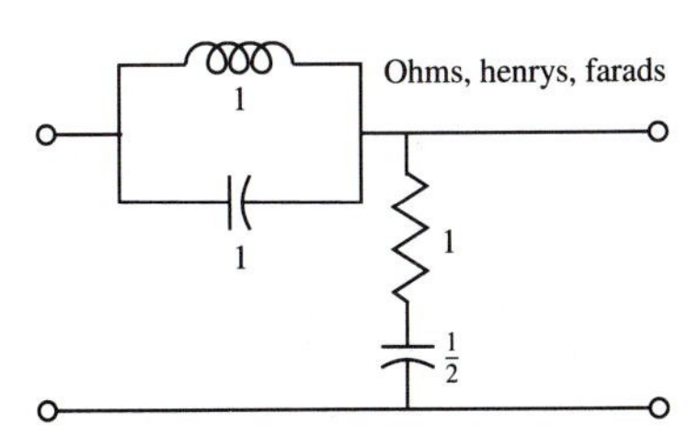

FIGURE 4.5-6
Ladder network for Example 4.5-3.

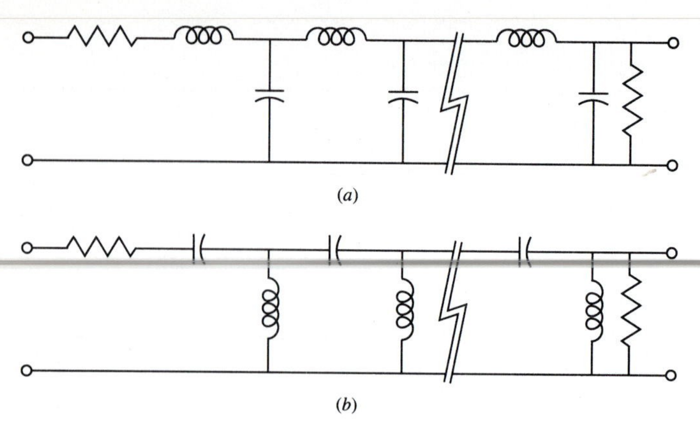

FIGURE 4.5-7
Common ladder networks: (*a*) low pass; (*b*) high pass.

illustrate these conclusions, the actual transfer functions for the circuit follow:

$$V_{OC}(s) = \frac{(s+2)(s^2+1)}{s^3+3s^2+s+2} \qquad y_{21}(s) = \frac{s^2+1}{s}$$

$$I_{SC}(s) = -1 \qquad z_{21}(s) = \frac{s+2}{s}$$

Another commonly encountered example of a ladder network is shown in Fig. 4.5-7(*a*). The poles of the series impedances and the zeros of the shunt impedances are all located at infinity. Thus this network has all its transmission zeros at infinity, i.e., it is a *low-pass* network. Similarly, the ladder network shown in Fig. 4.5-7(*b*) has all its transmission zeros at the origin, i.e., it is a *high-pass* network.

A summary of the conclusions regarding the pole and zero locations of network functions for various classes of networks is given in Table 4.5-1. In the following sections of this chapter we will present some methods for actually realizing transfer functions.

4.6 SYNTHESIS OF TRANSFER FUNCTIONS USING LOSSLESS LADDERS

In the preceding sections of this chapter we have discussed the properties of transfer functions. In this section we build on those properties to present a synthesis technique for their realization using a ladder network. The ladder itself will be restricted to the use of *LC* elements. Thus it is called a *lossless ladder*. As shown in Sec. 4.5, the transmission zeros for such a ladder must all be on the $j\omega$ axis (including the origin and infinity). For additional practicality, however, at the output pair of terminals of the ladder we shall connect a resistive termination that can act as a load. Thus, the natural frequencies of the entire network (the ladder plus the resistor) will be those of an *RLC* network, i.e., they may lie anywhere in the left-half plane.

TABLE 4.5-1
Locations of poles and zeros

	LC[1]		*RC* or *RL*[1]	
Network function	**Poles**	**Zeros**	**Poles**	**Zeros**
Driving point	Simple on $j\omega$ axis[2]	Simple on $j\omega$ axis[2]	Simple on negative real axis[3]	Simple on negative real axis[4]
Transfer functions of ladder networks	Simple on $j\omega$ axis[7]	Any order on $j\omega$ axis[2]	Simple on negative real axis[7]	Any order on negative real axis[2]
General transfer functions	Simple on $j\omega$ axis[7]	Any order on $j\omega$ axis[2]; quadrantal symmetry in right- and left-half planes	Simple on negative real axis[7]	Any order in right- or left-half plane[2,8]

	RLC[1]	
Network function	**Poles**	**Zeros**
Driving point	Any order in left-half plane[5]	Any order in left-half plane[6]
Transfer functions of ladder networks	Any order in left-half plane[5,7]	Any order in left-half plane or on $j\omega$ axis[2]
General transfer functions	Any order in left-half plane [5,7]	Any order in right- or left-half plane[2,8]

[1] Excluding mutual inductance.

[2] Including the origin and infinity.

[3] Including the origin for *RC* impedances and *RL* admittances and including infinity for *RC* admittances and *RL* impedances.

[4] Including infinity for *RC* impedances and *RL* admittances and including the origin for *RC* admittances and *RL* impedances.

[5] Any poles on the $j\omega$ axis must be simple.

[6] Any zeros on the $j\omega$ axis must be simple.

[7] Transfer immittances may have poles at the origin and infinity but dimensionless transfer functions may not.

[8] The positive real axis is excluded unless balanced networks are permitted.

Hurwitz Polynomials

As a preliminary step to describing the synthesis procedure, let us explore the properties of a polynomial whose roots all have negative real parts. Such a polynomial will have only positive coefficients. In addition, there will be no missing (or zero) coefficients. Such a polynomial is called a *Hurwitz polynomial*. As an example, the polynomial

$$P(s) = s^4 + 4s^3 + 7s^2 + 6s + 2 = (s^2 + 2s + 2)(s + 1)^2 \tag{1}$$

has simple roots at $s = -1 \pm j1$ and a second-order root at $s = -1$. Thus it is a Hurwitz polynomial.

Example 4.6-1 Hurwitz polynomials. Some examples of Hurwitz and non-Hurwitz polynomials follow:

Polynomials	**Remarks**
$P_1(s) = s^3 + 2s^2 + 2s + 1$	Hurwitz; has zeros at $s = -1$ and $s = -0.5 \pm j\sqrt{3}/2$
$P_2(s) = s^3 + s^2 + s + 1$	Not Hurwitz; has zeros on the $j\omega$ axis at $s = \pm j1$
$P_3(s) = s^2 + 3s^2 + 3s + 1$	Hurwitz; has a third-order zero at $s = -1$
$P_4(s) = s^3 + 0.8s^2 + 0.81s + 1.01$	Not Hurwitz; has zeros at $s = 0.1 \pm j1$

Now let

$$P(s) = m(s) + n(s) \tag{2}$$

where $P(s)$ is a Hurwitz polynomial and the polynomial $m(s)$ is defined as the even part of $P(s)$ (i.e., it is a polynomial containing only the even terms) and the polynomial $n(s)$ is defined as the odd part of $P(s)$. For example, for $P(s)$ of (1), we see that

$$m(s) = s^4 + 7s^2 + 2 \qquad n(s) = 4s^3 + 6s \tag{3}$$

More generally, we may write

$$m(s) = \tfrac{1}{2}[P(s) + P(-s)] \qquad n(s) = \tfrac{1}{2}[P(s) - P(-s)] \tag{4}$$

where $P(-s)$ is the polynomial $P(s)$ with the argument $-s$ substituted for s. For example, for the polynomial $P(s)$ of (1) we see that

$$P(-s) = s^4 - 4s^3 + 7s^2 - 6s + 2 \tag{5}$$

The zeros of $P(-s)$ are found by reflecting the zeros of $P(s)$ (which are all in the left-half plane) through the origin. An example for an arbitrary set of zeros is shown in Fig. 4.6-1. Thus, the zeros of $P(-s)$ will only occur in the right-half plane. As an example, it is readily verified that the polynomial given in (5) has simple roots at $s = 1 \pm j1$ and a second-order root at $s = 1$. Now let us define a function $N(s)$ as the ratio of the even

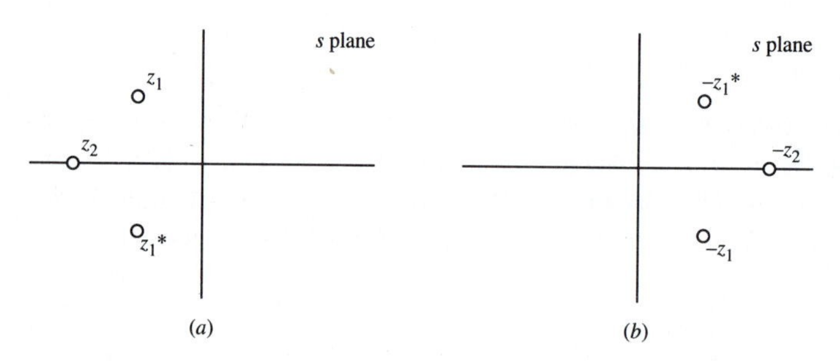

FIGURE 4.6-1
Zeros of $P(s)$ and $P(-s)$.

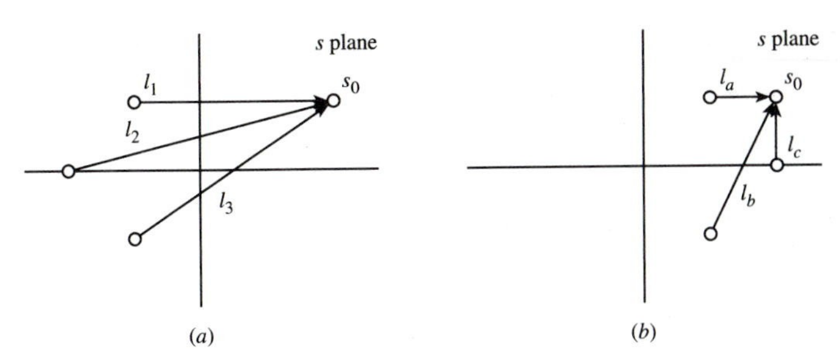

FIGURE 4.6-2
Magnitude of $P(s_0)$ and $P(-s_0)$.

and odd polynomials $m(s)$ and $n(s)$ of (2).[1] Thus we write

$$N(s) = \frac{m(s)}{n(s)} = \frac{P(s) + P(-s)}{P(s) - P(-s)} = \frac{P(s)/P(-s) + 1}{P(s)/P(-s) - 1} \tag{6}$$

If we evaluate the polynomials $P(s)$ and $P(-s)$ at some point s_0 in the right-half plane, then, as illustrated in Fig. 4.6-2(*a*), the magnitude of $P(s_0)$ is given by the product of the vector lengths l_1, l_2, and l_3 from the zeros of $P(s)$ to s_0 [assuming that the coefficient of the highest degree term in $P(s)$ is unity], while the magnitude of $P(-s_0)$ is similarly given by the product of the vector lengths l_a, l_b, and l_c. Comparing these lengths, we see that

$$\left| \frac{P(s)}{P(-s)} \right| \geq 1 \qquad \text{for Re } s \geq 0 \tag{7}$$

This is readily shown to be true for any Hurwitz polynomial. Now let $P(s)/P(-s) = U + jV$. The real part of the function $N(s)$ defined in (6) may be written

$$\text{Re } N(s) = \text{Re } \frac{U + jV + 1}{U + jV - 1} = \frac{U^2 + V^2 - 1}{(U - 1)^2 + V^2} \tag{8}$$

In the right member of this relation, the term $U^2 + V^2$ is simply $|P(s)/P(-s)|^2$, which, from (7), is always greater than or equal to 1 for $\text{Re } s \geq 0$. Applying this result to (8), we see that

$$\text{Re } N(s) \geq 0 \qquad \text{for Re } s \geq 0 \tag{9}$$

It may be shown that this result is sufficient to guarantee the realizability of $N(s)$ as a driving-point function.[2] In addition, since $N(s)$ is also an odd rational function, as shown in Sec. 4.2, it must be an *LC* driving-point function. We conclude that *the ratio of the even and odd parts (or odd and even parts) of a Hurwitz polynomial is realizable as an LC driving-point immittance.*

---

[1] A similar result is obtained if we consider the ratio of the odd and even polynomials.

[2] Formally, such a function is called a *positive real* one. See Wai-Kai Chen, *Passive and Active Filters—Theory and Implementation*, John Wiley and Sons, New York, 1986.

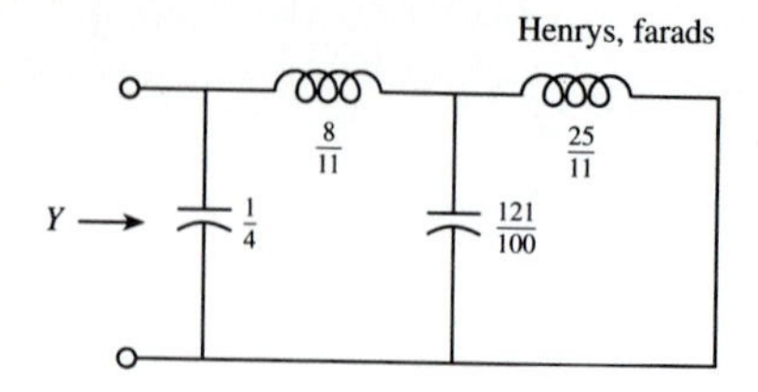

FIGURE 4.6-3
LC driving-point realization for Example 4.6-2.

Example 4.6-2 The *LC* driving-point realization of the ratio of the even and odd parts of a Hurwitz polynomial. As an example of this property of a Hurwitz polynomial, let us use the polynomial $P(s)$ given in (1) with even and odd parts given in (3). Arbitrarily choosing the ratio of even to odd parts to be an admittance and making a continued fraction expansion, we obtain[3]

$$Y(s) = \frac{m(s)}{n(s)} = \frac{s^4 + 7s^2 + 2}{4s^3 + 6s} = \frac{s}{4} + \cfrac{1}{8s/11 + \cfrac{1}{121s/100 + \cfrac{1}{25s/11}}}$$

which has the realization shown in Figure 4.6-3.

Exercise 4.6-2. Expand the odd/even parts of the polynomials defined in Example 4.6-1 in the form of a continued-fraction expansion with removals at infinity.

Answers

$$N_1(s) = \frac{s^3 + 2s}{2s^2 + 1} = \frac{s}{2} + \cfrac{1}{4s/3 + \cfrac{1}{3s/2}} \qquad N_2(s) = \frac{s^3 + s}{s^2 + 1} = s \qquad N_3(s) = \frac{s^3 + 3s}{3s^2 + 1} = \frac{s}{3} + \cfrac{1}{9s/8 + \cfrac{1}{8s/3}}$$

$$N_4(s) = \frac{s^3 + 0.81s}{0.8s^2 + 1.01} = 1.25s + \cfrac{1}{-1.76796s + \cfrac{1}{-0.44802s}}$$

Note that non-Hurwitz polynomials can terminate prematurely (for $j\omega$-axis zeros) as in $N_2(s)$ or will result in an expansion with negative coefficients as in $N_4(s)$.

Realization of $Y_T(s)$

The result obtained in the preceding paragraphs is of direct use in our primary topic in this section, namely the use of a resistance-terminated lossless ladder to synthesize

[3] This procedure is also useful in testing a polynomial to see if it is Hurwitz, i.e., to see if it has all its roots in the left-half plane, without actually having to find the roots. If the continued-fraction expansion of the ratio of the even and odd parts of the polynomial has n positive coefficients, then the polynomial is Hurwitz.

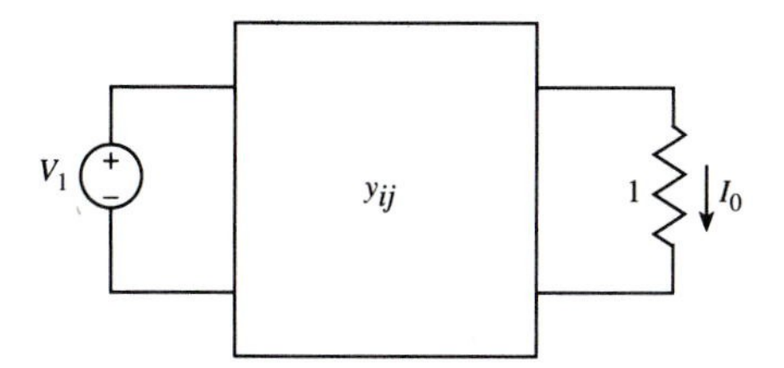

FIGURE 4.6-4
Resistance-terminated lossless ladder network.

transfer functions. We shall treat the realization of a terminated transfer admittance $Y_T(s) = I_0(s)/V_1(s) = N(s)/D(s)$, where $I_0(s)$ is the current in the terminating resistor, as shown in Fig. 4.6-4. The poles of this function will be those of an *RLC* network, i.e., they may be anywhere in the left-half plane. Thus, the denominator polynomial $D(s)$ will be Hurwitz. Since the ladder elements are *LC*, however, the zeros of $Y_T(s)$ will be restricted to the $j\omega$ axis. Thus, the numerator polynomial $N(s)$ will be even (if there is no zero at the origin) or odd. As shown in Fig. 4.6-4, we will use $y_{ij}(s)$ as the parameters of the lossless ladder (*not* including the termination), and for convenience, we will normalize the terminating resistor to a value of 1 Ω. For this configuration we may easily derive

$$Y_T(s) = \frac{I_0(s)}{V_1(s)} = \frac{N(s)}{D(s)} = \frac{N(s)}{m(s)+n(s)} = \frac{-y_{21}(s)}{y_{22}(s)+1} \tag{10}$$

where $m(s)$ and $n(s)$ are respectively the even and odd parts of the Hurwitz polynomial $D(s)$. There are now two cases to be considered depending on whether $N(s)$ is even or odd. If $N(s)$ is even, from (10), we obtain (Case 1)

$$-y_{21}(s) = \frac{N(s)}{n(s)} \qquad y_{22}(s) = \frac{m(s)}{n(s)} \tag{11}$$

where, since $m(s)$ and $n(s)$ are the even and odd parts of a Hurwitz polynomial, $y_{22}(s)$ is *LC* realizable. For the other case, namely, for $N(s)$ odd, from (10) we may write (Case 2)

$$-y_{21}(s) = \frac{N(s)}{m(s)} \qquad y_{22}(s) = \frac{n(s)}{m(s)} \tag{12}$$

and, as before, $y_{22}(s)$ is *LC* realizable. The synthesis procedure for the transfer function is now accomplished by synthesizing $y_{22}(s)$ in the form of a ladder network. The elements of the ladder network must be chosen so that the desired transmission zeros are realized. In selecting the elements, recall, from the discussion of the properties of ladder networks given in Sec. 4.5, that the poles of series impedances and the poles of shunt admittances must be located at the desired transmission zero locations. Thus, if we remove a pole at some frequency from $y_{22}(s)$ and produce a shunt admittance in our realization, such a removal produces a transmission zero at the same frequency. Similarly, at any frequency where $1/y_{22}(s)$ has a pole, the removal of such a pole to produce a series impedance also produces a transmission zero. For the case where transmission zeros are to be realized only at the origin and at infinity, the removals result in single L or C elements, and the realization for $y_{22}(s)$ will be a Cauer form.

Example 4.6-3 Synthesis of a low-pass network. Consider the realization of the third-order Butterworth function

$$Y_T(s) = \frac{H}{s^3 + 2s^2 + 2s + 1}$$

where H is a positive constant. This is a low-pass transfer function with a third-order transmission zero at infinity with $m(s) = 2s^2 + 1$ and $n(s) = s^3 + 2s$. Since the numerator of $Y_T(s)$ is even, we use (11) to obtain

$$-y_{21}(s) = \frac{H}{s^3 + 2s} \qquad y_{22}(s) = \frac{2s^2 + 1}{s^3 + 2s}$$

At this point we note that $y_{21}(s)$ and $y_{22}(s)$ have the same poles, i.e., $y_{22}(s)$ does *not* have any private poles. As a result, the first element in the realization of $y_{22}(s)$ must be a series element (*not* a shunt one). To produce one of the desired transmission zeros at infinity, this element must have a pole at infinity. The realization of the second and third transmission zeros at infinity is done by continuing the synthesis of $y_{22}(s)$ with a shunt element (with a pole of admittance at infinity) and another series element (with a pole of impedance at infinity). The element values are easily found using a continued-fraction expansion (a Cauer form). Thus we obtain

$$\frac{1}{y_{22}(s)} = \frac{s^3 + 2s}{2s^2 + 1} = \frac{s}{2} + \frac{1}{4s/3 + 1/(3s/2)}$$

The resulting realization for the transfer function is shown in Fig. 4.6-5. The final step in the synthesis process is to find the value of the multiplicative constant H.[4] Since the same constant H appears both in $Y_T(s)$ and $-y_{21}(s)$, this is most easily done by direct analysis of the lossless portion of the circuit. Doing this for the element values shown in Fig. 4.6-5, we obtain $-y_{21}(s) = 1/(s^3 + 2s)$. Thus, we see that $H = 1$.

Exercise 4.6-3. Find a realization for $Y_T(s)$ for the case where the denominator polynomial has the form given as $P_3(s)$ in Example 4.6-1 (see also Exercise 4.6-2). Also find the value of the multiplicative constant H.

Answer. The configuration is the same as the one shown in Fig. 4.6-5. The element values (from right to left) are $L_1 = \frac{1}{3}$ H, $C_2 = \frac{9}{8}$ F, and $L_3 = \frac{8}{3}$ H. The multiplicative constant $H = 1$.

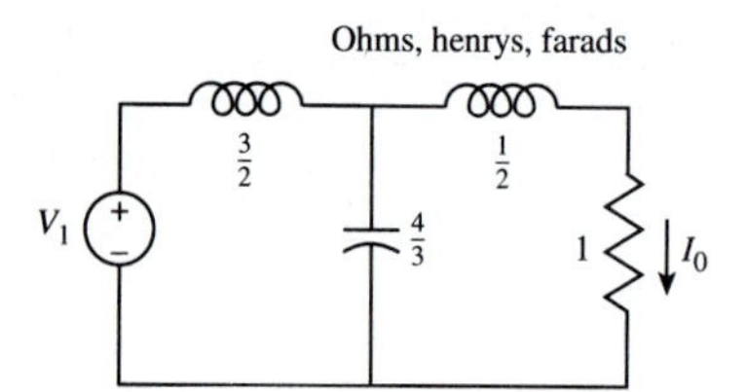

FIGURE 4.6-5
Realization of $Y_T(s)$ where $y_{22}(s)$ has no private poles.

[4] Note that this synthesis procedure will not realize a specific value of H. Thus we say that $Y_T(s)$ is realized *within a multiplicative constant*. Any desired value of H, of course, can be obtained by applying an impedance denormalization to the network.

Private Poles of $y_{22}(s)$

In the example given above, both $Y_T(s)$ and $-y_{21}(s)$ had the same transmission zeros, i.e., a third-order transmission zero at infinity. This will always be the case when $y_{22}(s)$ does not have private poles. An example of the synthesis procedure when $Y_T(s)$ has transmission zeros that are not present in $-y_{21}(s)$ follows.

Example 4.6-4 Synthesis when private poles are present. Consider the function

$$Y_T(s) = \frac{Hs^2}{s^3 + 2s^2 + 2s + 1}$$

Since the numerator of this function is even, we apply (11) to obtain

$$-y_{21}(s) = \frac{Hs^2}{s^3 + 2s} = \frac{Hs}{s^2 + 2} \qquad y_{22}(s) = \frac{2s^2 + 1}{s^3 + 2s}$$

Here we see that $-y_{21}(s)$ has simple transmission zeros at the origin and at infinity, while $Y_T(s)$ has a *second-order* transmission zero at the origin and a simple zero at infinity. In addition, we note that the poles of $-y_{21}(s)$ and $y_{22}(s)$ are not the same, i.e., $y_{22}(s)$ has poles at the origin and at $s = \pm j\sqrt{2}$. The pole of $y_{22}(s)$ at the origin is thus a private pole, which must be realized as a shunt element, in this case an inductor. The inductor produces the extra transmission zero at the origin needed for $Y_T(s)$ that is not present in $y_{21}(s)$. The complete expansion of $y_{22}(s)$ may now be put in the form

$$y_{22}(s) = \frac{1}{2s} + \frac{1}{4/3s + 2s/3}$$

Note that in the denominator of the right term in this expansion the quantity $4/3s$ represents the removal of a pole of impedance at the origin from $1/[y_{22}(s) - 1/2s]$; thus it realizes the second transmission zero at the origin. Similarly, the quantity $2s/3$ represents the removal of a pole of impedance at infinity and thus realizes the transmission zero at infinity. The resulting realization for the network function is shown in Fig. 4.6-6. It is readily shown that $H = \frac{3}{2}$.

Exercise 4.6-4. Find a realization for $Y_T(s)$ for the case where the numerator polynomial has the form Hs^2 and the denominator has the form given as $P_3(s)$ in Example 4.6-1 (see also Exercise 4.6-2). Also find the value of the multiplicative constant H.

Answer. The configuration is the same as the one shown in Fig. 4.6-6. The element values (from right to left) are $L_1 = 3$ H, $C_2 = \frac{8}{9}$ F, and $L_3 = \frac{3}{8}$ H. The multiplicative constant $H = \frac{8}{3}$.

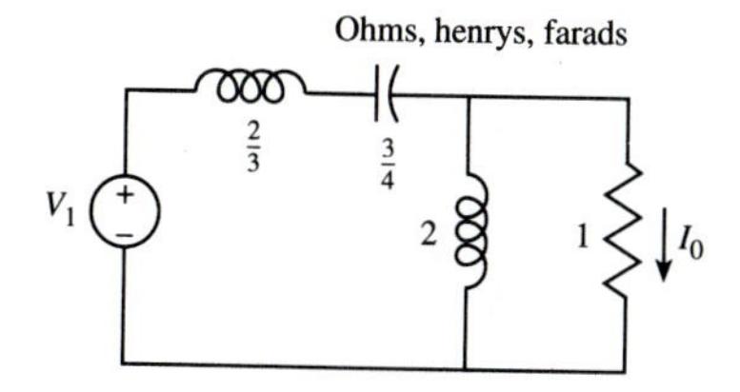

FIGURE 4.6-6
Realization of $Y_T(s)$ where $y_{22}(s)$ has a private pole.

A procedure dual to that described in this section may be used to realize open-circuit transfer impedances using a resistance-terminated lossless ladder network. In this case it is more convenient to work with the z parameters of the lossless network rather than with the y parameters. The details are left to the reader as an exercise. Using the procedures outlined in this section, a table of element values for filters that realize the low-pass Butterworth, Chebyshev, and Thomson approximations introduced in Chap. 2 is readily constructed. The results are given in App. A.

The realization procedure developed in this section has only been applied to the synthesis of transmission zeros at the origin and at infinity. In the following section, however, we shall show how the procedure may be extended to realize zeros anywhere on the $j\omega$ axis.

4.7 ZERO SHIFTING IN TRANSFER FUNCTION SYNTHESIS

In the preceding section we showed how a resistance-terminated lossless ladder network could be used to realize a transfer function with arbitrarily located left-half-plane poles, i.e., with a Hurwitz denominator polynomial, and with transmission zeros located at the origin or infinity. In this section we shall extend the technique so that we can also realize transmission zeros located anywhere on the $j\omega$ axis. Such zeros must, of course, occur in conjugate pairs. The synthesis procedure will make use of a technique called *zero shifting*.

Partial Pole Removals

To understand how zero shifting occurs, let us consider a driving-point function $y_{22}(s)$ having the form

$$y_{22}(s) = \frac{2s(s^2+4)}{s^2+1} = 2s + \frac{6s}{s^2+1} \tag{1}$$

The function has poles at $s = \pm j1$ and infinity, and zeros at $s = \pm j2$ and the origin. A plot of $B_{22}(\omega) = y_{22}(j\omega)/j$, the imaginary part of the admittance function $y_{22}(j\omega)$, is shown in Fig. 4.7-1(a). Now let us consider subtracting a term $Y_0(s) = Ks$, with an imaginary part $B_0(\omega) = K\omega$, from $y_{22}(s)$. From the right member of (1) we see that $y_{22}(s)$ has a residue of 2 at its pole at infinity. Thus if we assume $0 < K < 2$, the subtraction may be referred to as a *partial removal* of the pole of $y_{22}(s)$ at infinity. To make our example specific, we shall choose $K = \frac{5}{4}$. Note that for this value, as shown in Fig. 4.7-1(a), the plot of $B_0(\omega)$ intersects the plot of $B_{22}(\omega)$ at $\omega = 3$. If we let the difference be $y'_{22}(s)$, we obtain

$$y'_{22}(s) = y_{22}(s) - Y_0(s) = \frac{3s}{4} + \frac{6s}{s^2+1} = \frac{3s(s^2+9)}{4(s^2+1)} \tag{2}$$

A plot of the function $B'_{22}(\omega) = y'_{22}(j\omega)/j$ is shown in Fig. 4.7-1(b). From (2) we see that $y'_{22}(s)$ has zeros at $s = \pm j3$, whereas $y_{22}(s)$ had zeros at $s = \pm j2$. Clearly, the partial

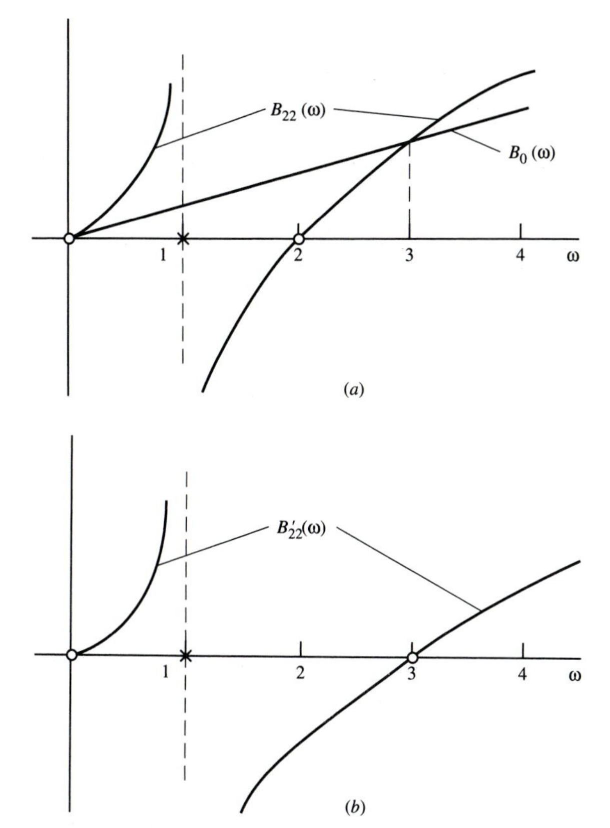

FIGURE 4.7-1
Partial removal of a pole at infinity.

removal of the pole at infinity from $y_{22}(s)$ has shifted the complex-conjugate zeros toward that pole. The operation is summarized in the upper half-plane pole-zero plots of Fig. 4.7-2. The choice of any other value of K is readily shown to produce a corresponding shift in the zero, with larger shifts produced by larger values of K until, for the choice $K = 2$, the zero is shifted all the way to the pole location, i.e., to infinity. The value of K required for any particular shift, i.e., for moving the zero of $y_{22}(s)$ to any value ω_0, where $2 < \omega_0 \leq \infty$, is readily found by noting that

$$Ks\bigg|_{s=j\omega_0} = y_{22}(s)\bigg|_{s=j\omega_0} \tag{3}$$

This relation is more easily evaluated by dividing both sides by s. Thus for the example here we have

$$K = \frac{y_{22}(s)}{s}\bigg|_{s=j3} = \frac{2(s^2+4)}{s^2+1}\bigg|_{s^2=-9} = \frac{5}{4} \tag{4}$$

Many variations of this procedure are possible. For example, starting with the $y_{22}(s)$ given in (1), if zeros were desired at $1 < \omega < 2$, then a partial removal of the pair of poles

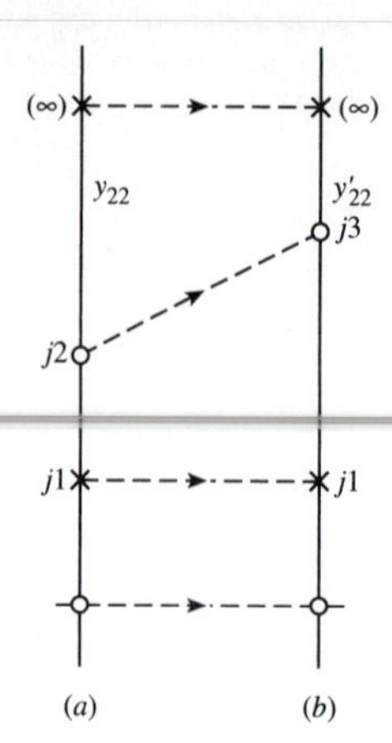

FIGURE 4.7-2
Zero shifting by partial removal of a pole at infinity.

at $s = \pm j1$ would shift the zeros at $s = \pm j2$ toward the poles at $\pm j1$. The term subtracted in this case would have the form $Ks/(s^2 + 1)$ where $0 < K < 6$. For example, for zeros at $s = \pm j\sqrt{2}$, we must choose K so that

$$\left.\frac{Ks}{s^2 + 1}\right|_{s=j\sqrt{2}} = \left.y_{22}(s)\right|_{s=j\sqrt{2}} \tag{5}$$

This evaluation is more readily accomplished by multiplying both sides of (5) by $(s^2 + 1)/s$. Thus, we find that

$$K = \left.\frac{s^2 + 1}{s} y_{22}(s)\right|_{s=j\sqrt{2}} = \left.2(s^2 + 4)\right|_{s^2=-2} = 4 \tag{6}$$

It is readily verified that subtracting $4s/(s^2 + 1)$ from $y_{22}(s)$ of (1) results in a function with zeros at $s = \pm j\sqrt{2}$. The operation is summarized in Fig. 4.7-3. The following example shows how the technique of shifting the zeros of driving-point functions described above can be applied to the realization of transfer functions.

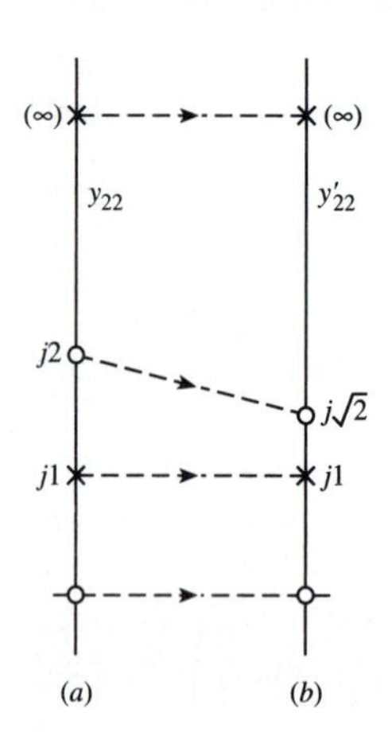

FIGURE 4.7-3
Zero shifting by partial removal of a pole pair at $\pm j1$.

Example 4.7-1 Zero shifting in transfer function synthesis. Consider the transfer function

$$Y_T(s) = \frac{Hs(s^2+9)}{2s^3+s^2+8s+1}$$

Applying (12) of Sec. 4.6, we obtain

$$-y_{21}(s) = \frac{Hs(s^2+9)}{s^2+1} \qquad y_{22}(s) = \frac{2s(s^2+4)}{s^2+1}$$

Note that $-y_{21}(s)$ has zeros at $s = \pm j3$ as well as at the origin. From the basic ladder synthesis techniques developed in the preceding section, we know that removal of poles from $y_{22}(s)$ or $1/y_{22}(s)$ will produce transmission zeros in $-y_{21}(s)$. However, neither of these functions have any poles at the desired transmission zero locations $s = \pm j3$. We therefore proceed to shift the zeros at $s = \pm j2$ to $s = \pm j3$ by partially removing the pole at infinity using the result given in (2). We see that $1/y'_{22}(s)$, the reciprocal of the resulting function, may be written as

$$\frac{1}{y'_{22}(s)} = \frac{4(s^2+1)}{3s(s^2+9)} = \frac{4/27}{s} + \frac{32s/27}{s^2+9}$$

The poles and zeros of the various functions are shown in Fig. 4.7-4. If we now remove the pole pair at $s = \pm j3$ from $1/y'_{22}(s)$, we produce the desired transmission zeros at $s = \pm j3$. Similarly, removing the pole at the origin from $1/y'_{22}(s)$ produces the transmission zero at the origin. Thus, the complete expansion used in realizing $y_{22}(s)$ for this example is

$$y_{22}(s) = \frac{5s}{4} + \frac{1}{4/27s + (32s/27)/(s^2+9)}$$

The network realization is shown in Fig. 4.7-5.

Exercise 4.7-1. Find a filter realization for the function $Y_T(s)$ given in Example 4.7-1 for each of the values of the transmission zero (specified as 3 rad/s in the example) given below. The element values are defined in Fig. 4.7-5.

Zeros of transmission, rad/s	Answers—element values			
	C_1, F	C_2, F	C_3, F	L_3, H
4	1.6000	6.4000	0.4267	0.1465
5	1.7500	6.2500	0.2604	0.1536
6	1.8286	6.1714	0.1763	0.1575
7	1.8570	6.1250	0.1276	0.1599

In the example given above, it should be noted that the shunt $\frac{5}{4}$-F capacitor does *not* produce a transmission zero at infinity since it represents a *partial removal* of the pole of $y_{22}(s)$ at infinity and only *complete pole removals* produce transmission zeros. Another way of seeing this is to note that as the frequency approaches infinity, the three capacitors shown in Fig. 4.7-5 provide a low-impedance path from the positive terminal

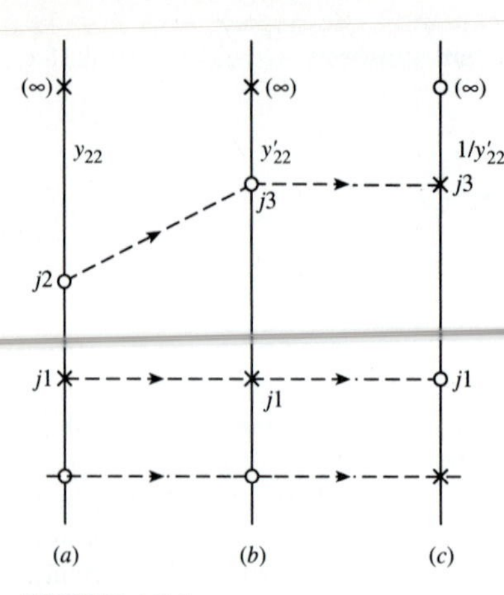

FIGURE 4.7-4
Zero shifting operations in Example 4.7-1.

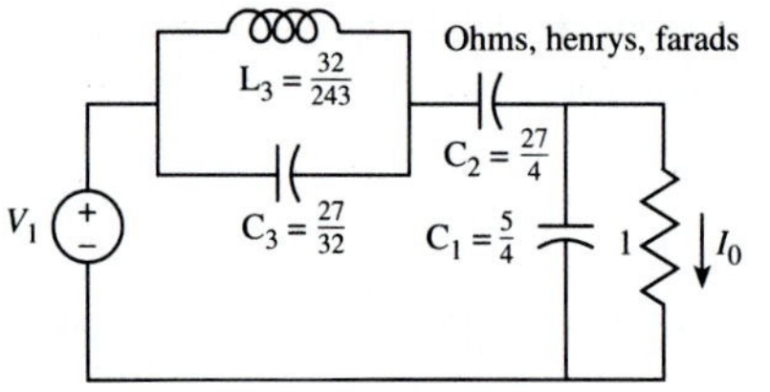

FIGURE 4.7-5
Network realization for Example 4.7-1.

of the voltage source to ground. Thus, to infinite frequencies they act as a voltage divider, and part of any infinite-frequency input voltage will appear across the output resistor.

Removals from Poles of Impedance

Zero shifting may be accomplished by partially removing poles on an impedance basis as well as on an admittance basis. An example follows.

Example 4.7-2 Zero shifting in impedance functions. Consider the realization of the function

$$Y_T(s) = \frac{Hs(s^2 + \frac{1}{4})}{2s^3 + s^2 + 8s + 1}$$

which has transmission zeros at $s = \pm j/2$ and the origin. Applying (12) of Sec. 4.6, we obtain

$$-y_{21}(s) = \frac{Hs(s^2 + \frac{1}{4})}{s^2 + 1} \qquad y_{22}(s) = \frac{2s(s^2 + 4)}{s^2 + 1}$$

Now consider $y_{22}(s)$. A partial removal of poles at $s = \pm j1$ is not suitable for producing zeros of $y'_{22}(s)$ at $s = \pm j/2$, since it can only be used to move zeros over the range of ω specified as $1 < \omega < 2$. Similarly, a partial removal of the pole at infinity cannot be used, since here the possible range is $2 < \omega < \infty$. Therefore, we must consider the reciprocal function

$$\frac{1}{y_{22}(s)} = \frac{\frac{1}{2}(s^2 + 1)}{s(s^2 + 4)} = \frac{\frac{1}{8}}{s} + \frac{3s/8}{s^2 + 4}$$

A partial removal of the pole at the origin of this function will shift the zeros from $s = \pm j1$

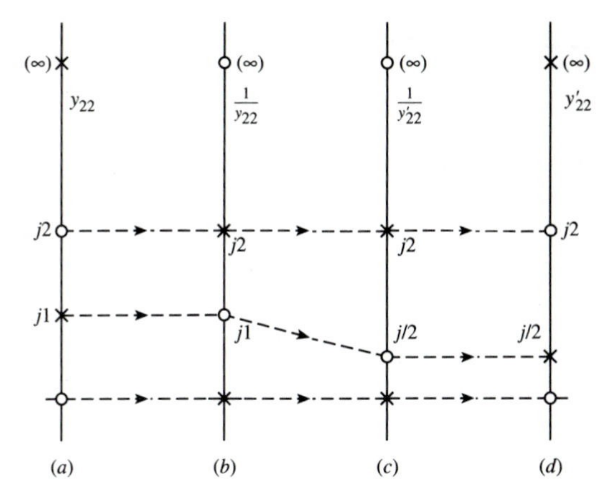

FIGURE 4.7-6
Zero shifting operations in Example 4.7-2.

to the desired $\pm j/2$ positions. Specifically, we must remove a term K/s, where

$$K = s\frac{1}{y_{22}(s)}\bigg|_{s^2=-1/4} = \frac{\frac{1}{2}(s^2+1)}{s^2+4}\bigg|_{s^2=-1/4} = \frac{1}{10}$$

Thus we obtain

$$\frac{1}{y'_{22}(s)} = \frac{1}{y_{22}(s)} - \frac{\frac{1}{10}}{s} = \frac{\frac{2}{5}(s^2+\frac{1}{4})}{s(s^2+4)}$$

$$y'_{22}(s) = \frac{(5/2s)(s^2+4)}{s^2+\frac{1}{4}} = \frac{5s}{2} + \frac{75s/8}{s^2+\frac{1}{4}}$$

The steps are summarized in Fig. 4.7-6. The removal of the term $(75s/8)/(s^2+\frac{1}{4})$ produces the desired transmission zeros at $s = \pm j/2$, and the removal of the impedance $2/5s$ produces a transmission zero at the origin and completes the realization of $y_{22}(s)$. The overall network is shown in Fig. 4.7-7.

Application Example

The zero shifting techniques introduced in this section provide a synthesis method that may be used to find realizations for the inverse-Chebyshev and elliptic approximations described in Chap. 2 since these approximations require $j\omega$-axis zeros.

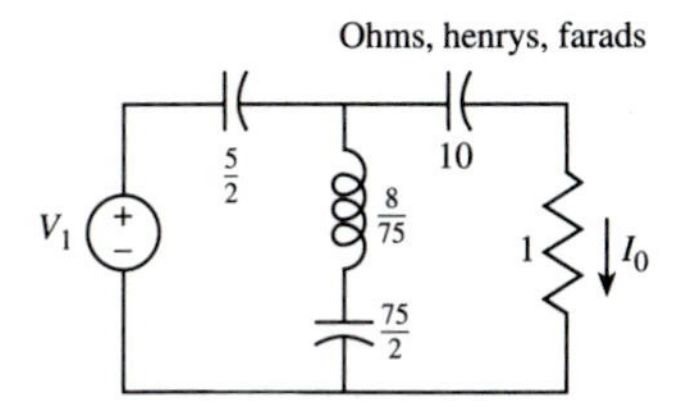

FIGURE 4.7-7
Network realization for Example 4.7-2.

Example 4.7-3 Realization of an inverse-Chebyshev approximation. A third-order inverse-Chebyshev approximation that provides a minimum of 20 dB attenuation throughout a normalized stopband having a frequency range of $1 \le \omega \le \infty$ radians per second was shown in Table 2.3-1 to have the form

$$Y(s) = H\frac{s^2 + 1.33333}{(s + 0.853447)(s^2 + 0.551936s + 0.471048)}$$

$$= H\frac{s^2 + 1.33333}{s^3 + 1.40538s^2 + 0.94200s + 0.40202}$$

The desired transmission zeros are at $s = \pm j1.15470$ (the square root of -1.33333) and infinity. From (12) of Sec. 4.6 we find

$$-y_{21}(s) = H\frac{s^2 + 1.33333}{s(s^2 + 0.94200)}$$

$$y_{22}(s) = \frac{1.40538(s^2 + 0.28605)}{s(s^2 + 0.94200)}$$

The locations of the poles and zeros of $y_{22}(s)$ and the following functions that we will define are shown in Fig. 4.7-8. The first step in the realization is to partially remove the pole at infinity from $1/y_{22}(s)$ in order to shift its zero from $j0.97062$ to $j1.15470$. The amount to be removed from the residue at infinity is

$$k_{\text{shift}} = \left.\frac{1/y_{22}(s)}{s}\right|_{s^2=-1.33333}$$

$$= \left.\frac{s^2 + 0.94200}{1.40538(s^2 + 0.28605)}\right|_{s^2=-1.33333} = 0.26582$$

Defining $y'_{22}(s)$ as the new function, we obtain

$$\frac{1}{y'_{22}(s)} = \frac{1}{y_{22}(s)} - 0.26582s = \frac{0.44573s(s^2 + 1.33333)}{s^2 + 0.28605}$$

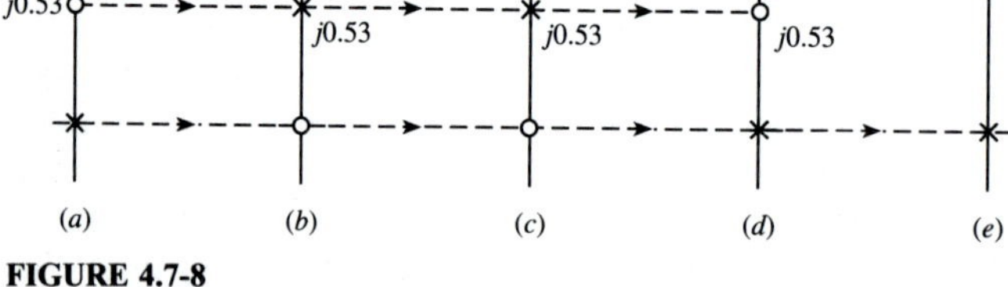

FIGURE 4.7-8
Zero shifting operations for Example 4.7-3.

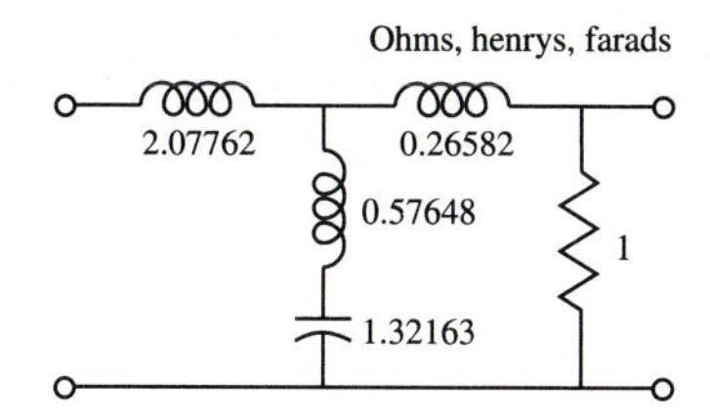

FIGURE 4.7-9
Network realization for Example 4.7-3.

Thus,

$$y'_{22}(s) = \frac{2.24349(s^2 + 0.28605)}{s(s^2 + 1.33333)} = \frac{0.48132}{s} + \frac{1.76218s}{s^2 + 1.33333}$$

Removal of the complex-conjugate pole pair produces a shunt element consisting of a series-connected inductor and capacitor and realizes the specified transmission zeros. Defining

$$y''_{22}(s) = y'_{22}(s) - \frac{1.76218s}{s^2 + 1.33333} = \frac{0.48132}{s}$$

we see that $1/y''_{22}(s)$ has a pole at infinity that may be removed as a series inductor to produce the required transmission zero at infinity. The final network configuration is shown in Fig. 4.7-9.

Appendix A gives realizations for some other inverse-Chebyshev realizations found using a method similar to that of the above example.

In the synthesis of a given transfer function $Y_T(s)$ using zero shifting techniques, there are usually many variations, based on the order in which the complete pole removals and the partial pole removals are made and depending on the poles from which partial removals are made. In addition, extra partial pole removals may frequently be made if it is desired to generate a certain network topology, for example, to place a shunt capacitor at the output or input of the network. Finally, in some situations, partial removals from poles may not be capable of producing a desired shift in a remotely located zero. Examples of some of the above cases may be found in the exercises at the end of this chapter.

4.8 DOUBLE-RESISTANCE-TERMINATED LOSSLESS LADDER NETWORKS

The synthesis techniques presented in the preceding sections of this chapter are based on a filter having the form of a lossless ladder with a single resistor as a load termination. A more practical network configuration is a lossless ladder with resistive terminations at both the input and the output ports. Not only does such a configuration provide the means for accommodating the internal resistance of the driving source as well as that of the load resistance, but also it has excellent sensitivity properties. As might be expected, the synthesis procedure is more complex than that for the single-terminated configuration. Because of this, in this section we will only present an introductory treatment of the method.

Transducer and Characteristic Functions

The synthesis of the transfer function for a double- resistance-terminated lossless ladder may be accomplished by realizing a related driving-point function. Note that such a procedure is similar to the one used in Sec. 4.6 for a single-resistance-terminated filter. To start our derivation, consider the network configuration consisting of a source resistance R_1 and a load resistance R_2 separated by a lossless (LC) ladder network N, as shown in Fig. 4.8-1. Under sinusoidal steady-state conditions, the input impedance $Z_{\text{in}}(s)$ may be written

$$Z_{\text{in}}(j\omega) = R_{\text{in}}(j\omega) + jX_{\text{in}}(j\omega) \tag{1}$$

where $R_{\text{in}}(j\omega)$ and $X_{\text{in}}(j\omega)$ are the real and imaginary parts. The sinusoidal steady-state average power $P_{\text{in}}(j\omega)$ transmitted from the left portion of the circuit (consisting of $\mathcal{V}_{\text{in}}$ and R_1) into the combination of N and R_2 is

$$P_{\text{in}}(j\omega) = |\mathcal{I}_{\text{in}}|^2 R_{\text{in}}(j\omega) = \left|\frac{\mathcal{V}_{\text{in}}}{R_1 + Z_{\text{in}}(j\omega)}\right|^2 R_{\text{in}}(j\omega) \tag{2}$$

where $\mathcal{I}_{\text{in}}$ and $\mathcal{V}_{\text{in}}$ are the rms-valued phasor current and voltage defined in Fig. 4.8-1. The power $P_{\text{out}}(j\omega)$ out of N into R_2 is

$$P_{\text{out}}(j\omega) = \left|\frac{\mathcal{V}_{\text{out}}}{R_2}\right|^2 R_2 \tag{3}$$

where $\mathcal{V}_{\text{out}}$ is the phasor output voltage. Since N is a lossless network, the power into it must equal the power out of it. Thus, $P_{\text{in}}(j\omega) = P_{\text{out}}(j\omega)$ and, from (2) and (3),

$$\left|\frac{\mathcal{V}_{\text{in}}}{R_1 + Z_{\text{in}}(j\omega)}\right|^2 R_{\text{in}}(j\omega) = \left|\frac{\mathcal{V}_{\text{out}}}{R_2}\right|^2 R_2 \tag{4}$$

Solving for the sinusoidal steady-state voltage transfer function magnitude squared, from (4) we obtain

$$\left|\frac{\mathcal{V}_{\text{out}}}{\mathcal{V}_{\text{in}}}\right|^2 = \frac{R_2 R_{\text{in}}(j\omega)}{|R_1 + Z_{\text{in}}(j\omega)|^2} \tag{5}$$

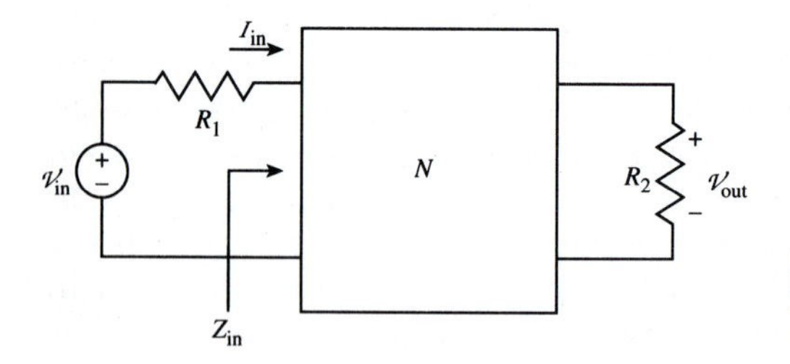

FIGURE 4.8-1
Double-resistance-terminated lossless ladder network.

To reduce this equation to a more useful form, we note that

$$|R_1 + Z_{in}(j\omega)|^2 - |R_1 - Z_{in}(j\omega)|^2 = 4R_1R_{in}(j\omega) \tag{6}$$

Solving this result for $R_{in}(j\omega)$ and substituting in (5), we obtain a relation for the voltage transfer function magnitude squared of the overall filter as determined by the terminating resistors and the specified driving-point impedance $Z_{in}(j\omega)$:

$$\left|\frac{\mathcal{V}_{out}}{\mathcal{V}_{in}}\right|^2 = \frac{R_2}{4R_1}\left(1 - \left|\frac{R_1 - Z_{in}(j\omega)}{R_1 + Z_{in}(j\omega)}\right|^2\right) \tag{7}$$

To simplify the subsequent derivation, at this point we define a function $H(j\omega)$, called the *transducer function*, by the relation

$$|H(j\omega)|^2 = \frac{R_2}{4R_1}\left|\frac{\mathcal{V}_{in}}{\mathcal{V}_{out}}\right|^2 \tag{8}$$

This function is proportional to the reciprocal of the voltage transfer function. It is also sometimes referred to as a *loss function*. We now also define a second function $K(j\omega)$, called the *characteristic function*, by

$$|K(j\omega)|^2 = |H(j\omega)|^2 - 1 \tag{9}$$

The characteristic function is a measure of the closeness to unity of the value of the transducer function. From (8) and (9) we obtain a new expression for the voltage transfer function magnitude squared:

$$\left|\frac{\mathcal{V}_{out}(j\omega)}{\mathcal{V}_{in}(j\omega)}\right|^2 = \frac{R_2}{4R_1}\frac{1}{|H(j\omega)|^2} = \frac{R_2}{4R_1}\left(1 - \left|\frac{K(j\omega)}{H(j\omega)}\right|^2\right) \tag{10}$$

Equating this to (7), we find that the ratio of the characteristic and transducer functions can be defined in terms of R_1 and $Z_{in}(j\omega)$:

$$\left|\frac{K(j\omega)}{H(j\omega)}\right|^2 = \left|\frac{R_1 - Z_{in}(j\omega)}{R_1 + Z_{in}(j\omega)}\right|^2 \tag{11}$$

By using a technique from complex-variable theory called *analytical continuation*, we can extend (11) over the entire s domain:

$$\frac{K(s)K(-s)}{H(s)H(-s)} = \frac{R_1 - Z_{in}(s)}{R_1 + Z_{in}(s)}\frac{R_1 - Z_{in}(-s)}{R_1 + Z_{in}(-s)} \tag{12}$$

In this expression, the zeros of $H(s)$ must lie in the left-half plane since they are the same as the poles of the voltage transfer function. The zeros of $K(s)$, however, may be shown to be in either the right- or the left-half plane. Since the transducer and characteristic functions both appear twice in (12), once with the argument s and once with $-s$, we may factor both sides of this expression to obtain

$$\frac{K(s)}{H(s)} = \frac{R_1 - Z_{in}(s)}{R_1 + Z_{in}(s)} \tag{13}$$

Solving this for $Z_{in}(s)$, we obtain

$$Z_{in}(s) = R_1 \frac{1 - K(s)/H(s)}{1 + K(s)/H(s)} \tag{14}$$

We may also apply analytical continuation to (9) to show that the poles of $K(s)$ and $H(s)$ are the same. If we now express the numerators of both of these functions in terms of their even (subscript e) and odd (subscript o) parts, then

$$H(s) = \frac{H_e + H_o}{D(s)} \qquad K(s) = \frac{K_e + K_o}{D(s)} \tag{15}$$

where $D(s)$ is the common denominator polynomial and, for convenience, we have dropped the (s) notation in H_e and H_o. Substituting these in (14), $Z_{in}(s)$ may be written

$$Z_{in}(s) = R_1 \frac{H_e + H_o - K_e - K_o}{H_e + H_o + K_e + K_o} \tag{16}$$

Input Impedance in Terms of Two-Port Parameters

Now let us digress slightly to derive another expression for $Z_{in}(s)$, this one as a function of R_2 and the z parameters of the two-port network N. By conventional circuit analysis we find

$$Z_{in}(s) = \frac{\Delta z/R_2 + z_{11}}{z_{22}/R_2 + 1} \tag{17}$$

where Δz is the determinant of the z-parameter matrix, namely,

$$\Delta z = z_{11}z_{22} - z_{12}z_{21} \tag{18}$$

Equation (16) may now be rearranged so that the z parameters in (17) are determined by the quantities H_e, H_o, K_e, and K_o. Note that since $z_{11}(s)$ and $z_{22}(s)$ are driving-point functions of an LC network, they can only be *odd* rational functions. There are two possible arrangements of terms in (16). The first is

$$Z_{in}(s) = \frac{R_1 \dfrac{H_o - K_o}{H_o + K_o} + R_1 \dfrac{H_e - K_e}{H_o + K_o}}{\dfrac{H_e + K_e}{H_o + K_o} + 1} \tag{19}$$

for which

$$z_{11} = R_1 \frac{H_e - K_e}{H_o + K_o} \qquad z_{22} = R_2 \frac{H_e + K_e}{H_o + K_o} \tag{20}$$

The second rearrangement of terms in (16) is

$$Z_{\text{in}}(s) = \frac{R_1 \dfrac{H_e - K_e}{H_e + K_e} + R_1 \dfrac{H_o - K_o}{H_e + K_e}}{\dfrac{H_o + K_o}{H_e + K_e} + 1} \tag{21}$$

For this case

$$z_{11} = R_1 \frac{H_o - K_o}{H_e + K_e} \qquad z_{22} = R_2 \frac{H_o + K_o}{H_e + K_e} \tag{22}$$

For either of these choices we obtain the same result; namely, the voltage transfer function for the overall network, as specified by $H(s)$ and $K(s)$, can be realized by synthesizing a related pair of driving-point functions, namely, z_{11} and z_{22}. The form of the driving-point realizations, of course, must be done in such a way as to obtain the desired transmission zeros of the overall transfer function. For network functions in which the transmission zeros are not at the origin or infinity, it is usually necessary to use zero shifting techniques to do this.

Scaling the Transducer Function

To obtain the best transfer properties for a double-resistance-terminated lossless ladder network, it is desirable to scale, i.e., multiply by a suitable constant, the transducer function defined in (8). Such a scaling can be used to improve the sensitivity and the signal gain. The scaling factor is determined so that at the frequencies where the transfer function has its maximum magnitude, the maximum power will be transferred from the source to the load. In general, this occurs when the impedance seen by the source impedance (R_1) is real. This requires that the input impedance of the LC network and its terminating resistance be real and equal to R_1. When this condition is satisfied, the power $P_{\text{in}}(j\omega)$ from the source into the network is

$$P_{\text{in}}(j\omega)_{\max} = \left| \frac{\mathcal{V}_{\text{in}}}{R_1 + R_1} \right|^2 R_1 = \frac{|\mathcal{V}_{\text{in}}|^2}{4R_1} \tag{23}$$

while the power $P_{\text{out}}(j\omega)$ delivered to the load is

$$P_{\text{out}}(j\omega)_{\max} = \left| \frac{\mathcal{V}_{\text{out}}}{R_2} \right|^2 R_2 = \frac{|\mathcal{V}_{\text{out}}|^2}{R_2} \tag{24}$$

At any frequency of maximum gain, (23) and (24) must be equal; thus

$$\frac{|\mathcal{V}_{\text{in}}|^2}{4R_1} = \frac{|\mathcal{V}_{\text{out}}|^2}{R_2} \tag{25}$$

The condition on the transfer function magnitude squared at any of these frequencies is

$$\left| \frac{\mathcal{V}_{\text{in}}}{\mathcal{V}_{\text{out}}} \right|^2 \frac{R_2}{4R_1} = 1 \tag{26}$$

From (8), the corresponding requirement on the transducer function at these frequencies is

$$|H(j\omega)|^2 = 1 \tag{27}$$

This requirement is satisfied by including a scaling constant C in the definition of the transducer function. Thus, we define

$$|H(j\omega)| = C\left|\frac{\mathcal{V}_{\text{in}}}{\mathcal{V}_{\text{out}}}\right| \tag{28}$$

where C is chosen so that (27) is satisfied at the frequencies of maximum gain. The resulting scaling, of course, also affects the overall gain of the realization.

The derivations given in the preceding paragraphs provide the necessary theory for synthesizing double-resistance-terminated lossless ladder filters. As an application of the theory, let us consider the case defined by the following specifications on the network function and the resulting filter realization.

1. The network function is a voltage transfer function with transmission zeros located only at the origin and infinity.
2. The source and load resistors are equal valued and normalized to unity.

It may be shown that these specifications exclude the even-order Chebyshev realizations. The following example illustrates the steps in the synthesis procedure.

Example 4.8-1 A second-order Butterworth double-resistance-terminated filter. It is desired to find a double-resistance-terminated realization with unity-valued resistors for a second-order Butterworth voltage transfer function. For this approximation

$$\frac{V_{\text{out}}(s)}{V_{\text{in}}(s)} = \frac{A}{s^2 + 1.414s + 1} \tag{29}$$

where A is an arbitrary multiplicative constant. The steps in the synthesis are:

1. Find the frequency of maximum gain (minimum loss). For the Butterworth functions this occurs at $\omega = 0$.
2. Generate the transducer function with a scaling constant C. From (29),

$$|H(j\omega)|^2 = C^2[(1 - \omega^2)^2 + (1.414\omega)^2] = C^2(\omega^4 + 1) \tag{30}$$

   Note that we have assumed that the effect of the arbitrary constant A is included in C.
3. Determine the value of C so that the transducer function has unity magnitude at the frequency of minimum loss, thus satisfying (27). From (30),

$$|H(j\omega)|^2\,|_{\omega=0} = C^2 \tag{31}$$

   Thus, the correct value of C is unity, and from analytic continuation,

$$H(s) = s^2 + 1.414s + 1 \tag{32}$$

4. Find the characteristic function using (9). From (30), with $C = 1$,

$$|K(j\omega)|^2 = |H(j\omega)|^2 - 1 = \omega^4 + 1 - 1 = \omega^4 \tag{33}$$

and from analytic continuation,

$$K(s)K(-s) = (s)^2(-s)^2 \tag{34}$$

Thus

$$K(s) = s^2 \tag{35}$$

5. Obtain the even and odd parts of $H(s)$ and $K(s)$. From (32) and (35), these are

$$\begin{array}{ll} H_e = s^2 + 1 & K_e = s^2 \\ H_o = 1.414s & K_o = 0 \end{array} \tag{36}$$

6. Find the two-port driving-point functions z_{11} and z_{22} using (20) or (22). In this case, we may apply (20) to the relations of (36) to obtain

$$\begin{aligned} z_{11} &= \frac{H_e - K_e}{H_o + K_o} = \frac{1}{1.414s} \\ z_{22} &= \frac{H_e + K_e}{H_o + K_o} = \frac{2s^2 + 1}{1.414s} = 1.414s + \frac{1}{1.414s} \end{aligned} \tag{37}$$

7. Realize the network two-port parameters z_{11} and z_{22} in such a way as to realize the desired transmission zeros. In this case, the transfer function has two transmission zeros at infinity and the realization has the form shown in Fig. 4.8-2.

Explicit Formulas for Butterworth Filters

Explicit formulas have been derived for double-resistance-terminated lossless ladder filters for various approximations. As an example of these, consider the equal-resistance-terminated realization of the Butterworth approximation. For the case where $R_1 = R_2 = 1$ and a 3-dB frequency of 1 rad/s, the element values are found by first defining the angle

$$\phi_k = \frac{k\pi}{2n} \tag{38}$$

where n is the desired order. The element values are given by

$$L_{2i-1} = 2 \sin \phi_{4i-3} \qquad C_{2i} = 2 \sin \phi_{4i-1} \qquad i = 1, 2, 3, \ldots, M \tag{39}$$

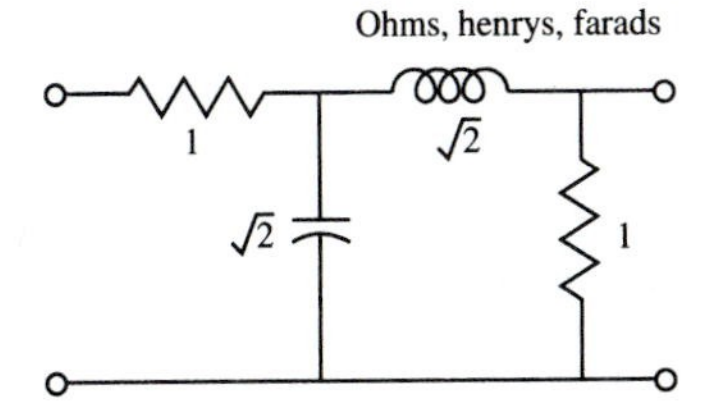

FIGURE 4.8-2
Filter for Example 4.8-1.

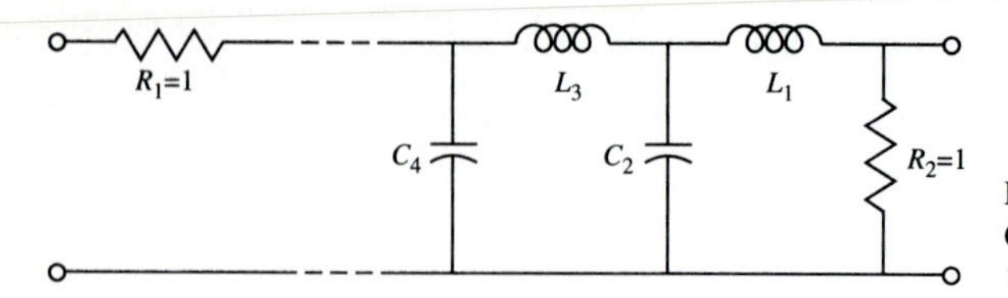

FIGURE 4.8-3
Configuration for Butterworth filter realizations.

where the elements are defined in Fig. 4.8-3 and where M is the highest value integer not larger than $\frac{1}{2}n$. This formulation takes account of the symmetry of the L and C values produced by the equal terminations.

Example 4.8-2 Determination of Butterworth filter element values. To find the normalized element values for a fifth-order Butterworth filter with equal 1-Ω terminating resistors, we first choose $i = 1$ in (39) and, using (38), find

$$L_1 = 2 \sin \phi_1 = 2 \sin \left(\tfrac{1}{10}\pi\right) = 0.6180 \text{ H} \qquad C_2 = 2 \sin \phi_3 = 2 \sin \left(\tfrac{3}{10}\pi\right) = 1.6180 \text{ F}$$

Next we choose $i = 2$ and obtain

$$L_3 = 2 \sin \phi_5 = 2 \sin \left(\tfrac{5}{10}\pi\right) = 2.0000 \text{ F}$$

The remaining element values, found from symmetry, are $C_4 = 1.6180$ F and $L_5 = 0.6180$ H. All these values are readily verified by the entries in the tables given in App. A.

Exercise 4.8-2. Repeat Example 4.8-2 for the following cases: (*a*) $n = 2$, (*b*) $n = 3$, (*c*) $n = 4$, (*d*) $n = 6$.

Answers. The answers are given in the tables in App. A.

Explicit relations for the element values of Butterworth functions using unequal-resistance-terminated filters, as well as for many cases of Chebyshev approximations, may be found in the references given in the Bibliography. For example, see W. K. Chen, *Passive and Active Filters*, Chap. 4.

In this section we have presented an introduction to the theory of double-resistance-terminated lossless ladder filters. The techniques developed may be extended to include the realization of a wide range of filter functions. Some of the results are given in the tables in App. A. The details of such an extension, however, are beyond the scope of our treatment here. The reader is referred to the more advanced texts listed in the Bibliography for a more detailed discussion of the methods.

4.9 TABLES FOR FILTER DESIGN

In the preceding sections of this chapter we have presented some of the simplest and most well-known techniques for the passive synthesis of transfer functions. In this section we shall illustrate the results of applying these techniques to various approximations to develop tables of filter realizations. To emphasize the application and use of tables, the treatment will be limited to low-order cases and a limited number of approximation types. An extensive set of tables covering most of the commonly

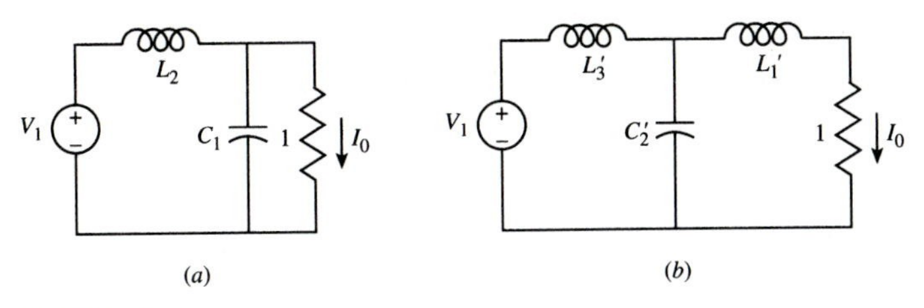

FIGURE 4.9-1
Low-pass realizations for $Y_T(s)$.

encountered orders and approximations is given in App. A. The material presented in this section will assist the reader in developing a wide range of applications for these tables.

Tables for Single-Resistance-terminated Filters

As an example of a filter design table, consider the single-resistance-terminated Butterworth filter that realizes a transfer admittance $Y_T(s) = I_0(s)/V_1(s)$. For the second-order case the filter has the form shown in Fig. 4.9-1(*a*). An impedance normalization has been made so that the output resistor has a value of 1 Ω. A frequency normalization is used so that the transfer function has a 3-dB frequency of 1 rad/s. The values for C_1 and L_2 are given in Table 4.9-1 using the legend at the top of the table. For the third-order case the filter has the form shown in Fig. 4.9-1(*b*). The values for L_1', C_2', and L_3' are given using the lower legend in Table 4.9-1. The filter configurations shown in Fig. 4.9-1 can be converted to realize a transfer impedance $Z_T(s) = V_0(s)/I_1(s)$ by a technique called *duality.* A description of the method is given later in this section. Here we note that duality produces the alternate second- and third-order filter realizations given in Figs. 4.9-2(*a*) and 4.9-2(*b*), respectively. The element values come from the same table (4.9-1)

TABLE 4.9-1
Element values for low-pass single-resistance-terminated Butterworth filters (1 rad/s bandwidth)

Elements in Figs. 4.9-1(*a*) ($n' = 2$) and 4.9-2(*b*) ($n = 3$)

n	C_1	L_2	C_3
2	0.7071	1.4141	
3	0.5000	1.3333	1.5000
n	L_1'	C_2'	L_3'

Elements in Figs. 4.9-2(*a*) ($n = 2$) and 4.9-1(*b*) ($n = 3$)

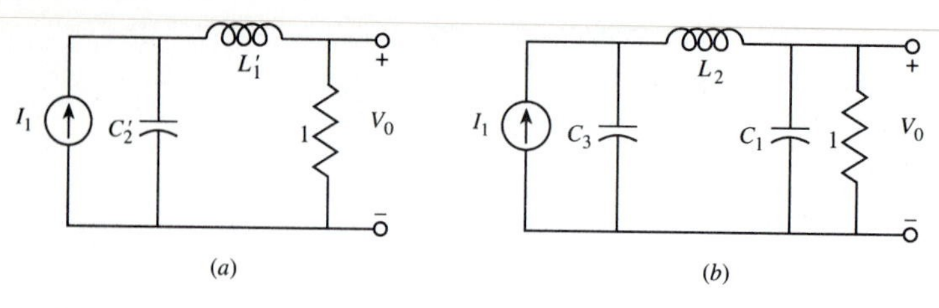

FIGURE 4.9-2
Low-pass realizations for $Z_T(s)$.

used for $Y_T(s)$. For the second-order case they are given by L_1' and C_2'. For the third-order case they are given by C_1, L_2, and C_3. Note that we have used the same table to produce different filter realizations. We shall see more examples of this versatility in the remainder of this section.

Tables for Double-Resistance-terminated Filters

Another example of passive filter design tables is for the double-resistance-terminated Butterworth filter. For the second- and third-order cases, the voltage transfer function $V_2(s)/V_1(s)$ realizations have the forms shown in Figs. 4.9-3(*a*) and 4.9-3(*b*), respectively. The element values are given in Table 4.9-2 using the top legend. Alternate realizations in which the first reactive element is an inductor rather than a capacitor are shown in Fig. 4.9-4. The element values for these realizations are determined by the legend at the bottom of Table 4.9-2. From Fig. 4.9-4 note that for the case where the terminating resistors are not equal ($R \neq 1$), the value of the source resistor is the inverse of the value used in Fig. 4.9-3.

Filter Conversion Methods

The filter realizations described above may be converted to realize many different types of transfer functions and to produce several alternate filter configurations. To describe these conversions, we begin by considering that the general form of a single-resistance-terminated lossless ladder network is as shown in Fig. 4.9-5(*a*). Similarly, we consider that the general form of a double-resistance-terminated lossless ladder network is as

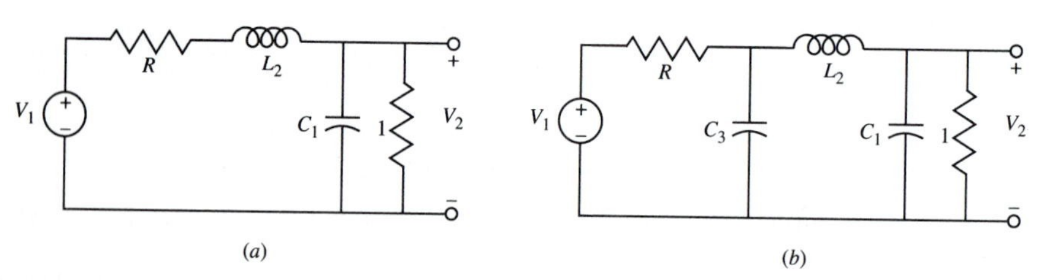

FIGURE 4.9-3
Double-resistance-terminated realizations for $V_2(s)/V_1(s)$.

TABLE 4.9-2
Element values for low-pass double-resistance-terminated Butterworth filters (1 rad/s bandwidth)

Elements in Figs. 4.9-3(a) ($n=2$) and 4.9-3(b) ($n=3$)

n	R	C_1	L_2	C_3
2	1.0	1.4141	1.4141	
2	0.5	3.3461	0.4483	
3	1.0	1.0000	2.0000	1.0000
3	0.5	3.2612	0.7789	1.1811
n	R	L_1'	C_2'	L_3'

Elements in Figs. 4.9-4(a) ($n=2$) and 4.9-4(b) ($n=3$)

shown in Fig. 4.9-6(a). We now show how the basic filter configurations given in these figures can be used as the starting point for the application of four conversion methods. These methods are in addition to the usual impedance and frequency denormalization that may be applied to the filters. The methods apply equally well to all the filter realizations catalogued in App. A.

CONVERSION METHOD 1: CHANGE OF VARIABLES. The first conversion method simply involves a change of variable. For example, consider the transfer admittance $Y_T(s) = I_0(s)/V_1(s)$ for a network N_1 with the variables defined in Fig. 4.9-5(a). Now consider the voltage transfer ratio $V_{OC}(s) = V_2(s)/V_1(s)$ for N_1. The variables are defined in Fig. 4.9-5(b), where we see that $V_2(s) = R_2 I_0(s)$. Thus we obtain $V_{OC}(s) = R_2 Y_T(s)$, that is, the two transfer functions are identical except for a multiplicative constant. Similarly, consider the transfer admittance $Y_T(s) = I_0(s)/V_1(s)$ of N_2 with the variables defined in Fig. 4.9-6(b). Comparing these variables with the ones defined in Fig. 4.9-6(a), we see that $I_0(s) = V_2(s)/R_2$. Thus we obtain $Y_T(s) = V_{OC}(s)/R_2$, that is, the two transfer functions are again identical except for a multiplicative constant. In both of the examples, however, note that an impedance denormalization will change the multiplicative

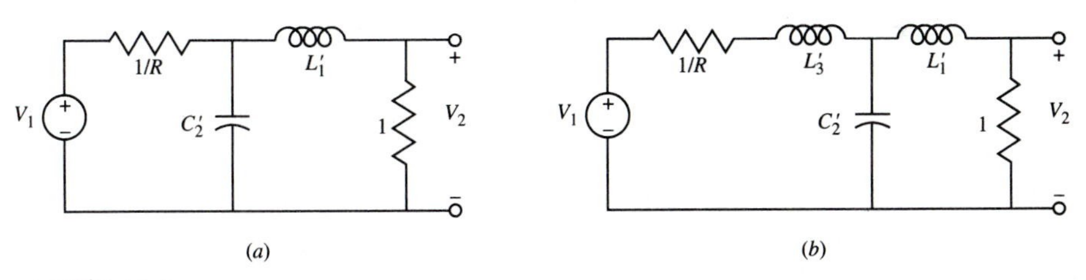

FIGURE 4.9-4
Double-resistance-terminated realizations for $V_2(s)/V_1(s)$.

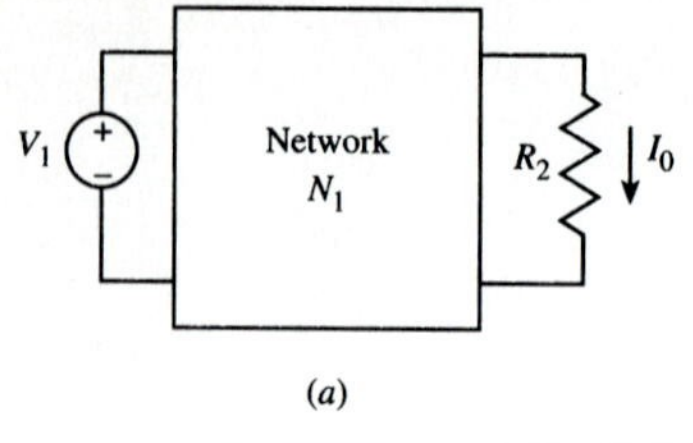

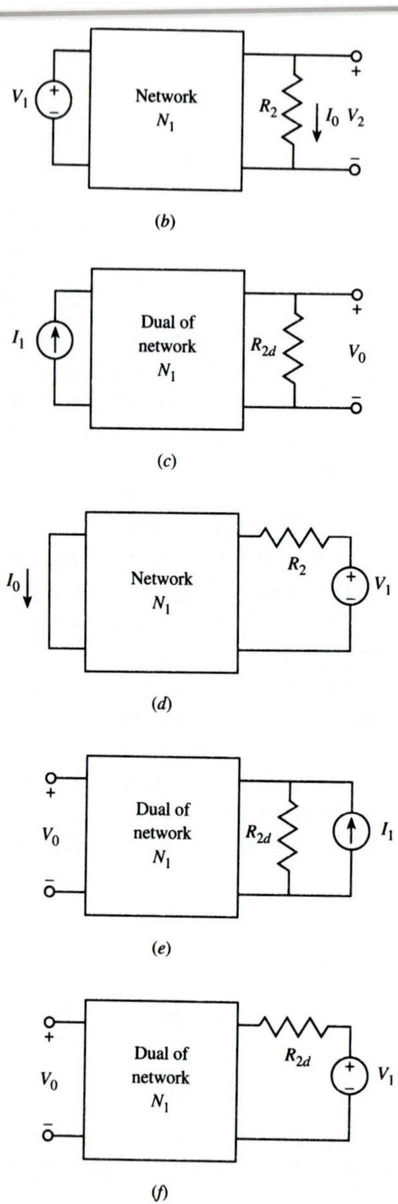

FIGURE 4.9-5
Conversions of a single-resistance-terminated filter.

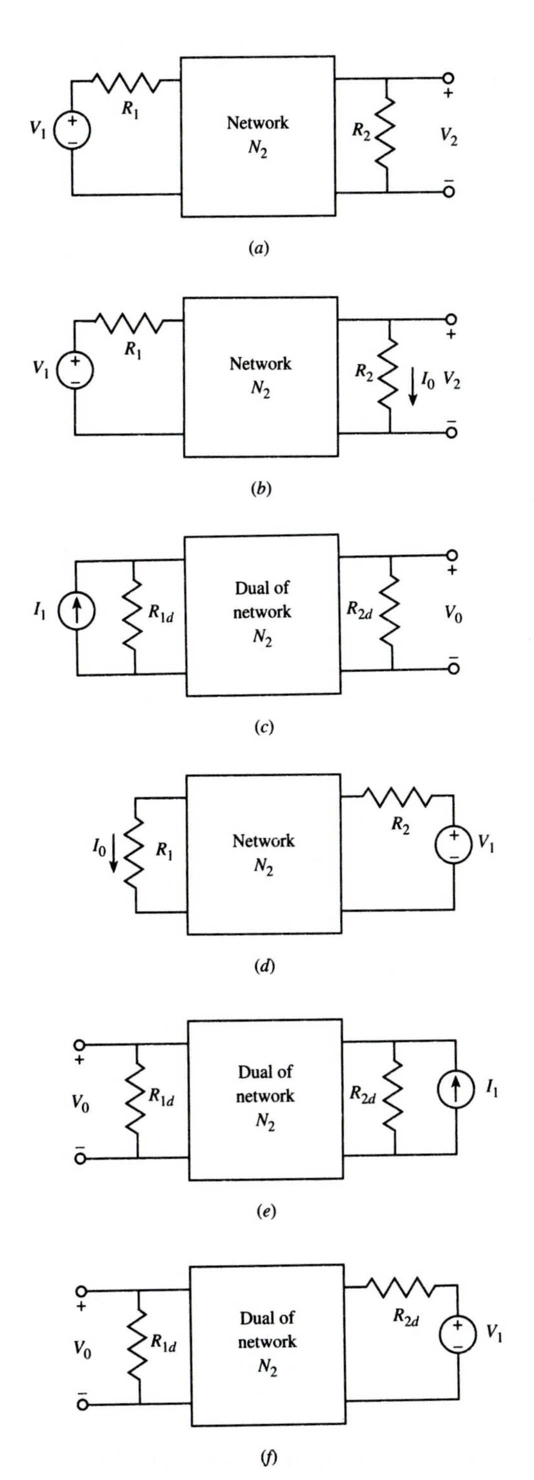

FIGURE 4.9-6
Conversions of a double-resistance-terminated filter.

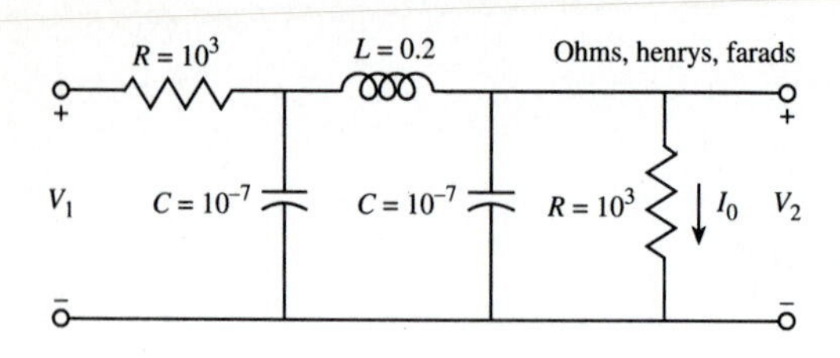

FIGURE 4.9-7
Example of change of variable.

constant in a transfer admittance $Y_T(s)$, but the constant in $V_{OC}(s)$ remains invariant under such an operation.

Example 4.9-1 Change of Variables. The filter shown in Fig. 4.9-7 has a low-pass Butterworth characteristic with a 3-dB frequency of 10^4 rad/s. Its voltage transfer function is

$$V_{OC}(s) = \frac{V_2(s)}{V_1(s)} = \frac{0.5 \times 10^{12}}{s^3 + 2 \times 10^4 s^2 + 2 \times 10^8 s + 10^{12}}$$

To find the transfer admittance $Y_T(s) = I_0(s)/V_1(s)$, we substitute $V_2(s) = I_0(s)\,R = I_0(s) \times 10^3$ to obtain

$$Y_T(s) = \frac{I_0(s)}{V_1(s)} = \frac{0.5 \times 10^9}{s^3 + 2 \times 10^4 s^2 + 2 \times 10^8 s + 10^{12}}$$

Note that the multiplicative constant in this latter expression will be changed if a different impedance denormalization is chosen for the filter.

Exercise 4.9-1. For the filter shown in Fig. 4.9-7, an additional frequency denormalization of 10 is made as well as an impedance denormalization of 10. The new element values are $R = 10^4\ \Omega$, $C = 10^{-9}$ F, and $L = 0.2$ H. Find the expressions for $V_{OC}(s)$ and $Y_T(s)$.

Answers.

$$V_{OC}(s) = \frac{V_2(s)}{V_1(s)} = \frac{0.5 \times 10^{15}}{s^3 + 2 \times 10^5 s^2 + 2 \times 10^{10} s + 10^{15}}$$

$$Y_T(s) = \frac{I_0(s)}{V_1(s)} = \frac{0.5 \times 10^{11}}{s^3 + 2 \times 10^5 s^2 + 2 \times 10^{10} s + 10^{15}}$$

CONVERSION METHOD 2: DUALITY. The second conversion method involves the use of duality, a fundamental property of circuits. For any given planar network,[1] using duality, we are able to construct a *dual network* that has the property that the form of the describing equations, and thus of the network functions, is the same as that of the original network. In addition, the elements have the same numerical values. The difference is in the type of elements, the way the elements are connected, and the type of variables used to define the equations. The specific relations are given in Table 4.9-3.

[1] A planar network is one that can be drawn in such a way that there are no crossing lines required.

TABLE 4.9-3
Relations between dual networks

Original network	Dual network
Kirchhoff voltage law (KVL) equations	Kirchhoff current law (KCL) equations
Mesh current	Node voltage
Series branch	Parallel branch
Inductor	Capacitor
Capacitor	Inductor
Resistor (Ω)	Resistor (S)
ICVS	VCIS
Controlling mesh current for an ICVS	Controlling node voltage for a VCIS
Output voltage for an ICVS	Output current for a VCIS

The dual of a given network may be found by the following graphical procedure:

1. In the center of each mesh of the given network place a node symbol. Number these symbols sequentially. In addition, place a ground reference node symbol outside of all the network meshes. Finally, draw a duplicate set of node symbols exterior to the network to provide the framework for the construction of the dual network.
2. Draw lines between the node symbols in such a way that each line crosses one network element. As each line is drawn, add an element to the dual network between corresponding nodes. Use the relations of Table 4.9-3 to determine the type of dual element. The reference polarity of sources in the dual network is readily observed by noting that a voltage source with a reference polarity that produces a positive mesh reference current has as its dual a current source whose reference direction is toward the numbered node symbol and away from the ground node symbol.

Example 4.9-2 The dual of a low-pass network. As an example of the construction of a dual network, consider the low-pass filter shown in Fig. 4.9-8(*a*). For this network, we readily find

$$Y_T(s) = \frac{I_0(s)}{V_1(s)} = \frac{1}{24s^3 + 42s^2 + 21s + 6}$$

Corresponding with step 1 given above, nodes have been identified and redrawn as shown in Fig. 4.9-8(*b*). Corresponding with step 2, the construction lines have been drawn and numbered in Fig. 4.9-8(*c*). The corresponding elements of the dual network are shown in Fig. 4.9-8(*d*). Since this is a dual network, it will have the same transfer function but with dual variables. Thus we obtain

$$Z_T(s) = \frac{V_2(s)}{I_1(s)} = \frac{1}{24s^3 + 42s^2 + 21s + 6}$$

The results of applying duality to the filter configurations shown in Figs. 4.9-5(*a*) and 4.9-6(*b*) are shown in Figs. 4.9-5(*c*) and 4.9-6(*c*), respectively. Note that the duality conversion has changed resistors R_1 and R_2 (ohms) to resistors $R_{1d} = 1/R_1$ and $R_{2d} = 1/R_2$ where R_{1d} and R_{2d} are also given in ohms. In most cases, the duality conversion

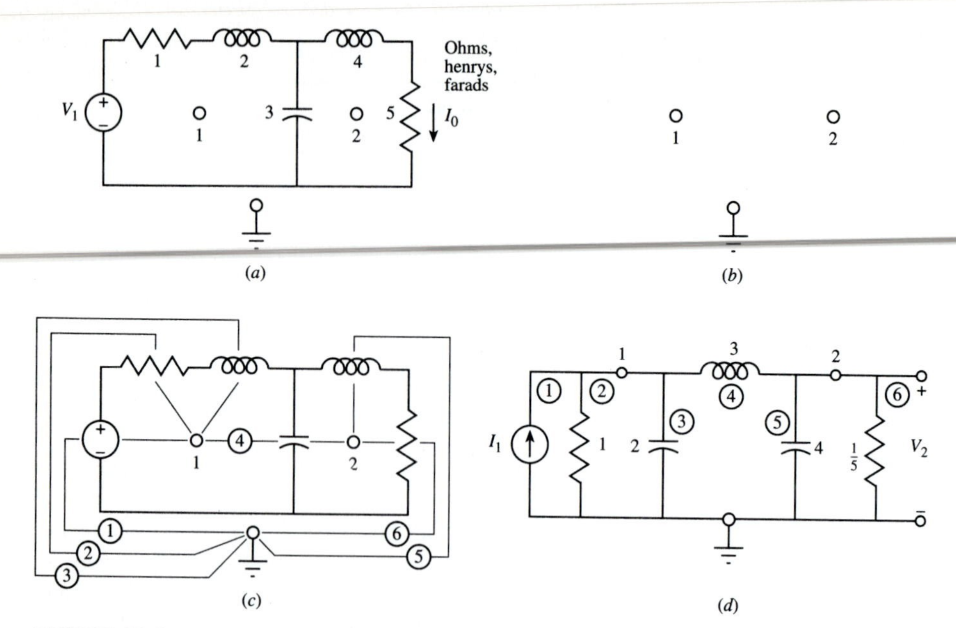

FIGURE 4.9-8
Construction of a dual network.

method is "built in" to filter realization tables. This is done by adding a second set of headings at the bottom of the table giving the identifiers for the dual elements. Examples of this may be found in Tables 4.9-1 and 4.9-2 and in the tables included in App. A.

CONVERSION METHOD 3: RECIPROCITY. A third method for the conversion of filter configurations is based on the reciprocity of the passive networks used in the realizations. In general, in any reciprocal network, interchanging the locations of the excitation source and the response variable leaves the transfer *immittance* (but *not* the dimensionless transfer ratios) of the network invariant. Thus, if we define the overall $z_{ij}(s)$ and $y_{ij}(s)$ as the two-port parameters of the entire filter [including the terminating resistor(s)], we have $z_{12}(s) = z_{21}(s)$ and $y_{12}(s) = y_{21}(s)$. For the networks shown in Figs. 4.9-5(a) and 4.9-6(b), application of reciprocity leads to the network configurations shown in Figs. 4.9-5(d) and 4.9-6(d), respectively. For each of these cases, the network function $Y_T(s) = I_0(s)/V_1(s)$ is unchanged by the indicated interchange of the locations of excitation and response. Similarly, for the networks shown in Figs. 4.9-5(c) and 4.9-6(c), application of reciprocity leads to the networks shown in Figs. 4.9-5(e) and 4.9-6(e), respectively. For these conversions the functions $Z_T(s) = V_0(s)/I_1(s)$ stay the same for each of the networks. Note that for the single-resistance-terminated filter shown in Fig. 4.9-5, this conversion method allows us to change from an ideal excitation source to a nonideal one.

Example 4.9-3 Use of reciprocity. From Table 4.9-2, we find that a normalized (1 rad/s 3-dB frequency) Butterworth filter with source resistance $\frac{1}{2}\ \Omega$ and load resistance $1\ \Omega$ has

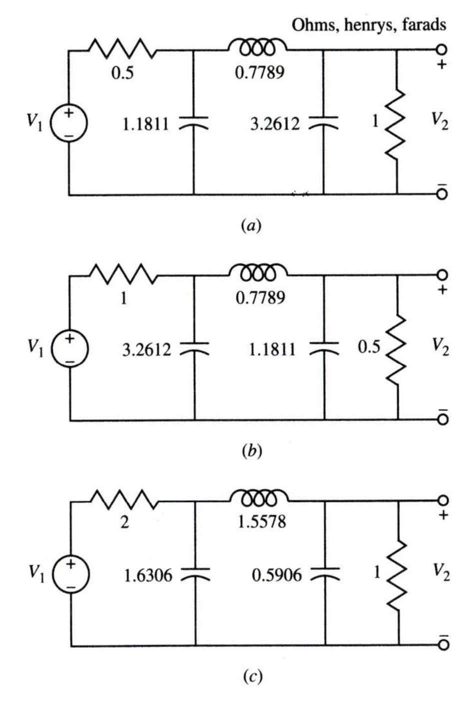

FIGURE 4.9-9
Example of the use of reciprocity.

the form shown in Fig. 4.9-9(*a*). Its voltage transfer function is

$$\frac{V_2(s)}{V_1(s)} = \frac{\frac{2}{3}}{s^3 + 2s^2 + 2s + 1}$$

It is desired to obtain a filter in which the source resistance is 2 Ω and the load resistance is 1 Ω. To modify the given circuit to obtain this result, we first reverse the network, break the connection between the 1-Ω resistor and ground to insert the excitation $V_1(s)$ and define $V_2(s)$ across the $\frac{1}{2}$-Ω resistor as shown in Fig. 4.9-9(*b*). Finally, we impedance denormalize the network by 2.0 to obtain the result shown in Fig. 4.9-9(*c*). For this, the voltage transfer function is

$$\frac{V_2(s)}{V_1(s)} = \frac{\frac{1}{3}}{s^3 + 2s^2 + 2s + 1}$$

CONVERSION METHOD 4: THEVENIN–NORTON TRANSFORMATIONS. A final method for the conversion of filters is the use of Norton–Thevenin and Thevenin–Norton source conversions. Thus, the Norton equivalent source consisting of $I_1(s)$ and R_{2d} for V_1 in Fig. 4.9-5(*e*) may be changed to the Thevenin equivalent source of $V_1(s)$ and R_{2d} shown in Fig. 4.9-5(*f*). Similarly, the Norton equivalent source of $I_1(s)$ and R_{2d} of Fig. 4.9-6(*e*) can be converted as shown in Fig. 4.9-6(*f*). Since $V_1(s) = R_{2d}I_1(s)$, the voltage transfer function for each of these circuits is $V_{OC}(s) = V_0(s)/V_1(s) = V_0(s)/[I_1(s)R_{2d}] = Z_T(s)/R_{2d}$.

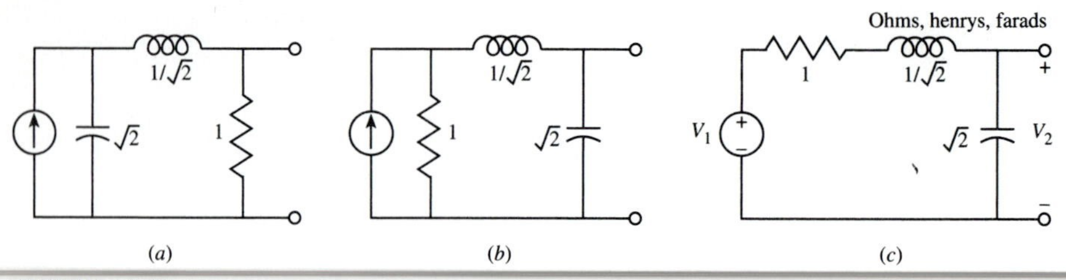

FIGURE 4.9-10
Realization of a voltage transfer function.

The four conversion methods described above may be applied in any combination to obtain a wide variety of filter configurations starting from a single prototype circuit.

Example 4.9-4 Conversion of a filter. As an example of the use of the techniques described above, consider the use of the tables to realize the following three filter functions, all of which realize a second-order low-pass Butterworth characteristic.

(*a*) *A voltage transfer function with a source resistance of unity and no load resistance.* From Table 4.9-1, using the heading at the bottom of the table (duality), we obtain the first network shown in Fig. 4.9-10. Applying reciprocity and a Norton–Thevenin conversion, we obtain the second and third networks, respectively. The third network satisfies the specifications.

(*b*) *A transfer impedance function with source and load resistors of unit value.* From Table 4.9-2, using the heading at the top of the table, we obtain the first network shown in Fig. 4.9-11. Applying a Thevenin–Norton conversion, we obtain the second network, which meets the specifications. Alternately, we can use the heading at the bottom of the table to directly obtain the dual realization shown as the third network of the figure.

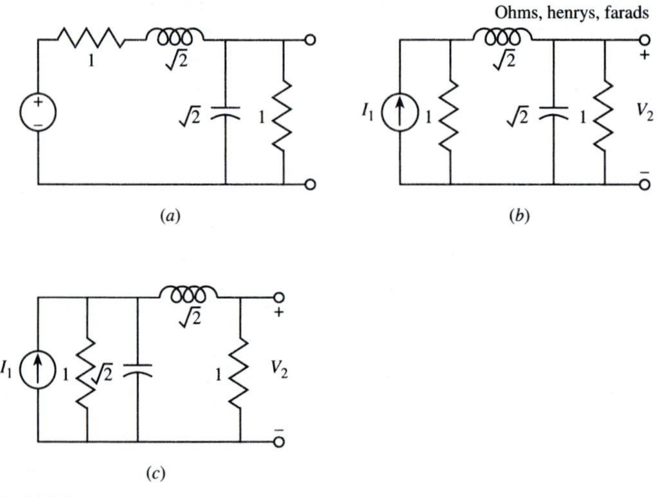

FIGURE 4.9-11
Realization of a transfer impedance.

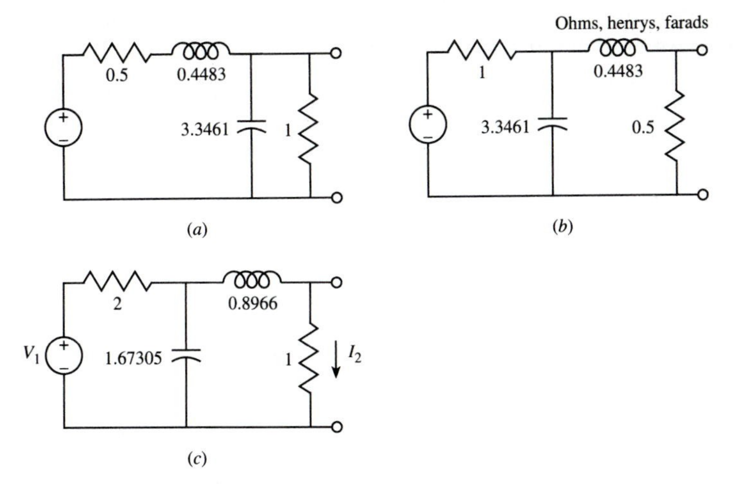

FIGURE 4.9-12
Realization of a transfer admittance.

(*c*) *A transfer admittance function with a source resistance of 2 Ω and a load resistance of 1 Ω*. From Table 4.9-2, using the heading at the top of the table, we obtain the first network shown in Fig. 4.9-12. Applying reciprocity and an impedance normalization, we obtain the second and third networks, respectively. The third network meets the specifications. Alternately, we can use the heading at the bottom of the table to obtain the network shown in Fig. 4.9-13(*a*). If we apply a Norton–Thevenin transformation, we obtain the network shown in Fig. 4.9-13(*b*), which meets the specifications.

In this section we have presented a set of conversion methods that allow us to greatly extend the range of applications of the tables that are a standard part of the literature of filters. The methods are of great importance, and the reader is strongly urged to become familiar with their use.

Sensitivity

One of the important reasons for the popularity of the passive filter realizations defined in the tables in this section and in App. A is their low sensitivities. In general, the Q and

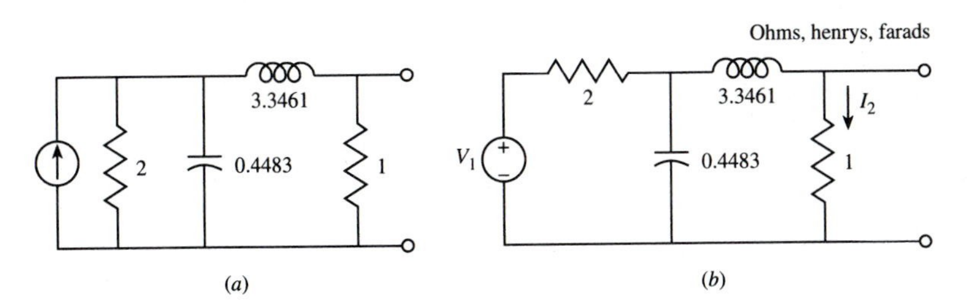

FIGURE 4.9-13
Alternate realization of a transfer admittance.

ω_n sensitivities have a maximum magnitude of 1.0, and the function sensitivities are also correspondingly low. An example follows.

Example 4.9-5 Sensitivity of passive low-pass filter. As an example of the sensitivities found in a passive filter realization, consider the third-order low-pass function realized by the double-resistance-terminated filter shown in Fig. 4.9-3(*b*). The general voltage transfer function is

$$N(s) = \frac{V_2(s)}{V_1(s)}$$

$$= \frac{\dfrac{1}{C_1 L_2 C_3 R_i}}{s^3 + s^2\left(\dfrac{1}{C_1 R_o} + \dfrac{1}{C_3 R_i}\right) + s\left(\dfrac{1}{C_1 C_3 R_i R_o} + \dfrac{1}{C_1 L_2} + \dfrac{1}{L_2 C_3}\right) + \dfrac{1}{C_1 L_2 C_3}\left(\dfrac{1}{R_i} + \dfrac{1}{R_o}\right)}$$

where R_i is the input resistance (on the left) and R_o is the output resistance (on the right). From Table 4.9-2, for a Butterworth characteristic and equal-resistance terminations ($R_i = R_o = 1$), the element values are $C_1 = C_3 = 1$ F and $L_2 = 2$ H. For these values, the transfer function can be written as

$$N(s) = \frac{V_2(s)}{V_1(s)} = \frac{0.5}{(s+1)(s^2+s+1)} = \frac{0.5}{(s+g)(s^2 + s\omega_n/Q + \omega_n^2)}$$

The Q, ω_n, and g sensitivities for the various network elements may be found using the methods of Sec. 3.5. The results are summarized in Table 4.9-4. Note that all the sensitivities are small; the largest magnitude is 0.5. The low-sensitivity properties of passive networks may also be demonstrated using function sensitivities. For the network realization described above, the function sensitivities are given by the plots in Fig. 4.9-14. The loci are defined over a frequency range $0 \le \omega \le 2$ rad/s. Note that in the passband $0 \le \omega \le 1$ rad/s, the magnitude of the function sensitivities is never greater than unity.

Analysis

In this section we have introduced tables giving ladder network realizations for the various approximations introduced in Chap. 2. These realizations, as well as the filters synthesized by the methods given in the earlier sections of the chapter, must sometimes be analyzed to determine the actual network function that they implement. There are

TABLE 4.9-4
Sensitivities for Example 4.9-5

Element	S^Q	S^{ω_n}	S^g
R_o	0.50	0	−0.50
C_1	0.25	−0.25	−0.50
L_2	−0.50	−0.50	0
C_3	0.25	−0.25	−0.50
R_i	0.50	0	−0.50

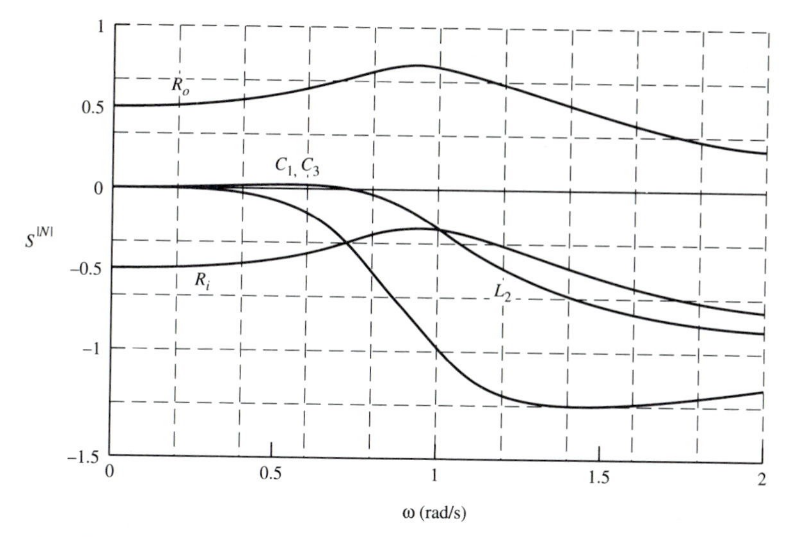

FIGURE 4.9-14
Function sensitivities for Example 4.9-5.

several situations in which such analysis is required: (1) to verify the correctness and accuracy of the element values determined from tables or from synthesis procedures, (2) to find the value of the multiplicative constant in the single-resistance-terminated ladder realizations for $Y_T(s)$ given in Secs. 4.6 and 4.7, and (3) to develop a symbolic form for the network function that can be used for the determination of sensitivity information. The algebraic steps required in the analysis procedure may be simplified by the use of a prototype ladder network with the elements defined topologically by blocks and identified by their impedance and admittance. Example prototype networks are shown for the four-element case in Figs. 4.9-15(*a*) and (*b*) and for the six-element case in Figs. 4.9-15(*c*) and (*d*). For each case, prototypes are given for the specification of the output (right) element as an admittance or an impedance. The admittance specification is most useful for low-pass and band-pass networks, while the impedance one is for high-pass and band-elimination situations. The voltage transfer functions for these prototype networks are given in Table 4.9-5. The use of these results is outlined in the following summary.

Summary 4.9-1 Analysis of ladder networks. To analyze a ladder network using the prototype networks shown in Fig. 4.9-15 and the transfer functions defined in Table 4.9-5, the steps are as follows:

1. For low-pass and band-pass networks choose the networks with the output (right) element defined as Y_1 in Figs. 4.9-15(*a*) and (*c*). For high-pass and band-elimination networks choose the networks with the output element defined as Z_1 in Figs. 4.9-15(*b*) and (*d*).
2. For 2nd- and 3rd-order low-pass and high-pass networks and 4th- and 6th-order band-pass and band-elimination ones choose the four-element prototypes shown in Figs.

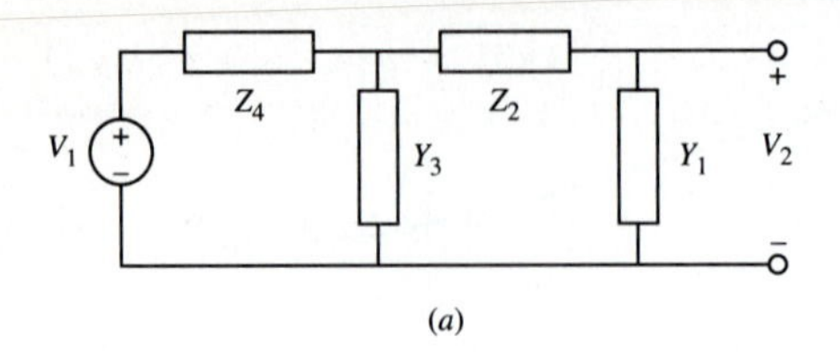

(a)

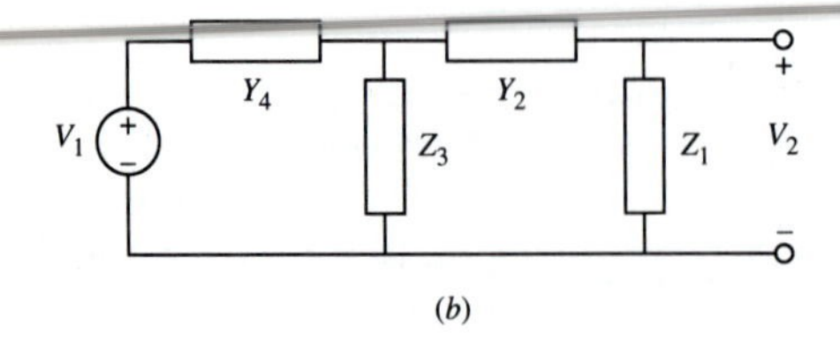

(b)

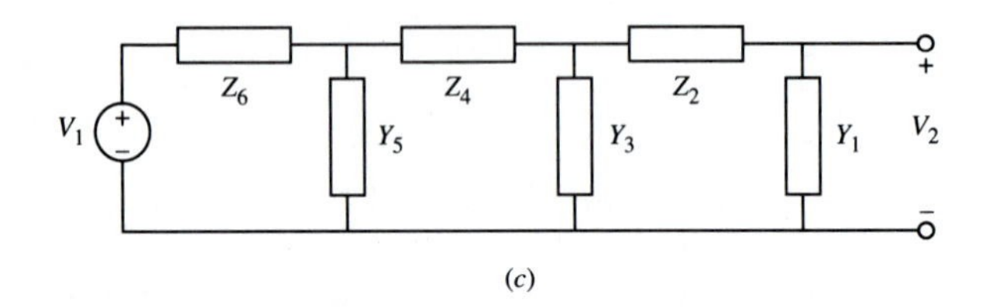

(c)

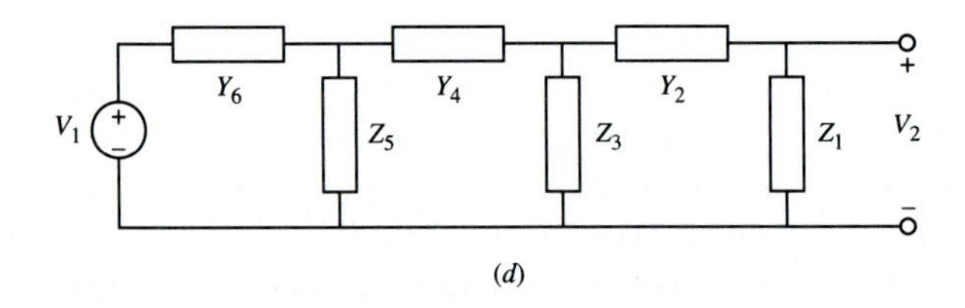

(d)

FIGURE 4.9-15
Prototype ladder networks for analysis.

TABLE 4.9-5
Voltage transfer functions for the networks shown in Fig. 4.9-15

Four elements—output Y
  Numerator: 1
  Denominator: $1 + Y_3Z_4 + Y_1Z_2 + Y_1Z_4 + Y_1Y_3Z_2Z_4$

Four elements—output Z
  Numerator: $Y_2Y_4Z_1Z_3$
  Denominator: $1 + Y_4Z_3 + Y_2Z_1 + Y_2Z_3 + Y_2Y_4Z_1Z_3$

Six elements—output Y
  Numerator: 1
  Denominator: $1 + Y_5Z_6 + Y_3Z_4 + Y_3Z_6 + Y_1Z_2 + Y_1Z_4 + Y_1Z_6 + Y_3Y_5Z_4Z_6 + Y_1Y_5Z_2Z_6 + Y_1Y_5Z_4Z_6$
  $+ Y_1Y_3Z_2Z_4 + Y_1Y_3Z_2Z_6 + Y_1Y_3Z_2Z_6 + Y_1Y_3Y_5Z_2Z_4Z_6$

Six elements—output Z
  Numerator: $Y_2Y_4Y_6Z_1Z_3Z_5$
  Denominator: $1 + Y_6Z_5 + Y_4Z_3 + Y_4Z_5 + Y_2Z_1 + Y_2Z_3 + Y_4Y_6Z_3Z_5 + Y_2Y_6Z_1Z_5 + Y_2Y_6Z_3Z_5 + Y_2Y_4Z_1Z_3$
  $+ Y_2Y_4Z_1Z_5 + Y_2Y_4Z_3Z_5 + Y_2Y_4Y_6Z_1Z_3Z_5$

4.9-15(*a*) and (*b*). For 4th- and 5th-order low-pass and high-pass networks and 8th- and 10th-order band-pass and band-elimination ones, choose the six-element prototypes shown in Figs. 4.9-15(*c*) and (*d*).

3. Identify the elements of the actual filter by assigning the symbolic names G_i, R_i, L_i, and C_i $(i = 1, 2, \ldots)$.
4. Replace the quantities Z_i and Y_i $(i = 1, 2, \ldots)$ in the expression given in Table 4.9-5 with the corresponding impedances and admittances of the filter elements using the symbolic names from step 3.
5. The resulting symbolic analysis can be changed to a numeric one by substituting numeric values for the quantities G_i, R_i, L_i, and C_i.

Example 4.9-6 Analysis of a third-order ladder network. As an example of the use of the analysis procedure given in the preceding summary, consider the third-order low-pass filter shown in Fig. 4.9-16. Subscripts corresponding with the ones used in the prototype networks of Fig. 4.9-15 are used to identify the *RLC* components. A comparison of the two figures gives

$$Y_1(s) = sC_1 + G_1 \qquad Z_2(s) = sL_2 \qquad Y_3(s) = sC_3 \qquad Z_4(s) = R_4$$

If these expressions are substituted in the expression under "Four elements—output *Y*" in Table 4.9-5, we obtain

$$\frac{V_2(s)}{V_1(s)} = \frac{1}{1 + (sC_3)(R_4) + (sC_1 + G_1)(sL_2) + (sC_1 + G_1)(R_4) + (sC_1 + G_1)(sC_3)(sL_2)(R_4)}$$

Rearranging terms, we obtain

$$\frac{V_2(s)}{V_1(s)} = \frac{1}{s^3C_1L_2C_3R_4 + s^2(C_1L_2 + G_1L_2C_3R_4) + s(G_1L_2 + C_1R_4 + C_3R_4) + (G_1R_4 + 1)}$$

For the element values for the Butterworth double-resistance- (1-Ω resistor) terminated filter in Table 4.9-2, we find $G_1 = 1$ S, $C_1 = 1$ F, $L_2 = 2$ H, $C_3 = 1$ F, and $R_4 = 1\ \Omega$. Inserting these values, the preceding symbolic expression becomes

$$\frac{V_2(s)}{V_1(s)} = \frac{1}{2s^3 + 4s^2 + 4s + 2}$$

This is the third-order Butterworth function with a dc value of 0.5. As an example of sensitivity determination, consider the case where all the numeric element values given above except the one for C_1 are used. We obtain

$$N(s) = \frac{V_2(s)}{V_1(s)} = \frac{1}{s^3(2C_1) + s^2(2C_1 + 2) + s(C_1 + 3) + 2}$$

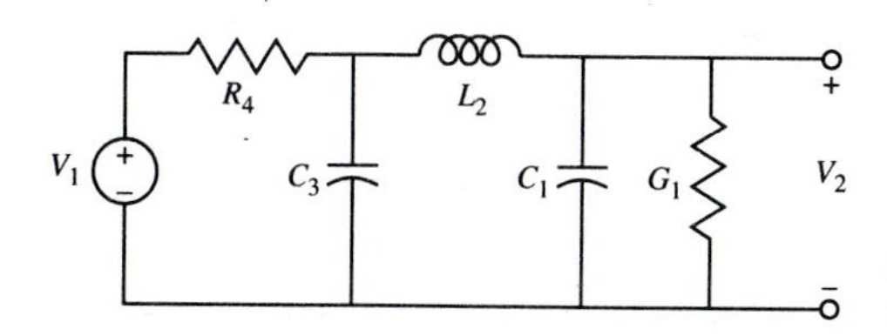

FIGURE 4.9-16
Third-order low-pass network for Example 4.9-6.

This is easily differentiated to obtain (assuming the nominal value $C_1 = 1$)

$$S_{C_1}^{N(s)} = \frac{\partial N(s)}{\partial C_1}\frac{C_1}{N(s)} = -\frac{2s^3 + 2s^2 + s}{2s^3 + 4s^2 + 4s + 2}$$

The methods given in Chap. 3 are readily applied to this expression to obtain the magnitude or phase sensitivity.

Exercise 4.9-6. Use the element values given in Table 4.9-2 for a third-order Butterworth filter with an input resistor of 0.5 Ω in the symbolic expression derived in the preceding example and verify that the resulting function is correct.

Answer

$$G_1 = 1\text{ S} \qquad C_1 = 3.2612\text{ F} \qquad L_2 = 0.7789\text{ H} \qquad C_3 = 1.1811\text{F}$$

$$\frac{V_2(s)}{V_1(s)} = \frac{1}{1.5s^3 + 3s^2 + 3s + 1.5}$$

4.10 COUPLED RESONATOR BAND-PASS FILTERS

In the preceding section we presented a discussion of the use of tables to realize low-pass filters. The techniques presented in Sec. 2.5 are readily applied to such filters to develop high-pass, band-pass, or band-elimination (notch) realizations. When these techniques are used to realize band-pass functions with a narrow bandwidth, however, the resulting filters present some practical problems. One of these is the large spread of element values that results. Another is the difficulty of using standard values for the network components. In this section we present a technique that is useful in obtaining narrow-band band-pass filters that have a more practical configuration.

Use of Coupled Resonators

One solution to the problems encountered in narrow-band filter realizations is the use of single reactive elements to couple the resonant circuits that provide the filtering capability of the band-pass realizations. The results are called *coupled resonator* filters. The coupling elements may be of three types: (1) capacitors, (2) inductors, or (3) mutual inductance. The determination of their element values may be made by developing coupled resonator circuits that are equivalent to the circuits obtained by the low-pass to band-pass transformation. The equivalence is established by using a node analysis or a mesh analysis. Although the equivalence is not exact, it provides satisfactory results in most applications. In this section we shall limit our discussion to the node-based capacitor-coupled resonator realization. It has several practical advantages including the minimization of the number of inductors required, the capability of choosing standard values for the inductors, and the fact that the inductors occur as grounded elements. In Chap. 7 we shall see that this latter feature permits us to apply inductance simulation methods in which passive filters are realized in an active *RC* form. A treatment of the other coupling methods may be found in the literature.

Low-Pass Prototype Network

The node-based capacitor-coupled resonator band-pass filter can be developed by first considering a low-pass prototype having the form shown in Fig. 4.10-1. In the figure, the quantities x_i $(i = 1, 3, 5, \ldots)$ are expressed in farads, and the quantities x_i $(i = 2, 4, 6, \ldots)$ in henrys. *Quality factors* q may be defined for the elements as

$$q = \omega_0 R_C C \qquad \text{and} \qquad q = \frac{\omega_0 L}{R_L} \tag{1}$$

where R_C is the value of a resistance in parallel with a capacitor and R_L is the value of a resistance in series with an inductor. For a normalized prototype filter with unity bandwidth, these factors are evaluated for $\omega_0 = 1$ rad/s. At this frequency, the quality factors for the individual elements defined in Fig. 4.10-1 are

$$\begin{gathered} q_1 = x_1 R_1 \qquad q_2 = q_3 = \cdots = q_{n-1} = \infty \\ q_n = x_n R_2 \qquad \text{or} \qquad q_n = \frac{x_n}{R_2} \end{gathered} \tag{2}$$

We may now define *coupling coefficients* k_{ij}, which represent the geometric mean value of adjacent reactive elements at the low-pass bandwidth frequency:

$$k_{ij} = \left. \frac{1}{\omega_0 \sqrt{x_i x_j}} \right|_{\omega_0 = 1} = \frac{1}{\sqrt{x_i x_j}} \tag{3}$$

Since the prototype network has a ladder structure, the coupling coefficients only exist for adjacent nodes. For the network shown in Fig. 4.10-1, we have

$$k_{12} = \frac{1}{\sqrt{x_1 x_2}} \qquad k_{23} = \frac{1}{\sqrt{x_2 x_3}} \qquad k_{34} = \frac{1}{\sqrt{x_3 x_4}} \qquad \text{etc.} \tag{4}$$

Now consider the effect of denormalizing the low-pass prototype in order to obtain the desired band-pass bandwidth BW. The band-pass center frequency will be 1 rad/s. The

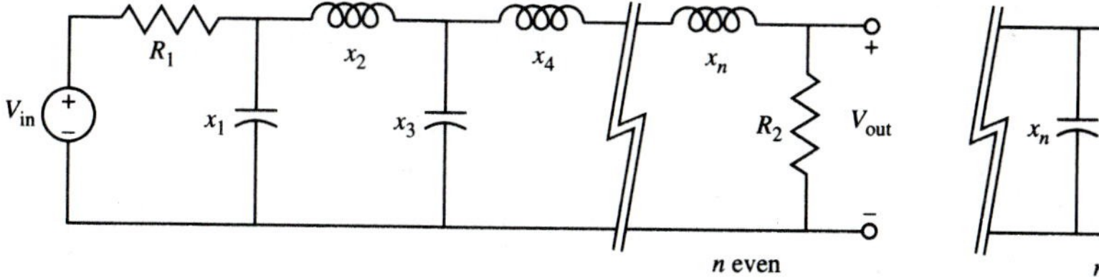

FIGURE 4.10-1
Low-pass prototype network.

low-pass element values become

$$X_i = \frac{x_i}{\mathrm{BW}} \tag{5}$$

The denormalized quality factors and coupling coefficients become

$$Q_i = \frac{q_i}{\mathrm{BW}} \qquad K_{ij} = \frac{1}{\sqrt{\dfrac{x_i}{\mathrm{BW}}\dfrac{x_j}{\mathrm{BW}}}} = \mathrm{BW}\,k_{ij} \tag{6}$$

These quantities will be used in the synthesis procedure given later in this section.

Coupled Resonator Filter

The general form of the narrow-band capacitor-coupled resonator band-pass filter is shown in Fig. 4.10-2. The resonators are defined by the L-C pairs: $L_{\mathrm{I}}, C_{\mathrm{I}}; L_{\mathrm{II}}, C_{\mathrm{II}}; \ldots$. The inductor values will shortly be shown to be a free parameter in the design process. Typically, they are all chosen to have the same value. The coupling capacitors are shown as $C_{12}, C_{23}, \ldots$. In addition, we define the total nodal capacitance C_i at the nodes designated as i ($i = 1, 2, \ldots, n$) in the figure. These represent the capacitance seen at the ith node when the adjacent nodes are both grounded. For this to be true, from Fig. 4.10-2 we see that

$$C_1 = C_{\mathrm{I}} + C_{12} \qquad C_2 = C_{12} + C_{\mathrm{II}} + C_{23} \qquad \text{etc.} \tag{7}$$

The total nodal capacitance values are the values that effectively resonate with the inductors of the resonators at the (normalized) band-pass center frequency of 1 rad/s. Thus, they must also obey the relations

$$C_1 = \frac{1}{L_{\mathrm{I}}} \qquad C_2 = \frac{1}{L_{\mathrm{II}}} \qquad \text{etc.} \tag{8}$$

Finally, the values of the coupling capacitors C_{ij} are related to the total nodal capacitance and the coupling coefficients by the relation

$$C_{ij} = K_{ij}\sqrt{C_i C_j} \tag{9}$$

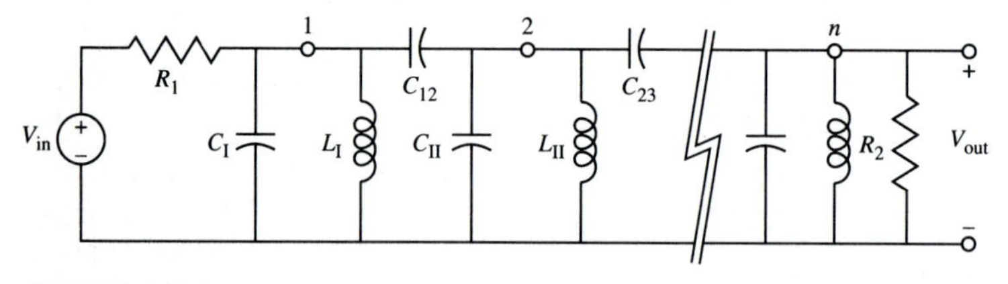

FIGURE 4.10-2
Capacitor-coupled resonator band-pass network.

The total nodal capacitance is also used to determine the values of the input and output resistors R_1 and R_2 in Fig. 4.10-2. These must be chosen so that quantities Q_1 and Q_n defined by the first equation of (6) have the correct values. For these we find

$$Q_1 = \omega_0 C_1 R_1\big|_{\omega_0=1} = C_1 R_1 \qquad Q_n = \omega_0 C_n R_2\big|_{\omega_0=1} = C_n R_2 \tag{10}$$

The quantities defined above can be used in a synthesis procedure to find the element values of a coupled resonator filter. The design procedure follows.

Summary 4.10-1 Design of a coupled resonator band-pass filter. A narrow-band band-pass filter with a (normalized) center frequency of 1 rad/s and a bandwidth BW may be designed as a coupled-capacitor resonator filter as follows:

1. Determine the element values x_i $(i = 1, 2, \ldots, n)$ of an appropriate low-pass prototype network with a (normalized) passband of 0 to 1 rad/s following the form given in Fig. 4.10-1.
2. Compute the normalized quality factors q_1 and q_n (the others will have an infinite value) and the normalized coupling coefficients k_{ij} using (2) and (3).
3. Compute the denormalized quality factors Q_i and coupling coefficients K_{ij} using (6).
4. Choose convenient values for the resonator inductors $L_{\mathrm{I}}, L_{\mathrm{II}}, \ldots$ shown in Fig. 4.10-2.
5. Determine the total nodal capacitance C_i for each of the nodes shown in Fig. 4.10-2 using the relations of (8).
6. Find the values of the coupling capacitors C_{ij} from (9).
7. Find the values of the terminating resistors R_1 and R_2 from (10).
8. Find the values of the shunt capacitors $C_{\mathrm{I}}, C_{\mathrm{II}}, \ldots$ from (7).

An example follows.

Example 4.10-1 Fourth-order coupled resonator band-pass filter. It is desired to synthesize a fourth-order narrow-band band-pass filter with a bandwidth of 0.1 rad/s and a (normalized) center frequency of 1 rad/s. The filter is to have a Butterworth characteristic. The low-pass prototype (with a 1-rad/s bandwidth) is found from App. A to have the form shown in Fig. 4.10-3(*a*). Let us first consider using the low-pass to band-pass transformation given in Sec. 2.5 to realize the band-pass filter. The initial step is to denormalize the low-pass filter so that it has a bandwidth of 0.1 rad/s. The result is shown in Fig. 4.10-3(*b*). If we apply the low-pass to band-pass transformation to this filter, we obtain the fourth-order band-pass filter shown in Fig. 4.10-3(*c*), which will meet the specifications. Note that this realization has a very large spread of element values for the inductor, namely $14.142/0.70711 \approx 200$. To obtain better control over the element values, let us consider a second realization, namely the use of a capacitor-coupled resonator design. The basic form of the realization is shown in Fig. 4.10-2. Using the procedure defined in Summary 4.10-1, from Fig. 4.10-3(*a*) we find (steps 1 and 2)

$$x_1 = 1.4142 \qquad x_2 = 1.4142 \qquad q_1 = 1.4142 \qquad q_2 = 1.4142 \qquad k_{12} = 0.70711$$

After applying the constant $\mathrm{BW} = 0.1$, the denormalized quality factors and coupling coefficients become (step 3)

$$Q_1 = 14.142 \qquad Q_2 = 14.142 \qquad K_{12} = 0.070711$$

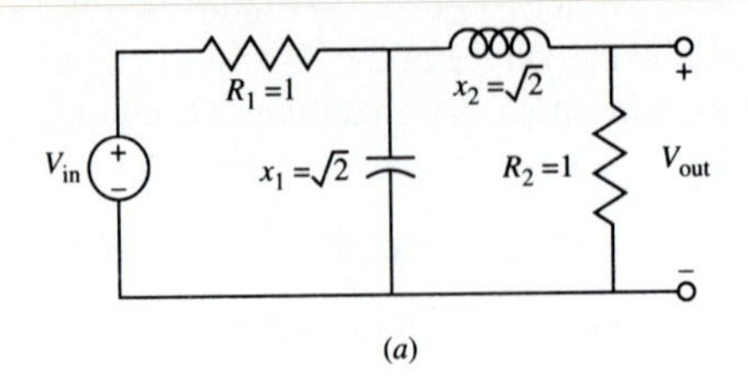

(a)

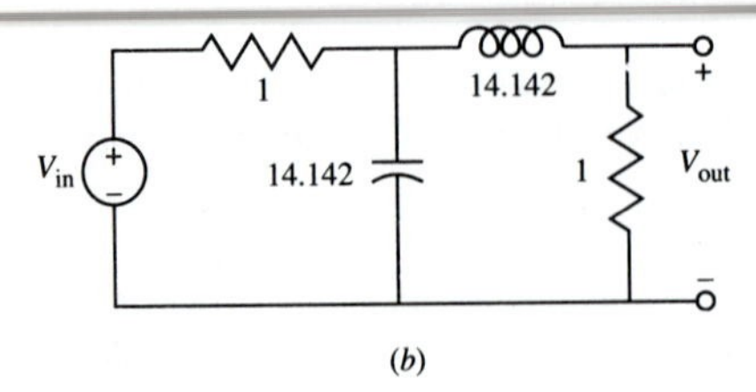

(b)

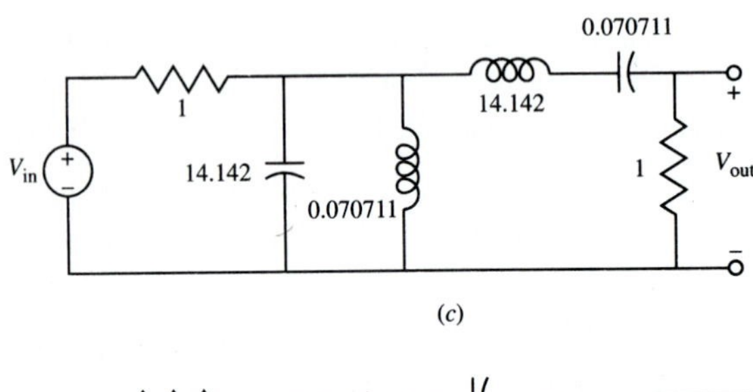

(c)

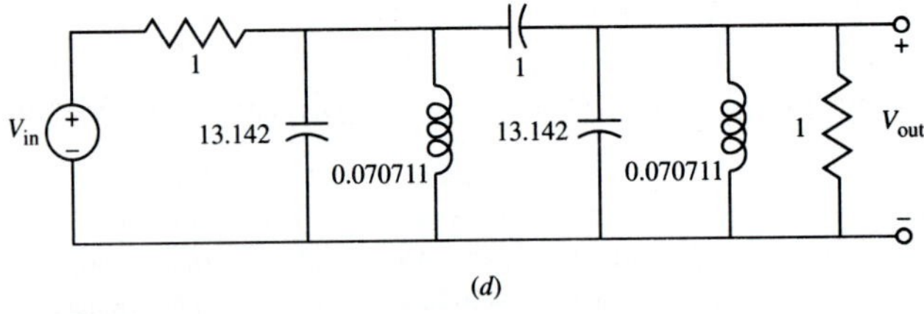

(d)

FIGURE 4.10-3
Development of fourth-order band-pass filters.

We now choose the inductors of the resonators as $L_{\mathrm{I}} = L_{\mathrm{II}} = 0.070711$ H (step 4), the same value as was obtained for the grounded inductor in the circuit shown in Fig. 4.10-3(*c*). The total nodal capacitance values are chosen to resonate with the inductors at the band-pass center frequency; thus $C_1 = C_2 = 14.142$ F (step 5). The coupling capacitor is found as $C_{12} = 1$ F (step 6). The values of the shunt capacitors are (step 7)

$$C_{\mathrm{I}} = C_1 - C_{12} = 13.142 \text{ F} \qquad C_{\mathrm{II}} = C_2 - C_{12} = 13.142 \text{ F}$$

The terminating resistance values are (step 8)

$$R_1 = \frac{Q_1}{C_1} = 1\ \Omega \qquad R_2 = \frac{Q_2}{C_2} = 1\ \Omega$$

The final circuit realization is shown in Fig. 4.10-3(*d*). Plots of the magnitude of the voltage transfer function for the low-pass to band-pass transformed filter and the coupled resonator

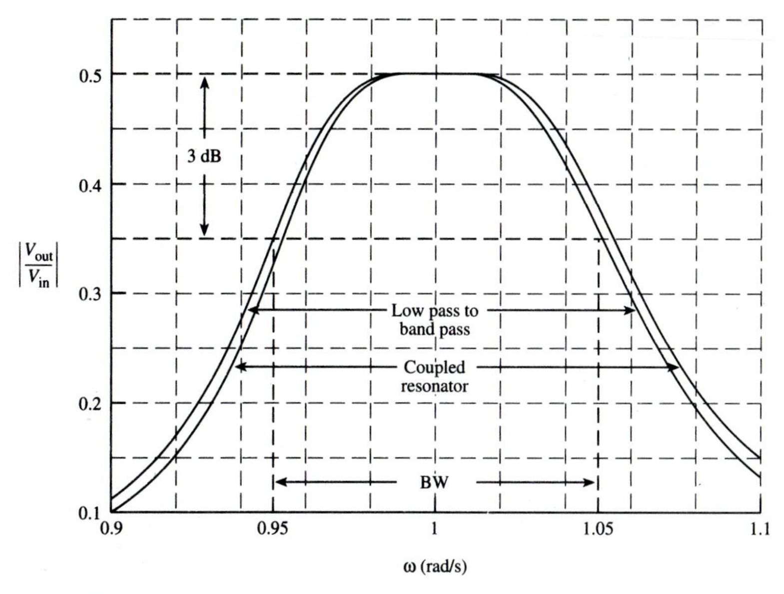

FIGURE 4.10-4
Magnitude plots for band-pass filters.

one are shown in Fig. 4.10-4. The plots illustrate the typical close agreement between the characteristics of the two types of realizations.

Exercise 4.10-1. Repeat the design given in Example 4.10-1 for each of the choices of the resonator inductors L_{I} and L_{II} given below:

	Answers		
L_{I} and L_{II}, H	**C_{12}, F**	**C_{I} and C_{II}, F**	**R_1 and R_2, Ω**
0.1	0.70711	9.2929	1.4142
0.5	0.14142	1.8586	7.0711

4.11 COMPUTER SIMULATION OF PASSIVE FILTER DESIGNS

One of the most important operations in the design of passive filters is the simulation of the final design. This provides the verification of the various steps of table look-up, normalizations, and other computational steps that are mandatory prior to the actual commitment of time, resources, and money required in the actual filter fabrication. Simulation operations can cover a wide range of goals and complexities. At the most basic end of the scale is a simple sinusoidal steady-state analysis that produces output plots of the magnitude and/or the phase and delay of the filter's transfer function. Other more specialized simulations can be used to provide some of the following:

Sensitivity analysis in which critical element tolerances and accuracies are determined

Symbolic analysis in which the complex-frequency plane pole and zero locations realized are identified and compared with the ones specified by the design

Monte Carlo analysis in which the effect of selected component tolerances can be estimated using statistical techniques

Application of optimization strategies to compensate for the effect of parasitics and other deviations from the ideal behavior of the components

Algorithms for any of the simulation operations described above are readily implemented on a PC (personal computer). Many shareware and commercial PC programs are available that provide implementations of the methods. In this section we illustrate the use of one of the most well known of these, the program PSpice, developed by MicroSim Corporation of Laguna Hills, California. It has a wide range of capabilities and is available in reasonably priced student editions as well as in more full-featured professional versions. Implementations of the program are also available for various workstation hosts. Some examples of the use of PSpice follow.

Example 4.11-1 Simulation of an elliptic filter design. As an example of the use of PSpice, it is desired to verify the design of a low-pass filter with a 1-dB ripple in the passband from 0 to 1 kHz and a minimum stopband attenuation of 55 dB in a stopband starting at 1.05 kHz. From App. A and Table A-6 we find a ninth-order elliptic filter will meet the specifications. Using the lower legend in the table, the basic design is shown in Fig. 4.11-1 (for convenience, the prime notation of the table has been omitted). The steps in the determination of the element values are summarized in Table 4.11-1.

The PSpice input file for plotting the overall magnitude characteristic from 0 to 2 kHz is shown in Fig. 4.11-2. The node numbers used are the ones shown in Fig. 4.11-1. Figure 4.11-3 shows plots of the simulation of the magnitude of the voltage transfer function of the filter. In Fig. 4.11-3(a) the magnitude characteristic corresponding with the

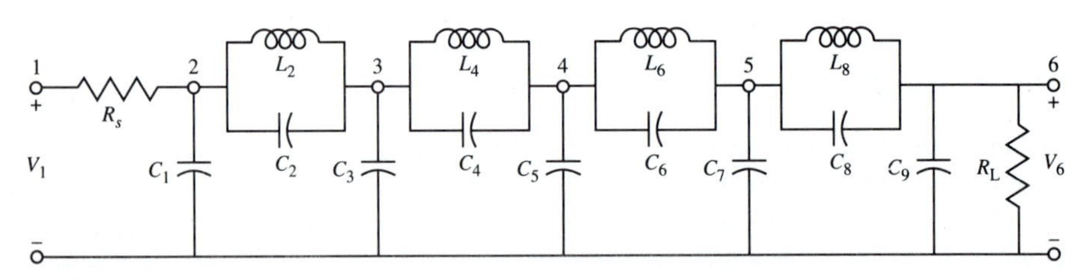

FIGURE 4.11-1
Ninth-order elliptic filter.

TABLE 4.11-1
Realization of a ninth-order elliptic filter

Elements	Normalized values from Table A-6 (Ω, H, F)	Values after frequency denormalization $\Omega_n = 2\pi \times 10^3$ (Ω, mH, mF)	Values after impedance denormalization $z_n = 10^3$ (kΩ, H, μF)
R_s	1.0	1.0	1.0
C_1	1.95471	0.311102	0.311102
L_2	0.95672	0.152267	0.152267
C_2	0.26172	0.041654	0.041654
C_3	1.94887	0.310172	0.310172
L_4	0.46951	0.074725	0.074725
C_4	1.76694	0.281217	0.281217
C_5	1.12605	0.179216	0.179216
L_6	0.35392	0.056328	0.056328
C_6	2.54224	0.404610	0.404610
C_7	1.40979	0.224375	0.224375
L_8	0.63099	0.100425	0.100425
C_8	0.99407	0.158211	0.158211
C_9	1.47897	0.235385	0.235385
R_L	1.0	1.0	1.0

input file shown in Fig. 4.11-2 is shown. In Fig. 4.11-3(*b*) the frequency and plotting limits have been changed to show an expanded plot of the passband from 0 to 1 kHz. This plot verifies the 1-dB equal-ripple specification for the passband. In Fig. 4.11-3(*c*) an expanded plot of the frequency range 1.05 to 2 kHz is shown. The ordinate scale limits have been chosen to verify the equal-ripple behavior of the stopband.

```
NINTH-ORDER ELLIPTIC LOW-PASS FILTER
VIN 1 0 AC 1
RS  1  2  1K
C1  2  0  0.311102U
          0.152267
L2  2  3
          0.041654U
C2  2  3  0.310172U
C3  3  0  0.074725
L4  3  4  0.281217U
          0.179216U
C4  3  4
          0.056328
C5  4  0  0.404610U
L6  4  5  0.224375U
C6  4  5  0.100425
          0.158211U
C7  5  0
          0.235385U
L8  5  6  1K
C8  5  6
C9  6  0
RL  6  0
- AC LIN 501 0.0001 2K
- PLOT AC VM(6)
- PROBE
- END
```

FIGURE 4.11-2
PSpice input file for ninth-order elliptic filter.

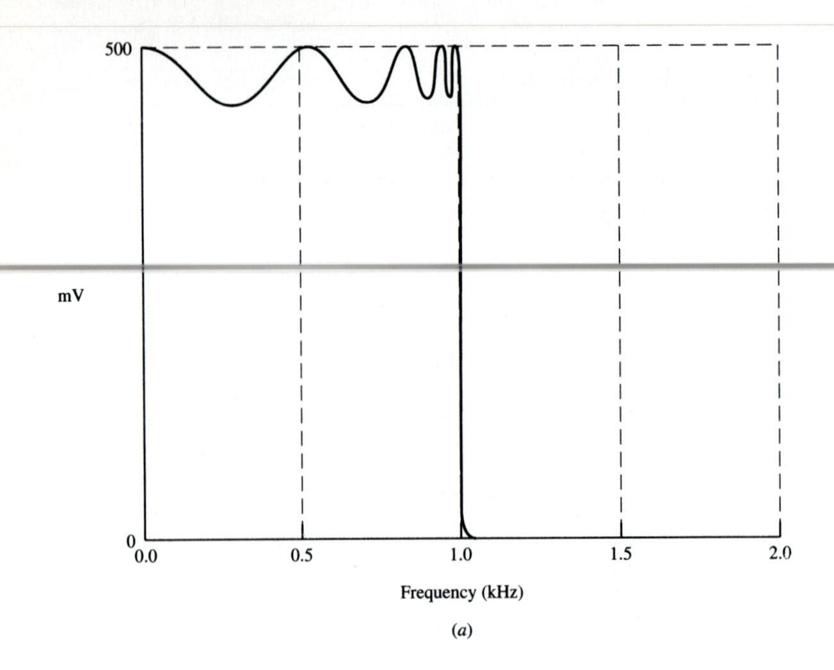

500
mV
0
0.0
0.5
1.0
1.5
2.0
Frequency (kHz)

(a)

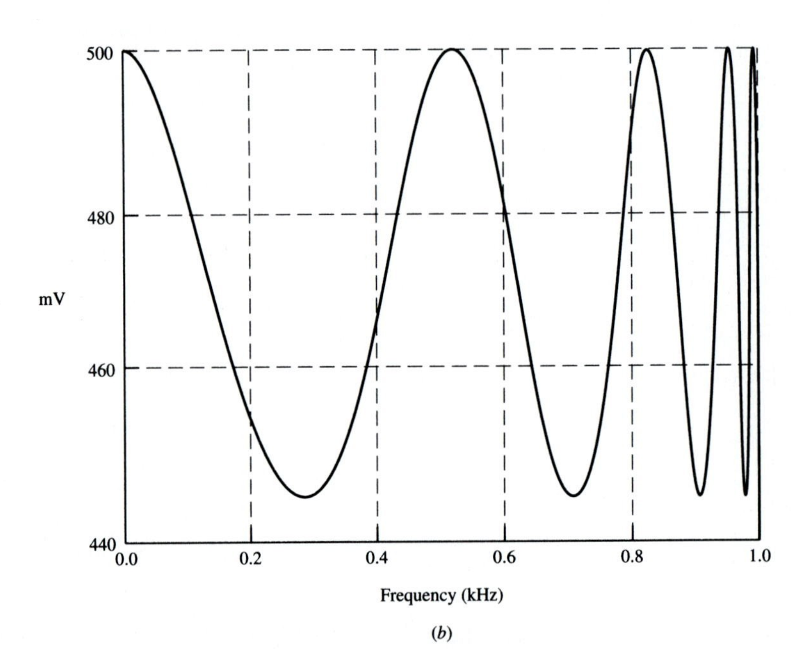

500
480
mV
460
440
0.0
0.2
0.4
0.6
0.8
1.0
Frequency (kHz)

(b)

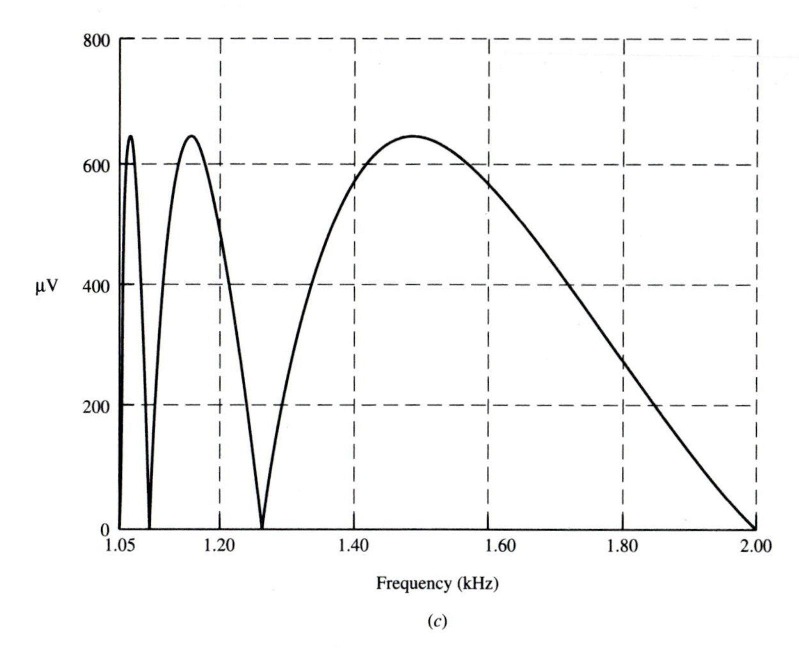

FIGURE 4.11-3
PSpice output magnitude plots.

PROBLEMS

Section 4.1

1. Find the incidence matrix for the network shown in Fig. P4.1-1. Show that it can be used to express the nodal voltages in terms of the branch voltages.

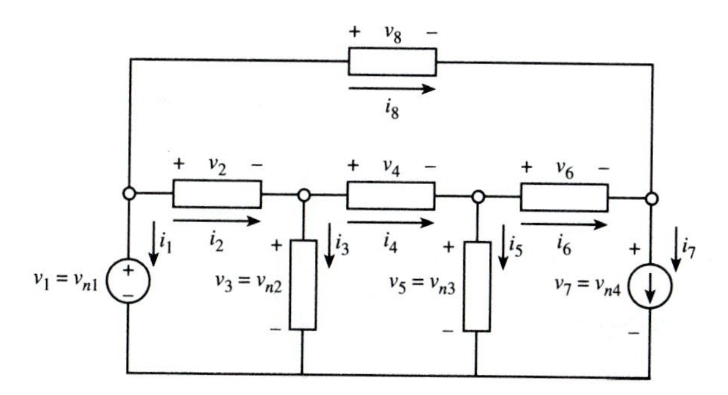

FIGURE P4.1-1

2. Apply (21) to find the admittance $Y(s)$ for the network shown in Fig. P4.1-2.

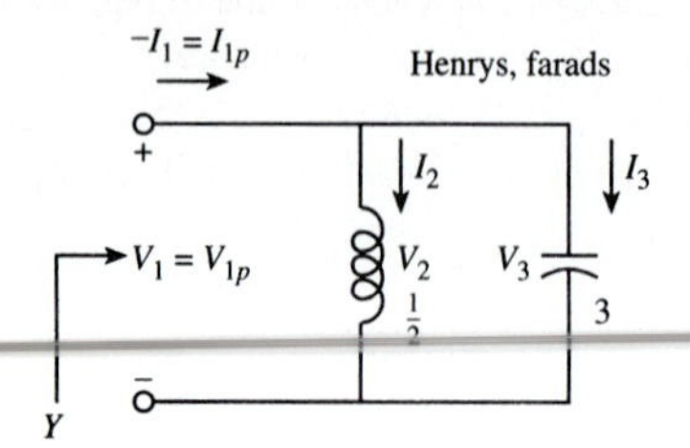

FIGURE P4.1-2

Section 4.2

1. Assume that an arbitrary network is composed of positive- and negative-valued capacitors and positive-valued inductors. Use the energy function method to determine in what areas of the complex plane the poles and zeros of driving-point functions may lie.
2. Use the Cauchy–Riemann conditions to prove that the function $B(\omega)$ defined in (8) has a derivative that is always positive.
3. (*a*) Sketch reactance plots of $X(\omega)$ versus ω for each of the functions given in Exercise 4.2-2. Assume that the functions are defined as impedances.
   (*b*) Sketch susceptance plots of $B(\omega)$ versus ω for each of the functions given in Exercise 4.2-2. Assume that the functions are defined as admittances.
4. Determine which of the following functions are *LC* realizable. For those that are not, explain why.

$$(a)\ F_a(s) = \frac{s^4 + 4s^2 + 3}{s^3 + 4s} \qquad (b)\ F_b(s) = \frac{s^3 + s^2 + 2s + 2}{s(s+1)(s^2+3)} \qquad (c)\ F_c(s) = \frac{s(s^2+2)}{s^4 + 2s^2 + 1}$$

$$(d)\ F_d(s) = \frac{s(s^2+1)(s^2+5)}{(s^2+3)(s^2+6)}$$

5. Determine which of the following functions are *LC* realizable. For those that are not, explain why.

$$(a)\ Z_a(s) = 3\frac{s^2+1}{s} \qquad (b)\ Z_b(s) = \frac{2s}{s^2+7} \qquad (c)\ Y_c(s) = \frac{7.6s}{s^2+s+1}$$

$$(d)\ Y_d(s) = \frac{s(s^2+1)}{s^4+2.5s^2+1} \qquad (e)\ Z_e(s) = \frac{s^2+2.5s+1}{s^2+1} \qquad (f)\ Z_f(s) = \frac{s(s^2+2)}{s^4+4s^2+3}$$

$$(g)\ Y_g(s) = \frac{s^4+s^3+2s^2+2s}{s^5+s^4+4s^3+4s^2+3s+3}$$

6. Derive an example function for each of the cases defined in Table 4.2-1.
7. Find partial-fraction expansions having the form given in (10) for each of the functions defined in Example 4.2-2.
8. Repeat the development given in (11) and (12) to prove that the residues for poles at the origin and infinity of an *LC*-realizable driving-point impedance are real and positive.

Section 4.3

1. (*a*) Find series Foster form realizations for each of the functions defined in Example 4.2-2. Assume that the functions are defined as impedances.
   (*b*) Repeat for the case where the functions are defined as admittances.
2. (*a*) Find shunt Foster form realizations for each of the functions defined in Example 4.2-2. Assume that the functions are defined as impedances.

(*b*) Repeat for the case where the functions are defined as admittances.

3. (*a*) Find removal-of-poles-at-infinity Cauer form realizations for each of the functions defined in Example 4.2-2. Assume that the functions are defined as impedances.
(*b*) Repeat for the case where the functions are defined as admittances.

4. (*a*) Find removal-of-poles-at-the-origin Cauer form realizations for each of the functions defined in Example 4.2-2. Assume that the functions are defined as impedances.
(*b*) Repeat for the case where the functions are defined as admittances.

5. Find the two Foster form and the two Cauer form realizations for the function given in Example 4.3-1 assuming that the function is a driving-point admittance rather than (as shown) a driving-point impedance.

6. Find two Foster form and two Cauer form realizations for the following driving-point admittance:

$$Y(s) = \frac{(s^2+1)(s^2+3)}{s(s^2+2)(s^2+4)}$$

7. (*a*) An *LC* driving-point function can be synthesized in the canonical form shown in Fig. P4.3-7. Draw a reactance plot of $X(\omega)$ versus ω.
(*b*) Sketch the circuit configuration that the two Foster form and the other Cauer form realization of the same driving-point function will have.

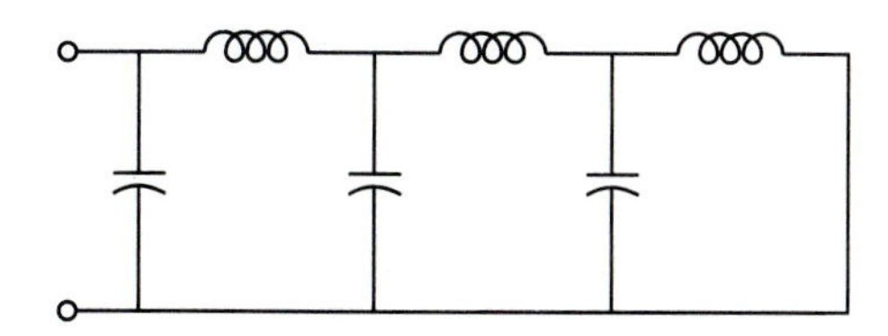

FIGURE P4.3-7

8. Determine the values of the elements if the network function given below is synthesized in the canonic form shown in Fig. P4.3-8:

$$Z(s) = \frac{(s^2+1)(s^2+3)}{s(s^2+2)(s^2+4)}$$

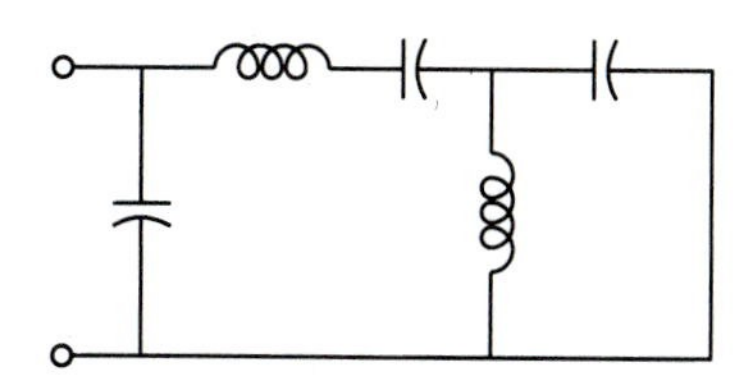

FIGURE P4.3-8

9. Realize the driving-point admittance given below in the *non*-canonic network configuration shown in Fig. P4.3-9, with the specified value of the input capacitor:

$$Y(s) = \frac{2s^5+8s^3+7s}{s^4+3s^2+2}$$

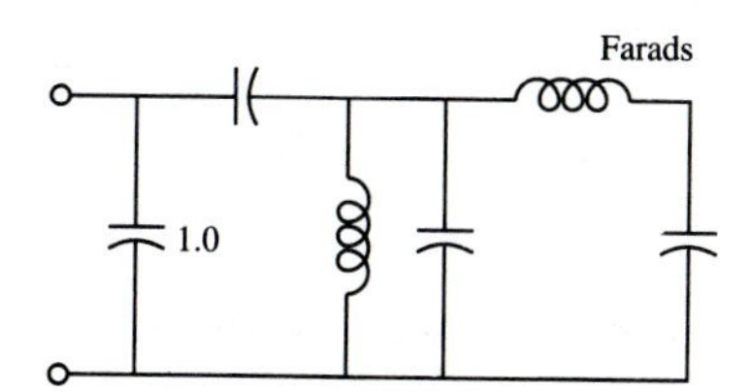

FIGURE P4.3-9

10. A new canonic form for the realization of an LC driving-point function is shown in Fig. P4.3-10. Sketch the circuit configurations that will realize the two Foster form and the two Cauer form realizations of the same driving-point function.

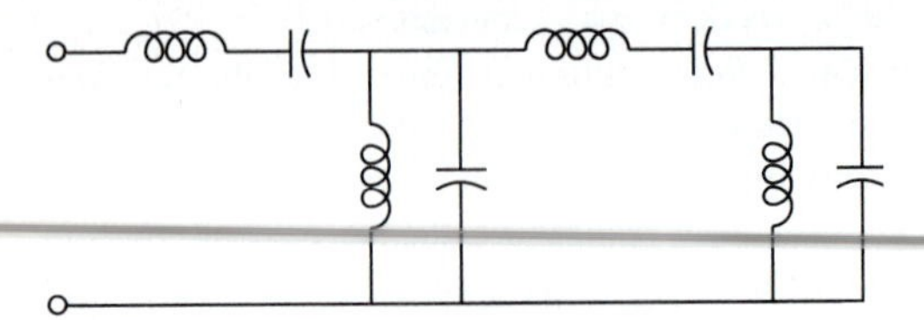

FIGURE P4.3-10

11. Fill in the table shown below for each of the indicated functions $Z_a(s)$ and $Z_b(s)$:

$$Z_a(s) = \frac{s(s^2+2)(s^2+4)}{(s^2+1)(s^2+3)} \qquad Z_b(s) = \frac{(s^2+1)(s^2+3)(s^2+5)}{s(s^2+2)(s^2+4)}$$

	Cauer removals at infinity			**Cauer removals at origin**		
	First element		**Last element**	**First element**		**Last element**
Circuit function	**Shunt or series?**	***L* or *C*?**	***L* or *C*?**	**Shunt or series?**	***L* or *C*?**	***L* or *C*?**
a						
b						

12. For the canonic network realization shown in Fig. P4.3-12(a), which of the realizations shown in Fig. P4.3-12(b) are also capable of realizing the same driving-point function?

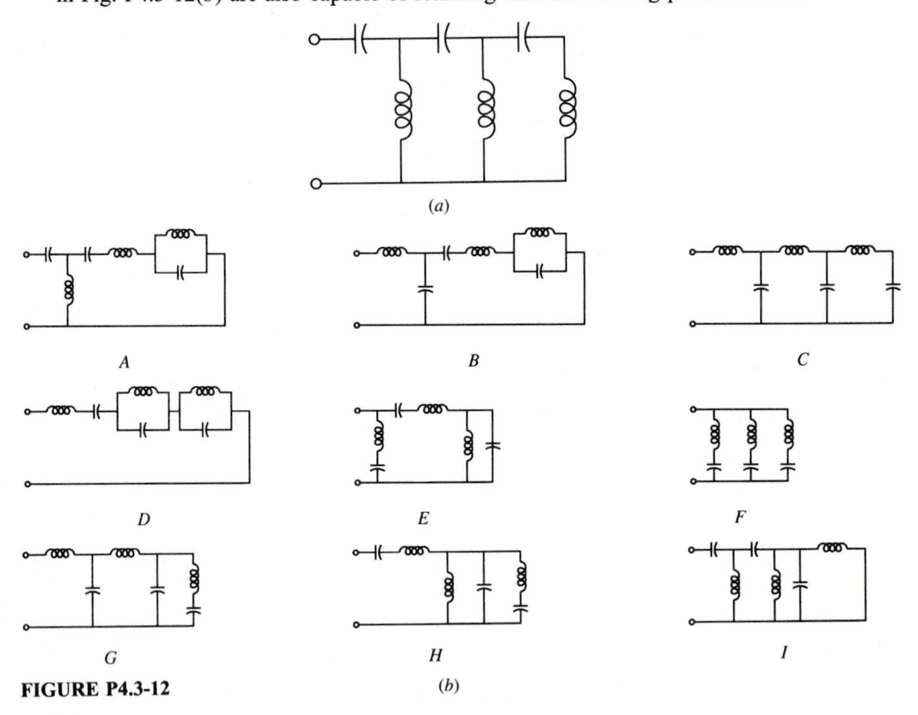

FIGURE P4.3-12

13. A reactance plot of $X(\omega)$ versus ω is shown in Fig. P4.3-13. Sketch the form of the circuits that provide the two Foster form and the two Cauer form realizations.

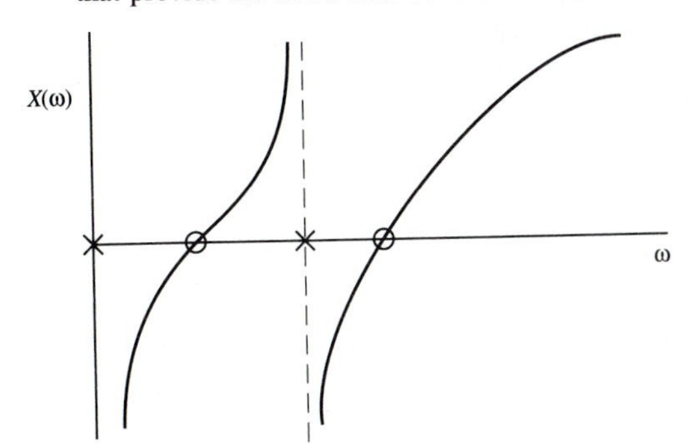

FIGURE P4.3-13

Section 4.4

1. Find the z parameters for the networks shown in Fig. P4.4-1.

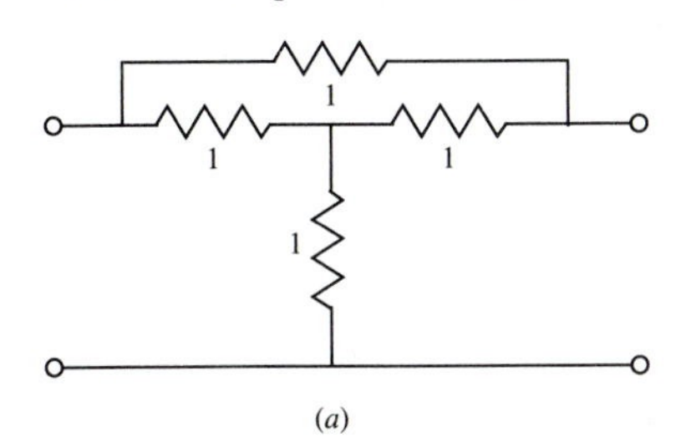

(*a*)

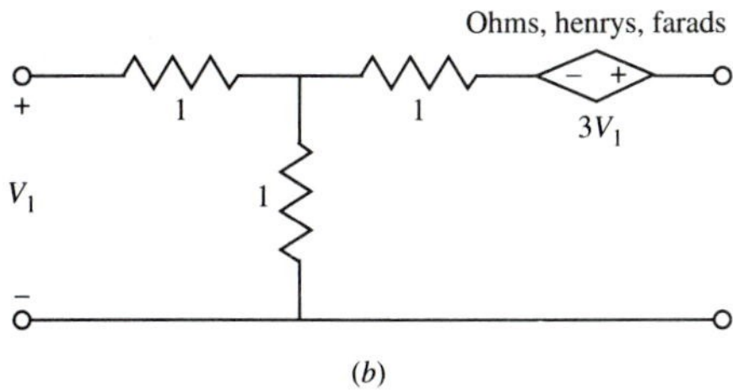

(*b*)

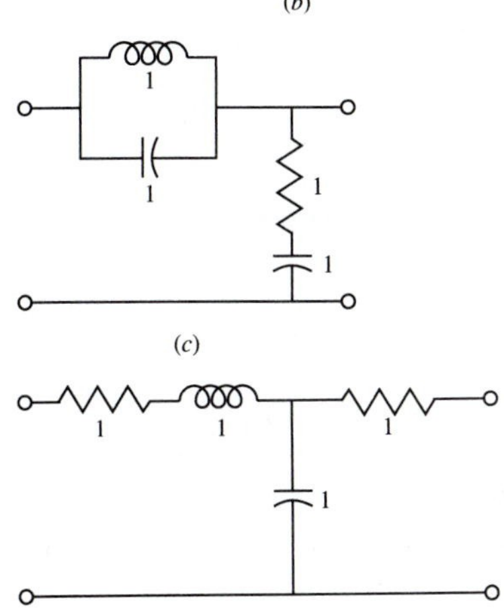

(*c*)

(*d*)

FIGURE P4.4-1

2. Find the y parameters for the networks shown in Fig. P4.4-2.

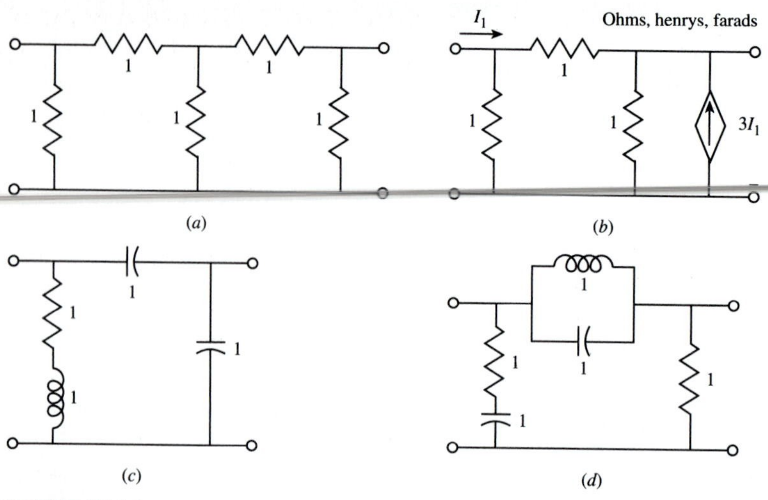

FIGURE P4.4-2

3. Find the y parameters for the network shown in Fig. P4.4-3 and show that the residue condition is satisfied. Assume that all the elements have unity value.

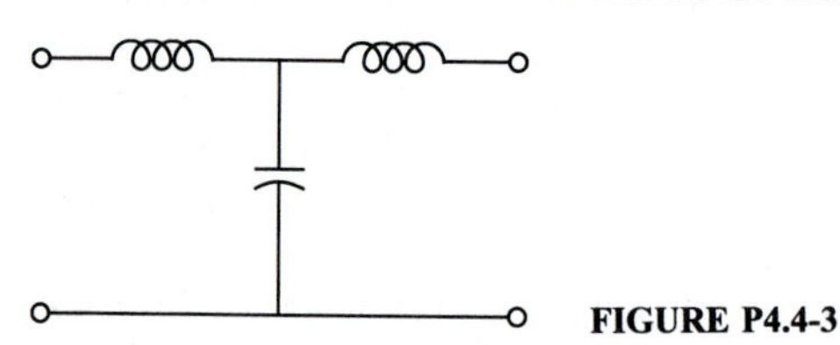

FIGURE P4.4-3

4. Find the z parameters for the network shown in Fig. P4.4-4 and show that the residue condition is satisfied. Assume that all the elements have unity value.

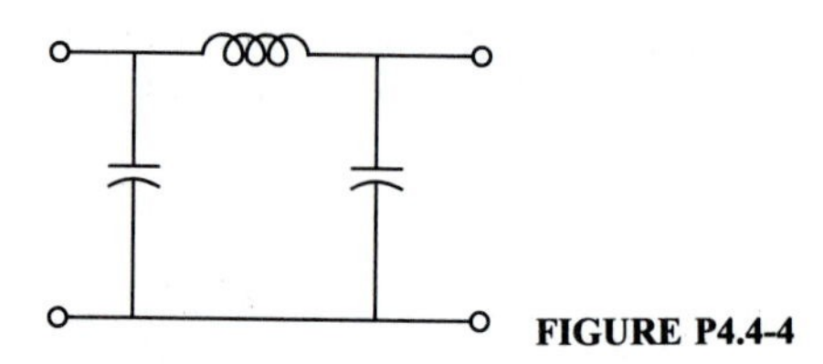

FIGURE P4.4-4

5. Identify any private poles in the z parameters for the networks shown in Figs. P4.4-1(c) and (d).
6. Identify any private poles in the y parameters for the networks shown in Figs. P4.4-2(c) and (d).
7. Find the open-circuit voltage transfer function for the two networks shown in Fig. 4.4-6 and identify any elements that do not affect the transfer function. Assume that the elements have unity value.
8. Find the short-circuit current transfer function for the two networks shown in Fig. 4.4-6 and identify any elements that do not affect the transfer function. Assume that the elements have unity value.

Section 4.5

1. Find the y parameters of the resistance network (assume that all the resistors have unity value) shown in Fig. P4.5-1 by applying wye-delta and delta-wye transformations to the network until a delta network of three resistors is obtained.

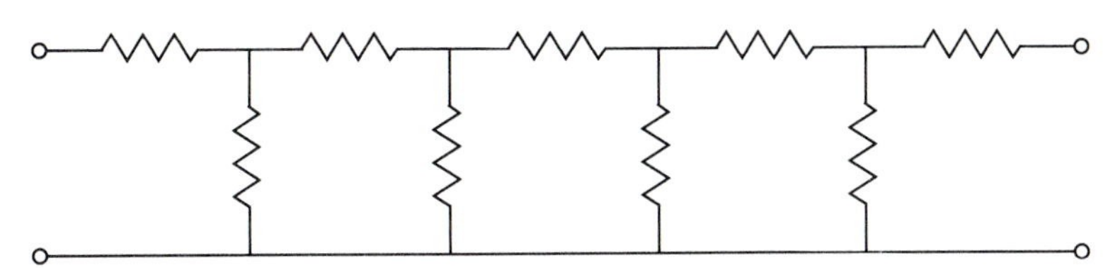

FIGURE P4.5-1

2. Show that the Fialkow condition holds for the z parameters of the network shown in Fig. 4.4-6(a). Assume that the elements have unity value.
3. Show that the Fialkow condition holds for the y parameters of the network shown in Fig. 4.4-6(b). Assume that the elements have unity value.
4. Show that the real-part condition holds for the z parameters of the network shown in Fig. 4.4-6(a). Assume that the elements have unity value.
5. Show that the real-part condition holds for the y parameters of the network shown in Fig. 4.4-6(b). Assume that the elements have unity value.
6. Find the open-circuit voltage transfer function for the network shown in Fig. 4.5-7(a). Assume that all elements are unity valued and that the network is connected across the break.
7. Find the open-circuit voltage transfer function for the network shown in Fig. 4.5-7(b). Assume that all elements are unity valued and that the network is connected across the break.
8. Find the short-circuit current transfer function for the network shown in Fig. 4.5-7(a). Assume that all elements are unity valued and that the network is connected across the break.
9. Find the short-circuit current transfer function for the network shown in Fig. 4.5-7(b). Assume that all elements are unity valued and that the network is connected across the break.
10. Apply the theory of ladder networks to find the number and location of the transmission zeros for the network functions (a) $y_{21}(s)$, (b) $z_{21}(s)$, (c) $V_{OC}(s)$, and (d) $I_{SC}(s)$ for the network shown in Fig. P4.5-10.

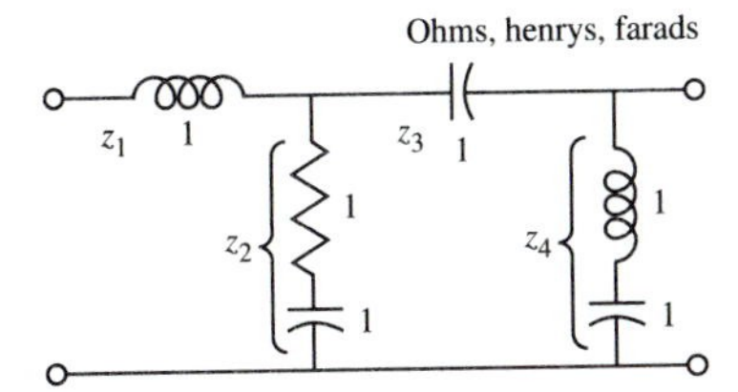

FIGURE P4.5-10

11. Verify the conclusions obtained in Prob. 10 by actually analyzing the network to find the four transfer functions.
12. Find example circuits that illustrate each of the LC cases listed in Table 4.5-1.
13. Find example circuits that illustrate each of the RC cases listed in Table 4.5-1.
14. Find example circuits that illustrate each of the RLC cases listed in Table 4.5-1.

Section 4.6

1. For each of the following polynomials expand the ratio of odd/even or even/odd parts in the form of a removals-at-infinity continued-fraction expansion to determine if they are Hurwitz:
   (*a*) $P_a(s) = s^3 + 0.2s^2 + 0.2s + 1$ (*b*) $P_b(s) = s^4 + 4s^3 + 6s^2 + 4s + 1$
   (*c*) $P_c(s) = s^4 + 2s^3 + 2s^2 + 2s + 1$
   (*d*) $P_d(s) = s^4 + 2.613126s^3 + 3.414214s^2 + 2.613126s + 1$
2. Find the conditions on the coefficients a, b, c, and d so that the following polynomial is Hurwitz:
$$P(s) = s^4 + as^3 + bs^2 + cs + d$$
3. Find a single-resistance-terminated (1-Ω) lossless ladder realization for each of the following transfer admittances. For each realization, find the actual value of the multiplicative constant H.
   (*a*) $Y_T(s) = \dfrac{H}{s^4 + s^3 + 3s^2 + s + 1}$ (*b*) $Y_T(s) = \dfrac{Hs}{s^4 + s^3 + 3s^2 + s + 1}$
   (*c*) $Y_T(s) = \dfrac{Hs^2}{s^4 + s^3 + 3s^2 + s + 1}$ (*d*) $Y_T(s) = \dfrac{Hs^3}{s^4 + s^3 + 3s^2 + s + 1}$
   (*e*) $Y_T(s) = \dfrac{Hs^4}{s^4 + s^3 + 3s^2 + s + 1}$
4. Find a single-resistance-terminated (1-Ω) lossless ladder realization for the following transfer admittance:
$$Y_T(s) = \frac{Hs^4}{(s^2 + s + 1)^2}$$
5. The following low-pass transfer functions are to be realized as lossless ladder networks with a single-resistance termination. The lossless ladder consists only of series L and shunt C elements. Answer the questions in the table that follows by filling in the squares with L, C, *yes*, or *no*, as appropriate.

Function realized	**First reactive element (at left)?**	**Last reactive element (at right)?**	**Does y_{22} have a private pole?**
Even-order voltage transfer function Odd-order voltage transfer function			
Even-order transfer admittance Odd-order transfer admittance			
Even-order transfer impedance Odd-order transfer impedance			

6. Show the *form* of the lossless ladder single-resistance-terminated network realization for each of the following network functions. *Do not* find the element values.
   (*a*) $Y_T(s) = Hs/D(s)$ (*b*) $Y_T(s) = Hs^2/D(s)$ (*c*) $Y_T(s) = Hs^3/D(s)$
$$D(s) = as^5 + bs^4 + cs^3 + ds^2 + es + f$$
   where a, b, c, d, e, and f are coefficients whose values are such that $D(s)$ is Hurwitz.
7. Repeat Prob. 6 for the case where the (Hurwitz) denominator polynomial is
$$D(s) = as^6 + bs^5 + cs^4 + ds^3 + es^2 + fs + g$$

8. Develop a synthesis procedure for a single-resistance-terminated lossless ladder network similar to that described in this section but based on the use of z parameters to realize the network function $Z_T(s) = V_2(s)/I_1(s)$, where $V_2(s)$ is the voltage across a 1-Ω output resistor and $I_1(s)$ is the input provided by a current source. Apply the procedure to the transfer function [redefined as $Z_T(s)$] given in Example 4.6-3.

Section 4.7

1. Realize the following transfer admittance as a single-resistance-terminated lossless ladder network using a total of seven elements (including the 1-Ω termination):

$$Y_T(s) = \frac{H(s^4 + 11s^2 + 24)}{s^5 + s^4 + 8s^3 + 4s^2 + 12s + 3}$$

2. Find another network that will meet the specifications given in Prob. 1.
3. The transfer admittance of a single-resistance-terminated lossless ladder network is given below. Show as many different configurations as possible that use six network elements (including the terminating 1-Ω resistor) to realize the function. *Do not* find the values of the network elements:

$$Y_T(s) = \frac{Hs(s^2 + 4)}{s^4 + s^3 + 4s^2 + 2s + 3}$$

4. Realize the following transfer function as a single-resistance-terminated lossless ladder network in a configuration in which a partial removal of an *interior* pole is used for zero shifting purposes:

$$Y_T(s) = \frac{H(s^2 + 4)}{s^3 + s^2 + 9s + 1}$$

5. Find another network that will meet the specifications given in Prob. 4.
6. Assume that the three network configurations shown in Fig. P4.7-6 all realize the same transfer admittance $Y_T(s)$. In each realization identify the element that has been used for zero shifting. Also identify the general form of y_{12} and y_{22} using simple numbers like 1, 2, 3, ... to show the correct *relative* positions of their poles and zeros.

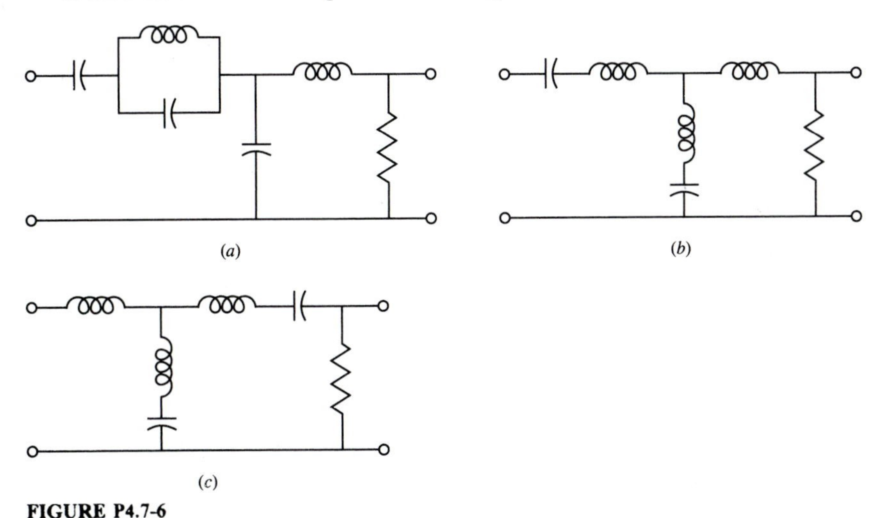

FIGURE P4.7-6

7. Obtain a lossless ladder network that has the form shown in Fig. P4.7-7 and the following y parameters:

$$y_{11}(s) = \frac{(s^2+4)(s^2+16)}{s(s^2+9)(s^2+25)} \qquad -y_{12}(s) = \frac{H(s^2+1)(s^2+36)}{s(s^2+9)(s^2+25)}$$

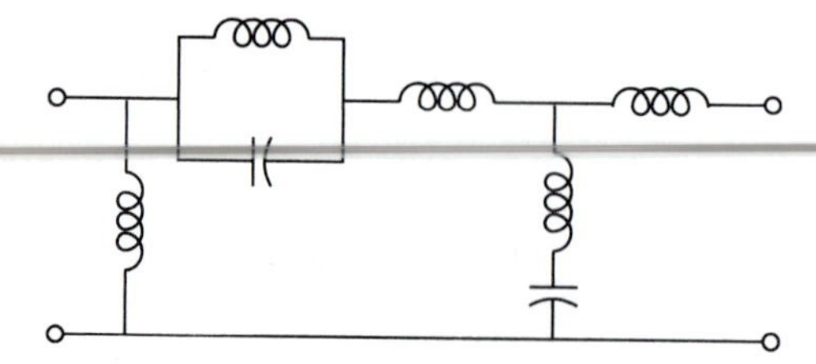

FIGURE P4.7-7

Section 4.8

1. Realize the following voltage transfer function as a double-resistance-terminated lossless ladder using 1-Ω terminating resistors:

$$\frac{V_{\text{out}}}{V_{\text{in}}} = \frac{H}{s^3 + 2s^2 + 2s + 1}$$

2. Find a double-resistance-terminated lossless ladder realization for a 1-dB ripple Chebyshev third-order voltage transfer function with a (normalized) 1-rad/s cutoff frequency. Use 1-Ω resistors for the terminations.
3. Find a double-resistance-terminated lossless ladder realization for a 1-dB ripple Chebyshev second-order voltage transfer function. If it is not possible to have both resistors with a value of 1 Ω, then make $R_1 = 1\ \Omega$ and use any convenient value for R_2.
4. Find a double-resistance-terminated lossless ladder realization using 1-Ω terminating resistors for the following voltage transfer function:

$$\frac{V_{\text{out}}}{V_{\text{in}}} = \frac{H(s^2 + 5.153354)}{s^3 + 1.45380s^2 + 1.81121s + 1.07538}$$

5. Find a double-resistance-terminated lossless ladder realization using 1-Ω terminating resistors for a voltage transfer function with an elliptic magnitude characteristic in which the ripple is 1 dB over a passband of 0 to 1 rad/s and that has a minimum of 34 dB attenuation at all frequencies greater than 2 rad/s.

Section 4.9

1. Find a realization for a fourth-order low-pass Butterworth transfer admittance having 10 kHz cutoff frequency and source and terminating resistors of 1000 Ω.
2. Find a realization for a fourth-order low-pass Butterworth transfer impedance having 10 kHz cutoff frequency and source and terminating resistors of 1000 Ω.
3. Find a realization for a fourth-order low-pass Butterworth transfer admittance having a source resistor of 500 Ω, a load resistor of 1000 Ω, and a cutoff frequency of 10 kHz.
4. Find a realization for a fourth-order low-pass Butterworth transfer admittance having a source resistor of 1000 Ω, a load resistor of 500 Ω, and a cutoff frequency of 10 kHz.
5. Find a realization for each of the following functions. They should all be low-pass third-order functions with a Butterworth magnitude characteristic. The bandwidth of all functions is 1 rad/s. Draw a sketch of each realization, clearly indicating where the excitation and

response variables are located. In each of the cases the network function will have the form

$$\frac{H}{s^3 + 2s^2 + 2s + 1}$$

Determine the value of the multiplicative constant H for each of the realizations.

(*a*) A voltage transfer function $V_2(s)/V_1(s)$. The realization should have a source resistor of $1\ \Omega$ and no load resistor.

(*b*) A transfer impedance function $V_2(s)/I_1(s)$. Both source and load resistors should be present and both should have a value of $1\ \Omega$.

(*c*) A transfer admittance function $I_0(s)/V_1(s)$. The source resistor should have a value of $2\ \Omega$ and the load resistor should be $1\ \Omega$.

6. Find a realization for each of the following functions. They should all be low-pass fourth-order functions with a Butterworth magnitude characteristic. The bandwidth of all functions is 1 rad/s. Draw a sketch of each realization, clearly indicating where the excitation and response variables are located. In each of the cases the network function will have the form

$$\frac{H}{s^4 + 2.613126s^3 + 3.414214s^2 + 2.613126s + 1}$$

Determine the value of the multiplicative constant H for each of the realizations.

(*a*) A voltage transfer function $V_2(s)/V_1(s)$. The realization should have a source resistor of $1\ \Omega$ and no load resistor.

(*b*) A transfer impedance function $V_2(s)/I_1(s)$. Both source and load resistors should be present and both should have a value of $1\ \Omega$. For this case, find two different realizations that meet the specifications.

(*c*) A transfer impedance function $V_2(s)/I_1(s)$. The source resistor should have a value of $2\ \Omega$ and the load resistor a value of $1\ \Omega$. For this case, find two different realizations that meet the specifications.

(*d*) A transfer impedance function $V_2(s)/I_1(s)$. The source resistor should have a value of $1\ \Omega$ and the load resistor a value of $2\ \Omega$. For this case, find two different realizations that meet the specifications.

Section 4.10

1. Design a normalized (1 rad/s center frequency) fourth-order coupled resonator band-pass filter that has a Butterworth magnitude characteristic and a bandwidth of 0.1 rad/s. Choose a low-pass prototype design in which $R_1 = 2\ \Omega$ and $R_2 = 1\ \Omega$. Let the inductors have a value of 0.1 H.

2. Design a normalized (1 rad/s center frequency) fourth-order coupled resonator band-pass filter that has a Chebyshev magnitude characteristic with a $\frac{1}{2}$-dB ripple and a bandwidth of 0.1 rad/s. Choose a low-pass prototype design in which $R_1 = 2\ \Omega$ and $R_2 = 1\ \Omega$. Let the inductors have a value of 0.1 H.

3. Design a normalized (1 rad/s center frequency) sixth-order coupled resonator band-pass filter that has a Butterworth magnitude characteristic and a bandwidth of 0.1 rad/s. Let the inductors have a value of 0.07071 H.

4. Design a normalized (1 rad/s center frequency) sixth-order coupled resonator band-pass filter that has a Butterworth magnitude characteristic and a bandwidth of 0.1 rad/s. Choose a low-pass prototype design in which $R_1 = \frac{1}{2}\ \Omega$ and $R_2 = 1\ \Omega$. Let the inductors have a value of 0.07071 H.

CHAPTER 5

SINGLE-AMPLIFIER ACTIVE *RC* FILTERS

In this and the following chapters we will consider the general subject of active filters. More specifically, we will present methods for realizing all types of network functions through the use of filter circuits comprising both active and passive elements, the latter being restricted exclusively to resistors and capacitors. Such filters are referred to collectively as *active RC* or *inductorless* filters. There are many reasons why active *RC* filters are attractive and indeed may be preferable to their purely passive *RLC* counterparts. For example, active *RC* filters usually weigh less and require less space than do passive ones. This is an important consideration in satellite and airborne applications. As another example, active filters can be fabricated in microminiature form using integrated circuit techniques. Thus they can be mass produced inexpensively. On the other hand, since it is not possible to "integrate" an inductor, passive circuits can only be produced using discrete components. This is usually far more expensive. For these and many other reasons, many traditional passive filtering applications, notably in the telecommunications industry, have now been modified so as to use active filters exclusively. As a result, annual production of active filters is now in the millions, and many companies offer these items as an off-the-shelf standard item. In this chapter we shall introduce one of the most well-known types of active *RC* filters, the single-amplifier one.

5.1 THE SINGLE-AMPLIFIER FILTER

In this section we introduce the simplest of the active *RC* filter configurations. It uses a single VCVS (voltage-controlled voltage source) or *amplifier* as an active element. In later sections we shall describe how this configuration may be used to realize various filter functions.

Cascade and Direct Realizations

There are two general methods of using active *RC* filters to realize network functions. The first of these is the *cascade* method. The method is so named because the network function to be realized is first factored into a product of second-order terms.[1] Each term is then individually realized by an active *RC* circuit, and a cascade or series connection of the circuits is then used to realize the overall network function. The individual active *RC* circuits, of course, must be designed in such a manner that they do not interact with each other when the cascade connection is made; i.e., they must be isolated from each other. The second general method of using active *RC* circuits to realize network functions is the *direct* method, in which a single circuit is used to realize the entire network function. A discussion of this method will be given in later chapters.

The cascade method of using active *RC* circuits to realize network functions, as described above, offers many advantages to the network designer. First, since any given active *RC* circuit is only required to realize a second-order network function, its configuration is usually relatively simple and the number of elements required is usually small. As a result, the synthesis procedure needed for determining the values of the elements is straightforward, and the implementation of additional constraints, such as the use of standard element values or the minimization of sensitivity, is usually easy to achieve. As another advantage, each active *RC* circuit may be individually tuned to make certain that it exactly realizes its specific second-order factor. This, of course, is far easier than trying to tune a circuit in which all the elements interact, as is the case when the direct method of realization is used.

The VCVS Amplifier

Now let us consider the active element. Although theoretically any type of controlled source may be used as the active element, in practice, the VCVS has proven to be the preferred one. Ideally, the VCVS is a two-port device characterized by the following properties: (1) infinite input impedance, (2) zero output impedance, and (3) an output voltage that is linearly proportional to the input voltage, the constant of proportionality being referred to as the *gain*. A model and a circuit symbol for a VCVS are given in Fig. 5.1-1. The VCVS is also referred to as a *voltage amplifier*, or more simply just as an *amplifier*. The gain constant may be positive, in which case the VCVS is said to be *noninverting*, or negative, in which case it is said to be *inverting*. Among the reasons for

[1] If an odd-order function is to be realized, either a passive first-order circuit or a third-order active *RC* circuit will be needed in the cascade. This will be discussed further in Sec. 5.6.

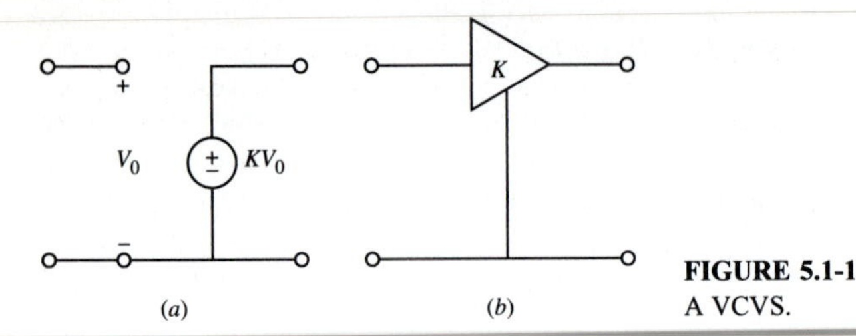

FIGURE 5.1-1
A VCVS.

the popularity of the VCVS as the active element of active *RC* filters is the ease with which it can be realized using an operational amplifier. As an example, the noninverting VCVS can be realized using a differential-input operational amplifier in the circuit shown in Fig. 5.1-2*a*. The gain of the resulting VCVS is given by the relation

$$\frac{V_2}{V_1} = \frac{R_1 + R_2}{R_1} = K \tag{1}$$

Obviously the gain will always be greater than unity. A unity-gain noninverting VCVS can be realized as shown in Fig. 5.1-2*b*. This circuit is also referred to as a *voltage follower.* The inverting VCVS can be realized as shown in Fig. 5.1-2*c*. For this circuit the gain is

$$\frac{V_2}{V_1} = \frac{-R_2}{R_1} = K \tag{2}$$

Example 5.1-1 Noninverting amplifier. To realize a VCVS with a gain $K = 1.5858$, from (1) we require $(R_1 + R_2)/R_1 = 1.5858$ for the resistors in Fig. 5.1-2*a*. Suitable impedance-normalized values for the resistors would be $R_1 = 1\ \Omega$ and $R_2 = 0.5858\ \Omega$. In determining an appropriate impedance denormalization for these components, we must take into account the actual properties of the operational amplifier. For example, the popular μA741 has input impedances that are greater than 1 MΩ, thus it is desirable that the values of R_1 and R_2 be considerably less than this value. Similarly, the output impedance of a μA741 is typically less than 100 Ω, so the resistor values should be much greater than

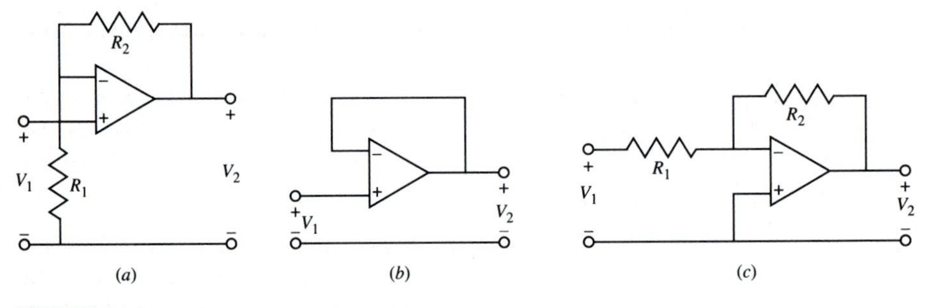

FIGURE 5.1-2
VCVS realizations: (*a*) noninverting; (*b*) voltage follower; (*c*) inverting.

this. In most applications, the choices $R_1 = 100\ \text{k}\Omega$ and $R_2 = 58.58\ \text{k}\Omega$ would be satisfactory. For applications at frequencies where wiring parasitic capacities cause problems, these values could be reduced by a factor of 10.

Exercise 5.1-1. Find the actual gain realized in the VCVS design given in Example 5.1-1 for each of the sets of values of the input impedance R_i, the output impedance R_o, and the dc open-loop gain A_d given below. *Hint:* The results given in App. C may be used to simplify the analysis.

R_i, Ω	R_o, Ω	A_d	Answers: Grain
10^6	10^2	10^5	1.58577
10^6	10^2	10^4	1.58554
10^5	10^3	10^5	1.58577
10^5	10^3	10^4	1.58545
10^4	10^4	10^5	1.58567
10^4	10^4	10^4	1.58455

Analysis of Circuits Containing VCVSs

On many occasions in this text we will encounter circuits comprised of passive components and VCVSs. The analysis of such circuits to determine a specific transfer function is facilitated by using a technique called the *method of constraints*. It is summarized as follows.

Summary 5.1-1 Analysis of VCVS circuits using the method of constraints. To analyze a circuit containing passive components and one or more VCVSs, we may proceed as follows:

1. Write the admittance equations describing the circuit *without* the VCVSs. This may be done either on an external terminal basis, in which case the equations describe the multiport parameters of the circuit, or on an internal node basis, in which case the usual node equations result.
2. Apply the *voltage constraint* introduced by each VCVS to reduce the number of unknown voltage variables. For a VCVS with gain K relating an output voltage V_i to an input voltage V_j the equation $V_i = KV_j$ or $V_j = V_i/K$ may be used to eliminate either V_i or V_j.
3. Apply the *current constraint* for each VCVS. If the output of a VCVS is at node i, the zero output impedance of the VCVS effectively shorts any input current to this terminal (or node) to ground. As a result, the corresponding ith equation is not an independent one, and it may be deleted to reduce the number of equations.

As a result of the application of these steps, we see that the constraints imposed by each VCVS reduces the order of the equations describing the circuit by one degree. When the constraints imposed by all the VCVSs have been implemented, the remaining equations may be solved for any choice of input and output variables.

The method of constraints can be applied to develop two equations that will be of considerable use in much of the following treatment of active filters. The development of these equations is given in the following two examples.

Example 5.1-2 Voltage transfer function for a general single-amplifier filter. As an example of the use of the method of constraints, consider the general single-amplifier filter configuration shown in Fig. 5.1-3. The passive portion of the network may be considered as a three-port network and defined by a set of short-circuit admittance (or y) parameters. Thus, for the passive network portion of the filter (*without* the VCVS) we may write (for step 1)

$$\begin{bmatrix} I_1(s) \\ I_2(s) \\ I_3(s) \end{bmatrix} = \begin{bmatrix} y_{11}(s) & y_{12}(s) & y_{13}(s) \\ y_{21}(s) & y_{22}(s) & y_{23}(s) \\ y_{31}(s) & y_{32}(s) & y_{33}(s) \end{bmatrix} \begin{bmatrix} V_1(s) \\ V_2(s) \\ V_3(s) \end{bmatrix} \tag{3}$$

where, due to the passive nature of the network, $y_{ij}(s) = y_{ji}(s)$; that is, the y matrix is symmetric. Now let us apply the method of Summary 5.1-1 to include the effect of the VCVS. From step 2 we obtain $V_3(s) = V_2(s)/K$. If this is substituted in (3), we find

$$\begin{bmatrix} I_1(s) \\ I_2(s) \\ I_3(s) \end{bmatrix} = \begin{bmatrix} y_{11}(s) & y_{12}(s) + y_{13}(s)/K \\ y_{21}(s) & y_{22}(s) + y_{23}(s)/K \\ y_{31}(s) & y_{32}(s) + y_{33}(s)/K \end{bmatrix} \begin{bmatrix} V_1(s) \\ V_2(s) \end{bmatrix} \tag{4}$$

From step 3 of Summary 5.1-1, we may eliminate the second equation of the matrix array given in (4) to obtain

$$\begin{bmatrix} I_1(s) \\ I_3(s) \end{bmatrix} = \begin{bmatrix} y_{11}(s) & y_{12}(s) + y_{13}(s)/K \\ y_{31}(s) & y_{32}(s) + y_{33}(s)/K \end{bmatrix} \begin{bmatrix} V_1(s) \\ V_2(s) \end{bmatrix} \tag{5}$$

Since $I_3(s) = 0$ (due to the infinite input impedance of the VCVS), the second equation of the array given in (5) may be set to zero and solved to obtain

$$\frac{V_2(s)}{V_1(s)} = \frac{-Ky_{31}(s)}{y_{33}(s) + Ky_{32}(s)} \tag{6}$$

This equation will be of considerable use in many of the developments that follow.

Example 5.1-3 Voltage transfer function for a specific network configuration. In many of the single-amplifier active *RC* filters to be discussed in this chapter we will restrict the passive network so that it has the form shown in Fig. 5.1-4. The method of constraints may be applied to find the voltage transfer function for this network. From step 1 of

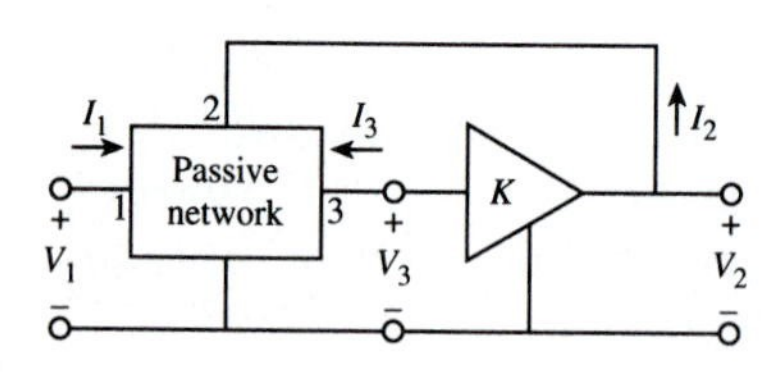

FIGURE 5.1-3
The single-amplifier filter configuration.

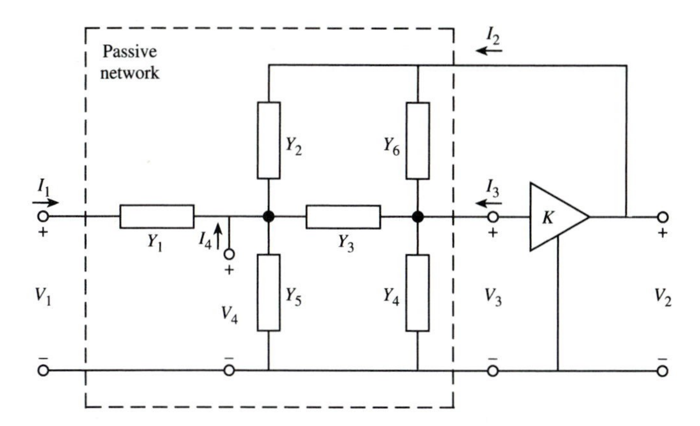

FIGURE 5.1-4
Single-amplifier filter with a specific form of passive network.

Summary 5.1-1 we may write

$$\begin{bmatrix} I_1 \\ I_2 \\ I_3 \\ I_4 \end{bmatrix} = \begin{bmatrix} Y_1 & 0 & 0 & -Y_1 \\ 0 & Y_2+Y_6 & -Y_6 & -Y_2 \\ 0 & -Y_6 & Y_3+Y_4+Y_6 & -Y_3 \\ -Y_1 & -Y_2 & -Y_3 & Y_1+Y_2+Y_3+Y_5 \end{bmatrix} \begin{bmatrix} V_1 \\ V_2 \\ V_3 \\ V_4 \end{bmatrix} \tag{7}$$

where, for convenience, we have deleted the (s) notation. From steps 2 and 3 we set $V_3 = V_2/K$ and delete the second equation in the above array to obtain

$$\begin{bmatrix} I_1 \\ I_3 \\ I_4 \end{bmatrix} = \begin{bmatrix} Y_1 & 0 & -Y_1 \\ 0 & -Y_6+(Y_3+Y_4+Y_6)/K & -Y_3 \\ -Y_1 & -Y_2-Y_3/K & Y_1+Y_2+Y_3+Y_5 \end{bmatrix} \begin{bmatrix} V_1 \\ V_2 \\ V_4 \end{bmatrix} \tag{8}$$

In this expression the node currents $I_3 = 0$ and $I_4 = 0$ since no external currents flow into these nodes. The corresponding equations may be solved for V_4 and equated to each other to obtain an expression involving only V_1 and V_2. From this we obtain the result

$$\frac{V_2(s)}{V_1(s)} = \frac{KY_1Y_3}{(Y_1+Y_2+Y_5)(Y_3+Y_4+Y_6)+Y_3(Y_4+Y_6)-K[Y_6(Y_1+Y_2+Y_3+Y_5)+Y_2Y_3]} \tag{9}$$

This result will be of considerable use in the developments that follow.

General Single-Amplifier Filter Configuration

The general single-amplifier filter configuration shown in Fig. 5.1-3 forms the basis for much of our study of cascade filters. In this circuit, the use of a VCVS as a (zero-impedance) output element satisfies the requirement that the filter's properties are not changed when it is loaded by being connected to another element of the cascade. When the passive network portion of the filter is comprised of R and C elements, a wide range of filter functions can be produced by the overall structure. The feedback provided by the VCVS effectively converts the negative real poles that characterize the RC network into the complex-conjugate ones required for effective filtering. The mechanism for accomplishing this conversion is determined by (6). In this equation, we recall from Chap. 4 that the functions $-y_{31}$, $-y_{32}$, and y_{33} cannot have negative-valued coefficients in their numerator and denominator polynomials. Thus, if we write

$$-y_{31}(s) = \frac{N_{31}(s)}{D(s)} \qquad -y_{32}(s) = \frac{N_{32}(s)}{D(s)} \qquad y_{33}(s) = \frac{N_{33}(s)}{D(s)} \tag{10}$$

the polynomials $N_{31}(s)$, $N_{32}(s)$, and $N_{33}(s)$ will only have coefficients that are positive (or zero). From (10) we can write (6) in the form

$$\frac{V_2(s)}{V_1(s)} = \frac{A(s)}{B(s)} = \frac{KN_{31}(s)}{N_{33}(s) - KN_{32}(s)} \tag{11}$$

For the network shown in Fig. 5.1-4, comparing (9) and (11), we find

$$\begin{aligned} N_{31}(s) &= Y_1 Y_3 \\ N_{32}(s) &= Y_6(Y_1 + Y_2 + Y_3 + Y_5) + Y_2 Y_3 \\ N_{33}(s) &= (Y_1 + Y_2 + Y_5)(Y_3 + Y_4 + Y_6) + Y_3(Y_4 + Y_6) \end{aligned} \tag{12}$$

For a positive value of the VCVS gain K, the denominator $B(s)$ of (11) represents a difference of positive-valued polynomials $N_{33}(s)$ and $KN_{32}(s)$. This is referred to as a *difference decomposition* of $B(s)$. For the case where $B(s)$ is second order, $y_{33}(s)$ will typically also be of second order and, as an RC driving-point function, can only have zeros on the negative real axis. Thus, typically, $N_{33}(s) = (s + \sigma_1)(s + \sigma_2)$. For most applications, the passive RC network is chosen so that $-y_{32}(s)$ has a simple zero at the origin, and thus $N_{32}(s) = \alpha s$, where α is a positive constant. The resulting second-order denominator polynomial in (11) can be written as

$$B(s) = s^2 + a_1 s + a_0 = (s + \sigma_1)(s + \sigma_2) - K\alpha s \tag{13}$$

where a_1 and a_0 determine the resulting pole locations and σ_1, σ_2, and α are functions of the various passive network elements. Such a decomposition produces a circular root locus as a function of K. Its general form is shown in Fig. 5.1-5. From the figure we see that when K is increased beyond the value $(\sigma_1 + \sigma_2)/\alpha$, the poles of the network function

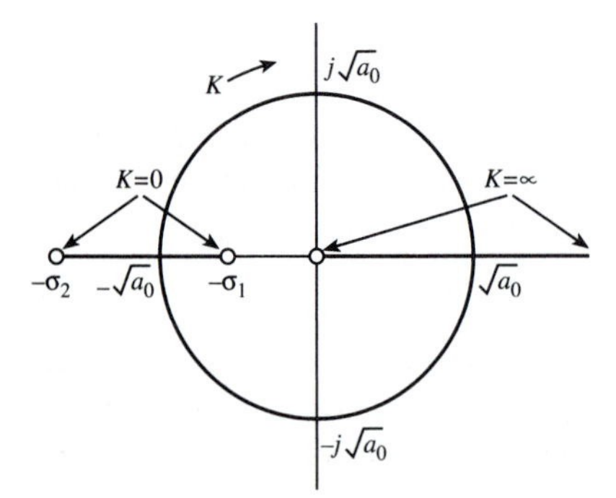

FIGURE 5.1-5
A difference decomposition.

move into the right-half plane and the network becomes unstable. For a frequency-normalized denominator polynomial having the form $P(s) = s^2 + (1/Q)s + 1$, the optimum difference decomposition is $(s + 1)^2 - K\alpha s$. This produces a lower bound for S_K^Q of $2Q - 1$. Since this value is relatively large, *RC* amplifier realizations are in general suitable only for realizing network functions with low values of Q.

5.2 LOW-PASS SINGLE-AMPLIFIER FILTERS

In the preceding section we presented a basic *RC* amplifier filter configuration suitable for use in the realization of second-order voltage transfer functions. The circuit is given in Fig. 5.1-4. In this section we will show how the basic structure shown in the figure can be used to realize low-pass functions. The general procedure that we shall use will be extended to include other types of filter functions in the following sections.

General Low-Pass Function

The general form of the second-order low-pass voltage transfer function can be written as

$$\frac{V_2(s)}{V_1(s)} = \frac{H_0\omega_n^2}{s^2 + (\omega_n/Q)s + \omega_n^2} = \frac{KN_{31}(s)}{N_{33}(s) - KN_{32}(s)} \tag{1}$$

where H_0 is the gain at direct current, ω_n is the undamped natural frequency, and Q is the quality factor originally defined in Sec. 3.5. The right member of (1) is taken from (11) of Sec. 5.1. If the poles of the network function of (1) are located at $p_0 = \sigma_0 \pm j\omega_0$, then the relation between the quantities ω_n and Q and the pole locations are given as

$$p_0 = \sigma_0 \pm j\omega_0 = -\frac{\omega_n}{2Q} \pm j\frac{\omega_n}{2Q}\sqrt{4Q^2 - 1} \tag{2}$$

These relations are further illustrated in Fig. 5.2-1. Plots of $|V_2/V_1(j\omega)|$ and Arg $V_2/V_1(j\omega)$ for various values of Q for the normalized values $H_0 = 1$ and $\omega_n = 1$ rad/s are shown in Fig. 5.2-2. The polynomial $s^2 + (\omega_n/Q)s + \omega_n^2$ in the denominator of (1) is called the *standard form* of the second-order polynomial.

Low-Pass Single-Amplifier Filter Realization

Let us now compare (1) with the relations given in (12) of Sec. 5.1. Looking first at the numerators, we see that the quantities Y_1 and Y_3 must be constants; i.e., they must represent the admittances of resistors. As such they may be written as

$$Y_1 = G_1 \qquad Y_3 = G_3 \tag{3}$$

Now consider the denominators of (1). The denominator of the right member may be regarded as specifying the decomposition of the denominator polynomial $s^2 + (\omega_n/Q)s + \omega_n^2$ of the center member. The remaining elements of the filter are

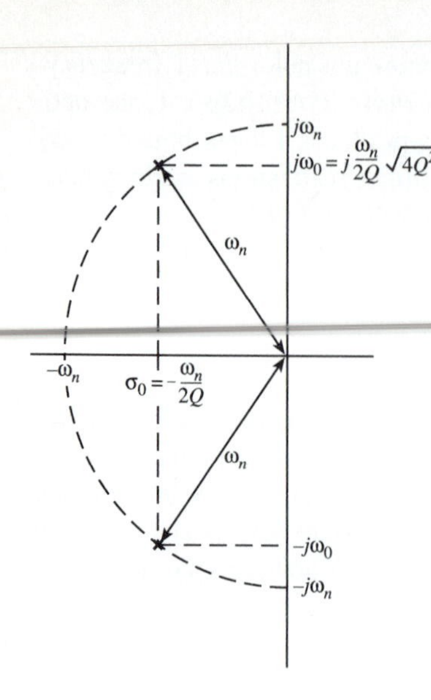

FIGURE 5.2-1
Relations between parameters defining poles.

determined by noting that, in (12) of Sec. 5.1, to satisfy the decomposition given in (13) of Sec. 5.1, $N_{33}(s)$ must be of second order with negative real zeros. In addition, $N_{32}(s)$ must have a simple zero at the origin. Using (3), the expression for $N_{32}(s)$ becomes

$$N_{32}(s) = Y_6(G_1 + Y_2 + G_3 + Y_5) + Y_2G_3 = \alpha s$$

This relation is satisfied by

$$Y_2 = sC_2 \qquad Y_6 = 0 \tag{4}$$

Using these results in $N_{33}(s)$, we obtain

$$N_{33}(s) = (G_3 + Y_4)(G_1 + sC_2 + Y_5) + G_3Y_4$$

The desired form of $N_{33}(s)$ is obtained by letting

$$Y_4 = sC_4 \qquad Y_5 = 0 \tag{5}$$

The resulting realization for the low-pass *RC* amplifier filter is shown in Fig. 5.2-3. The transfer function for this is readily shown to be

$$\frac{V_2(s)}{V_1(s)} = \frac{G_1G_3K}{s^2C_2C_4 + s(G_3C_2 + G_1C_4 + G_3C_4 - KG_3C_2) + G_1G_3} \tag{6}$$

A more conventional form of (6) is achieved by dividing the numerator and denominator by C_2C_4 and using $R_i = 1/G_i$. The result is

$$\frac{V_2(s)}{V_1(s)} = \frac{K/R_1R_3C_2C_4}{s^2 + s(1/R_3C_4 + 1/R_1C_2 + 1/R_3C_2 - K/R_3C_4) + 1/R_1R_3C_2C_4} \tag{7}$$

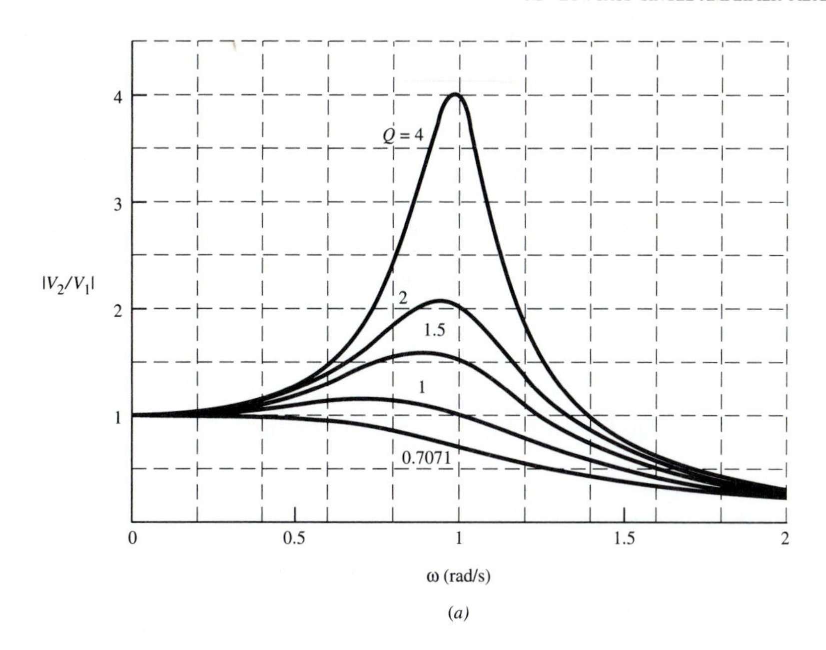

(*a*)

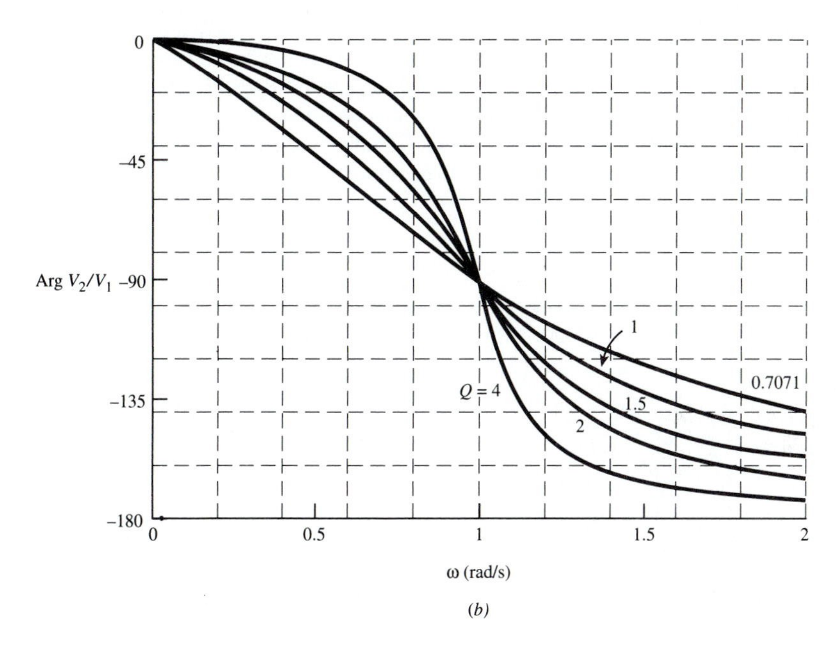

(*b*)

FIGURE 5.2-2
Magnitude and phase loci for low-pass functions.

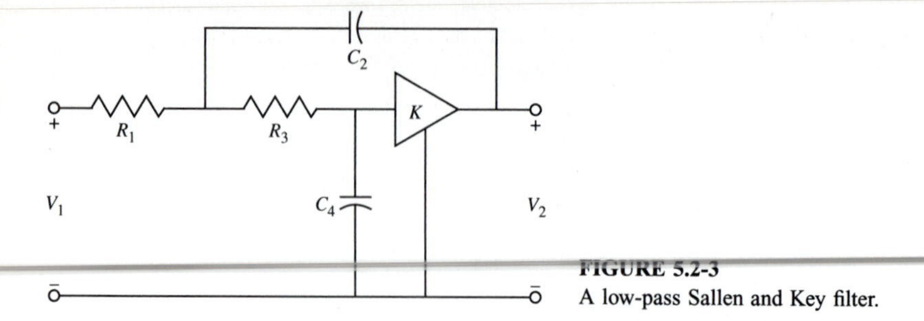

FIGURE 5.2-3
A low-pass Sallen and Key filter.

The realization shown in Fig. 5.2-3 is called a *Sallen and Key low-pass filter configuration.*[1]

Example 5.2-1 Root locus for low-pass single-amplifier filter. The low pass filter defined in Fig. 5.2-3 is readily shown to be an example of the difference decomposition defined in Sec. 5.1 with a root locus shown in Fig. 5.1-5. To illustrate this, let $R_1 = R_3 = 1$ and $C_2 = C_4 = 1$ as normalized values. This produces ω_n normalized to unity, $N_{33}(s) = s^2 + 3s + 1 = (s + 0.382)(s + 2.618)$, and $N_{32}(s) = s$. Equating the denominators of (1), we see that

$$s^2 + (1/Q)s + 1 = (s + 0.382)(s + 2.618) - Ks$$

The locus of the roots of this expression [the poles of the network function defined in (1)] are plotted as a function of the VCVS gain K in Fig. 5.2-4. In the figure we see that as the gain K is increased, the poles of (1) start from the s-plane locations $s = -0.382, -2.618$ $(K = 0)$. They then move toward each other, becoming coincident at $s = -1$ $(K = 1)$. Next, they break apart and become complex conjugate, finally crossing the $j\omega$ axis at $s = \pm j1$ $(K = 3)$. Further increases in the gain produce an unstable network.

[1] R. P. Sallen and E. L. Key, "A Practical Method of Designing RC Active Filters," *IRE Trans. Circuit Theory,* vol. CT-2, March 1955, pp. 74–85. Their paper is one of the most referenced in the field of active filters.

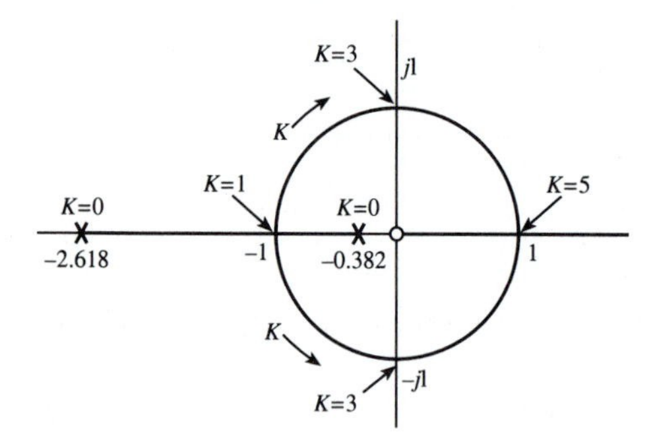

FIGURE 5.2-4
Locus of the poles for Example 5.2-1.

DESIGN 1. Comparing (1) and (7) gives the following set of equations:

$$\omega_n = \frac{1}{\sqrt{R_1 R_3 C_2 C_4}} \tag{8a}$$

$$\frac{1}{Q} = \sqrt{\frac{R_3 C_4}{R_1 C_2}} + \sqrt{\frac{R_1 C_4}{R_3 C_2}} + (1 - K)\sqrt{\frac{R_1 C_2}{R_3 C_4}} \tag{8b}$$

$$H_0 = K \tag{8c}$$

Since there are five unknowns, namely, R_1, C_2, R_3, C_4, and K, and only three equations, a unique solution may be obtained by specifying

$$R_1 = R_3 = R \tag{9a}$$

$$C_2 = C_4 = C \tag{9b}$$

Substituting (9) into (8) yields

$$\omega_n = \frac{1}{RC} \qquad \frac{1}{Q} = 3 - K \qquad H_0 = K \tag{10}$$

A set of design equations is now easily derived as (design 1)

$$RC = \frac{1}{\omega_n} \qquad K = 3 - \frac{1}{Q} \tag{11}$$

In this case H_0 is no longer a free parameter but is constrained to the value K.

Example 5.2-2 An equal-resistance, equal-capacitance low-pass filter (design 1). A low-pass filter is desired in which $\omega_n = 6283$ rad/s (1 kHz) and $Q = 0.7071$ (a Butterworth characteristic). From (11) we obtain

$$RC = \frac{1}{6283} = 1.59 \times 10^{-4} \qquad K = 1.586$$

If we select $C = 0.1\ \mu\text{F}$, then $R = 1.59\ \text{k}\Omega$. The resulting realization has $H_0 = 1.586$.

DESIGN 2. A different design procedure (design 2) from the one using (9) occurs if we set $K = 1$. This value of gain is achieved with a high degree of precision by the voltage follower circuit shown in Fig. 5.1-2*b*. From (8) we obtain

$$\frac{1}{Q} = \sqrt{\frac{R_3 C_4}{R_1 C_2}} + \sqrt{\frac{R_1 C_4}{R_3 C_2}} \tag{12}$$

We now define two parameters n and m as

$$n = \frac{R_3}{R_1} \qquad m = \frac{C_4}{C_2} \tag{13}$$

It should be noted that n and m are the ratios of the resistor and capacitor values, respectively. Furthermore, let us define

$$R_1 = R \qquad C_2 = C \tag{14}$$

Equation (7) can now be written as

$$\frac{V_2(s)}{V_1(s)} = \frac{1/mnR^2C^2}{s^2 + (1/RC)[(n+1)/n]s + 1/mnR^2C^2} \tag{15}$$

Equating (1) and (15) results in the following design formulas:

$$\omega_n = \frac{1}{\sqrt{mn}\,RC} \qquad \frac{1}{Q} = (n+1)\sqrt{\frac{m}{n}} \tag{16}$$

From the expression for $1/Q$, it can be shown that for any given value of m the maximum value of Q will occur when $n = 1$. This case is not optimum, however, for if $n = 1$, Q is equal to $1/(2\sqrt{m})$. For most values of Q, this will produce excessively high capacitor ratios. A more practical approach is to select a value of m compatible with standard capacitor values such that

$$m \le \frac{1}{4Q^2} \tag{17}$$

Then n can be calculated from

$$n = \left(\frac{1}{2mQ^2} - 1\right) \pm \frac{1}{2mQ^2}\sqrt{1 - 4mQ^2} \tag{18}$$

This equation provides two values of n for any given Q and m. The values are readily shown to be reciprocal; thus, the use of either one produces the same element spread.

Example 5.2-3 A unity-gain low-pass filter (design 2). It is desired to use a unity-gain VCVS to realize the voltage transfer function

$$\frac{V_2(s)}{V_1(s)} = \frac{0.988}{s^2 + 0.179s + 0.988}$$

where the complex-frequency variable has been normalized by a radian frequency of 10^{-4} rad/s. From (1) we obtain $\omega_{n(\text{normalized})} = 0.994$ rad/s and $Q = 5.553$. From (17) we must select $m \le 0.0081$. Choosing $m = 0.001$, from (18) we find $n = 0.0329, 30.397$. If we select $n = 30.397$, then from (16) with $\omega_n = 10^4\omega_{n(\text{normalized})}$ we find $RC = 5.7703 \times 10^{-4}$. If we select $C_2 = C = 0.1\ \mu\text{F}$, then $C_4 = 100$ pF, $R_1 = R = 5.77\ \text{k}\Omega$, and $R_3 = 175.4\ \text{k}\Omega$.

DESIGN 3. Another practical choice for the low-pass single-amplifier active filter is the one in which both capacitors are equal valued and the gain of the VCVS is set to 2.0. The equal-valued capacitors are convenient because they can be impedance normalized to a "stock" value. The gain of 2.0 is attractive because it can be precisely obtained by using equal-valued feedback resistors around an operational amplifier. From (1) and (7) for $C_2 = C_4 = C$ and $K = 2$ we readily derive (design 3)

$$R_1 = \frac{Q}{\omega_n C} \qquad R_3 = \frac{1}{R_1\omega_n^2C^2} \quad \text{or} \quad R_3 = \frac{R_1}{Q^2} \tag{19}$$

Example 5.2-4 An equal-capacitance, gain-of-2 low-pass filter (design 3). To realize the normalized (Butterworth) voltage transfer function

$$\frac{V_2(s)}{V_1(s)} = \frac{2}{s^2 + \sqrt{2}s + 1}$$

where a normalization of 10^4 rad/s has been made, from (19) using $C = 1$, $\omega_n = 1$, and $Q = 1/\sqrt{2}$, we find $R_1 = 1/\sqrt{2}$ and $R_3 = \sqrt{2}$. Frequency denormalizing and applying an additional impedance denormalization of 10^3, we obtain the design values $R_1 = 0.707$ kΩ, $R_3 = 1.414$ kΩ, $C_2 = C_4 = 0.1$ μF, and $K = 2$.

Since as pointed out in connection with (8) there are only three network function specifications, namely, ω_n, Q, and H_0, while there are five network parameters, there are obviously many other design procedures than those given above that may be specified for the Sallen and Key low-pass filter. The three described in the preceding paragraphs are typical of the majority of these. Some of the other procedures may be found in the Problems.

Sensitivity

Now let us consider the sensitivity aspects of the Sallen and Key filter. For added generality we will use the circuit shown in Fig. 5.2-5, in which the gain-determining resistors used in connection with the operational amplifier are shown explicitly as R_A and R_B and where the VCVS gain $K = 1 + R_B/R_A$. Using the sensitivity definitions of Sec. 3.5 and the relations of (8) and (1), we obtain

$$S_{R_1}^{Q} = -\tfrac{1}{2} + Q\sqrt{\frac{R_3C_4}{R_1C_2}} = -S_{R_3}^{Q} \tag{20a}$$

$$S_{C_2}^{Q} = -\tfrac{1}{2} + Q\left(\sqrt{\frac{R_1C_4}{R_3C_2}} + \sqrt{\frac{R_3C_4}{R_1C_2}}\right) = -S_{C_4}^{Q} \tag{20b}$$

$$S_{K}^{Q} = QK\sqrt{\frac{R_1C_2}{R_3C_4}} \tag{20c}$$

$$S_{R_A}^{Q} = -Q(K-1)\sqrt{\frac{R_1C_2}{R_3C_4}} = -S_{R_B}^{Q} \tag{20d}$$

$$S_{R_1,R_3,C_2,C_4}^{\omega_n} = -\tfrac{1}{2} \tag{20e}$$

$$S_{K,R_A,R_B}^{\omega_n} = 0 \tag{20f}$$

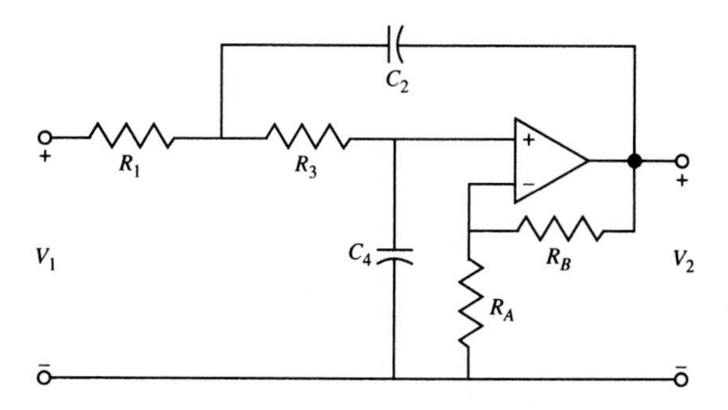

FIGURE 5.2-5
Operational amplifier implementation of low-pass Sallen and Key filter.

The values of these sensitivities for designs 1, 2, and 3 are given in Table 5.2-1. This table emphasizes the fact that the ω_n sensitivities are identical for all three designs but that the Q sensitivities vary widely.

It is frequently possible to reduce the sensitivities cataloged in Table 5.2-1, although a price is usually exacted for such a reduction. Frequently this price is an increase in the spread of element values. For example, choosing ratios of 1:10 for the resistors R_1 and R_3 and 10:1 for the capacitors C_2 and C_4, rather than the 1:1 ratios specified in design 1, considerably lowers many of the Q sensitivities. A similar result is obtained by choosing R_1 and R_3 equal in design 2. Finally the value of the gain K may be chosen in such a way as to obtain the best compromise between the active and passive sensitivities. Examples of many of these situations may be found in the Problems.

The Q-sensitivity information given in Table 5.2-1 for the single-amplifier filter may be verified and extended by determining the function sensitivities for the circuit. This is done in the following example.

Example 5.2-5 Function sensitivities for a single-amplifier filter. The voltage transfer function given for a low-pass single-amplifier filter in (7) can be evaluated to obtain the function sensitivities for a specific design of the filter. As an example, consider the use of design 1 to realize a normalized realization in which $\omega_n = 1$ rad/s and $Q = 3$. The element values are $R_1 = R_3 = 1\ \Omega$, $C_2 = C_4 = 1$ F, and $K = 2.6667$. The resulting voltage transfer function is

$$N(s) = \frac{V_2(s)}{V_1(s)} = \frac{2.6667}{s^2 + s/3 + 1}$$

The results of determining the function sensitivities and evaluating them over the range $0 \le \omega \le 2$ rad/s are shown in Fig. 5.2-6. In the figure, note that the peak value of $S_K^{|N(j\omega)|}$ is approximately 9. This is in good agreement with the (approximate) dependence $S_K^Q \approx 3Q$ indicated in Table 5.2-1. Similarly, the peaks values of $S_{C_2}^{|N(j\omega)|}$ and $S_{R_1}^{|N(j\omega)|}$ are approximately 6 and 3, respectively, in good agreement with the $2Q$ and Q dependencies indicated for the

TABLE 5.2-1
Sensitivities for three designs of Sallen and Key low-pass single-amplifier filters

Sensitivity	Design 1	Design 2	Design 3
$S_{R_1}^Q = -S_{R_3}^Q$	$-\frac{1}{2} + Q$	$-\frac{1}{2} + Q\sqrt{mn}$	$\frac{1}{2}$
$S_{C_2}^Q = -S_{C_4}^Q$	$-\frac{1}{2} + 2Q$	$-\frac{1}{2} + Q\left(\sqrt{\frac{m}{n}} + \sqrt{mn}\right)$	$\frac{1}{2} + Q^2$
S_K^Q	$3Q - 1$	$\frac{Q}{\sqrt{mn}}$	$2Q^2$
$S_{R_A}^Q = -S_{R_B}^Q$	$1 - 2Q$	0	$-Q^2$
$S_{R_1, R_3, C_2, C_4}^{\omega_n}$	$-\frac{1}{2}$	$-\frac{1}{2}$	$-\frac{1}{2}$
$S_{K, R_A, R_B}^{\omega_n}$	0	0	0

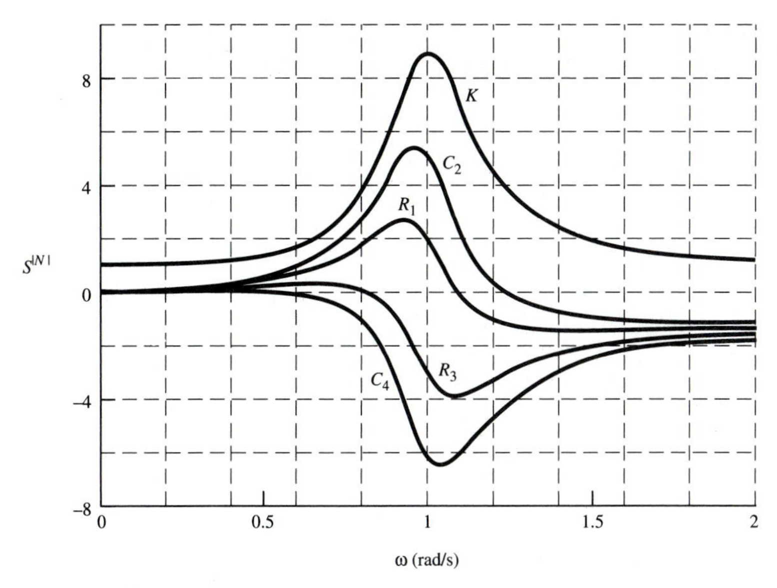

FIGURE 5.2-6
Sensitivities of a low-pass single-amplifier filter.

corresponding Q sensitivities in the table. Finally, note that $S_{C_4}^{|N(j\omega)|}$ and $S_{R_3}^{|N(j\omega)|}$ have negative values, as do the corresponding Q sensitivities given in the table.

5.3 BAND-PASS AND HIGH-PASS SINGLE-AMPLIFIER FILTERS

In the preceding section we used the general second-order *RC*-amplifier configuration described at the beginning of this chapter to develop a low-pass filter circuit. In this section we will use similar procedures to develop band-pass and high-pass circuits. We first consider the band-pass case. The general second-order band-pass voltage transfer function has the form

$$\frac{V_2(s)}{V_1(s)} = \frac{H_0(\omega_n/Q)s}{s^2 + (\omega_n/Q)s + \omega_n^2} = \frac{KN_{31}(s)}{N_{33}(s) - KN_{32}(s)} \tag{1}$$

In this equation H_0 is the maximum magnitude of the network function in the passband. It is also referred to as the *gain at the resonant frequency*. The quantities ω_n and Q have the same significance as they did in the low-pass case, namely, ω_n is the *undamped natural frequency* and Q is the *quality factor*. These quantities are related to the poles of the network function as described in (2) of Sec. 5.2 and in Fig. 5.2-1. The right member of (1) is taken from (11) of Sec. 5.1. Plots of $|V_2/V_1(j\omega)|$ and Arg $V_2/V_1(j\omega)$ for various values of Q for the case where normalized values $H_0 = 1$ and $\omega_n = 1$ rad/s are used are shown in Fig. 5.3-1.

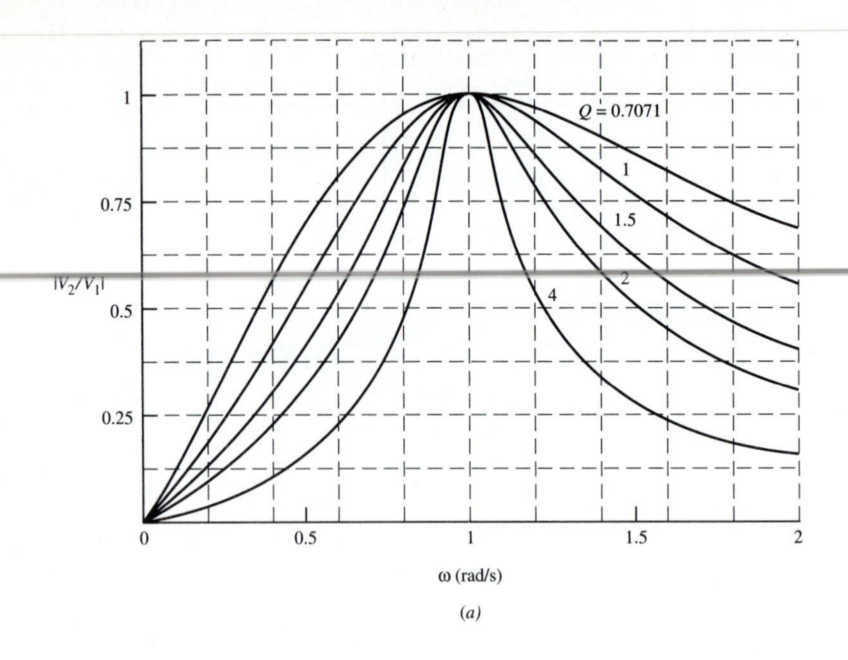

(a)

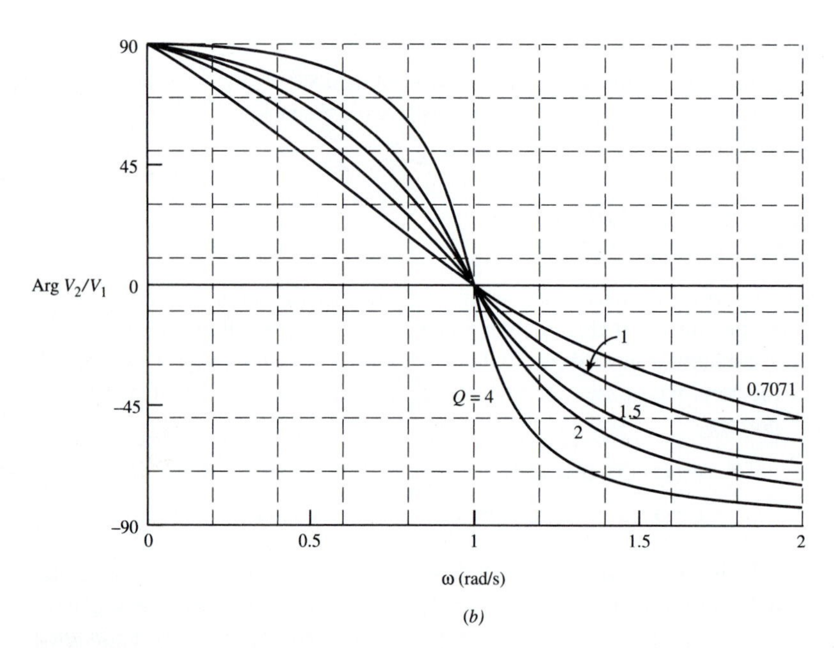

(b)

FIGURE 5.3-1
Magnitude and phase loci for band-pass functions.

Band-Pass Single-Amplifier Filter Realization

Comparing the numerator of (1) with (12) of Sec. 5.1 we see that either Y_1 or Y_3 must be a capacitor and the other a resistor. Let us first choose

$$Y_1 = G_1 \qquad Y_3 = sC_3 \tag{2}$$

Thus $N_{31}(s) = sG_1C_3$. Following the same procedure used on connection with (4) and (5) of Sec. 5.2, we obtain

$$N_{32}(s) = Y_6(G_1 + Y_2 + sC_3 + Y_5) + sC_3Y_2$$
$$N_{33}(s) = (sC_3 + Y_4 + Y_6)(G_1 + Y_2 + Y_5) + sC_3(Y_4 + Y_6)$$

The denominator decomposition must be a difference one having the same general form as used for the low-pass filter. To obtain this, we choose

$$Y_2 = G_2 \qquad Y_6 = 0 \qquad \text{thus} \qquad N_{32}(s) = sC_3G_2 \tag{3}$$

For these choices we obtain

$$N_{33}(s) = (sC_3 + Y_4)(G_1 + G_2 + Y_5) + sC_3Y_4$$

Finally, to make $N_{33}(s)$ a second-order polynomial, as required by the decomposition, we choose $Y_4 = G_4$ and $Y_5 = sC_5$. Thus we obtain

$$N_{33}(s) = s^2C_3C_5 + s(G_1C_3 + G_2C_3 + G_4C_3 + G_4C_5) + G_4(G_1 + G_2) \tag{4}$$

The overall voltage transfer function now becomes

$$\frac{V_2(s)}{V_1(s)} = \frac{sKG_1C_3}{s^2C_3C_5 + s(G_1C_3 + G_2C_3 + G_4C_3 + G_4C_5 - KG_2C_3) + G_4(G_1 + G_2)} \tag{5}$$

This may also be put in the form

$$\frac{V_2(s)}{V_1(s)} = \frac{sK/R_1C_5}{s^2 + s(1/R_1C_5 + 1/R_2C_5 + 1/R_4C_5 + 1/R_4C_3 - K/R_2C_5) + (1/R_4C_3C_5)(1/R_1 + 1/R_2)} \tag{6}$$

The final circuit configuration is shown in Fig. 5.3-2. It is usually referred to as a *Sallen and Key band-pass filter circuit.*

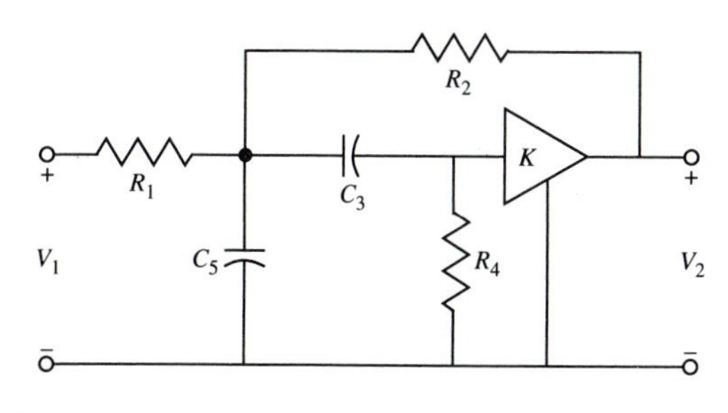

FIGURE 5.3-2
A band-pass Sallen and Key filter.

DESIGN 1. If we compare (1) and (6), we obtain the following equations:

$$\omega_n = \left(\frac{1 + R_1/R_2}{R_1R_4C_3C_5}\right)^{1/2} \tag{7a}$$

$$\frac{1}{Q} = \frac{[1 + (R_1/R_2)(1 - K)]\sqrt{R_4C_3/R_1C_5} + \sqrt{R_1C_3/R_4C_5} + \sqrt{R_1C_5/R_4C_3}}{\sqrt{1 + R_1/R_2}} \tag{7b}$$

$$H_0 = \frac{K/R_1C_5}{1/R_1C_5 + 1/R_2C_5 + 1/R_4C_5 + 1/R_4C_3 - K/R_2C_5} \tag{7c}$$

One design procedure for obtaining a unique solution for the values of the network elements is to choose an equal-valued resistor–equal-valued capacitor one. Thus (design 1)

$$R_1 = R_2 = R_4 = R \qquad C_3 = C_5 = C \tag{8}$$

In this case, the relations of (7) become

$$\omega_n = \frac{\sqrt{2}}{RC} \qquad Q = \frac{\sqrt{2}}{4 - K} \qquad H_0 = \frac{K}{4 - K} \tag{9}$$

Solving for RC and K, we obtain

$$RC = \frac{\sqrt{2}}{\omega_n} \qquad K = 4 - \frac{\sqrt{2}}{Q} \tag{10}$$

Note that for K to be positive, Q must be greater than $\sqrt{2}/4$.

Example 5.3-1 An equal-resistance, equal-capacitance band-pass filter (design 1). It is desired to realize a band-pass filter in which $Q = 10$ and $\omega_n = 10^4$ rad/s. From (10) we find that $RC = \sqrt{2} \times 10^{-4}s$ and $K = 3.8586$. Choosing C as 0.1 μF, we obtain $R = 1.414$ kΩ. The resulting value of H_0 is 27.289.

DESIGN 2. There are many possible choices for the element values of the filter shown in Fig. 5.3-2 other than those given by (10). Another practical design results if the VCVS gain is constrained to a value of 2, which, as has been pointed out, is a value readily obtained with great stability by using two equal-valued precision resistors as feedback around an operational amplifier. In this case we may select (design 2)

$$R_1 = C_3 = C_5 = 1 \qquad K = 2 \tag{11}$$

For these choices, equating (1) and (6), we obtain

$$\omega_n^2 = \left(1 + \frac{1}{R_2}\right)\frac{1}{R_4} \tag{12a}$$

$$\frac{\omega_n}{Q} = 1 + \frac{2}{R_4} - \frac{1}{R_2} \tag{12b}$$

This set of simultaneous nonlinear equations is readily solved to satisfy specific values of ω_n and Q.

Example 5.3-2 An equal-capacitance, gain-of-2 band-pass filter (design 2). It is desired to design an impedance and frequency normalized band-pass filter in which $\omega_n = 1$ and $Q = 2$. The capacitors are to be of unity value, and the VCVS is to have a gain of 2. Using (12) and for convenience letting $G_i = 1/R_i$, we obtain

$$1 = (1 + G_2)G_4$$
$$0.5 = 1 + 2G_4 - G_2$$

Solving the first equation for G_4, substituting the result in the second equation, and rearranging the terms, we obtain $G_2^2 + 0.5G_2 - 2.5 = 0$. Solving this equation for G_2 and substituting the result in the first equation, we find $R_2 = 0.7403\ \Omega$ and $R_4 = 2.3508\ \Omega$.

A different band-pass filter circuit from the one described above is obtained by letting $Y_1 = sC_1$ and $Y_3 = G_3$ rather than using the choices of (2). Following through the remaining steps of the realization procedure yields a circuit configuration with three capacitors. As such it is of less practical interest than the one described above. Further details concerning this circuit may be found in the Problems.

High-Pass Single-Amplifier Filter Realization

The second type of filter to be considered in this section is the high-pass one. The general second-order high-pass voltage transfer function has the form

$$\frac{V_2(s)}{V_1(s)} = \frac{H_0 s^2}{s^2 + (\omega_n/Q)s + \omega_n^2} = \frac{KN_{31}(s)}{N_{33}(s) - KN_{32}(s)} \tag{13}$$

In this equation, H_0 is the gain at infinite frequency and ω_n and Q are the undamped natural frequency and quality factor, respectively. The right member of (13) is taken from (11) of Sec. 5.1. Plots of $|V_2/V_1(j\omega)|$ and Arg $V_2/V_1(j\omega)$ for the normalized values $H_0 = 1$ and $\omega_n = 1$ rad/s are shown in Fig. 5.3-3.

The filter circuit that realizes this network function could be developed in a manner similar to that used for the low-pass and band-pass functions. Instead, we here select a simpler approach. We first note that this type of network function is readily related to the low-pass one by using the low-pass to high-pass transformation introduced in Sec. 2.5; namely, we let $s = 1/p$, where s is the original low-pass complex-frequency variable and p is the new high-pass complex-frequency variable. We may now directly use low-pass filter realizations to derive high-pass filter realizations by applying this frequency transformation to the elements. In addition, we then apply an impedance denormalization of $1/p$. This latter, of course, leaves the voltage transfer function invariant since it is dimensionless. For any (low-pass) resistor of value R, the procedure is as follows:

$$\boxed{Z_{LP}(s) = R} \xrightarrow[\text{transformation}]{\text{Low-pass to high-pass}} \boxed{Z_{HP}(p) = R} \xrightarrow[\text{of } 1/p]{\text{Impedance denormalization}} \boxed{Z_{z\ \text{denormalized}}(p) = R/p} \tag{14}$$

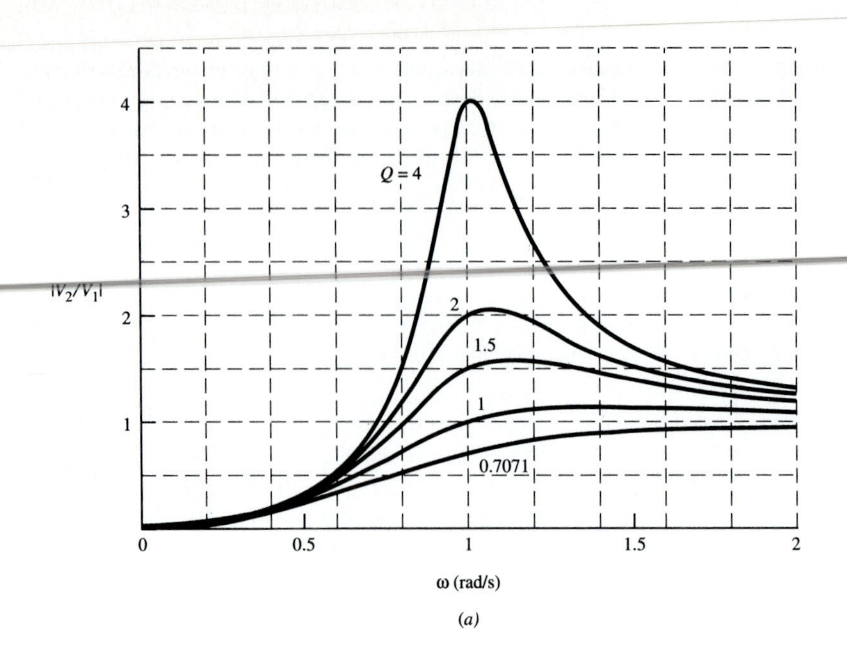

(a)

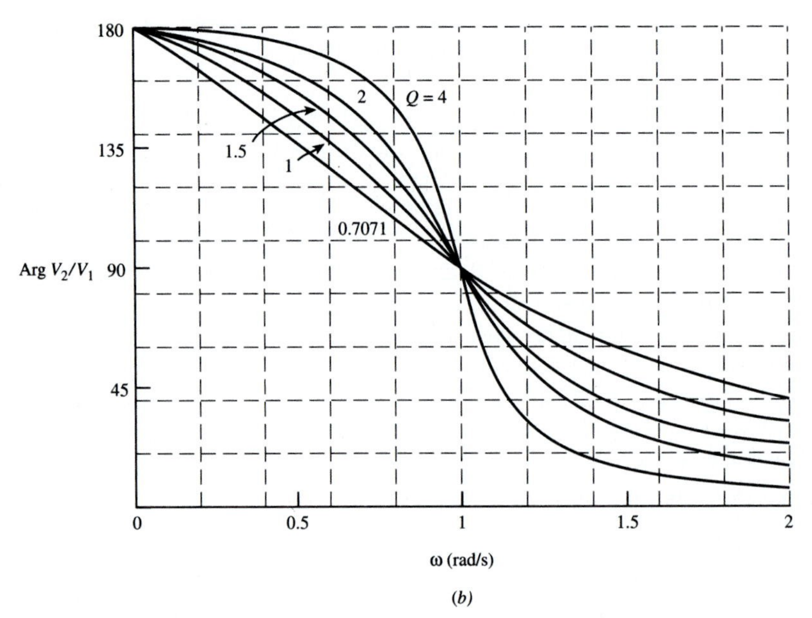

(b)

FIGURE 5.3-3
Magnitude and phase loci for high-pass functions.

The result of the two transformations is a capacitor of value $1/R$ farads in the high-pass realization. Similarly, for a (low-pass) capacitor of value C we obtain

$$\boxed{Y_{LP}(s) = sC} \xrightarrow[\substack{\text{Low-pass to}\\\text{high-pass}\\\text{transformation}}]{} \boxed{Y_{HP}(p) = C/p} \xrightarrow[\substack{\text{Impedance}\\\text{denormalization}\\\text{of } 1/p}]{} \boxed{Y_{z\,\text{denormalized}}(p) = C} \tag{15}$$

Thus, the result of the two transformations in this case is a resistor of value $1/C$ ohms. Since the gain of the VCVS is both dimensionless and not a function of frequency, the VCVS remains invariant under the two transformations. These results are summarized in Table 5.3-1. Applying this table to the network of Fig. 5.2-3 gives the positive-gain high-pass realization shown in Fig. 5.3-4. Thus the choices for the passive elements of Fig. 5.1-4 are seen to be $Y_1 = sC_1$, $Y_2 = G_2$, $Y_3 = sC_3$, $Y_4 = G_4$, and $Y_5 = Y_6 = 0$. Substitution of these values into $N_{31}(s)$, $N_{32}(s)$, and $N_{33}(s)$ as defined in (12) of Sec. 5.1 permits the transfer function of (13) to be found as

$$\frac{V_2(s)}{V_1(s)} = \frac{s^2C_1C_3K}{s^2C_1C_3 + s(G_2C_3 + G_4C_1 + G_4C_3 - KG_2C_3) + G_2G_4} \tag{16}$$

Dividing the numerator and denominator of this by C_1C_3, we obtain

$$\frac{V_2(s)}{V_1(s)} = \frac{s^2K}{s^2 + s(1/R_2C_1 + 1/R_4C_3 + 1/R_4C_1 - K/R_2C_1) + 1/R_2R_4C_1C_3} \tag{17}$$

TABLE 5.3-1

A transformation of network elements under the low-pass to high-pass transformation with an additional impedance transformation $1/p$

Low pass	High pass
Resistor R →	Capacitor $C = 1/R$
Capacitor C →	Resistor $R = 1/C$
VCVS: input V_0 (+, −), output KV_0 →	VCVS: input V_0 (+, −), output KV_0

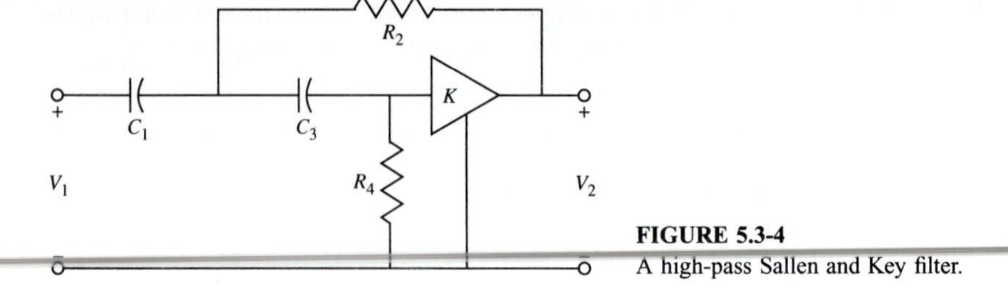

FIGURE 5.3-4
A high-pass Sallen and Key filter.

DESIGN 1. If we compare (13) and (17), we obtain the equations

$$\omega_n = \frac{1}{\sqrt{R_2 R_4 C_1 C_3}} \tag{18a}$$

$$\frac{1}{Q} = \sqrt{\frac{R_4 C_3}{R_2 C_1}} + \sqrt{\frac{R_2 C_1}{R_4 C_3}} + \sqrt{\frac{R_2 C_3}{R_4 C_1}} - K\sqrt{\frac{R_4 C_3}{R_2 C_1}} \tag{18b}$$

$$H_0 = K \tag{18c}$$

Following the design 1 procedure for the low-pass network in Sec. 5.2 by choosing $R_2 = R_4 = R$ and $C_1 = C_3 = C$ gives (design 1)

$$\omega_n = \frac{1}{RC} \tag{19a}$$

$$\frac{1}{Q} = 3 - K \tag{19b}$$

$$H_0 = K \tag{19c}$$

which are identical with (10) of Sec. 5.2. Thus a high-pass realization with $Q = 0.7071$ and $\omega_n = 6283$ rad/s can be obtained by using the R, C, and K values of Example 5.2-2 as $1/C$, $1/R$, and K, respectively, in the circuit of Fig. 5.3-4.

DESIGN 2. As with the positive-gain low-pass realization of Sec. 5.2, a second high-pass filter design may be obtained by setting $K = 1$. Assuming that $m = C_3/C_1$ and $n = R_4/R_2$ and letting $C_1 = C$ and $R_2 = R$, (17) becomes

$$\frac{V_2}{V_1} = \frac{s^2}{s^2 + \dfrac{s}{RC}\dfrac{1}{n}\dfrac{m+1}{m} + \dfrac{1}{mn(RC)^2}} \tag{20}$$

Comparing this with (13), we get the design formulas (design 2)

$$\omega_n = \frac{1}{\sqrt{mn}\,RC} \qquad \frac{1}{Q} = \frac{m+1}{\sqrt{mn}} \tag{21}$$

From these it can be shown that for any given value of n the minimum value of $1/Q$ will occur when $m = 1$. Since it is usually desirable to have a minimum $1/Q$ for any given n, we will let $m = 1$. In this case (21) simplifies to

$$\omega_n = \frac{1}{\sqrt{n}RC} \qquad Q = \frac{\sqrt{n}}{2} \tag{22}$$

Example 5.3-3 A unity-gain high-pass filter (design 2). It is desired to use a unity-gain VCVS to realize the voltage transfer function

$$\frac{V_2(s)}{V_1(s)} = \frac{s^2}{s^2 + 0.179s + 0.988}$$

where the complex-frequency variable has been normalized by a value of 10^{-4} rad/s. The terms Q and $\omega_{n(\text{normalized})}$ have the same values found in Example 5.2-3, namely, 5.553 and 0.994 rad/s. From (22) we find $n = 4Q^2 = 123.34$ and $RC = 1/\omega_n\sqrt{n} = 0.09058$. Denormalizing, $RC = 0.09058 \times 10^{-4}$. If we select $C = 0.01\ \mu\text{F}$, then $R = R_2 = 905.8\ \Omega$ and $R_4 = 111.72\ \text{k}\Omega$.

DESIGN 3. Another practical set of values for the high-pass single-amplifier positive-gain RC amplifier filter shown in Fig. 5.3-4 is the one in which both capacitors are equal valued and the gain of the VCVS is set to 2.0. Thus, for the normalized case we have $C_1 = C_3 = C$ and $K = 2$. From (17) for this case we find that (design 3)

$$\omega_n = \frac{1}{C\sqrt{R_2R_4}} \qquad \frac{\omega_n}{Q} = \frac{1}{C}\left(\frac{2}{R_4} - \frac{1}{R_2}\right) \tag{23}$$

These equations can be solved for values of R_2 and R_4 in a manner similar to that used in Example 5.3-2.

Sensitivity

Now let us consider some of the sensitivity aspects of the band-pass and high-pass filter circuit realizations given in this section. A comparison of the denominator of the low-pass voltage transfer function given in (7) of Sec. 5.2 with the denominators of the band-pass and high-pass functions given in (6) and (17) of this section readily shows that the type of polynomial decomposition used to obtain the complex-conjugate poles is the same in all three cases. Thus, the root loci for all three filters have the general form shown in Fig. 5.2-4. As such, sensitivity analyses of the band-pass and high-pass designs will yield results similar to those given in Table 5.2-1 for the low-pass types. For example, the ω_n sensitivities to K are zero for all three types of filters. Further details of such analyses are covered in the Problems.

In this section and the preceding one, we have presented a discussion of the single-amplifier Sallen and Key structures for low-pass, band-pass, and high-pass network functions. The advantages of these structures in general are that they are characterized by easy-to-use design relations; the designer has good control over the element values and their spread; and it is possible to use low values of VCVS gain, which is convenient since such values are readily stabilized. There are also some disadvantages, of which the major one is the high sensitivities that result when these circuits are used to realize high-Q functions. This is readily apparent not only from sensitivity analyses such as those given in Table 5.2-1 but also from the shape of their root loci, as shown in Fig. 5.2-4. Obviously, in this figure, when the poles are "high Q," i.e., located close to the $j\omega$ axis, a very small increase in the gain will move them into the right half of the complex-frequency plane and thus make the circuit unstable. One solution to this sensitivity problem that has been proposed is the use of negative-gain filter structures, i.e., ones using inverting VCVSs. Such filters have root loci that do not cross into the right-half plane. Not only are such filters always stable, but their sensitivities are usually lower than those of the positive-gain ones. Unfortunately, the negative-gain filters also have several disadvantages: Their element value spreads are frequently large; their design methods are often more complex; and they usually require large values of VCVS gain, which are more difficult to stabilize than low ones. In general, these negative points outweigh the positive ones. As a result, negative-gain filters are not widely used. Some examples of them may be found in the Problems.

5.4 INFINITE-GAIN SINGLE-AMPLIFIER FILTERS

In the preceding sections of this chapter we have considered the use of a low-gain noninverting amplifier as the gain element of a single-amplifier filter. The amplifier was realized by using two gain-determining resistors as the feedback elements for an operational amplifier. In this section we show that an inverting operational amplifier (with an ideal open-loop gain of infinity) can be used directly as a gain element for an active filter. The resulting realization is referred to as an *infinite-gain single-amplifier filter*. We shall see that this type of filter has many advantages (and some disadvantages) when compared with the type described in the preceding sections.

General Infinite-Gain Single-Amplifier Configuration

The general form of the infinite-gain single-amplifier filter is shown in Fig. 5.4-1 where the passive portion of the network is represented by its y parameters. The derivation of the voltage transfer function is done using the *method of constraints*. The procedure is similar to the one given in Summary 5.1-1. It is described as follows:

Summary 5.4-1 Analysis of operational amplifier circuits using the method of constraints. To analyze a circuit containing passive components and one or more operational amplifiers, we may proceed as follows:

1. Write the admittance equations describing the circuit *without* the operational amplifiers. This may be done either on an external terminal basis, in which case the equations describe the multiport parameters of the circuit, or on an internal node basis, in which case the usual node equations result.
2. Apply the *voltage constraint* introduced by each operational amplifier to reduce the number of unknown voltage variables. For an operational amplifier with one input

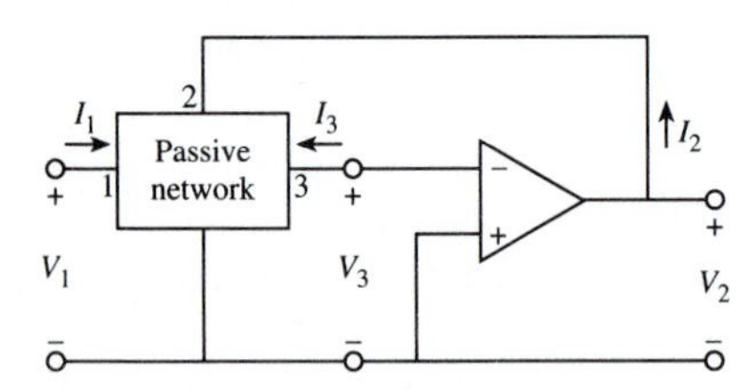

FIGURE 5.4-1
An infinite-gain single-amplifier filter configuration.

terminal grounded and the ungrounded input terminal connected to the terminal (or node) where V_i is defined, the constraining equation is $V_i = 0$. For an operational amplifier with both input terminals ungrounded and connected to the terminals (or nodes) at which V_i and V_j are defined, the constraining equation is $V_i = V_j$. In either case the number of unknown node voltages is reduced by 1.

3. Apply the *current constraint* for each operational amplifier. If the output of an operational amplifier is at node i, the zero output impedance of the amplifier effectively shorts any input current to this terminal (or node) to ground. As a result, the corresponding ith equation is not an independent one and it may be deleted to reduce the number of equations.

As a result of the application of these steps, we see that the constraints imposed by each operational amplifier reduce the order of the equations describing the circuit by one degree. When the constraints imposed by all the operational amplifiers have been implemented, the remaining equations may be solved for any choice of input and output variables.

The method of constraints can be applied to develop two equations that will be of considerable use in much of the following treatment of active filters. The development of these equations is given in the following two examples.

Example 5.4-1 Voltage transfer function for a general infinite-gain single-amplifier filter. As an example of the use of the method of constraints, consider the general infinite-gain single-amplifier filter configuration shown in Fig. 5.4-1. The passive portion of the network may be considered as a three-port network and defined by a set of short-circuit admittance (or y) parameters. Thus, for the passive network portion of the filter (*without* the operational amplifier) we may write (for step 1)

$$\begin{bmatrix} I_1(s) \\ I_2(s) \\ I_3(s) \end{bmatrix} = \begin{bmatrix} y_{11}(s) & y_{12}(s) & y_{13}(s) \\ y_{21}(s) & y_{22}(s) & y_{23}(s) \\ y_{31}(s) & y_{32}(s) & y_{33}(s) \end{bmatrix} \begin{bmatrix} V_1(s) \\ V_2(s) \\ V_3(s) \end{bmatrix} \tag{1}$$

where, due to the passive nature of the network, $y_{ij}(s) = y_{ji}(s)$; that is, the y matrix is symmetric. Now let us apply the method of Summary 5.4-1 to include the effect of the operational amplifier. From step 2 we obtain $V_3(s) = 0$. If this is substituted in (1), we find

$$\begin{bmatrix} I_1(s) \\ I_2(s) \\ I_3(s) \end{bmatrix} = \begin{bmatrix} y_{11}(s) & y_{12}(s) \\ y_{21}(s) & y_{22}(s) \\ y_{31}(s) & y_{32}(s) \end{bmatrix} \begin{bmatrix} V_1(s) \\ V_2(s) \end{bmatrix} \tag{2}$$

From step 3 of Summary 5.4-1, we may eliminate the second equation of the matrix array given in (2) to obtain

$$\begin{bmatrix} I_1(s) \\ I_3(s) \end{bmatrix} = \begin{bmatrix} y_{11}(s) & y_{12}(s) \\ y_{31}(s) & y_{32}(s) \end{bmatrix} \begin{bmatrix} V_1(s) \\ V_2(s) \end{bmatrix} \tag{3}$$

Since $I_3(s) = 0$ (due to the infinite input impedance of the operational amplifier), the second equation of the array given in (3) may be set to zero and solved to obtain

$$\frac{V_2(s)}{V_1(s)} = \frac{-y_{31}(s)}{y_{32}(s)} \tag{4}$$

Note that the zeros of the transfer function are determined by the zeros of $y_{31}(s)$, and the poles are determined by the *zeros* of $y_{32}(s)$. This equation will be of considerable use in many of the developments that follow.

Example 5.4-2 Voltage transfer function for a specific passive network. In many of the single-amplifier *RC* filters to be discussed in this chapter we will restrict the passive network so that it has the form shown in Fig. 5.4-2. The method of constraints may be applied to find the voltage transfer function for this network. From step 1 of Summary 5.4-1 we may write

$$\begin{bmatrix} I_1 \\ I_2 \\ I_3 \\ I_4 \end{bmatrix} = \begin{bmatrix} Y_1 & 0 & 0 & -Y_1 \\ 0 & Y_2 + Y_6 & -Y_6 & -Y_2 \\ 0 & -Y_6 & Y_3 + Y_4 + Y_6 & -Y_3 \\ -Y_1 & -Y_2 & -Y_3 & Y_1 + Y_2 + Y_3 + Y_5 \end{bmatrix} \begin{bmatrix} V_1 \\ V_2 \\ V_3 \\ V_4 \end{bmatrix} \tag{5}$$

where, for convenience, we have deleted the (s) notation. From steps 2 and 3 we set $V_3 = 0$ and delete the second equation in the above array to obtain

$$\begin{bmatrix} I_1 \\ I_3 \\ I_4 \end{bmatrix} = \begin{bmatrix} Y_1 & 0 & -Y_1 \\ 0 & -Y_6 & -Y_3 \\ -Y_1 & -Y_2 & Y_1 + Y_2 + Y_3 + Y_5 \end{bmatrix} \begin{bmatrix} V_1 \\ V_2 \\ V_4 \end{bmatrix} \tag{6}$$

In this expression the node currents $I_3 = 0$ and $I_4 = 0$ since no external currents flow into these nodes. The corresponding equations may be solved for V_4 and equated to each other to obtain an expression involving only V_1 and V_2. From this we obtain the result

$$\frac{V_2(s)}{V_1(s)} = \frac{-Y_1 Y_3}{Y_6(Y_1 + Y_2 + Y_3 + Y_5) + Y_2 Y_3} \tag{7}$$

This result will be of considerable use in the developments that follow. By comparing (7) with the various general second-order filter functions defined in the previous sections of this chapter, we can use the configuration of Fig. 5.4-2 to develop various types of filter realizations.

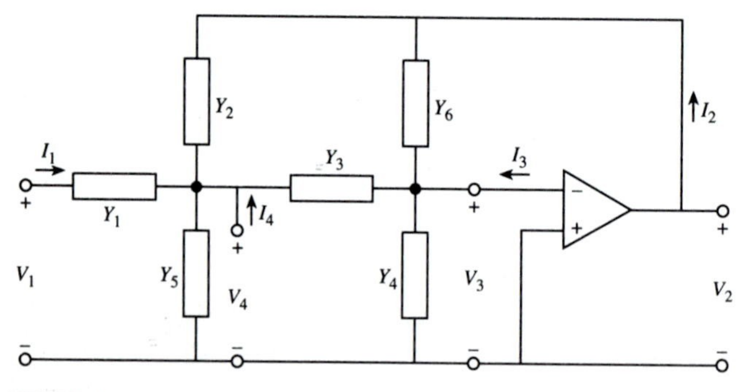

FIGURE 5.4-2
Infinite-gain single-amplifier filter with a specific form of passive network.

Low-Pass Infinite-Gain Single-Amplifier Filter

As a first infinite-gain single-amplifier filter realization, consider the low-pass function described by the general form

$$\frac{V_2(s)}{V_1(s)} = \frac{H_0\omega_n^2}{s^2 + (\omega_n/Q)s + \omega_n^2} \tag{8}$$

If we compare this with the voltage transfer function of the filter shown in Fig. 5.4-2 as given in (7), we see that it is necessary that $Y_1 = G_1$ and $Y_3 = G_3$. The denominator of (7) can thus be written as

$$D(s) = Y_6(G_1 + Y_2 + G_3 + Y_5) + Y_2G_3 \tag{9}$$

For this polynomial to be of second order, it is necessary that $Y_2 = G_2$, $Y_5 = sC_5$, and $Y_6 = sC_6$. The resulting realization for a low-pass filter is shown in Fig. 5.4-3. Its transfer function is

$$\frac{V_2(s)}{V_1(s)} = \frac{-G_1G_3}{s^2C_5C_6 + sC_6(G_1 + G_2 + G_3) + G_2G_3} \tag{10}$$

This may also be put in the form

$$\frac{V_2(s)}{V_1(s)} = \frac{-1/R_1R_3C_5C_6}{s^2 + s\dfrac{1}{C_5}\left(\dfrac{1}{R_1} + \dfrac{1}{R_2} + \dfrac{1}{R_3}\right) + \dfrac{1}{R_2R_3C_5C_6}} \tag{11}$$

Comparing (11) to (8) gives the equations

$$\omega_n = \frac{1}{\sqrt{R_2R_3C_5C_6}} \tag{12a}$$

$$\frac{1}{Q} = \sqrt{\frac{C_6}{C_5}}\left(\frac{\sqrt{R_2R_3}}{R_1} + \sqrt{\frac{R_3}{R_2}} + \sqrt{\frac{R_2}{R_3}}\right) \tag{12b}$$

$$|H_0| = \frac{R_2}{R_1} \tag{12c}$$

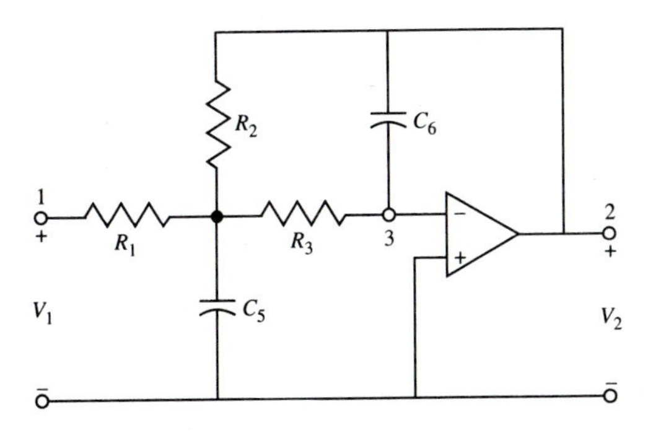

FIGURE 5.4-3
A low-pass infinite-gain single-amplifier filter.

A design procedure that allows the selection of standard values of capacitors is given as follows:

Given. H_0, Q, and ω_n.

Choose. $C_5 = C$ (a convenient value).

Calculate

$$C_6 = mC \qquad m \le \frac{1}{4Q^2(1+|H_0|)} \tag{13a}$$

$$R_2 = \frac{1}{2\omega_n CmQ}\left[1 \pm \sqrt{1 - 4mQ^2(1+|H_0|)}\right] \tag{13b}$$

$$R_1 = \frac{R_2}{|H_0|} \tag{13c}$$

$$R_3 = \frac{1}{\omega_n^2 C^2 R_2 m} \tag{13d}$$

Another approach that may be used for the design of the low-pass infinite-gain filter is given in the Problems.

Example 5.4-3 Low-pass infinite-gain single-amplifier filter. It is desired to realize a second-order low-pass Butterworth ($Q = 0.7071$) filter function in which $f_n = \omega_n/2\pi = 100$ Hz and $|H_0| = 1$. Using the design procedure of (13), we choose $C_5 = C = 0.1$ μF. Since from (13*a*), m must be less than $\frac{1}{4}$, we select $m = 0.1$ so that $C_6 = C/10 = 0.01$ μF. We then find $R_1 = R_2 = 199.7$ kΩ and $R_3 = 12.68$ kΩ.

Exercise 5.4-3. For each of the following sets of values of $|H_0|$, Q, and ω_n, find a design for a low-pass infinite-gain single-amplifier filter. For the values $\omega_n = 1$, choose $C = 1$ F. For other values of ω_n, choose $C = 0.1$ μF. Use the largest value of m permitted by the inequality of (13*a*).

Case	$\lvert H_0 \rvert$	Q	ω_n
1	1	1	1
2	1	5	1
3	10	0.7071	628.32
4	1	10	628.32
5	2	5	6283.2

Answers

Case	R_1	R_2	R_3	C_6
1	4 Ω	4 Ω	2 Ω	0.125 F
2	20 Ω	20 Ω	10 Ω	0.005 F
3	24.759 kΩ	247.59 kΩ	22.508 kΩ	4.545 nF
4	636.6 kΩ	636.6 kΩ	318.3 kΩ	0.1250 nF
5	23.87 kΩ	47.75 kΩ	15.92 kΩ	0.3333 nF

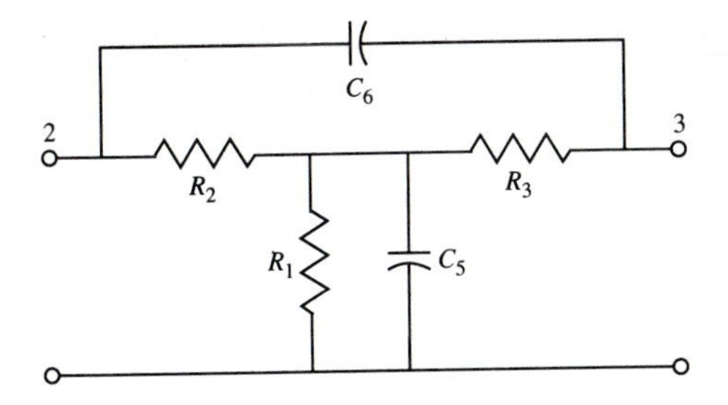

FIGURE 5.4-4
Configuration for determining $y_{32}(s)$ for the passive network in Fig. 5.4-3.

It is of interest to examine the circuit of Fig. 5.4-3 in more detail. From (4) we know that the zeros of the transfer admittance $y_{32}(s)$ create the complex pole pair of the overall voltage transfer function. To determine $y_{32}(s)$, it is necessary that terminal 1 be shorted to ground. The resulting passive *RC* network is shown in Fig. 5.4-4. Note that it is *not* a ladder network. As such, the zeros of the transfer admittance $y_{32}(s)$ can be located anywhere in the left half of the complex-frequency plane. Actually the network is simply the well-known bridged-T circuit, with an added shunt resistor. This same configuration will be the one that produces the complex natural frequencies of all the infinite-gain filter types discussed in this section.

Sensitivity

One of the major advantages of the infinite-gain single-amplifier filter is the low sensitivities that this circuit possesses. Using the definitions of Sec. 3.5, from (12) we obtain

$$S_{R_1}^{Q} = Q\left(\frac{1}{R_1}\sqrt{\frac{R_2R_3C_6}{C_5}}\right) \tag{14a}$$

$$S_{R_2}^{Q} = -\frac{Q}{2}\left(\frac{1}{R_1}\sqrt{\frac{R_2R_3C_6}{C_5}} - \sqrt{\frac{R_3C_6}{R_2C_5}} + \sqrt{\frac{R_2C_6}{R_3C_5}}\right) \tag{14b}$$

$$S_{R_3}^{Q} = -\frac{Q}{2}\left(\frac{1}{R_1}\sqrt{\frac{R_2R_3C_6}{C_5}} + \sqrt{\frac{R_3C_6}{R_2C_5}} - \sqrt{\frac{R_2C_6}{R_3C_5}}\right) \tag{14c}$$

$$S_{C_5}^{Q} = -S_{C_6}^{Q} = \tfrac{1}{2} \tag{14d}$$

$$S_{R_1}^{\omega_n} = 0 \tag{14e}$$

$$S_{R_2,R_3,C_5,C_6}^{\omega_n} = -\tfrac{1}{2} \tag{14f}$$

The various terms enclosed in parentheses for the Q sensitivities of R_1, R_2, and R_3, however, are all less in magnitude than the expression for $1/Q$ given in (12*b*). Thus the product of Q and any of these bracketed terms must be less than unity in magnitude. We conclude that

$$|S_{R_1}^{Q}| < 1 \qquad |S_{R_2}^{Q}| < \tfrac{1}{2} \qquad |S_{R_3}^{Q}| < \tfrac{1}{2} \tag{15}$$

The sensitivity of this circuit to the operational amplifier gain is also readily shown to be even lower than the values given above. Thus this circuit provides extremely low sensitivities with bounds on the sensitivities that are independent of Q. Similar observations hold for the band-pass and high-pass realizations that use this circuit. These will be discussed later in this section.

As predicted by the low Q sensitivities given in (15), the function sensitivities for the low-pass infinite-gain single-amplifier filter shown in Fig. 5.4-3 are also relatively low. An example follows.

Example 5.4-4 Function sensitivities for an infinite-gain single-amplifier filter. For the infinite-gain single-amplifer filter, the voltage transfer function given in (11) for the low-pass case can be evaluated to obtain the function sensitivities for a specific design of the filter. As an example, consider the design equations specified by (13) for the realization of a normalized function in which $|H_0| = 1$, $Q = 3$, and $\omega_n = 1$ rad/s. The element values are $R_1 = R_2 = 7.8475\ \Omega$, $R_3 = 12.7429\ \Omega$, $C_5 = 1$ F, and (for $m = 0.01$) $C_6 = 0.01$ F. The voltage transfer function is

$$\frac{V_2(s)}{V_1(s)} = \frac{-1}{s^2 + s/3 + 1}$$

The Q sensitivities for the resistors have the values

$$S_{R_1}^{Q} = 0.3823 \qquad S_{R_2}^{Q} = -0.1177 \qquad S_{R_3}^{Q} = -0.2646$$

The Q sensitivities for the capacitors are

$$S_{C_5}^{Q} = 0.5 \qquad S_{C_6}^{Q} = -0.5$$

The results of determining the function sensitivities and evaluating them over a range $0 \le \omega \le 2$ rad/s are shown in Fig. 5.4-5. A comparison of these curves with the ones determined for a similar function realized by a Sallen and Key filter in Example 5.2-5 readily demonstrate the superior sensitivity properties of the infinite-gain realization.

Band-Pass Infinite-Gain Single-Amplifier Filter

The second infinite-gain multiple-feedback filter we shall consider here is a band-pass one having a network function of the form

$$\frac{V_2(s)}{V_1(s)} = \frac{H_0(\omega_n/Q)s}{s^2 + (\omega_n/Q)s + \omega_n^2} \tag{16}$$

If we compare this with (7), we note that there are two possible choices for the realization of this function. Let us first consider the one where $Y_1 = G_1$ and $Y_3 = sC_3$. Thus the denominator of (7) becomes

$$D(s) = Y_6(G_1 + Y_2 + sC_3 + Y_5) + Y_2 sC_3 \tag{17}$$

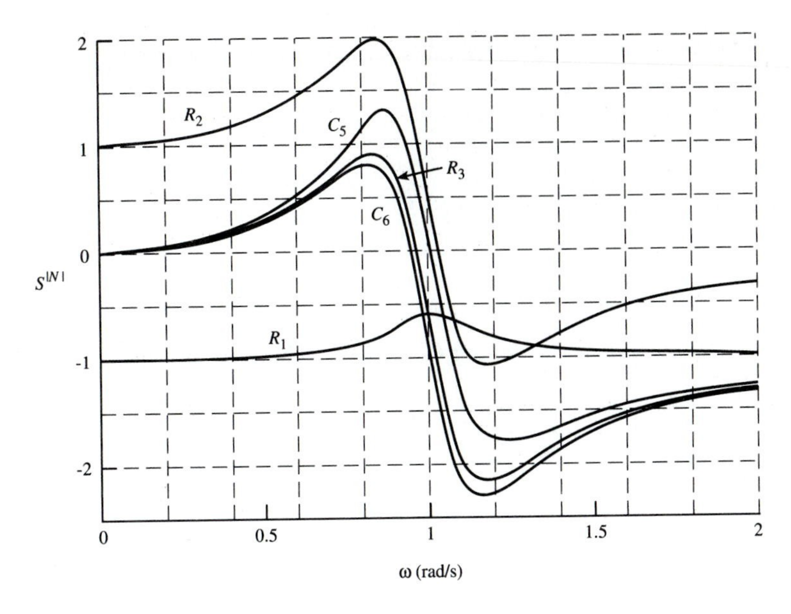

FIGURE 5.4-5
Sensitivity of an infinite-gain single-amplifier filter.

In order for $D(s)$ to be second order, $Y_2 = sC_2$, $Y_5 = G_5$, and $Y_6 = G_6$. The resulting realization is shown in Fig. 5.4-6. The transfer function is

$$\frac{V_2(s)}{V_1(s)} = \frac{-sG_1C_3}{s^2C_2C_3 + s(G_6C_2 + G_6C_3) + G_6(G_1 + G_5)} \tag{18}$$

This can also be written as

$$\frac{V_2(s)}{V_1(s)} = \frac{-\dfrac{s}{R_1C_2}}{s^2 + s\left(\dfrac{1}{R_6C_3} + \dfrac{1}{R_6C_2}\right) + \dfrac{1}{R_6C_2C_3}\left(\dfrac{1}{R_1} + \dfrac{1}{R_5}\right)} \tag{19}$$

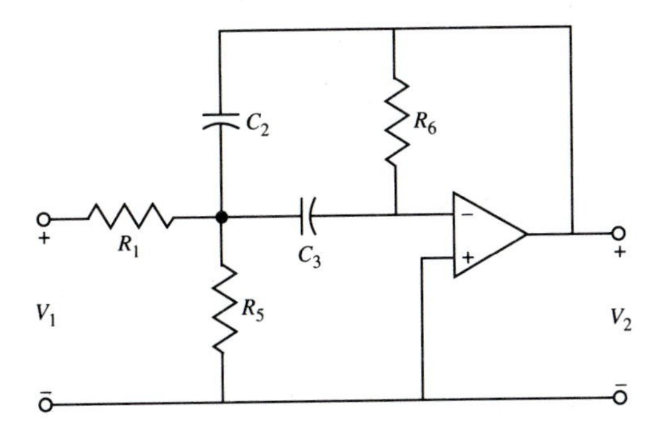

FIGURE 5.4-6
A band-pass infinite-gain single-amplifier filter.

Comparing this with (16) results in the equations

$$\omega_n = \frac{\sqrt{1 + R_5/R_1}}{\sqrt{R_5 R_6 C_2 C_3}} \tag{20a}$$

$$\frac{1}{Q} = \frac{\sqrt{R_5 C_2/R_6 C_3} + \sqrt{R_5 C_3/R_6 C_2}}{\sqrt{1 + R_5/R_1}} \tag{20b}$$

$$|H_0| = \frac{R_6/R_1}{1 + C_2/C_3} \tag{20c}$$

A design procedure can be obtained by letting $C_2 = C_3 = C$. Thus we can solve for R_1, R_5, and R_6 as follows:

$$R_1 = \frac{Q}{\omega_n C|H_0|} \tag{21a}$$

$$R_5 = \frac{Q}{(2Q^2 - |H_0|)\omega_n C} \tag{21b}$$

$$R_6 = \frac{2Q}{\omega_n C} \tag{21c}$$

Note that, for a positive value of R_5, $|H_0|$ must be less than $2Q^2$.

Example 5.4-5 Band-pass infinite-gain single-amplifier filter. It is desired to realize a second-order band-pass filter in which $|H_0| = 2$, $Q = 2$, and $\omega_n = 10$ krad/s. If we choose $C_2 = C_3 = C = 0.01$ μF, then from (21), $R_1 = 10$ kΩ, $R_5 = 3.33$ kΩ, and $R_6 = 40$ kΩ.

Exercise 5.4-5. For each of the following sets of values of $|H_0|$, Q, and ω_n, find a design for a band-pass infinite-gain single-amplifier filter. For the values $\omega_n = 1$, choose $C = 1$ F. For other values of ω_n, choose $C = 0.01$ μF.

			Answers, Ω		
$\|H_0\|$	Q	ω_n, (rad/s)	R_1	R_5	R_6
1	1	1	1	1	2
1	5	1	5	0.1020	10
5	2	1	0.4	0.6667	4
1	10	10 krad/s	100 kΩ	502.5	200 kΩ
2	5	628.38	397.8 kΩ	16.58 kΩ	1591 kΩ

It is possible to eliminate R_5 from the circuit shown in Fig. 5.4-6 and still achieve a band-pass realization. However, if this is done, $|H_0|$ is no longer a free parameter. For example, if $C_2 = C_3$ and $R_5 = \infty$, then (20*b*) and (20*c*) become

$$\frac{1}{Q} = 2\sqrt{\frac{R_1}{R_6}} \tag{22a}$$

$$|H_0| = \frac{R_6}{2R_1} \tag{22b}$$

Thus $|H_0|$ is equal to $2Q^2$. For large values Q, $|H_0|$ will be very large. This is undesirable since if similar stages were cascaded, the overall gain would become too large, and the allowable signal level of the resulting filter would be seriously degraded. Another band-pass realization is possible if we choose $Y_1 = sC_1$ and $Y_3 = G_3$. The resulting circuit requires three capacitors rather than the two needed in the one described above. A treatment of it is left to the reader as an exercise.

High-Pass Infinite-Gain Single-Amplifier Filter

The filter configuration shown in Fig. 5.4-2 can also be used to realize a high-pass function having the form

$$\frac{V_2(s)}{V_1(s)} = \frac{H_0 s^2}{s^2 + (\omega_n/Q)s + \omega_n^2} \tag{23}$$

To do this, we may use Table 5.3-1 to transform the low-pass circuit shown in Fig. 5.4-3 into a high-pass one. The result is shown in Fig. 5.4-7 in which $Y_1 = sC_1$, $Y_2 = sC_2$, $Y_3 = sC_3$, $Y_5 = G_5$, and $Y_6 = G_6$. The transfer function is

$$\frac{V_2(s)}{V_1(s)} = \frac{-s^2 C_1 C_3}{s^2 C_2 C_3 + s(G_6 C_1 + G_6 C_2 + G_6 C_3) + G_5 G_6} \tag{24}$$

This may be put in the form

$$\frac{V_2(s)}{V_1(s)} = \frac{-s^2 \dfrac{C_1}{C_2}}{s_2 + s\dfrac{1}{R_6}\left(\dfrac{C_1}{C_2 C_3} + \dfrac{1}{C_2} + \dfrac{1}{C_3}\right) + \dfrac{1}{R_5 R_6 C_2 C_3}} \tag{25}$$

Note that three capacitors are required; i.e., the realization is not canonic. By equating (25) to (23), we obtain

$$\omega_n = \frac{1}{\sqrt{R_5 R_6 C_2 C_3}} \tag{26a}$$

$$\frac{1}{Q} = \sqrt{\frac{R_5}{R_6}}\left(\frac{C_1}{\sqrt{C_2 C_3}} + \sqrt{\frac{C_3}{C_2}} + \sqrt{\frac{C_2}{C_3}}\right) \tag{26b}$$

$$|H_0| = \frac{C_1}{C_2} \tag{26c}$$

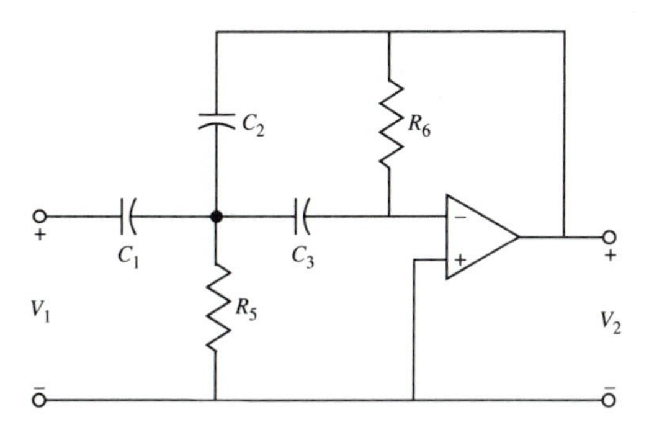

FIGURE 5.4-7
A high-pass infinite-gain single-amplifier filter.

A set of design equations can be developed if $C_1 = C_3 = C$, where C is selected as a convenient value. The equations are

$$R_5 = \frac{|H_0|}{\omega_n QC(2|H_0| + 1)} \tag{27a}$$

$$R_6 = \frac{(2|H_0| + 1)Q}{\omega_n C} \tag{27b}$$

$$C_2 = \frac{C}{|H_0|} \tag{27c}$$

Example 5.4-6 High-pass infinite-gain single-amplifier filter. It is desired to realize a second-order high-pass Butterworth ($Q = 1/\sqrt{2}$) transfer function where $f_n = \omega_n/2\pi =$ 100 Hz and $|H_0| = 1$. Choosing $C = C_1 = C_3 = 0.1$ μF, we find $C_2 = 0.1$ μF, $R_5 = 7.503$ kΩ, and $R_6 = 33.762$ kΩ.

5.5 POSITIVE- AND NEGATIVE-FEEDBACK INFINITE-GAIN SINGLE-AMPLIFIER FILTERS

The infinite-gain single-amplifier circuit introduced in Sec. 5.4 has the disadvantage that undesirably large spreads of resistance values are required in high-Q realizations. Such spreads may compromise basic assumptions of (relatively) low output impedance and high input impedance for the active elements of the circuit. As a result, they may cause serious errors in the predicted circuit performance. A modification of the basic design permits a reduction in these spreads. The method is to provide positive feedback to the positive terminal of the operational amplifier, in addition to the feedback already provided to the negative terminal by the passive network. Although this leads to an increase in sensitivity, the overall result is usually worthwhile.

Band-Pass Positive- and Negative-Feedback Filter

A circuit realization for a band-pass infinite-gain single-amplifier filter with positive and negative feedback is shown in Fig. 5.5-1. The positive feedback is provided by the voltage divider consisting of the two resistors R_A and R_B. To simplify the equations describing the circuit, we may represent the effect of the two resistors by defining a feedback constant K, where

$$K = \frac{R_A}{R_B} \tag{1}$$

The voltage transfer function for the circuit is

$$\frac{V_2(s)}{V_1(s)} = \frac{-s(K+1)/R_1C_2}{s^2 + s(1/R_2C_2 + 1/R_2C_1 - K/R_1C_2) + 1/R_1R_2C_1C_2} \tag{2}$$

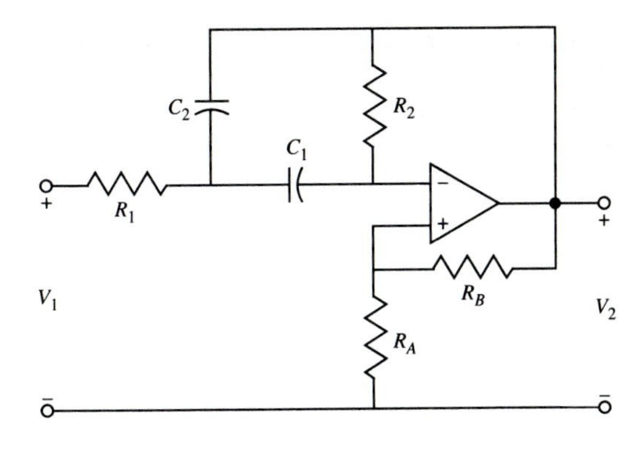

FIGURE 5.5-1
Band-pass positive- and negative-feedback infinite-gain single-amplifier filter.

For the standard second-order band-pass function

$$\frac{V_2(s)}{V_1(s)} = \frac{H_0(\omega_n/Q)s}{s^2 + (\omega_n/Q)s + \omega_n^2} \tag{3}$$

we obtain the relations

$$\omega_n = \frac{1}{\sqrt{R_1 R_2 C_1 C_2}} \tag{4a}$$

$$\frac{1}{Q} = \sqrt{\frac{R_1}{R_2}}\left(\sqrt{\frac{C_1}{C_2}} + \sqrt{\frac{C_2}{C_1}} - K\frac{R_2}{R_1}\sqrt{\frac{C_1}{C_2}}\right) \tag{4b}$$

$$|H_0| = \frac{(K+1)/R_1C_1}{1/R_2C_2 + 1/R_2C_1 - K/R_1C_1} \tag{4c}$$

A design procedure may be specified for an equal-valued capacitor realization. The steps are given in the following summary.

Summary 5.5-1 Band-pass positive- and negative-feedback infinite-gain single-amplifier filter realization. A filter with the form shown in Fig. 5.5-1 and the voltage transfer function given in (3) may be found by the following steps:

1. Choose a suitable value C for the capacitors and let $C_1 = C_2 = C$.
2. Determine the resistance ratio parameter m_0 that would be required if there were no positive feedback. This is given by the relation

$$m_0 = \frac{1}{4Q^2}$$

3. Select the desired resistor ratio m where

$$m_0 \le m \le 1$$

4. Use m to determine the desired value of the feedback constant K that determines the ratio of feedback resistors

$$K = \frac{R_A}{R_B} = 2m - \frac{\sqrt{m}}{Q}$$

Choose a convenient value of R_B and calculate $R_A = KR_B$.

5. Determine the value of R_2 and R_1 by the relations

$$R_2 = \frac{1}{\omega_n C\sqrt{m}} \qquad R_1 = mR_2$$

6. Note that the value of $|H_0|$, the gain at resonance, is determined by

$$|H_0| = \frac{K+1}{2m - K}$$

If this gain is excessively large, it may be reduced by adding a voltage divider at the input of the network to reduce the gain. This is accomplished by adding a resistor R_3 and changing the designation of R_1 to R_1' as shown in Fig. 5.5-2. If $|H_0'|$ is the desired gain (where $|H_0'| < |H_0|$), the values of the voltage-divider resistors are found as

$$R_1' = R_1\frac{|H_0|}{|H_0'|} \qquad R_3 = \frac{R_1}{1 - |H_0'|/|H_0|}$$

An example of the use of these relations follows.

Example 5.5-1 Band-pass filter with positive and negative feedback. As an example of the procedure given in Summary 5.5-1, consider the realization of a band-pass filter with a resonant frequency $\omega_n = 10^4$ rad/s and a Q of 10. The maximum resistance spread is to be 100. Following the steps in the summary, we obtain:

Step 1. $C_1 = C_2 = C = 0.01\ \mu\text{F}$

Step 2. $m_0 = \dfrac{1}{4Q^2} = \dfrac{1}{4 \times 10^2} = 0.0025$

Step 3. $0.0025 \le m \le 1, \qquad m = 0.01$

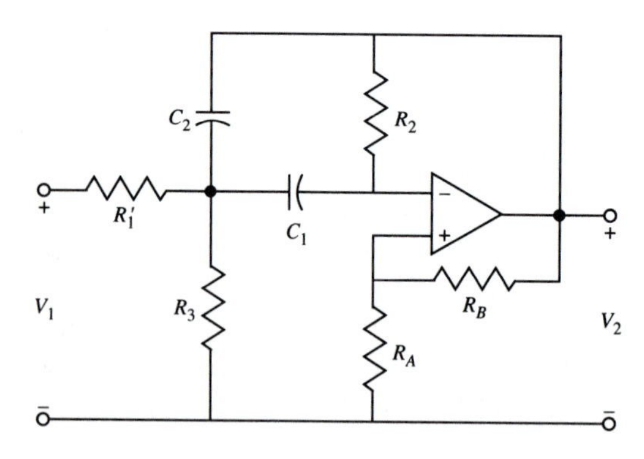

FIGURE 5.5-2
Band-pass filter with reduced gain.

Step 4. $K = 2m - \dfrac{\sqrt{m}}{Q} = 2 \times 0.01 - \dfrac{0.1}{10} = 0.01$

$$R_B = 100\ \text{k}\Omega \qquad R_A = KR_B = 0.01 \times 10^5 = 1\ \text{k}\Omega$$

Step 5. $R_2 = \dfrac{1}{\omega_n C \sqrt{m}} = \dfrac{1}{10^4 \times 0.01 \times 10^{-6} \times 0.1} = 100\ \text{k}\Omega$

$$R_1 = mR_2 = 0.01 \times 10^5 = 1\ \text{k}\Omega$$

Step 6. $|H_0| = \dfrac{K+1}{2m - K} = \dfrac{0.01 + 1}{2 \times 0.01 - 0.01} = 101$

Assuming a desired value of $|H_0|$ of 50, we find

$$R_1' = R_1 \frac{|H_0|}{|H_0'|} = 10^3 \frac{101}{50} = 2.02\ \text{k}\Omega$$

$$R_3 = \frac{R_1}{1 - |H_0'|/|H_0|} = \frac{10^3}{1 - 50/101} = 1.98\ \text{k}\Omega$$

Note that the adjustment to the gain provided by R_1' and R_3 is only approximate since the calculation assumes that the current flowing through the resistors to ground is much greater than the current flowing out of the junction of the resistors. For small values of m the error produced is usually negligible.

Exercise 5.5-1. For each of the following sets of values of Q and m, find a normalized design ($\omega_n = 1$ rad/s, $C = 1$ F, and $R_B = 1\ \Omega$) for the band-pass filter shown in Fig. 5.5-1. Also find the value of $|H_0|$.

		Answers			
Q	m	R_A	R_1	R_2	$\|H_0\|$
1	0.5	0.2929	0.7071	1.4142	1.8284
2	0.1	0.0419	0.3162	3.1623	6.5895
5	0.1	0.1368	0.3162	3.1623	17.974
10	0.1	0.1684	0.3162	3.1623	36.947
20	0.1	0.1842	0.3162	3.1623	74.895

Biquadratic Positive- and Negative-Feedback Filter

In the preceding discussion we used positive feedback in the band-pass infinite-gain single-amplifier filter to reduce the spread of element values. Positive feedback can also be used in connection with a more complex passive network to generalize the type of network function that can be realized by the infinite-gain single-amplifier filter configuration. Specifically, we extend the concept to the realization of a *biquadratic function*, that is, one in which the numerator as well as the denominator may be a second-order polynomial. Such a function includes the low-pass, band-pass, high-pass, and

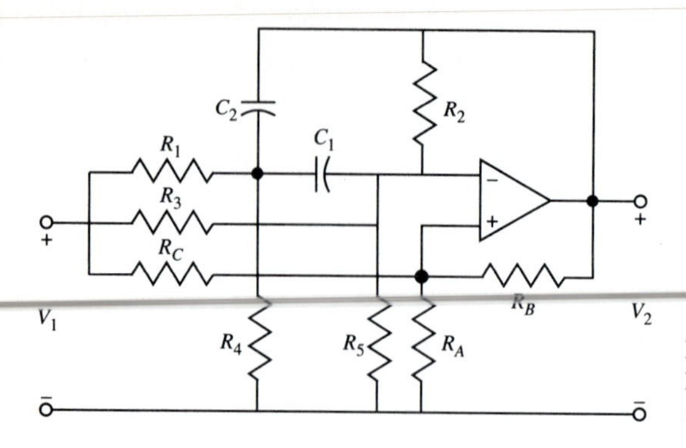

FIGURE 5.5-3
Biquadratic positive- and negative-feedback infinite-gain single-amplifier filter.

band-elimination ones as special cases. The circuit is shown in Fig. 5.5-3. It realizes a voltage transfer function having the form

$$\frac{V_2(s)}{V_1(s)} = \frac{cs^2 + ds + e}{s^2 + as + b} \tag{5}$$

The circuit was originally proposed by Friend, Harris, and Hilberman[1] for thin-film hybrid integrated circuit fabrication. It is also referred to as the SAB (Single-Amplifier Biquad) or the STAR (Standard Tantalum Active Resonator) circuit. It has been used extensively for voice-frequency signal processing applications. Due to the large number of circuit elements, the network function for this circuit is quite complex. A computer-generated version of the numerator and denominator polynomials for the voltage transfer function is given in Fig. 5.5-4. To use the circuit, the coefficients a, b, c, d, and e of the network function given in (5) are used to determine the values of the elements of the filter by the procedure given in the following summary.

Summary 5.5-2 Biquadratic positive- and negative-feedback infinite-gain single-amplifier filter. A filter with the form shown in Fig. 5.5-3 and the voltage transfer function given in (5) may be found by the following steps. In these steps we assume that a suitable frequency normalization of the data has been made so that the critical frequencies (such as bandwidth or center frequency) are approximately unity. In addition, the overall gain of the function must be such that $c \leq 1$. If this is not the case, the numerator coefficients c, d, and e must be multiplied by a constant of value less than unity to satisfy this relation. The procedure given below produces an impedance-normalized realization.

1. For convenience we choose an equal-valued capacitor realization. Thus we let $C_1 = C_2 = 1$ F. All the following steps assume that the value of the capacitors are unity.
2. Choose a value for the positive-feedback parameter K. The parameter is similar to the one defined in step 4 of Summary 5.5-1. As a first try, we choose $K = 0.1$.
3. Choose an impedance-normalized value for the resistor R_B. Let $R_B = 1\ \Omega$.

---

[1] J. J. Friend, C. A. Harris, and D. Hilberman, "STAR: An Active Biquadratic Section," *IEEE Trans. Circuits and Systems*, vol. CAS-22, 1975, pp. 115–121.

Numerator:

+ s2 C1 C2 R1 R2 R3 R4 R5 RA RB

+ s1 C2 R1 R2 R3 R4 RA RB
− s1 C2 R1 R2 R4 R5 RA RC
− s1 C2 R1 R2 R4 R5 RB RC
+ s1 C2 R1 R3 R4 R5 RA RB
+ s1 C1 R1 R2 R3 R4 RA RB
+ s1 C1 R1 R2 R3 R5 RA RB
− s1 C1 R1 R2 R4 R5 RA RC
− s1 C1 R1 R2 R4 R5 RB RC
+ s1 C1 R1 R3 R4 R5 RA RB
− s1 C1 R2 R3 R4 R5 RA RC
− s1 C1 R2 R3 R4 R5 RB RC

+ s0 R1 R2 R3 RA RB
− s0 R1 R2 R5 RA RC
− s0 R1 R2 R5 RB RC
+ s0 R1 R3 R5 RA RB
+ s0 R2 R3 R4 RA RB
− s0 R2 R4 R5 RA RC
− s0 R2 R4 R5 RB RC
+ s0 R3 R4 R5 RA RB

Denominator:

+ s2 C1 C2 R1 R2 R3 R4 R5 RA RB
+ s2 C1 C2 R1 R2 R3 R4 R5 RB RC

− s1 C2 R1 R2 R3 R4 RA RC
− s1 C2 R1 R2 R4 R5 RA RC
+ s1 C2 R1 R3 R4 R5 RA RB
+ s1 C2 R1 R3 R4 R5 RB RC
− s1 C1 R1 R2 R3 R4 RA RC
− s1 C1 R1 R2 R3 R5 RA RC
− s1 C1 R1 R2 R4 R5 RA RC
+ s1 C1 R1 R3 R4 R5 RA RB
+ s1 C1 R1 R3 R4 R5 RB RC
− s1 C1 R2 R3 R4 R5 RA RC

− s0 R1 R2 R3 RA RC
− s0 R1 R2 R5 RA RC
+ s0 R1 R3 R5 RA RB
+ s0 R1 R3 R5 RB RC
− s0 R2 R3 R4 RA RC
− s0 R2 R4 R5 RA RC
+ s0 R3 R4 R5 RA RB
+ s0 R3 R4 R5 RB RC

FIGURE 5.5-4
Network function for the filter of Fig. 5.5-3.

4. Determine the values of R_C and R_A:

$$R_C = \frac{K}{c} \qquad R_A = \frac{K}{1 - c}$$

Note that in the relation for R_A we require that $c \leq 1$.

5. At this point, to simplify future computational steps, we define two constants x_1 and x_2:

$$x_1 = \frac{2K}{-a + \sqrt{a^2 + 8Kb}} \qquad x_2 = \frac{c + 2ex_1^2 - dx_1}{1 + K}$$

6. Use the constants x_1 and x_2 to find the values of R_1 and R_4:

$$R_1 = \frac{x_1}{x_2} \qquad R_4 = \frac{R_1 x_1}{R_1 - x_1}$$

Note that to avoid having a negative value for R_4, if the value calculated for x_2 is greater than 1, the coefficients c, d, and e of (5) must be further reduced. Also note that if these coefficients are changed so that $x_2 = 1$, then $R_1 = x_1$ and $R_4 = \infty$, an open circuit. This makes it possible to reduce the number of elements in the circuit by eliminating one resistor.

7. Again, to simplify our computations, we define two additional constants y_1 and y_2. The constant y_1 must be selected so that it satisfies $0 \leq y_1 \leq 1$ and also so that the constant y_2 (defined below) is nonnegative:

$$y_2 = \frac{(1+K)(c-y_1)}{x_1 b(e/b - c)}$$

8. Use the constants y_1 and y_2 and other terms to find the values of the resistors R_2, R_3, and R_5:

$$R_2 = \frac{y_2}{x_1 y_2 b + K} \qquad R_3 = \frac{y_2}{y_1} \qquad R_5 = \frac{R_3 y_2}{R_3 - y_2}$$

These steps are illustrated in the example that follows.

Example 5.5-2 Elliptic filter with positive and negative feedback. As an example of the procedure given in Summary 5.5-2, consider the realization of a second-order low-pass filter with the following specifications:

a. A normalized passband from 0 to 1 rad/s in which the maximum deviation of the magnitude is 1 dB.

b. A minimum attenuation of 17 dB (from the maximum passband magnitude) at all frequencies greater than 2 rad/s.

c. A maximum gain in the passband of 1.0.

The specifications require the use of an elliptic approximation. From Table 2.4-1, after determining a multiplicative constant to satisfy requirement c given above, we find

$$\frac{V_2(s)}{V_1(s)} = \frac{0.1397s^2 + 1.0428}{s^2 + 0.9989s + 1.1701}$$

Comparing this with (5), we find

$$a = 0.9989 \qquad b = 1.1701 \qquad c = 0.1397 \qquad d = 0 \qquad e = 1.0428$$

Following the steps in the summary, we obtain:

Step 1. $C_1 = C_2 = 1$ F
Step 2. Try $K = 0.1$
Step 3. $R_B = 1\ \Omega$

Step 4. $R_C = \dfrac{K}{c} = \dfrac{0.1}{0.1397} = 0.7158\ \Omega$

$$R_A = \frac{K}{1-c} = \frac{0.1}{1 - 0.1397} = 0.1162\ \Omega$$

Step 5. $$x_1 = \frac{2K}{-a + \sqrt{a^2 + 8Kb}}$$

$$= \frac{2 \times 0.1}{-0.9989 + \sqrt{(0.9989)^2 + (8 \times 0.1 \times 1.1701)}} = 0.5105$$

$$x_2 = \frac{c + 2ex_1^2 - dx_1}{1 + K}$$

$$= \frac{0.1397 + [2 \times 1.0428 \times (0.5105)^2] - 0}{1 + 0.1} = 0.6212$$

Step 6. $$R_1 = \frac{x_1}{x_2} = \frac{0.5105}{0.6211} = 0.8219\ \Omega$$

$$R_4 = \frac{R_1 x_1}{R_1 - x_1} = \frac{0.8219 \times 0.5105}{0.8219 - 0.5105} = 1.3478\ \Omega$$

Step 7. $$y_1 = 0$$

$$y_2 = \frac{(1 + K)(c - y_1)}{x_1 b(e/b - c)} = \frac{(1 + 0.1)(0.1397 - 0)}{0.5105 \times 1.1701 \times (1.0428/1.1701 - 0.1397)}$$

$$= 0.3423$$

Step 8. $$R_2 = \frac{y_2}{x_1 y_2 b + K} = \frac{0.3423}{(0.5105 \times 0.3423 \times 1.1701) + 0.1} = 1.1242\ \Omega$$

$$R_3 = \frac{y_2}{y_1} = \frac{0.3423}{0} = \infty$$

$$R_5 = \frac{R_3 y_2}{R_3 - y_2} = \frac{y_2}{1 - y_2/R_3} = \frac{0.3423}{1 - 0} = 0.3423\ \Omega$$

Note that, in this design, choosing $y_1 = 0$ has resulted in the elimination of the resistor R_3.

Exercise 5.5-2. For each of the sets of normalized coefficient values given below, find a design for the biquadratic filter configuration shown in Fig. 5.5-3. The coefficient letters refer to the general biquadratic form given in (5). Use the design procedure given in Summary 5.5-2. Choose $R_B = 1$, $K = 0.1$, and y_1 as 0 or 1 as required by the specific case.

Case	*a*	*b*	*c*	*d*	*e*
1	0.1	1	0.9	0	1.089
2	0.1	1	0.9	0	0
3	2	2	0.9	0.9	0
4	0.1	1.21	0.9	0	0.9

Answers

Case	R_A	R_C	R_1	R_2	R_3	R_4	R_5
1	1	0.1111	0.2654	3.9251	∞	4.3053	20.952
2	1	0.1111	0.3056	2.2000	0.4889	1.3750	∞
3	1	0.1111	1.4687	0.5038	0.1120	0.8686	∞
4	1	0.1111	0.2497	3.2166	2.5869	2.2729	∞

5.6 HIGHER ORDER FILTERS

Most of the synthesis techniques presented in the previous sections of this chapter were restricted to second-order filter realizations. As pointed out in Sec. 5.1, however, such realizations may be cascaded to achieve higher order ones, since the use of a VCVS or an operational amplifier as the output element provides the necessary isolation. Such a cascade will have the form shown in Fig. 5.6-1. The overall voltage transfer function $T(s)$ will be given by

$$T(s) = T_1(s)\, T_2(s) \cdots T_{n/2}(s) \tag{1}$$

where n is the order of the overall filter and has been assumed to be even and $T_i(s)$ is the voltage transfer function of an individual section that has the form

$$T_i(s) = \frac{a_{2i}s^2 + a_{1i}s + a_{0i}}{s^2 + (\omega_{ni}/Q_i)s + \omega_{ni}^2} \tag{2}$$

There are several important variations that should be considered in applying the cascade method. These are treated briefly in this section.

Use of First-Order Sections

A first consideration in the application of the cascade method of realizing higher order functions occurs if the function to be realized is of odd order. In such a case, there are two possible techniques that may be applied. The first of these is to add a passive first-order circuit as one of the elements of the cascade. This will realize the negative real pole associated with the odd-order function. Depending on whether or not a zero at the origin is also to be realized, the passive network will have the form shown in Fig. 5.6-2*a*, for

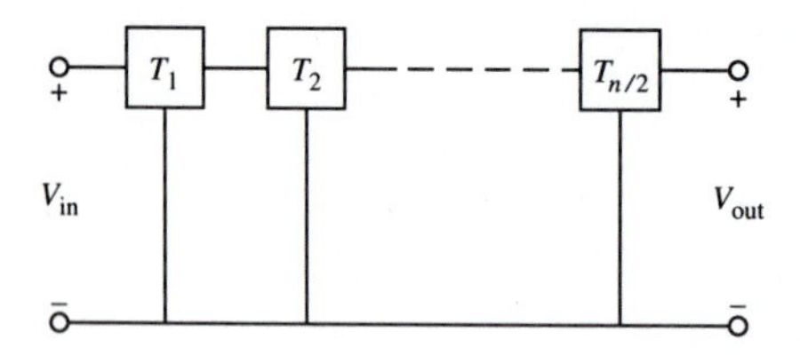

FIGURE 5.6-1
A cascade of second-order filter realizations.

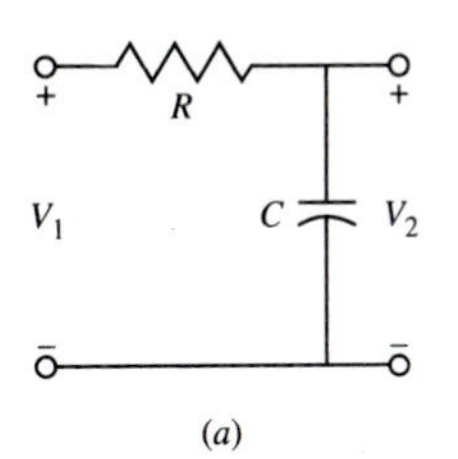

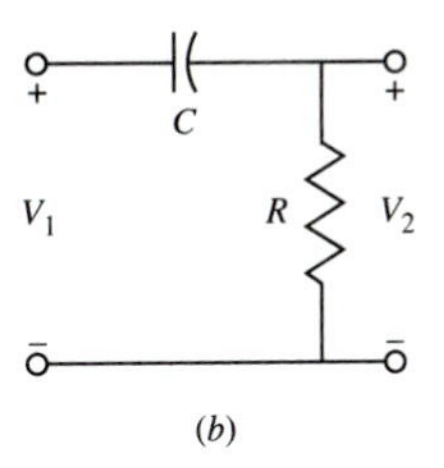

FIGURE 5.6-2
First-order filter sections.

which the voltage transfer function (with *no* zero at the origin) is

$$\frac{V_2(s)}{V_1(s)} = \frac{1/RC}{s + 1/RC} \tag{3}$$

or the form shown in Fig. 5.6-2*b*, which has the voltage transfer function (*with* a zero at the origin)

$$\frac{V_2(s)}{V_1(s)} = \frac{s}{s + 1/RC} \tag{4}$$

Example 5.6-1 Fifth-order cascade filter. As an example of the use of a first-order section in a cascade filter realization, consider the synthesis of a normalized fifth-order low-pass filter with a Butterworth magnitude characteristic. From Table 2.3-1*b* we may write

$$T(s) = T_1(s)T_2(s)T_3(s) = \frac{H_1}{s+1} \frac{H_2}{s^2 + 1.61803s + 1} \frac{H_3}{s^2 + 0.61803s + 1}$$

If we use the circuit of Fig. 5.6-2*a* to realize the first-order function and the circuit of Fig. 5.2-3 (design 1) to realize the second-order functions, we obtain the realization shown in Fig. 5.6-3. Note that a voltage follower is used to buffer the output of the first-order stage. The gain constants are $H_1 = 1$, $H_2 = 1.382$, and $H_3 = 2.382$. The dc gain of the cascade is $H_0 = 1 \times 1.382 \times 2.382 = 3.292$.

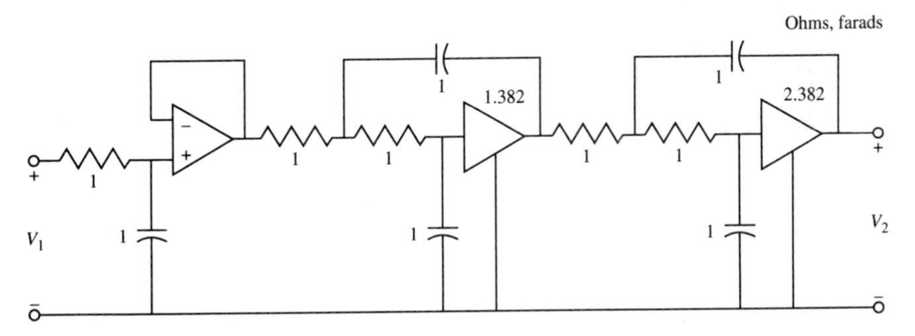

FIGURE 5.6-3
Fifth-order cascade filter.

In general, to maximize the dynamic range of signals that can be filtered using a cascade realization, the individual sections of the cascade should be ordered (from input to output) in order of increasing Q (with a first-order section considered as having a Q of zero). The preceding example illustrates this general rule.

Use of Third-Order Sections

A second technique that may be used when the function to be realized is of odd order (excluding the case where the order is 1) is the use of a single third-order active *RC* circuit of the Sallen and Key type to realize both a negative real pole and a pair of complex-conjugate ones. Such a circuit is shown in Fig. 5.6-4.[1] To make the circuit as practical as possible, all the capacitors have been chosen equal to a normalized value of unity, as shown. For this constraint, the voltage transfer function of the circuit is

$$\frac{V_2(s)}{V_1(s)} = \frac{K}{R_1R_2R_3s^3 + [2R_1R_3 + R_1R_2 + R_2R_3(2-K)]s^2 + [R_1 + R_3 + (R_2+R_3)(2-K)]s + 1} \tag{5}$$

To determine the values R_i of the resistors and the gain value K of the noninverting VCVS, we begin by noting that for a third-order low-pass voltage transfer function having the form

$$\frac{V_2(s)}{V_1(s)} = \frac{H}{a_3s^3 + a_2s^2 + a_1s + 1} \tag{6}$$

the following set of simultaneous (nonlinear) equations results:

$$a_3 = R_1R_2R_3 \tag{7a}$$
$$a_2 = 2R_1R_3 + R_1R_2 + R_2R_3(2-K) \tag{7b}$$
$$a_1 = R_1 + R_3 + (R_2+R_3)(2-K) \tag{7c}$$

[1] L. P. Huelsman, "An Equal-Valued-Capacitor Active *RC* Network Realization of a Third-Order Low-Pass Butterworth Characteristics," *Electronic Letters*, vol. 7, no. 10, May 20, 1971, pp. 271–272.

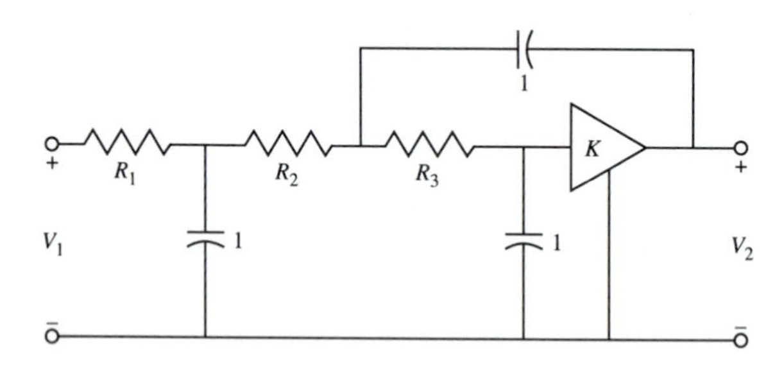

FIGURE 5.6-4
Third-order low-pass Sallen and Key filter.

In general, the sensitivities characterizing the third-order Sallen and Key filter are higher than those of the second-order one. As a result, this configuration is useful mainly for realizations in which the quadratic term has a low Q. The solution of the equations of (7) is usually accomplished by the application of numerical methods. A treatment of this is beyond the scope of this text. Values of the results obtained for realizing the negative real pole and the lowest Q quadratic factor of various approximations are given in Table 5.6-1.

Example 5.6-2 Fifth-order cascade filter. As an example of the use of a third-order section in a cascade filter realization, consider the fifth-order low-pass function with a Butterworth magnitude characteristic described in Example 5.6-1. From Table 2.1-3*b* we obtain

$$T(s) = T_1(s)T_2(s) = \frac{H_1}{(s+1)(s^2+1.61803s+1)} \frac{H_2}{s^2+0.61803s+1}$$

$$= \frac{H_1}{s^3+2.61803s^2+2.61803s+1} \frac{H_2}{s^2+0.61803s+1}$$

The circuit shown in Fig. 5.6-4 may be used to realize the third-order function. The resistor values (for $n = 5$) are taken from Table 5.6-1. The circuit of Fig. 5.2-3 (design 1) is used to realize the second-order function. The resulting realization is shown in Fig. 5.6-5. The gain constants are $H_1 = 2$ and $H_2 = 2.382$. The dc gain of the cascade is $H_0 = 2 \times 2.382 = 4.764$.

TABLE 5.6-1
Element values for Fig. 5.6-4 (with $K = 2$) that realize the negative real pole and the complex-conjugate pole pair with the lowest Q for various low-pass functions

Approximation	n	R_1	R_2	R_3
Butterworth	3	1.56520	1.46940	0.43480
	5	2.10994	0.93280	0.50809
	7	2.28449	0.84595	0.51745
	9	2.35892	0.81451	0.52046
Chebyshev 0.5-dB ripple	3	1.87657	2.77768	0.26806
	5	3.43569	3.04181	0.55393
	7	4.91761	3.88785	0.80423
	9	6.37762	4.82524	1.04759
Chebyshev 1.0-dB ripple	3	2.27516	3.64424	0.24549
	5	4.03696	3.92009	0.50845
	7	5.73554	4.98496	0.73885
	9	7.41686	6.17387	0.96277
Thomson	3	0.20824	1.66950	0.19176
	5	0.60601	0.22987	0.13792
	7	0.46861	0.16189	0.10330
	9	0.37859	0.12567	0.08223

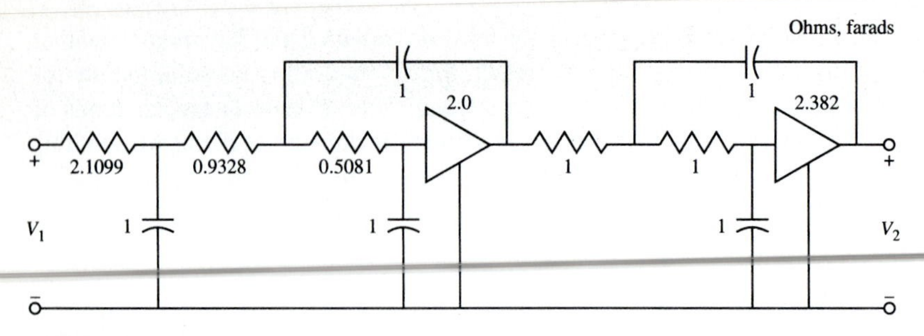

FIGURE 5.6-5
Fifth-order cascade filter.

Use of Higher Order Sections

At this point, the use of even higher order single-amplifier filter stages might be considered. In general, however, the sensitivity of such stages becomes quite large as compared with those of the cascade of lower order stages. An example follows that illustrates this.

Example 5.6-3 Fourth-order single-amplifier and cascaded filters. As an example of the large sensitivities that occur in high-order single-amplifier filters, consider the realization of a fourth-order low-pass function with a Butterworth magnitude characteristic. The voltage transfer function from Table 2.1-3*b* is

$$\frac{V_2(s)}{V_1(s)} = \frac{H}{(s^2 + 1.84776s + 1)(s^2 + 0.76537s + 1)}$$

A fourth-order single-amplifier Sallen and Key type of realization is shown in Fig. 5.6-6*a*. The element values were determined in a manner similar to that used for the third-order filter. A cascade realization of the same function is shown in Fig. 5.6-6*b*. The magnitude plots obtained from the two realizations, after adjustment to a common dc value of unity (0 dB), as shown in Fig. 5.6-7, are identical. Now consider the effect of a positive 5 percent change in the gains of the amplifiers. For the fourth-order realization, $K = 2.1$. For the cascaded realizations, $K_1 = 1.2098$ and $K_2 = 2.3463$. The resulting magnitude plots, as shown in the figure, readily illustrate the poorer sensitivity properties of the fourth-order realization.

Realization of Other Functions

Cascade sections can also be used to realize functions with complex-conjugate zeros. An example of an elliptic band-pass approximation follows.

Example 5.6-4 Elliptic band-pass filter. It is desired to realize a band-pass function with the following specifications:

Center frequency $f_0 = 1$ kHz

Passboard of 0.1 kHz with 1-dB ripple

Maximum passband magnitude 1.0

Stopbands with edges 0.2 kHz apart with a minimum of 17 dB attenuation

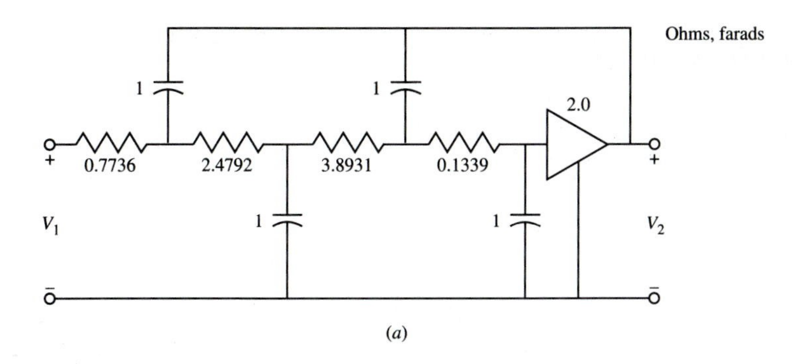

(a)

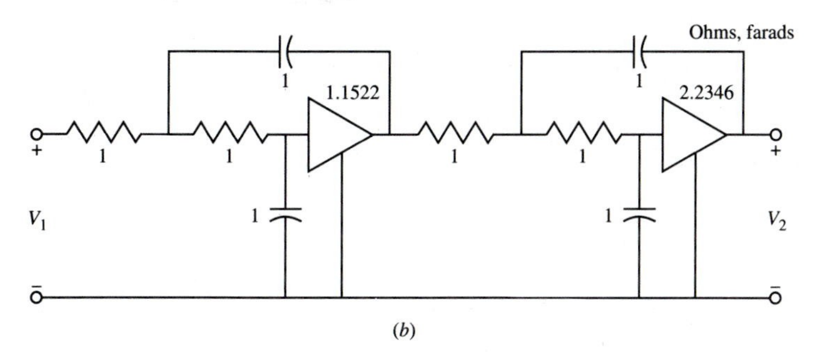

(b)

FIGURE 5.6-6
Alternate realizations of a fourth-order filter.

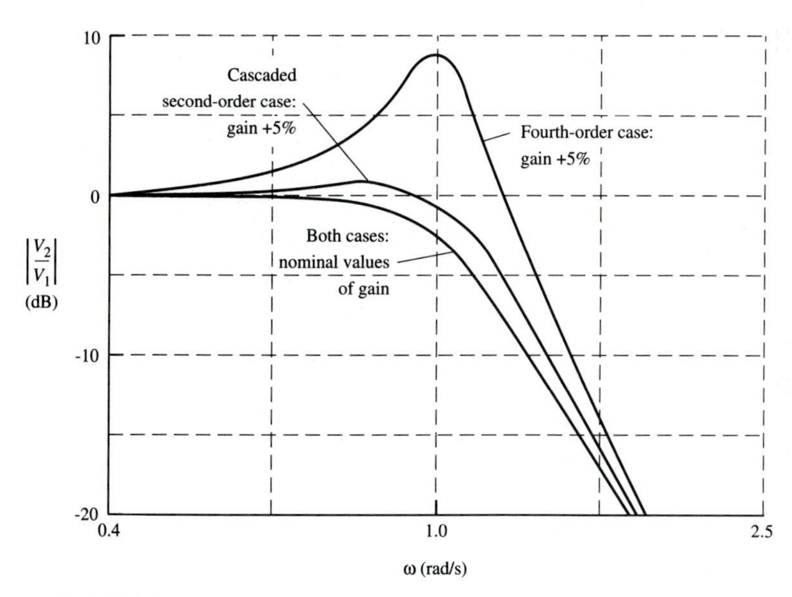

FIGURE 5.6-7
Effects of 5 percent gain changes on the filters of Fig. 5.6-6.

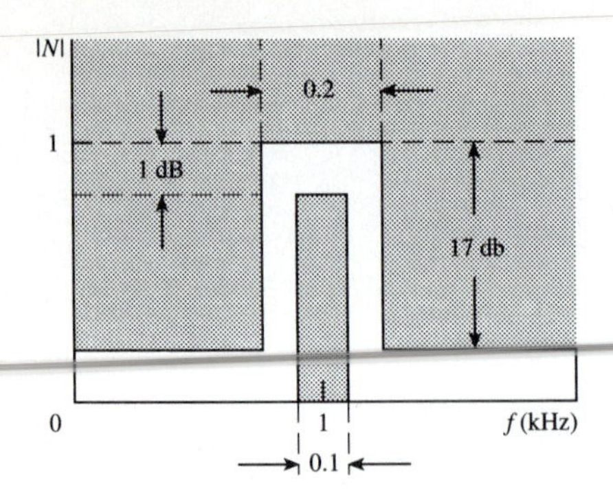

FIGURE 5.6-8
Band-pass function magnitude characteristic.

The magnitude characteristic should lie within the unshaded areas shown in Fig. 5.6-8. The first step is to convert the specifications to those of a normalized band-pass function. These become:

Center frequency $\omega_0 = 1$ rad/s
Passband of 0.1 rad/s with 1-dB ripple
Maximum passband magnitude 1.0
Stopbands with edges 0.2 rad/s apart with a minimum of 17 dB attenuation

The low-pass prototype specifications are now:

Passband of 0.1 rad/s with 1-dB ripple
Maximum passband magnitude 1.0
Stopband from 0.2 rad/s to ∞ with a minimum of 17 dB attenuation

The normalized low-pass function is characterized by:

Passband of 1.0 rad/s with 1-dB ripple
Maximum passband magnitude 1.0
Stopband from 2 rad/s to ∞ with a minimum of 17 dB attenuation

From Table 2.4-1, we find that these normalized specifications (except the passband magnitude) are met by a second-order elliptic function

$$N_{LP}(s) = \frac{H(s^2 + 7.464102)}{s^2 + 0.998942s + 1.170077}$$

After denormalizing by a factor of 0.1 to obtain the low-pass prototype and applying the low-pass to band-pass transformation, we obtain

$$N_{BP}(s) = \frac{H_1(s^2 + 0.76158)}{s^2 + 0.047553s + 0.908518} \frac{H_2(s^2 + 1.313063)}{s^2 + 0.052341s + 1.100694}$$

The constants H_1 and H_2 can now be determined to satisfy the passband specification by noting that (from Table 2.4-1) the stopband attenuation at 0 and ∞ frequencies of the band-pass function is 17.095 dB. This corresponds with a (linear) gain of 0.139717. Thus

$$|N_{BP}(0)| = |N_{BP}(\infty)| = H_1 \times H_2 = 0.139717$$

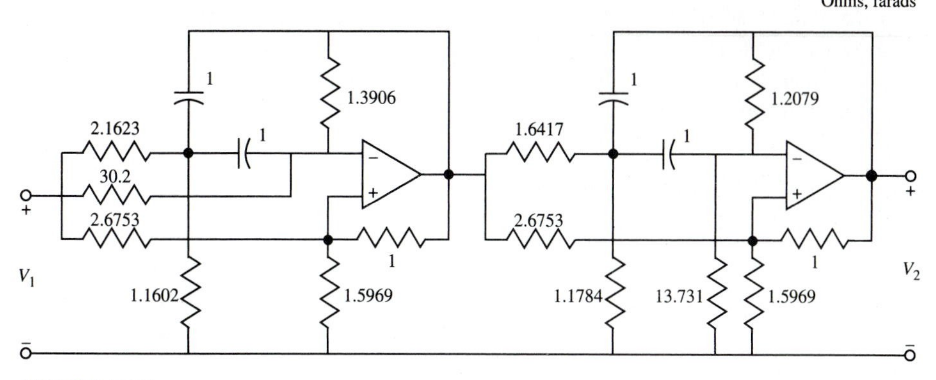

FIGURE 5.6-9
Elliptic band-pass filter realization.

For convenience we choose $H_1 = H_2 = \sqrt{0.139717} = 0.373788$. The cascaded functions to be realized are

$$N_1(s) = \frac{0.373788s^2 + 0.284668}{s^2 + 0.047553s + 0.908518} \qquad N_2(s) = \frac{0.373788s^2 + 0.490807}{s^2 + 0.052341s + 1.100694}$$

These functions can be realized by the biquadratic circuit shown in Fig. 5.5-3. For $N_1(s)$ we choose $y_1 = 1$ and for $N_2(s)$ we choose $y_1 = 0$. For both realizations we choose $K = 1.0$ and $R_B = 1$. The resulting normalized band-pass realization is shown in Fig. 5.6-9. Frequency and impedance denormalizations are easily applied to complete the design. A plot of the magnitude of the resulting function is shown in Fig. 5.6-10.

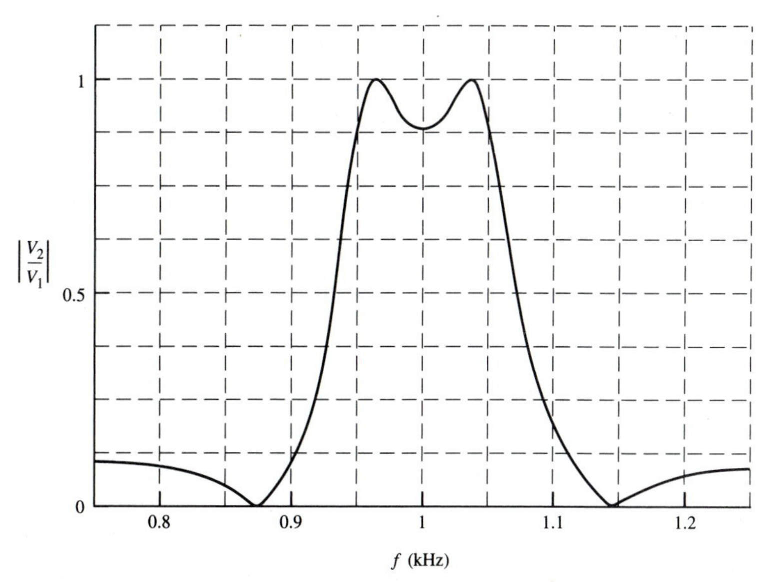

FIGURE 5.6-10
Magnitude plot of elliptic band-pass filter.

In general, to improve the dynamic range of a circuit with both finite poles and finite zeros, the quadratic factors should be arranged so that a given function contains poles and zeros that are as close to each other as possible. The preceding example illustrates this general rule.

5.7 OPERATIONAL AMPLIFIER PARASITICS—GAIN BANDWIDTH

The active *RC* realization techniques presented in the previous sections of this chapter are based on the use of ideal operational amplifiers. In this case the symbol shown in Fig. 5.7-1*a* is modeled by the circuit shown in Fig. 5.7-1*b* in which we assume that the gain A is infinite. The model is called an infinite-gain frequency-independent one. In it, the input impedances at the positive and negative terminals are also infinite, and the output impedance of the VCVS is zero. In practical operational amplifiers, the gain constant A is actually *non*infinite and frequency *de*pendent. These properties may pose significant limitations on the performance of active *RC* filters and may require modification of the design procedures. This is especially true for high-frequency filter applications. In this section we present an introductory study of these topics.

Gain Bandwidth

A practical (nonideal) operational amplifier has an open-loop gain that is frequency dependent and noninfinite in value at direct current. The frequency dependence is usually modeled by a single negative real pole. These characteristics may be described by an open-loop voltage gain $A(s)$ that has the form

$$A(s) = \frac{A_0 \omega_a}{s + \omega_a} \tag{1}$$

In this expression, A_0 is the open-loop dc gain and $-\omega_a$ is the location of the dominant pole. Typical values of these parameters would be $A_0 = 10^5$ and $\omega_a = 30$ rad/s. A Bode plot of $|A(j\omega)|$ is given in Fig. 5.7-2. We may now define a gain bandwidth product GB and express (1) in terms of it by

$$\text{GB} = A_0 \omega_a \qquad \text{for which} \qquad A(s) = \frac{\text{GB}}{s + \omega_a} \tag{2}$$

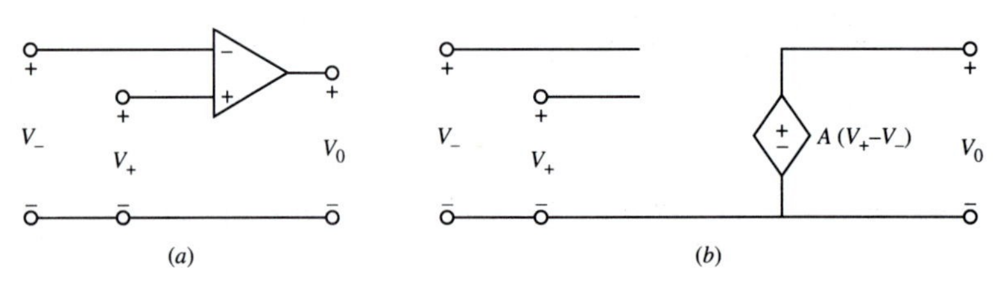

FIGURE 5.7-1
Operational amplifier symbol and ideal model.

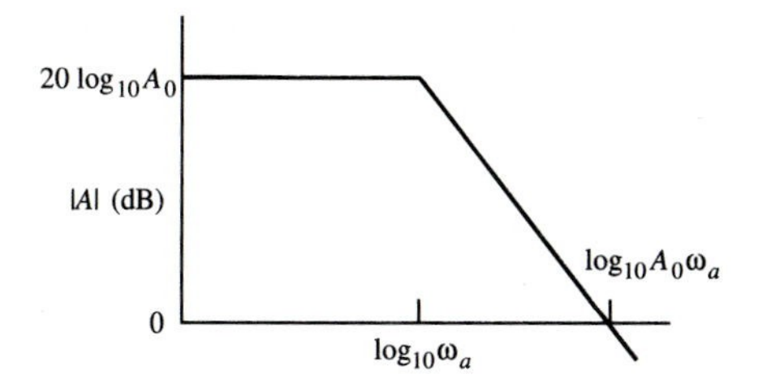

FIGURE 5.7-2
Bode plot of operational amplifier gain.

From (2), and from Fig. 5.7-2, we see that GB closely approximates the frequency at which the open-loop gain $|A(j\omega)| = 1$ (0 dB). In most applications, where the closed-loop gain is relatively small, the right equation of (2) may be approximated as

$$A(s) = \frac{\text{GB}}{s} \tag{3}$$

This equation will be the starting point for the discussions of gain bandwidth effects which follow.

Gain Bandwidth Effects in Noninverting Amplifiers

The operational amplifier realization of a noninverting VCVS shown in Fig. 5.1-2*a* may be analyzed to determine the effects of gain bandwidth. The circuit is shown symbolically in Fig. 5.7-3*a*. A model is shown in Fig. 5.7-3*b*. The gain $A(s)$ is defined in (3). For generality, the resistors R_1 and R_2 have been replaced by the impedances $Z_1(s)$ and $Z_2(s)$. The frequency-dependent noninverting amplifier gain $K(s)$ is given as

$$K(s) = \frac{V_{\text{out}}(s)}{V_{\text{in}}(s)} = \frac{A(s)}{1 + A(s)Z_1(s)/[Z_1(s) + Z_2(s)]} \tag{4}$$

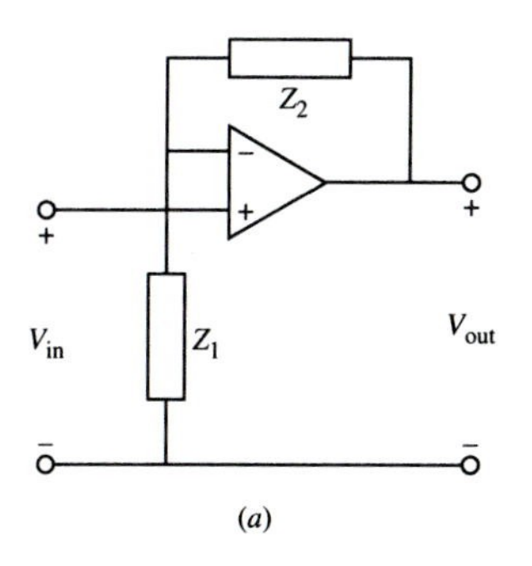

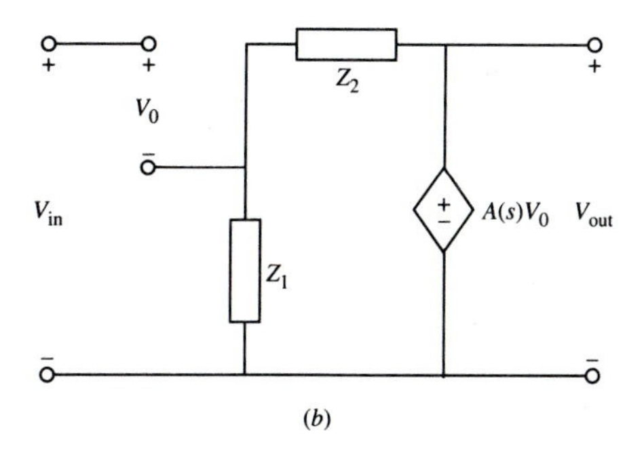

FIGURE 5.7-3
Non-inverting amplifier.

If we substitute for $A(s)$ from (3) we obtain

$$K(s) = \frac{V_{out}(s)}{V_{in}(s)} = \frac{GB}{s + GB\, Z_1(s)/[Z_1(s) + Z_2(s)]} \quad (5)$$

If we let $Z_i(s) = R_i(s)$ $(i = 1, 2)$, then for $GB = \infty$ we can define the infinite-gain-bandwidth gain as K_0. It corresponds with the quantity K defined in (1) of Sec. 5.1. From (5) we may write

$$K_0 = \frac{R_1 + R_2}{R_1} \quad \text{for which} \quad K(s) = \frac{V_{out}(s)}{V_{in}(s)} = \frac{GB}{s + GB/K_0} \quad (6)$$

Example 5.7-1 Gain bandwidth in a noninverting amplifier. A second-order low-pass Butterworth function is to be realized by a Sallen and Key design 1 filter. The gain K_0 required for the noninverting amplifier (for $Q = 1.414$) is 1.586. The operational amplifier used to implement the gain has $A_0 = 10^5$ and $\omega_a = 1$ rad/s. For these values we find $GB = A_0\omega_a = 10^5$. From (6), the resulting expression for the frequency-dependent amplifier gain is

$$K(s) = \frac{10^5}{s + 0.6305 \times 10^5}$$

Note that for $s = 0$, $K(0) = K_0 = 1.586$.

Gain Bandwidth Effects in the Sallen and Key Single-Amplifier Filter

Now let us see how gain bandwidth affects the Sallen and Key filter. We shall use the low-pass filter introduced in Sec. 5.2 to illustrate our results. The effects in band-pass and high-pass realizations are similar. Our goal is to determine what happens to the design pole locations due to the effects of gain bandwidth. The algebra is simplified by assuming a normalized design in which ω_n [see (1) of Sec. 5.2] is set to unity. To indicate this, we define the design values of the variables as follows:

$$\omega_{n(\text{design})} = \omega_{n0} = 1 \qquad Q_{(\text{design})} = Q_0 \qquad K_{(\text{design})} = K_0 \quad (7)$$

We now also choose design 1 and assume unity-valued resistors and capacitors. For these choices the design form of the voltage transfer function for the filter shown in Fig. 5.2-3 becomes

$$\frac{V_2(s)}{V_1(s)} = \frac{K_0}{s^2 + (1/Q_0)s + 1} = \frac{K_0}{s^2 + (3 - K_0)s + 1} \quad (8)$$

Note that

$$K_0 = \frac{3Q_0 - 1}{Q_0} \quad (9)$$

To determine the effects of gain bandwidth on the pole locations, we replace K_0 in the right member of (8) by the expression for $K(s)$ given in (6). The value GB is replaced by the frequency-normalized value $\text{GB}_n = \text{GB}/\omega_n$. The pole locations are now determined by the denominator polynomial

$$D(s) = s^2 + \left(3 - \frac{\text{GB}_n}{s + \text{GB}_n/K_0}\right)s + 1 \tag{10}$$

If we substitute the value of K_0 from (9), after some algebra we obtain

$$D(s) = \frac{1}{\text{GB}_n X(s)}\left[s^3 + s^2\left(3 + \frac{\text{GB}_n Q_0}{3Q_0 - 1}\right) + s\left(1 + \frac{\text{GB}_n}{3Q_0 - 1}\right) + \frac{\text{GB}_n Q_0}{3Q_0 - 1}\right] \tag{11}$$

The factor $X(s)$ in the above expression is used to indicate the extra terms that occur when (10) is rationalized. It has no effect on the pole locations. From (11) we see that the pole of the operational amplifier has changed the voltage transfer function from a second-order one to a third-order one. This may be verified by letting $\text{GB}_n = \infty$ in (11). If we do this and use (9), we obtain

$$D(s)\big|_{\text{GB}_n=\infty} = \frac{s^2}{K_0} + \frac{s}{K_0 Q_0} + \frac{1}{K_0} \tag{12}$$

which agrees with (8) [the factor $X(s)$ becomes negligible in the evaluation]. The denominator polynomial of (11) can be factored as

$$D(s) = \frac{1}{\text{GB}_n X(s)}\left[(s + g)\left(s^2 + \frac{\omega_n}{Q}s + \omega_n^2\right)\right] \tag{13}$$

where Q and ω_n are the values actually realized by the filter. In general, $\omega_n < \omega_{n0}$ while Q may be larger or smaller than Q_0. The negative real pole at $s = -g$ is usually located far out on the negative real axis and, as a result, has a negligible effect on the properties of the realization. The dominant complex-conjugate poles, however, as determined by the actual values of ω_n and Q, may be considerably different from the design values. For given values of Q_0 and GB_n, the complex-conjugate roots of (11) are usually found by applying root-solving techniques to the third-order polynomial. A graph showing the resulting pole positions as a function of a range of values of Q_0 and GB_n is given by the plot of the upper left-half of the complex plane in Fig. 5.7-4.[1] Note that the original design values appear on the outer circle (for $\text{GB}_n = \infty$). An example of the use of this graph follows.

[1] A. Budak and D. M. Petrela, "Frequency Limitations of Active Filters Using Operational Amplifiers," *IEEE Trans. Circuit Theory*, vol. CT-19, no. 4, 1972, pp. 322–328.

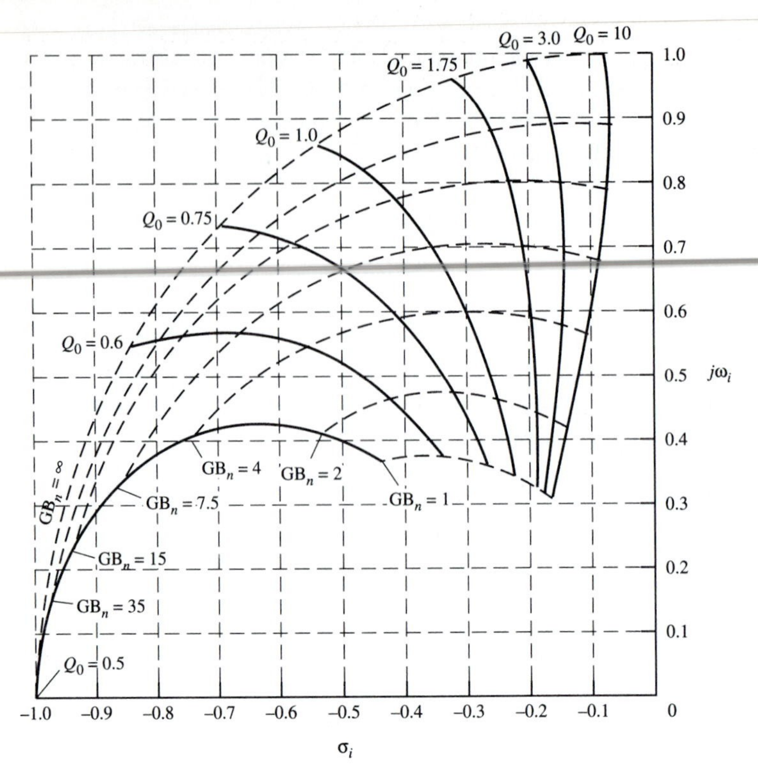

FIGURE 5.7-4
Effect of GB on the poles of the second-order low-pass or high-pass Sallen and Key equal-*R* equal-*C* (design 1) filter.

Example 5.7-2 The actual ω_n and Q of a single-amplifier low-pass filter. It is desired to find the actual ω_n and Q of a low-pass single-amplifier filter realization that uses design 1 of Sec. 5.2. The desired ω_n and Q are respectively $100 \times 2\pi$ krad/s and 1. The gain bandwidth of the amplifier is 1.5 MHz, and its frequency response has the form given in (1). The normalized gain bandwidth $GB_n = 15$. From Fig. 5.7-4 the actual normalized pole locations are found to be $\alpha_n \pm j\beta_n = -0.4 \pm j0.78$. From this result we calculate that

$$\omega_n = \sqrt{\alpha_n^2 + \beta_n^2} = \sqrt{0.4^2 + 0.78^2} = 0.877 \text{ rad/s}$$

$$Q = \frac{\omega_n}{2\alpha_n} = \frac{0.877}{2(.4)} = 1.096$$

Therefore the actual denormalized ω_n and Q are $87.7 \times 2\pi$ krad/s and 1.096. In this case the deviation is small enough so that the realization is readily tuned to the proper values. For some realizations, however, such a tuning procedure may be difficult to accomplish.

An analysis of the various other single-amplifier filter realizations is readily accomplished following the method described above. The results of a series of such analyses are given in Table 5.7-1. The shifts of the pole p_i for these analyses are illustrated in Figs. 5.7-4 through 5.7-6.

TABLE 5.7-1
Denominator polynomials for single-amplifier filters

Function	Design	Denominator polynomial
Low pass (Fig. 5.2-3)	$R_1 = R_3 = R$ $C_2 = C_4 = C$ (Fig. 5.7-4)	$s^3 + s^2\left(3 + \frac{\text{GB}_n Q_0}{3Q_0 - 1}\right) + s\left(1 + \frac{\text{GB}_n}{3Q_0 - 1}\right) + \frac{\text{GB}_n Q_0}{3Q_0 - 1}$
	$K = H_0 = 1$ $m = C_4/C_2$ $n = R_3/R_1 = 1$ (Fig. 5-7.5)	$s^3 + s^2\left(\text{GB}_n + \frac{1}{Q_0} + 2Q_0\right) + s\left(1 + \frac{\text{GB}_n}{Q_0}\right) + \text{GB}_n$
Band pass (Fig. 5.3-2)	$C_3 = C_5 = C$ $R_1 = R_2 = R_4 = R$ (Fig. 5.7-6)	$s^3 + s^2\left(2\sqrt{2} + \frac{\text{GB}_n Q_0}{4Q_0 - \sqrt{2}}\right) + s\left(1 + \frac{\text{GB}_n}{4Q_0 - \sqrt{2}}\right) + \frac{\text{GB}_n Q_0}{4Q_0 - \sqrt{2}}$
High pass (Fig. 5.3-4)	$R_2 = R_4 = R$ $C_1 = C_3 = C$ (Fig. 5.7-4)	Same as low-pass case for equal resistors, equal capacitors
	$K = H_0 = 1$ $n = R_4/R_2 = 1$ $m = C_3/C_1$ (Fig. 5.7-5)	Same as low-pass case for unity gain and $n = 1$

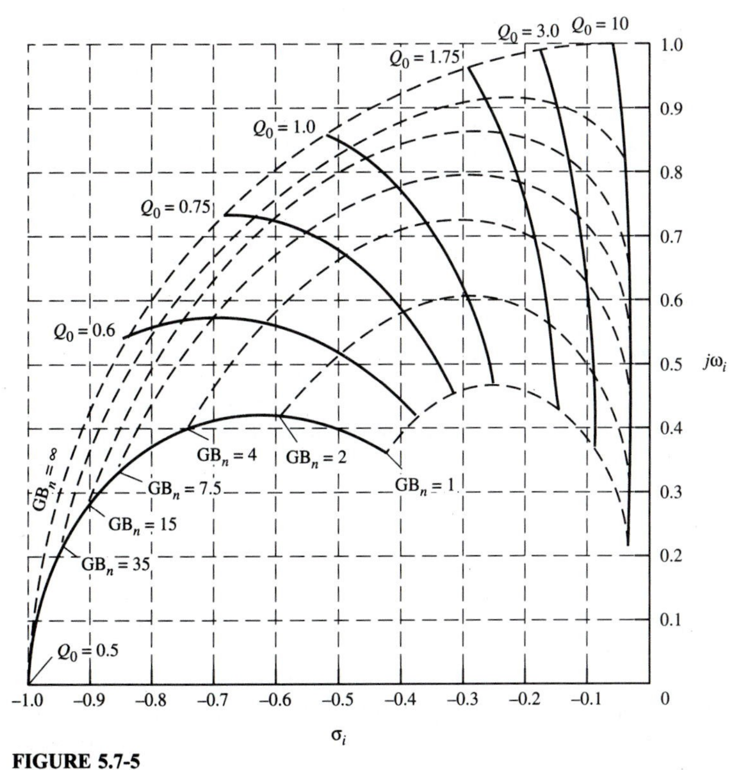

FIGURE 5.7-5
Effect of GB on the poles of the second-order low-pass or high-pass Sallen and Key unity-gain (design 2) filter.

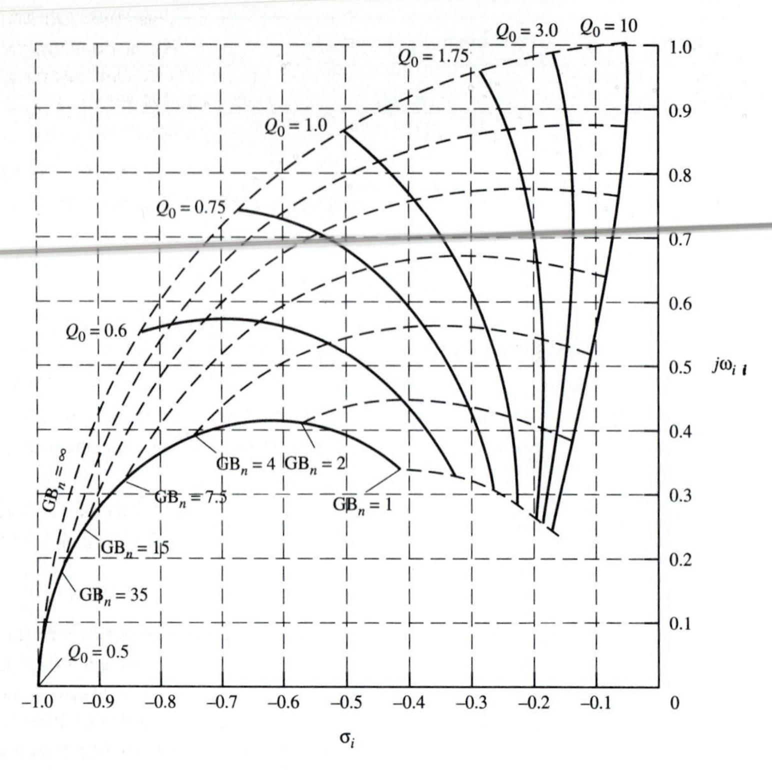

FIGURE 5.7-6
Effect of GB on the poles of the second-order band-pass Sallen and Key equal-R equal-C (design 1) filter.

Gain Bandwidth Effects in the Infinite-Gain Single-Amplifier Filter

The techniques used for showing the effects of gain bandwidth in the preceding discussion may also be used with the infinite-gain single-amplifier filter configuration. In this case, for the circuit shown in Fig. 5.4-2, we first treat the operational amplifier as having a gain $A(s) = -\text{GB}/s$. We next apply the method of constraints outlined in Summary 5.1-1 to analyze the circuit. The resulting transfer function is

$$\frac{V_2(s)}{V_1(s)} = \frac{-Y_1Y_3}{[Y_6(Y_1+Y_2+Y_3+Y_5)+Y_2Y_3]+(s/\text{GB})[(Y_1+Y_2+Y_5)(Y_3+Y_6)+Y_3Y_6]} \tag{14}$$

If we evaluate this for the low-pass, band-pass, and high-pass designs given in Sec. 5.4, we obtain the third-order denominator polynomials given in Table 5.7-2. These are for the normalized design values

$$\omega_{n(\text{design})} = \omega_{n0} = 1 \qquad Q_{(\text{design})} = Q_0 \tag{15}$$

TABLE 5.7-2
Denominator polynomials for infinite-gain single-amplifier filters

Low pass (Fig. 5.4-3)	$R_1 = R_2 = R_3 = R$ (Fig. 5.7-7)	$s^3 + s^2\left(\text{GB}_n + \dfrac{1}{Q_0} + 3Q_0\right) + s\left(\dfrac{\text{GB}_n}{Q_0} + 2\right) + \text{GB}_n$
Band pass (Fig. 5.4-6)	$C_2 = C_3 = C$ (Fig. 5.7-8)	$s^3 + s^2\left(\text{GB}_n + \dfrac{1}{Q_0} + 2Q_0\right) + s\left(\dfrac{\text{GB}_n}{Q_0} + 1\right) + \text{GB}_n$
High pass (Fig. 5.4-7)	$C_1 = C_2 = C_3 = 1$ (Fig. 5.7-9)	$2s^3 + s^2\left(\text{GB}_n + \dfrac{1}{Q_0} + 3Q_0\right) + s\left(\dfrac{\text{GB}_n}{Q_0} + 1\right) + \text{GB}_n$

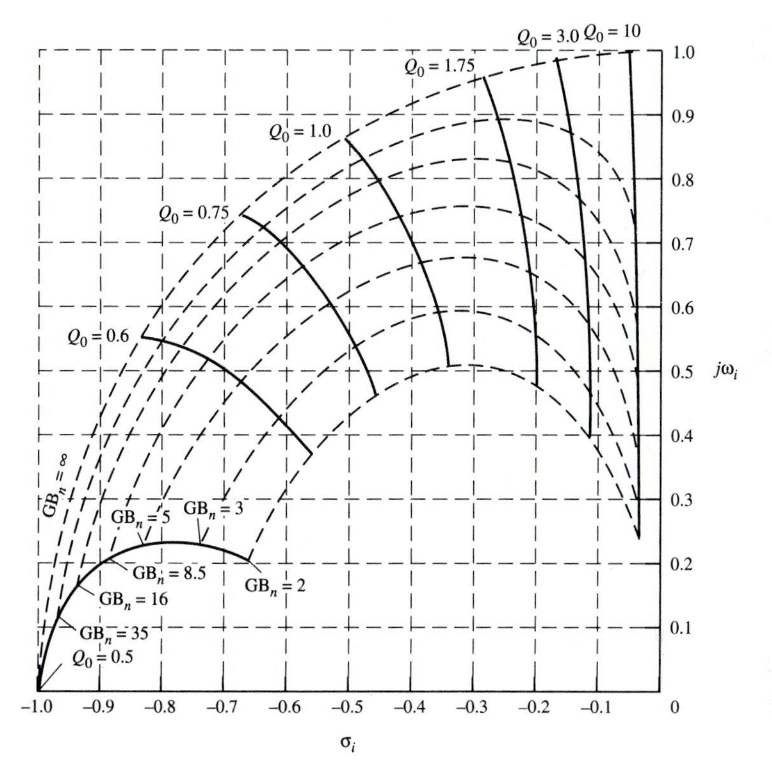

FIGURE 5.7-7
Effect of GB on the poles of the second-order low-pass infinite-gain single-amplifier filter.

The actual values of ω_n and Q realized by the filter are found by solving the polynomials. Plots of the dominant complex-conjugate upper half-plane pole locations for a range of values of Q_0 and GB_n are given in Figs. 5.7-7 to 5.7-9.

Compensation

In the preceding discussion we have illustrated the *effects* of operational amplifier gain bandwidth limitations. The resultant shift in the design locations of the pole positions caused by this parasitic property can greatly modify the characteristics of the filter. This is especially true when high values of the design Q_0 are called for and where low values of GB_n must be used. The effects are usually quite pronounced in cascade realizations of high-order filter functions due to the high-Q poles present. The effects produced by gain bandwidth, however, can be minimized by a process called *compensation*. This is a method for changing the filter parameters to compensate for the effects of gain bandwidth. The compensation is accomplished by applying a mathematical problem-solving method called *optimization*. This is the same procedure as was used in the determination of delay equalizers in Example 2.6-8. The process is best illustrated with

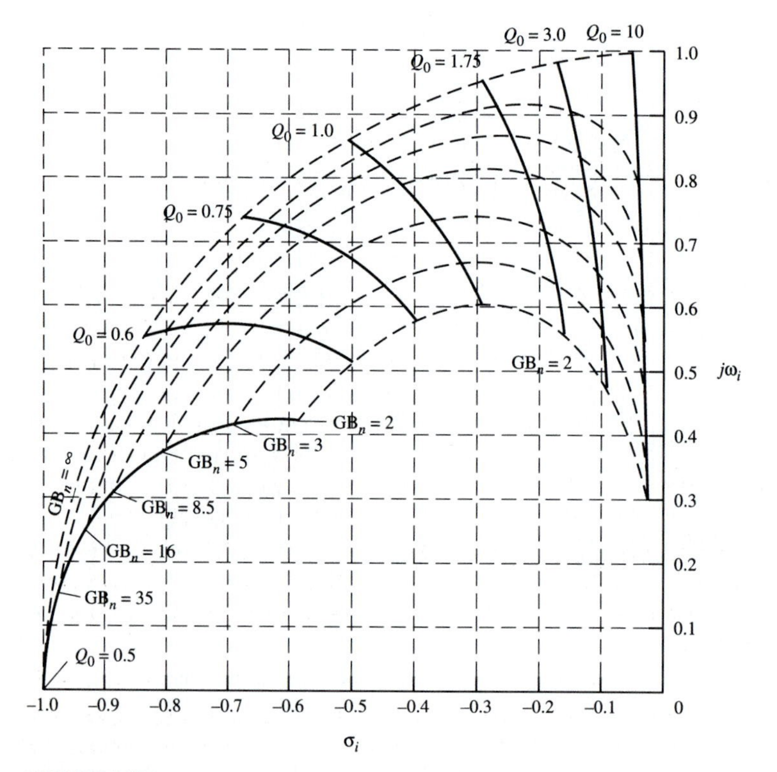

FIGURE 5.7-8
Effect of GB on the poles of the second-order band-pass infinite-gain single-amplifier filter.

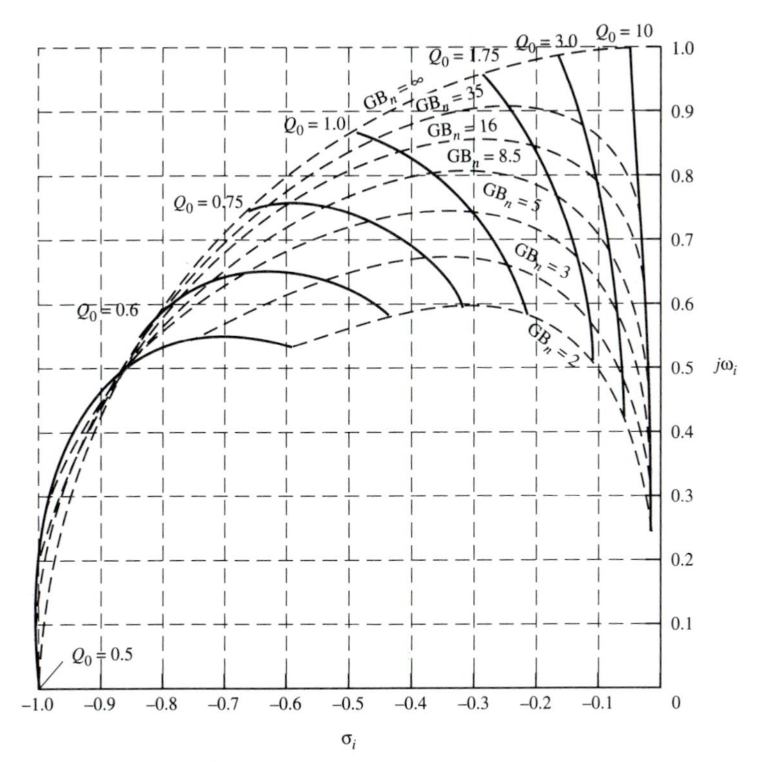

FIGURE 5.7-9
Effect of GB on the poles of the second-order high-pass infinite-gain single-amplifier filter.

respect to a specific filter configuration and design. Let us consider the Sallen and Key filter, design 1, as our example. For this filter, to compensate for the effects of gain bandwidth, we may vary the gain K from the value given by (9) and the capacitors C from their design value of unity. The goal is to restore ω_n and Q to $\omega_n = \omega_{n0} = 1$ and $Q = Q_0$. Note that it is not possible to apply a simple frequency denormalization as part of the compensation operation since the gain bandwidth has a fixed frequency that cannot be denormalized. The results of performing such an operation for a range of values of Q_0 and GB_n are given by the plot in Fig. 5.7-10. The appropriate intersection of the lines of constant Q_0 and GB_n determine the values of K and S ($= 1/C$) as given by the ordinate and abscissa scales to use in the filter circuit. Note that, as indicated by the "limit line," compensations are not available for some extreme combinations of values of Q_0 and GB_n.

Example 5.7-3 Compensation of gain bandwidth effects in a cascade Sallen and Key filter. As an example of the use of the compensation method, consider the synthesis of a sixth-order Butterworth low-pass filter using a cascade of three Sallen and Key, design 1, second-order stages. The operational amplifiers used in the stages have $GB_n = 5$. From

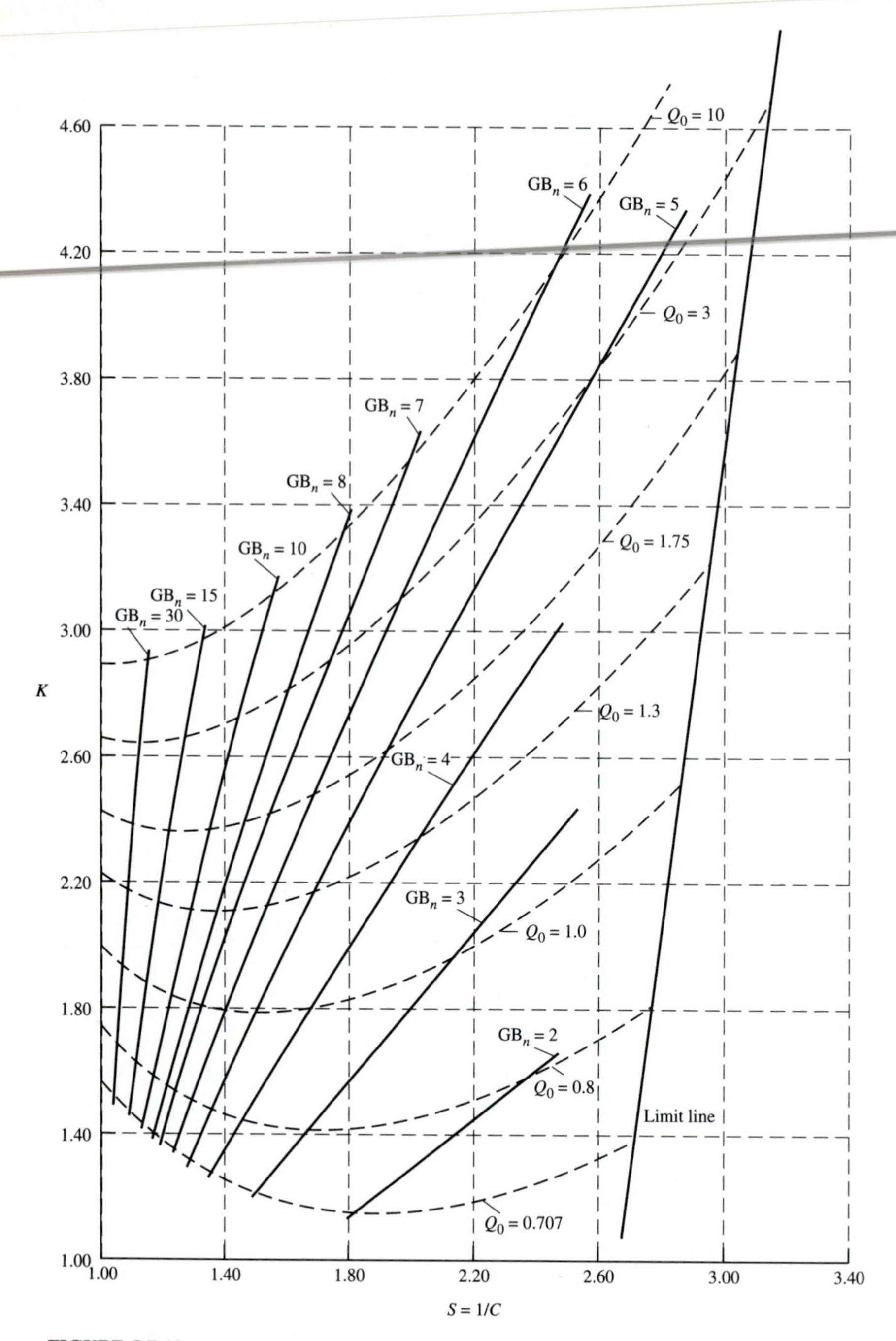

FIGURE 5.7-10
Compensation for the GB effects in a Sallen and Key equal-R equal-C (design 1) filter.

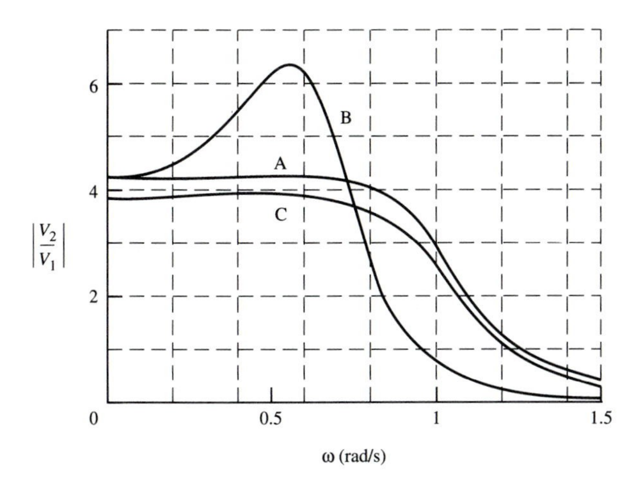

FIGURE 5.7-11
Effect of GB on a sixth-order Butterworth filter.

Table 2.1-3 and (9), the ideal design is as follows:

	Stage 1	Stage 2	Stage 3
Q_0	0.518	0.707	1.932
K_0	1.068	1.586	2.482
R	1	1	1
C	1	1	1

The dc gain is $H_0 = 1.068 \times 1.586 \times 2.482 = 4.204$. A plot of the magnitude of the voltage transfer function under the (ideal) conditions that the operational amplifiers have infinite gain bandwidth is shown as A in Fig. 5.7-11. Now let us reanalyze the circuit, keeping the design values given above but using operational amplifiers with a gain bandwidth $\text{GB}_n = 5$. For example, in SPICE this can be done by replacing the VCVS infinite-gain bandwidth model shown in Fig. 5.7-12*a* with a model that implements the expression for $K(s)$ given in (6) as shown in Fig. 5.7-12*b*. To do this, we let

$$\frac{R_{\text{GB}}}{L_K} = \frac{\text{GB}_n}{K_0} \qquad \text{for which} \qquad R_{\text{GB}} = \text{GB}_n \qquad L_K = K_0$$

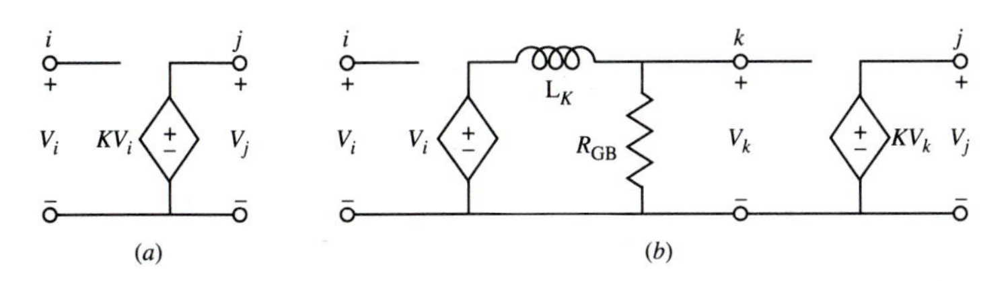

FIGURE 5.7-12
SPICE models for a noninverting VCVS: (*a*) without GB; (*b*) with GB.

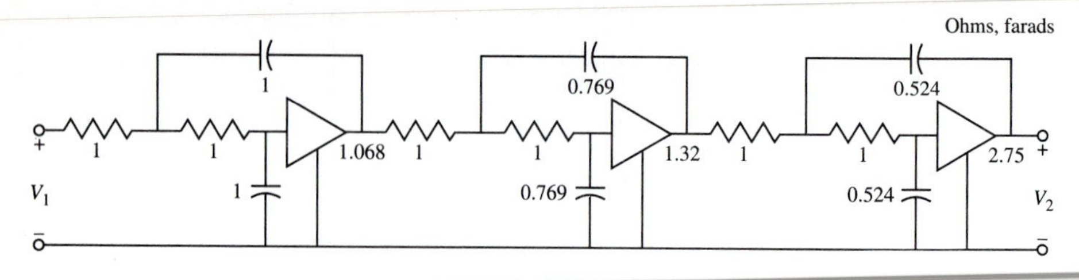

FIGURE 5.7-13
Compensated realization of a sixth-order Butterworth filter.

The effects of gain bandwidth are shown in the *B* magnitude plot of Fig. 5.7-11. The locus obviously bears little resemblance to a maximally flat magnitude one. If we now apply the compensation information given in Fig. 5.7-10, interpolating between the curves as necessary, we obtain the following compensated design specifications (due to the low value of Q in stage 1, no compensation is needed for that stage):

	Stage 1	**Stage 2**	**Stage 3**
K (dc)	1.068	1.32	2.75
R	1	1	1
S	1	1.30	1.91
C	1	0.769	0.524

The compensated normalized circuit is shown in Fig. 5.7-13. The VCVSs are realized with operational amplifiers with $\mathrm{GB}_n = 5$. The values shown are for their dc gains. The overall dc gain is $H_0 = 1.068 \times 1.32 \times 2.75 = 3.877$. The resulting magnitude characteristic is shown as C in Fig. 5.7-11. The close approximation to the specified Butterworth characteristic is quite evident.

5.8 USE OF COMPUTER PROGRAMS IN CASCADE SINGLE-AMPLIFIER SYNTHESIS

The computations required in designing the second-order active filter building blocks used as cascaded elements in this chapter are readily implemented on a PC (personal computer). Many shareware and commercial PC programs are available that provide a wide range of design capabilities. A useful program that illustrates many of the designs presented in this chapter is Active RC, a program developed by Linda C. Dalmas and the author. It implements the design of low-pass single-amplifier filters and also the design of the multiple-amplifier filters introduced in the following chapter. It includes approximation capabilities for low-pass Butterworth and Chebyshev designs. In addition

This is the second-order stage 1 of 2 for the Sallen and Key filter, design 1.

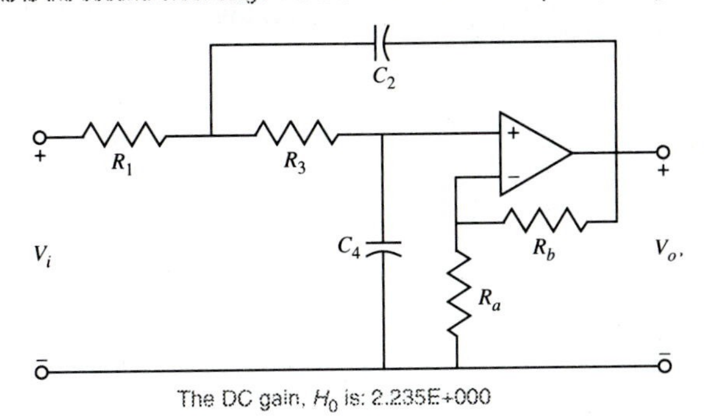

The DC gain, H_0 is: 2.235E+000

Element values X =		Sensitivity: S_X^Q	Sensitivity: $S_X^{w_n}$
R1	= 1.000E+004	8.066E-001	-5.000E-001
R3	= 1.000E+004	-8.066E-001	-5.000E-001
Ra	= 1.000E+004	-1.613E+000	0.000E+000
Rb	= 1.235E+004	1.613E+000	0.000E+000
C2	= 1.592E-009	2.113E+000	-5.000E-001
C4	= 1.592E-009	-2.113E+000	-5.000E-001

Frequency denorm = 1.000E+000 Impedance denorm = 1.000E+000

Press any key to continue

(*a*)

This is the second-order stage 2 of 2 for the Sallen and Key filter, design 1.

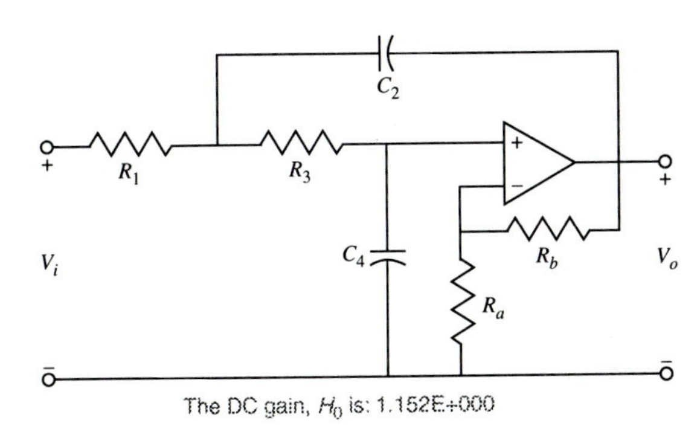

The DC gain, H_0 is: 1.152E+000

Element values X =		Sensitivity: S_X^Q	Sensitivity: $S_X^{w_n}$
R1	= 1.000E+004	4.120E-002	-5.000E-001
R3	= 1.000E+004	-4.120E-002	-5.000E-001
Ra	= 1.000E+004	-8.239E-002	0.000E+000
Rb	= 1.522E+003	8.239E-002	0.000E+000
C2	= 1.592E-009	5.824E-001	-5.000E-001
C4	= 1.592E-009	-5.824E-001	-5.000E-001

Frequency denorm = 1.000E+000 Impedance denorm = 1.000E+000

Press any key to continue

(*b*)

FIGURE 5.8-1
Output from the program Active RC.

to determining the network functions, it provides realizations for the following low-pass configurations:

Sallen and Key—design 1
Sallen and Key—design 2
Sallen and Key—design 3
Infinite gain single amplifier
State variable (to be covered in Chap. 6)
Tow-Thomas (to be covered in Chap. 6)

The program provides detailed sensitivity information for each filter realization. This permits interaction with the designer, allowing him or her to implement the low-Q stages of a cascade with simple (but high-sensitivity) single-amplifier structures, while using more complex (but low-sensitivity) multiple-amplifier structures for the high-Q stages. An example of the use of the program follows.

Example 5.8-1 Fourth-order low-pass active filter. As an example of the use of the Active RC program, consider the realization of a Butterworth low-pass filter with a passband (as determined by the −3.0103-dB frequency) of 0 to 10 kHz and with a stopband characterized by a minimum of 24 dB attenuation starting at 20 kHz. When these specifications are entered, the program determines that a fourth-order filter is required. The resulting component values for the two Sallen and Key design 1 stages are shown in Fig. 5.8-1.

The program Active RC is available for a small charge. Additional information may be obtained from Dr. L. P. Huelsman, Department of Electrical and Computer Engineering, University of Arizona, Tucson, AZ 85721.

PROBLEMS

Section 5.1

1. (*a*) Find the actual gain realized by the inverting VCVS circuit shown in Fig. 5.1-2*c* for the case where $R_1 = R_2 = 100\ \text{k}\Omega$. Assume that the input resistance $R_i = 10^6\ \Omega$, the output resistance $R_o = 100\ \Omega$, and the dc open-loop gain $A_d = 10^5$. *Hint:* The results given in App. C may be used to simplify the analysis.
(*b*) Repeat part (*a*) for $R_1 = 50\ \text{k}\Omega$ and $R_2 = 100\ \text{k}\Omega$.
(*c*) Repeat part (*a*) for the values $R_i = 10^4\ \Omega$ and $R_o = 10^4\ \Omega$.

2. (*a*) Apply the method of constraints given in Summary 5.1-1 to find the voltage transfer function for the circuit shown in Fig. P5.1-2. Assume that all the passive elements have unity value but keep K as a literal quantity. Use the indicated node numbers.
(*b*) Check the results obtained in part (*a*) by using (9).
(*c*) In order that the voltage transfer function for this circuit can realize complex-conjugate poles, should the VCVS be inverting or noninverting?

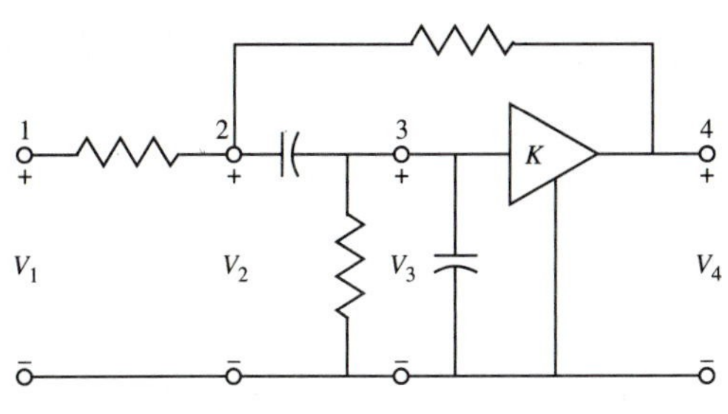

FIGURE P5.1-2

3. Find the polynomials $N_{31}(s)$, $N_{32}(s)$, and $N_{33}(s)$ defined in (12) for the network shown in Fig. P5.1-2. Assume that all the passive elements have unity value.
4. (*a*) Apply the method of constraints given in Summary 5.1-1 to analyze the circuit shown in Fig. P5.1-4. Assume that all the passive elements have unity value but keep K as a literal quantity.
   (*b*) In order that the voltage transfer function for this circuit can realize complex-conjugate poles, should the VCVS be inverting or noninverting?

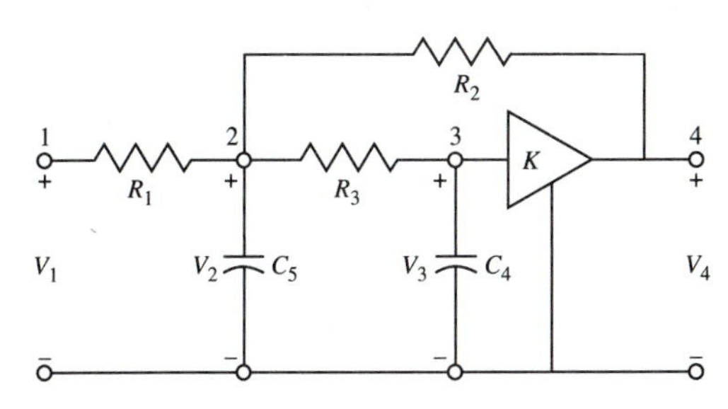

FIGURE P5.1-4

5. Find the polynomials $N_{31}(s)$, $N_{32}(s)$, and $N_{33}(s)$ defined in (12) for the network shown in Fig. P5.1-4. Assume that all the passive elements have unity value.
6. (*a*) In the filter shown in Fig. P5.1-4, what type of filter characteristic (for example, low pass or high pass) does the voltage transfer function realize?
   (*b*) Assume that the components have the value $R_1 = R_2 = R_3 = 1\ \Omega$ and $C_4 = 10$ F. Find the values of C_5 and K such that the denominator of the voltage transfer function has the form

$$D(s) = s^2 + 0.2s + 1$$

   (*c*) Draw a sketch of the locus of the poles of the network as the amplifier gain $|K|$ is varied.
7. (*a*) Derive an expression for S_K^Q for the case where a denominator polynomial has the form

$$D(s) = s^2 + \frac{1}{Q}s + 1 = (s+1)^2 - K\alpha s$$

   where α is not a function of K.

(*b*) Repeat part (*a*) for the case where the denominator polynomial has the form

$$D(s) = s^2 + \frac{1}{Q}s + 1 = s^2 + 3s + 1 - K\alpha s$$

8. Identify (by inspection) each of the filter configurations shown in Fig. P5.1-8 as low pass, high pass, or band pass. Do not attempt to analyze the circuits.

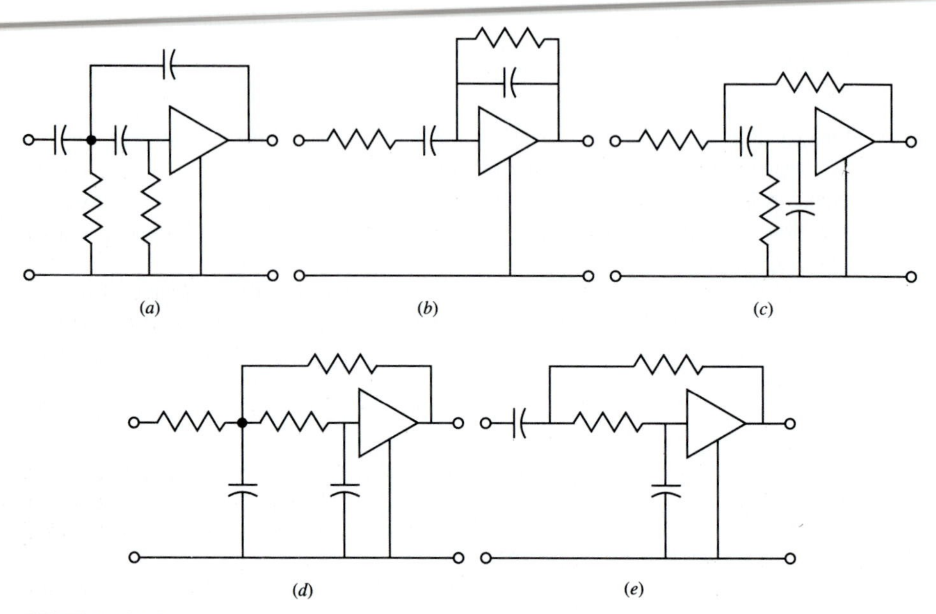

FIGURE P5.1-8

Section 5.2

1. (*a*) Use design 1 to realize a second-order Thomson filter with a delay (at direct current) of 1 ms. Choose $C = 10^{-7}$ F.

(*b*) Repeat part (*a*) using design 2. Choose the *equality* in (17) to determine the value of m.

(*c*) Repeat part (*a*) using design 3.

2. An alternate design procedure for the low-pass filter shown in Fig. 5.2-3 that allows the H_0 specifications as well as the ω_n and Q ones to be met is as follows: Choose $K = H_0$ and $C_2 = C_4 = C$; then

$$R_3 = \frac{1}{2\omega_n QC}[1 + \sqrt{1 + 4Q^2(H_0 - 2)}] \qquad R_1 = \frac{1}{\omega_n^2 C^2 R_3}$$

where $H_0 \geq 2$. Apply these relations to design a filter with $f_n = 30$ Hz, $Q = 1/\sqrt{2}$, $H_0 = 10$, and $C = 0.1$ μF.

3. Derive the design relations given in Prob. 2.

4. Determine the sensitivities of (20) for a realization of the low-pass Sallen and Key filter shown in Fig. 5.2-5 in which $R_3 = 10R_1$ and $C_4 = 0.1C_2$.

5. Determine the sensitivities of (20) for a realization of the low-pass Sallen and Key filter shown in Fig. 5.2-5 in which $R_1 = R_3 = 1\ \Omega$, $K = 1$, $C_2 = 2Q/\omega_n$ farads, and $C_4 = 1/2Q\omega_n$ farads.

6. (*a*) For a positive-gain low-pass realization having the form shown in Fig. 5.1-4 in which $Y_i = 1$ (for $i = 1, 3, 6$) and $Y_i = sC_i$ (for $i = 2, 4$) find the voltage transfer function.

(*b*) For the restriction $Q\omega_n > 1$, find an expression for the required gain K as a function of $Q\omega_n$.

7. (*a*) A filter has the following network function:

$$N(s) = \frac{V_2(s)}{V_1(s)} = \frac{|K|}{s^2 + 4s + (2 + |K|)}$$

Find the function sensitivity $S_{|K|}^{N(j\omega)}$.

(*b*) Will the gain sensitivity for such a filter be larger or smaller than the gain sensitivity for the Sallen and Key low-pass filter? Why?

8. (*a*) Analyze the Sallen and Key low-pass filter shown in Fig. 5.2-3 to find the voltage transfer function for the case where $R_1 = R_3 = 1\ \Omega$, $C_2 = C_4 = 1$ F, and a resistance R_i, representing the input resistance of the VCVS, is placed in shunt with the capacitor C_4.

(*b*) Define $G_i = 1/R_i$ in the transfer function found in part (*a*) and determine the range of values for K (as a function of G_i) that produces complex-conjugate left-half-plane poles.

9. For the filter defined in Prob. 8, find an expression for the loci of the poles as K is varied. The loci should be defined in terms of G_i.

10. (*a*) Analyze the Sallen and Key low-pass filter shown in Fig. 5.2-3 to find the voltage transfer function for the case where $R_1 = R_3 = 1\ \Omega$, $C_2 = C_4 = 1$ F, and a resistance R_o, representing the output resistance of the VCVS, is placed in series with the VCVS at its positive output terminal.

(*b*) For the network defined in part (*a*), find a value of K for which the real part of the pole positions is constant, independent of the value of R_o.

11. For the filter defined in Prob. 10, find an expression for the loci of the poles as K is varied. The loci should be defined in terms of R_o.

12. A normalized ($\omega_n = 1$ rad/s) low-pass voltage transfer function $N(s) = V_2(s)/V_1(s)$ is realized by a Sallen and Key design 1 filter. The element values are

$$R_1 = 1 \qquad R_3 = 1 \qquad C_2 = 1 \qquad C_4 = 1 \qquad K = 2.667$$

If the element values are changed to

$$R_1 = 1.01 \qquad R_3 = 1.01 \qquad C_2 = 1.01 \qquad C_4 = 1.01 \qquad K = 2.693$$

use sensitivity techniques to find

$$\frac{\Delta|N(j1)|}{|N(j1)|}$$

Section 5.3

1. (*a*) Design a band-pass Sallen and Key filter having the form shown in Fig. 5.3-2. Use design 1 to meet the specifications $f_n = 5$ kHz, $Q = 5$, and $C = 10^{-8}$ F.

(*b*) Repeat part (*a*) using design 2.

2. Prove whether or not the band-pass Sallen and Key filter shown in Fig. 5.3-2 can be used to realize complex-conjugate poles if the gain $K = 1$.
3. Develop a band-pass Sallen and Key filter following the method outlined in this section but with the initial choice of elements $Y_1 = sC_1$ and $Y_3 = G_3$. Obtain design relations similar to those given in (6) and (7).
4. (*a*) For the Sallen and Key band-pass filter whose voltage transfer function is given in (6), find expressions for the sensitivities similar to those given for the low-pass filter in (20) of Sec. 5.2. Use an operational amplifier realization for the VCVS as shown in Fig. 5.2-5.
   (*b*) Use the results obtained in part (*a*) to make a table of sensitivities for design 1 and design 2. It should be similar to Table 5.2-1.
5. (*a*) Design a high-pass Sallen and Key filter having the form shown in Fig. 5.3-4. Use design 1 to meet the specifications $f_n = 5$ kHz, $Q = 1/\sqrt{2}$, and $C = 10^{-8}$ F.
   (*b*) Repeat part (*a*) using design 2. Choose $m = 1$.
   (*c*) Repeat part (*a*) using design 3.
6. An alternate design procedure for the high-pass filter shown in Fig. 5.3-4 that allows the H_0 specification as well as the ω_n and Q ones to be met is as follows: Choose $C_1 = C_3 = C$; then

$$R_2 = \frac{1/2Q + \sqrt{2(H_0 - 1) + 1/4Q^2}}{2\omega_n C} \qquad R_4 = \frac{2/\omega_n C}{1/2Q + \sqrt{2(H_0 - 1) + 1/4Q^2}}$$

   Apply these relations to design a filter with $Q = 1/\sqrt{2}$, $f_n = 300$ Hz, $H_0 = 100$, and $C = 10^{-8}$ F.
7. Derive the design relations given in Prob. 6.
8. (*a*) For the Sallen and Key high-pass filter whose voltage transfer function is given in (17), find expressions for the sensitivities similar to those given for the low-pass filter in (20) of Sec. 5.2. Use an operational amplifier realization for the VCVS as shown in Fig. 5.2-5.
   (*b*) Use the results obtained in part (*a*) to make a table of sensitivities for design 1, design 2, and design 3. It should be similar to Table 5.2-1.

Section 5.4

1. Apply the method of constraints given in Summary 5.4-1 to find the voltage transfer function for the filter shown in Fig. P5.4-1. The normalized element values are $R_1 = 0.1\ \Omega$, $R_2 = 10\ \Omega$, $R_3 = 0.09901\ \Omega$, and $C = 1$ F. Use the node numbers shown in the figure.

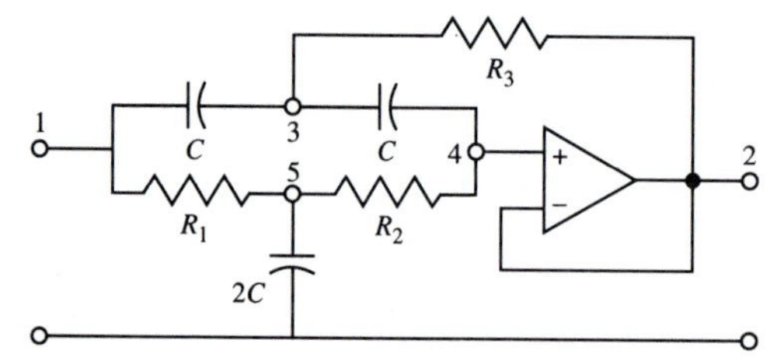

FIGURE P5.4-1

2. Apply the method of constraints given in Summary 5.4-1 to find the voltage transfer function for the filter shown in Fig. P5.4-2. Use the node numbers shown in the figure.

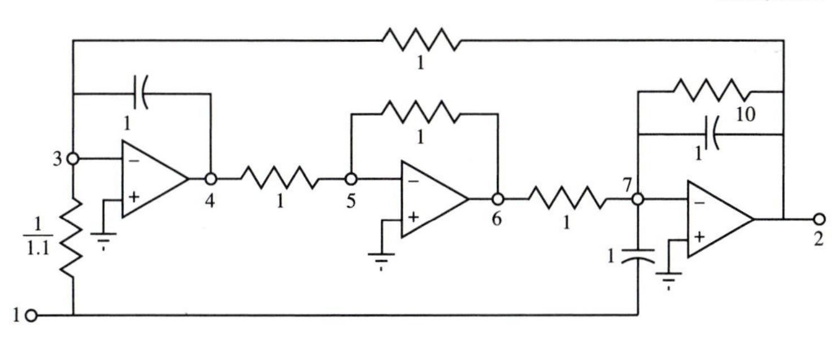

FIGURE P5.4-2

3. Use the infinite-gain single-amplifier low-pass filter shown in Fig. 5.4-3 to realize a second-order Thomson function having a delay of 500 μs and a unity gain at direct current. Choose $C = 0.05$ μF and $m = 0.1$ and use the higher of the two possible values for R_2.
4. For the infinite-gain single-amplifier low-pass filter shown in Fig. 5.4-3, assume a design in which we set $R_1 = R_2 = R_3 = R$ and find expressions for the quantities ω_n, $1/Q$, and $|H_0|$ given in (8).
5. For an infinite-gain single-amplifier low-pass filter with $m = 0.1$ and $C = 1$, realizing $Q = 1$, $\omega_n = 1$ rad/s, and $|H_0| = 1$, find the Q sensitivities to the resistors and capacitors.
6. An infinite-gain single-amplifier low-pass filter realizes a normalized voltage transfer function $N(s) = V_2(s)/V_1(s)$ in which $|H_0| = 1$, $Q = 3$, and $\omega_n = 1$ rad/s. The element values are

$$R_1 = R_2 = 7.8475\ \Omega \qquad R_3 = 12.7429\ \Omega \qquad C_5 = 1\text{ F} \qquad C_6 = 0.01\text{ F}$$

If the element values are all increased by 1 percent, use sensitivity techniques to find $\Delta|N(j1)|/|N(j1)|$.

7. Realize an infinite-gain single-amplifier band-pass filter with the specifications $f_n = 10$ kHz and $Q = 10$. Choose $|H_0|$ such that the resistor R_5 shown in Fig. 5.4-6 is infinite (an open circuit), and use $C = 1$ nF.
8. (*a*) What restrictions are placed on the pole positions realized by an infinite-gain single-amplifier low-pass filter constrained to be equal R and equal C?
   (*b*) Repeat part (*a*) for a band-pass filter.
9. (*a*) Find expressions for the sensitivities of an infinite-gain single-amplifier band-pass filter similar to those given in (14) for the low-pass case.
   (*b*) Evaluate the expressions given in part (*a*) for a realization of $Q = 1$, $\omega_n = 1$ rad/s, and $|H_0| = 1$ in which $C = 1$ F.
10. Realize an infinite-gain single-amplifier high-pass filter with the specifications $f_n = 1$ kHz, $Q = 1/\sqrt{2}$, and $|H_0| = 5$. Choose $C = 0.01$ μF.
11. Find expressions for the sensitivities of an infinite-gain single-amplifier high-pass filter similar to those given in (14) for the low-pass case.

Section 5.5

1. Design a band-pass filter having the form shown in Fig. 5.5-1 with a resonant frequency $f_n = 10$ kHz, $Q = 10$, and a maximum element spread of 100. Choose $C = 0.1$ μF.
2. (*a*) For a band-pass filter having the form shown in Fig. 5.5-1, the normalized specifications

are $\omega_n = 1$ rad/s and $Q = 10$, and C and R_B are chosen as unity. Find the element values and the Q sensitivity to each of the elements for the case where $m = 0.01$.

(*b*) For a band-pass filter having the form shown in Fig. 5.4-6 and designed using (21) of Sec. 5.4, find the element values and the Q sensitivity to each of the elements if the filter is to meet the specifications given in part (*a*) and also is to have the same value of H_0 as was obtained in that realization.

3. (*a*) For a band-pass filter having the form shown in Fig. 5.5-1, the normalized specifications are $\omega_n = 1$ rad/s and $Q = 10$, and C and R_B are chosen as unity. Find the element values and the Q sensitivity to each of the elements for the case where $m = 0.1$.

(*b*) For a band-pass filter having the form shown in Fig. 5.4-6 and designed using (21) of Sec. 5.4, find the element values and the Q sensitivity to each of the elements if the filter is to meet the specifications given in part (*a*) and also is to have the same value of H_0 as was obtained in that realization.

4. Use the biquadratic filter structure shown in Fig. 5.5-3 to realize a second-order all-pass filter with a Thomson (Bessel) approximation and a delay of 2 ms at direct current. Choose an impedance denormalization such that all the capacitors have a value of 0.01 μF. In the procedure given in Summary 5.5-2, choose $K = 0.1$ and $y_1 = 0$ and adjust the gain so that $x_2 = 1$.

Section 5.6

1. Realize a fourth-order low-pass active *RC* filter using the circuit shown in Fig. 5.2-3 and design 1. The magnitude characteristic is to be maximally flat with a 3-dB frequency of 2 rad/s. All of the capacitors used in the realization should have a value of 2 F.

2. Realize a fourth-order active *RC* high-pass filter using the circuit shown in Fig. 5.3-4 and design 1. The magnitude characteristic is to have a 1-dB ripple in the passband with a cutoff frequency of 1 rad/s. All the capacitors in the filter are to have a value of 2 F. Show the operational amplifiers and their gain-determining resistors.

3. Design a sixth-order active *RC* low-pass filter that has a Thomson characteristic with a delay at direct current of 1 ms. Use the circuit shown in Fig. 5.2-3 and design 1. All the capacitors should have a value of 0.01 μF.

4. (*a*) Design an eighth-order active *RC* low-pass filter with a 1-dB ripple in the passband and a cutoff frequency of 20 kHz. Use the filter shown in Fig. 5.2-3 for each of the component second-order realizations. Use design 1 of Sec. 5.2 to determine the element values. Choose $C = 10^{-9}$ F.

(*b*) Repeat the problem specified in part (*a*) using design 2. For the highest Q stage choose $m = 0.001$ and $C_4 = 10^{-10}$ F. For the second highest Q stage choose $m = 0.01$ and $C_4 = 10^{-10}$ F. For the third highest Q stage choose $m = 0.05$ and $C_4 = 0.5 \times 10^{-9}$ F. For the lowest Q stage choose $m = 0.2$ and $C_4 = 2 \times 10^{-9}$ F.

(*c*) Repeat the problem specified in part (*a*) using design 3. Choose $C = 10^{-9}$ F.

5. (*a*) Plot the component second-order and the overall magnitude characteristic of a fourth-order Butterworth band-pass filter function with a center frequency of 1 rad/s and a bandwidth of 1 rad/s, which is realized by a cascade of two second-order band-pass filters in which the H_0 constant of each filter is set to unity.

(*b*) Repeat part (*a*) for a cascade of a second-order low-pass filter and a second-order high-pass one. Arrange the design so that the low-pass filter realizes the poles closest to the origin and the H_0 constants of both filters are set to unity.

(*c*) Repeat part (*b*) for the case where the high-pass filter realizes the poles closest to the origin.

6. (*a*) Design a band-pass filter with a maximally flat magnitude characteristic and a 3-dB bandwidth from 725 to 800 Hz and with 0 dB gain at the resonant frequency. Use the narrow-band transformation on an appropriate second-order Butterworth function to find the band-pass one. Use the infinite-gain single-amplifier band-pass filter shown in Fig. 5.4-6 to realize the band-pass pole pairs. Choose the values of the second-order band-pass filter gain constants $|H_0|$ so that they are equal. Choose $C = 10^{-8}$ F.

(*b*) Repeat the problem assuming that the gain at the resonant frequency is to be 20 dB.

7. Use a cascade of two of the infinite-gain single-amplifier band-pass filters shown in Fig. 5.4-6 to realize a normalized maximally flat magnitude fourth-order band-pass network function with a center frequency of 1 rad/s and a bandwidth of 0.5 rad/s. All the capacitors should have a value of unity. The gain of the band-pass network at the resonant frequency should be unity. The gain constants $|H_0|$ of the individual band-pass sections should be the same.

8. (*a*) Design a cascade realization for a fourth-order Butterworth low-pass filter with a (normalized) bandwidth of 1 rad/s. Use the Sallen and Key circuit with design 3 (amplifier gain $K = 2$). Let $C_2 = C_4 = 1$ F.

(*b*) Find the function sensitivity $S_K^{N(s)}$ where $N(s)$ is the voltage transfer function and K is the gain constant for *both* amplifiers.

(*c*) What percentage change in the magnitude of the network function will occur at a frequency of 1 rad/s if a 1 percent change is made in the gain K of both amplifiers?

Section 5.7

1. (*a*) It is desired to use a Sallen and Key low-pass filter and design 1 to realize the following voltage transfer function:

$$\frac{V_2(s)}{V_1(s)} = \frac{H}{s^2 + 0.5714s + 1}$$

Find the design values for R, C, and K assuming that the operational amplifier has infinite gain bandwidth. What is the value of the numerator multiplicative constant H that is realized?

(*b*) If an operational amplifier in which $\text{GB}_n = 7.5$ is used with the design values found in part (*a*), find the network function (including the value of the numerator multiplicative constant H) that will actually be realized.

2. (*a*) It is desired to use a Sallen and Key low-pass filter and design 2 to realize the following voltage transfer function:

$$\frac{V_2(s)}{V_1(s)} = \frac{H}{s^2 + 0.5714s + 1}$$

Use $n = 1$ and $C = 1$ and find the design values for the other components assuming that the operational amplifier has infinite gain bandwidth. What is the value for the numerator multiplicative constant H that is realized?

(*b*) If an operational amplifier in which $\text{GB}_n = 7.5$ is used with the design values found in part (*a*), find the network function (including the value of the numerator multiplicative constant H) that will actually be realized.

3. (*a*) It is desired to use a Sallen and Key band-pass filter and design 1 to realize the following voltage transfer function:

$$\frac{V_2(s)}{V_1(s)} = \frac{Hs}{s^2 + 0.5714s + 1}$$

Find the design values for R, C, and K assuming that the operational amplifier has infinite gain bandwidth. What is the value for the numerator multiplicative constant H that is realized?

(*b*) If an operational amplifier in which $GB_n = 7.5$ is used with the design values found in part (*a*), find the network function (including the value of the numerator multiplicative constant H) that will actually be realized.

4. (*a*) For the design given in Example 5.7-2, find the actual network function realized when an ideal operational amplifier is used.

(*b*) For the specifications of part (*a*), find the design values of R and K if $C = 10^{-9}$.

(*c*) For the design given in Example 5.7-2, find the network function realized if the operational amplifier has $GB_n = 15$.

(*d*) Use the value of R found in part (*b*) and find new values of C and K that will meet the original design specifications of Example 5.7-2 if a $GB_n = 15$ operational amplifier is used.

(*e*) Find the network function actually realized with the compensated element values determined in part (*d*).

5. A Sallen and Key low-pass filter (design 1) was to realize $\omega_n = 10^4$ rad/s. When a nonideal operational amplifier ($GB_n \neq \infty$) of unknown GB is used in the realization, however, the actual voltage transfer function is found to be

$$\frac{V_2(s)}{V_1(s)} = \frac{1.37 \times 10^8}{s^2 + 0.74 \times 10^4 s + 0.685 \times 10^8}$$

Find the voltage transfer function that results if the operational amplifier of unknown GB is replaced with one for which $GB_n = 30$.

6. (*a*) It is desired to use an infinite-gain single-amplifier low-pass filter to realize a normalized network function in which the design values are $\omega_n = 1$ rad/s, $Q = 1$, and $|H_0| = 1$. Find the network function actually realized if an operational amplifier with $GB_n = 5$ is used.

(*b*) Repeat part (*a*) for the case where a design value $Q = 10$ is specified.

CHAPTER 6

MULTIPLE-AMPLIFIER ACTIVE *RC* FILTERS

The technology of making active devices has advanced to the point where certain traditional guidelines are no longer valid. One of these guidelines was that one should minimize the number of active devices. This viewpoint is reflected in the single-amplifier filters discussed in the preceding chapter. With respect to modern integrated circuit technology, however, there is frequently little advantage to minimizing the number of active elements. Therefore if multiple-amplifier realizations can provide better performance, such realizations may be preferable to single-amplifier filters. In this chapter we continue our discussion of the cascade realization method by developing several multiple-amplifier filter configurations.

6.1 THE STATE-VARIABLE (KHN) FILTER

The *state-variable filter* realization, also called the *KHN filter* from the names of its originators,[1] has extreme flexibility, good performance, and low sensitivities. These characteristics have made it the workhorse of commercially available active filters.

[1] W. J. Kerwin, L. P. Huelsman, and R. W. Newcomb, "State-Variable Synthesis for Insensitive Integrated Circuit Transfer Functions," *IEEE J. Solid-State Circuits*, vol. SC-2, September 1967, pp. 87–92. An earlier reference to this configuration was given in W. H. Schussler, *On the Representation of Transfer Functions and Networks on Analog Computers*, Westdeutscher Verlag, Cologne, 1961.

Theory of the State-Variable Filter

The name *state variable* is derived from the fact that state-variable methods of solving differential equations are used in the development of the realization. To see this, consider the inverting second-order band-pass transfer function

$$\frac{V_2(s)}{V_1(s)} = \frac{-|H|s}{s^2 + a_1 s + a_0} \tag{1}$$

In this equation we now introduce an arbitrary frequency-domain variable $X(s)$ and multiply both the numerator and denominator by $X(s)/s^2$ to obtain

$$\frac{V_2(s)}{V_1(s)} = \frac{-|H|\dfrac{X(s)}{s}}{X(s) + \dfrac{a_1 X(s)}{s} + \dfrac{a_0 X(s)}{s^2}} \tag{2}$$

If we separately equate the numerators and denominators of both members of (2), two equations result:

$$X(s) = V_1(s) - \frac{a_1 X(s)}{s} - \frac{a_0 X(s)}{s^2} \tag{3a}$$

$$V_2(s) = -|H|\frac{X(s)}{s} \tag{3b}$$

If we now take the inverse Laplace transform of both sides of the equations of (3), we get the following integral (time-domain) equations:

$$x(t) = v_1(t) - a_1 \int x(t)\,dt - a_0 \int \left[\int x(t)\,dt \right] dt \tag{4a}$$

$$v_2(t) = -|H| \int x(t)\,dt \tag{4b}$$

where $x(t) = \mathscr{L}^{-1}[X(s)]$. The quantities $x(t)$, $\int x(t)\,dt$, and $\int [\int x(t)\,dt]\,dt$ are called *state variables*, which accounts for the name of the filter. A block diagram for solving (4) is shown in Fig. 6.1-1.

Realization of the State-Variable Filter

The block diagram shown in Fig. 6.1-1 is readily converted to a filter circuit by using operational amplifiers to model the inverting integrators. The result is shown in Fig.

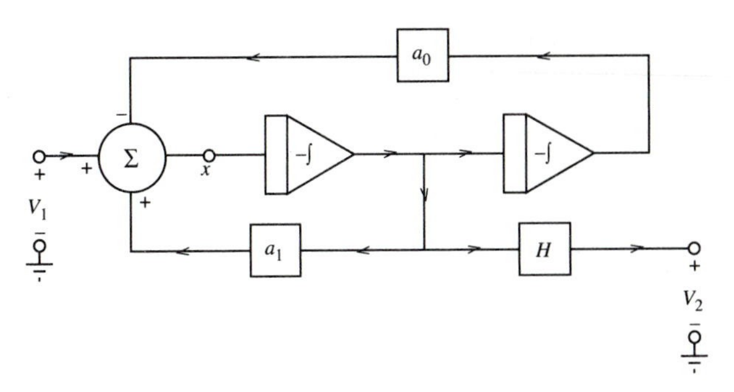

FIGURE 6.1-1
General state-variable filter configuration.

6.1-2. The band-pass transfer function for this circuit is developed by first observing that

$$V_{\mathrm{LP}}(s) = \frac{-V_{\mathrm{BP}}(s)}{sR_2C_2} \qquad V_{\mathrm{BP}}(s) = \frac{-V_{\mathrm{HP}}(s)}{sR_1C_1} \tag{5}$$

where $V_{\mathrm{LP}}(s)$, $V_{\mathrm{BP}}(s)$, and $V_{\mathrm{HP}}(s)$ are respectively the low-pass, band-pass, and high-pass outputs. In addition one can express $V_{\mathrm{HP}}(s)$ as

$$V_{\mathrm{HP}}(s) = -\frac{R_6}{R_5}V_{\mathrm{LP}}(s) + \frac{R_4}{R_3+R_4}\frac{R_5+R_6}{R_5}V_1(s) + \frac{R_3}{R_3+R_4}\frac{R_5+R_6}{R_5}V_{\mathrm{BP}}(s) \tag{6}$$

Using (5) to eliminate $V_{\mathrm{LP}}(s)$ and $V_{\mathrm{HP}}(s)$, we obtain

$$V_{\mathrm{BP}}(s) = -\frac{1}{sR_1C_1}\left[\frac{R_6}{R_5}\frac{V_{\mathrm{BP}}(s)}{sR_2C_2} + \frac{1+R_6/R_5}{1+R_3/R_4}V_1(s) + \frac{1+R_6/R_5}{1+R_4/R_3}V_{\mathrm{BP}}(s)\right] \tag{7}$$

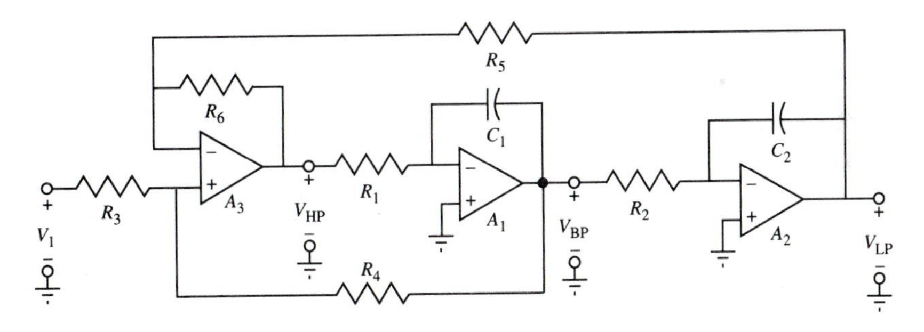

FIGURE 6.1-2
The state-variable filter.

Solving for $V_{BP}(s)/V_1(s)$ in the above and using (5), we obtain the band-pass, low-pass, and high-pass transfer functions

$$\frac{V_{BP}(s)}{V_1(s)} = \frac{-\left[\dfrac{1+R_6/R_5}{1+R_3/R_4}\dfrac{s}{R_1C_1}\right]}{D(s)} \tag{8a}$$

$$\frac{V_{LP}(s)}{V_1(s)} = \frac{\dfrac{1+R_6/R_5}{1+R_3/R_4}\dfrac{1}{R_1R_2C_1C_2}}{D(s)} \tag{8b}$$

$$\frac{V_{HP}(s)}{V_1(s)} = \frac{\dfrac{1+R_6/R_5}{1+R_3/R_4}s^2}{D(s)} \tag{8c}$$

$$D(s) = s^2 + \frac{s}{R_1C_1}\frac{1+R_6/R_5}{1+R_4/R_3} + \frac{R_6/R_5}{R_1R_2C_1C_2} \tag{8d}$$

Note that the low-pass and high-pass realizations are noninverting, whereas the band-pass one is inverting. The expressions for ω_n and Q are given as

$$\omega_n = \sqrt{\frac{R_6/R_5}{R_1R_2C_1C_2}} \tag{9a}$$

$$\frac{1}{Q} = \frac{1+R_6/R_5}{1+R_4/R_3}\sqrt{\frac{R_5R_2C_2}{R_6R_1C_1}} \tag{9b}$$

The term H_0 will be different for all three realizations. It is given as follows:

Low pass: $$H_0 = \frac{1+R_5/R_6}{1+R_3/R_4} \quad \text{[see (1) of Sec. 5.2]} \tag{10a}$$

Band pass: $$H_0 = -R_4/R_3 \quad \text{[see (1) of Sec. 5.3]} \tag{10b}$$

High pass: $$H_0 = \frac{1+R_6/R_5}{1+R_3/R_4} \quad \text{[see (13) of Sec. 5.3]} \tag{10c}$$

where, for the low-pass case, H_0 is the gain at direct current; for the band-pass case, H_0 is the gain at the resonant frequency; and for the high-pass case, H_0 is the gain at infinite frequency.

DESIGN 1. Just as was the case for the single-amplifier filters, several design procedures may be specified for the state-variable filter. A simple procedure is to choose $R_5 = R_6$, $R_1 = R_2 = R$, and $C_1 = C_2 = C$. Equation (9) becomes

$$\omega_n = \frac{1}{RC} \tag{11a}$$

$$\frac{1}{Q} = \frac{2}{1+R_4/R_3} \tag{11b}$$

For these choices the following design procedure (design 1) may be used:

1. Assume ω_n and Q are specified.
2. Choose convenient values for $C_1 = C_2 = C$ and $R_3 = R_5 = R_6$.
3. Calculate

$$R_1 = R_2 = \frac{1}{\omega_n C} \tag{12a}$$

$$R_4 = (2Q - 1)R_3 \tag{12b}$$

4. H_0 becomes

Low pass and high pass:
$$H_0 = \frac{2Q - 1}{Q} \tag{13a}$$

Band pass:
$$H_0 = 1 - 2Q \tag{13b}$$

For the frequency-normalized case in which $\omega_n = 1$ rad/s, this procedure reduces to a simple design in which all the resistors and all the capacitors are unity valued except R_4. This latter resistor is used to determine the Q using the relation $R_4 = 2Q - 1$.

Example 6.1-1 A state-variable (design 1) band-pass filter. It is desired to design a band-pass state-variable filter with $Q = 20$ and $f_n = 1$ kHz. Let us choose $C_1 = C_2 = 0.01$ μF and $R_3 = R_5 = R_6 = 10$ kΩ. Thus, from (12) we get $R_1 = R_2 = 15.9$ kΩ and $R_4 = 390$ kΩ. From (10) or (13), $H_0 = -39.0$. This same realization yields low-pass and high-pass transfer functions with $Q = 20$ and $f_n = 1$ kHz at the appropriate output terminals of Fig. 6.1-2. For these realizations, $H_0 = 1.95$.

One of the main characteristics of this design is that the reciprocals of the integrator time constants as specified by $\omega_1 = 1/R_1C_1$ and $\omega_2 = 1/R_2C_2$ are chosen to be equal to the undamped natural frequency ω_n. Thus, $\omega_1 = \omega_2 = \omega_n$.

DESIGN 2. As an example of a different design procedure we choose the case where the reciprocal of the time constants is greater than the undamped natural frequency. Specifically, for $\omega_1 = 1/R_1C_1$ and $\omega_2 = 1/R_2C_2$, we choose $\omega_1 = \omega_2 = \sqrt{10}\omega_n$. In a later section we shall illustrate some of the improvement in filter performance produced by this choice. A design procedure (design 2) follows:

1. Choose a convenient value for $C_1 = C_2 = C$.
2. Calculate

$$R = R_1 = R_2 = \frac{1}{\sqrt{10}\omega_n C} \tag{14a}$$

$$R_3 = R_6 = R \qquad \text{and} \qquad R_5 = 10R \tag{14b}$$

$$R_4 = R(1.1\sqrt{10}Q - 1) \tag{14c}$$

3. The resulting value of H_0 is

Low pass:
$$H_0 = \frac{11Q - \sqrt{10}}{Q} \tag{15a}$$

Band pass:
$$H_0 = 1 - 1.1\sqrt{10}Q \tag{15b}$$

High pass:
$$H_0 = \frac{1.1Q - 1/\sqrt{10}}{Q} \tag{15c}$$

For the frequency-normalized case in which $\omega_n = 1$ rad/s, this procedure reduces to the following:

$$C_1 = C_2 = 1 \qquad R_1 = R_2 = R_3 = R_6 = 1/\sqrt{10} \tag{16a}$$
$$R_5 = \sqrt{10} \qquad R_4 = 1.1Q - 1/\sqrt{10} \tag{16b}$$

Example 6.1-2 A state-variable (design 2) band-pass filter. A state-variable filter is to be designed (design 2) to meet the same specifications used in Example 6.1-1 ($Q = 20$ and $f_n = 1$ kHz). If we choose $C_1 = C_2 = 0.005\ \mu$F, from (14) we obtain $R_1 = R_2 = R_3 = R_6 = 10.1$ kΩ and $R_4 = 690$ kΩ. From (15) we find $H_0 = 68.57$. The low-pass output will have $H_0 = 10.84$ and the high-pass output will have $H_0 = 1.084$.

Sensitivity

One of the reasons for the popularity of the state-variable filter is the low sensitivities that characterize its performance. Using the definitions of Sec. 3.5 and the relations of (9), for the general case, we find

$$S^Q_{R_1,C_1} = -S^Q_{R_2,C_2} = \tfrac{1}{2} \tag{17a}$$

$$S^Q_{R_3} = -S^Q_{R_4} = \frac{-1}{1 + R_3/R_4} \tag{17b}$$

$$S^Q_{R_5} = -S^Q_{R_6} = -\frac{Q}{2}\frac{R_5 - R_6}{1 + R_4/R_3}\sqrt{\frac{R_2C_2}{R_5R_6R_1C_1}} \tag{17c}$$

$$S^{\omega_n}_{R_1,R_2,R_5,C_1,C_2} = -S^{\omega_n}_{R_6} = -\tfrac{1}{2} \qquad S^{\omega_n}_{R_3,R_4} = 0 \tag{17d}$$

These relations may be simplified for the two designs defined above. For design 1, (17*b*) and (17*c*) become

$$S^Q_{R_3} = -S^Q_{R_4} = \frac{1 - 2Q}{2Q} \approx -1 \tag{18a}$$

$$S^Q_{R_5} = -S^Q_{R_6} = 0 \tag{18b}$$

The approximation is for the high-Q case. Similarly, for design 2 we find

$$S^Q_{R_3} = -S^Q_{R_4} = \frac{1 - 1.1\sqrt{10}Q}{1.1\sqrt{10}Q} \approx -1 \tag{19a}$$

$$S^Q_{R_5} = -S^Q_{R_6} = 0.4091 \tag{19b}$$

Note that the sensitivities are all less in magnitude than 1, and most of them have a magnitude of $\frac{1}{2}$ or less. The Q sensitivities to the gains of the operational amplifiers are even smaller, being of the order of Q/A_0, where A_0 is the dc open-loop operational amplifier gain. As a result of its low sensitivities, the state-variable filter has been successfully used to realize functions with Q's in the hundreds.

The excellent sensitivity properties of this filter, as reflected by the low values of the Q and ω_n sensitivities, are further verified if the function sensitivities are evaluated. In general, for high-Q realizations, the magnitude sensitivities for frequencies in the center of the resonant peak are very low. The sensitivities at the band edges, however, may be larger. An example follows.

Example 6.1-3 Function sensitivities of a state-variable filter. The voltage transfer function given for a state-variable band-pass filter in (8) can be evaluated to obtain the function sensitivities for a specific design of the filter. As an example, consider the use of design 1 to realize a frequency- and impedance-normalized function in which $\omega_n = 1$ rad/s and $Q = 10$. The element values are $C_1 = C_2 = 1$ F, $R_1 = R_2 = R_3 = R_5 = R_6 = 1\ \Omega$, and $R_4 = 19\ \Omega$. The resulting voltage transfer function is

$$\frac{V_{BP}(s)}{V_1(s)} = \frac{-1.9s}{s^2 + 0.1s + 1}$$

Some of the results obtained from determining the function sensitivities and evaluating them over the range $0.75 \le \omega < 1.25$ rad/s are shown in Fig. 6.1-3. The figure shows plots of the magnitude sensitivity for R_3, R_4, R_6, and C_2 as well as a plot of the magnitude of the network function. The plots for the magnitude sensitivities of R_1, R_2, R_5, and C_1 are nearly identical with the one for C_2.

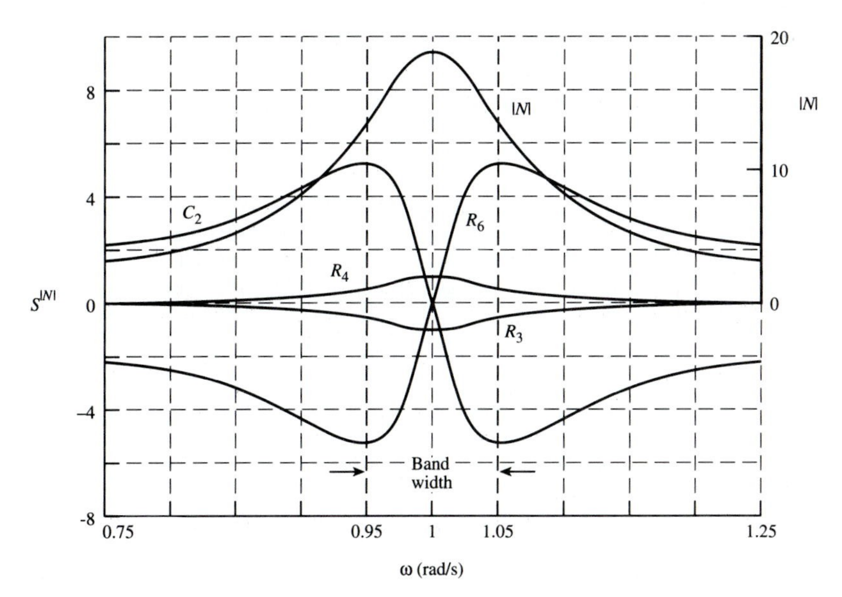

FIGURE 6.1-3
Function sensitivities for a state-variable filter.

The function sensitivity curves shown in Fig. 6.1-3 merit further discussion. Note that the sensitivities for R_3 and R_4 are very low and have a maximum magnitude of unity at the resonant frequency $\omega = 1$ rad/s. The sensitivities of the other filter elements, however, are quite different. They have a value of zero at the resonant frequency but have relatively large magnitudes at the band-edge frequencies. We shall see that similar large band-edge sensitivities characterize many of the multiple-amplifier filter configurations. From the curves of Fig. 6.1-3 we see that many possibilities exist for sensitivity cancellation. For example, $S_{R_3}^{|N(j\omega)|} = -S_{R_4}^{|N(j\omega)|}$ and $S_{R_5}^{|N(j\omega)|} = -S_{R_6}^{|N(j\omega)|}$; thus changes in the magnitude of $N(j\omega)$ produced by equal normalized changes in these resistors are canceled, leaving a net change of zero. As another example of a possible sensitivity cancellation, since the sensitivity curves for R_1, R_2, C_1, and C_2 are nearly the same, the effects of changes in the resistor values can be canceled by opposite changes in the capacitor values for these elements.

Example 6.1-4 Sensitivity cancellation in a state-variable filter. A state-variable band-pass filter is designed using the specifications of Example 6.1-3. The function sensitivities are given in Fig. 6.1-3. The network elements that are to be used in the realization are found to have the following values:

$$R_1 = R_2 = R_3 = R_5 = R_6 = 1.01\ \Omega \qquad R_4 = 19.19\ \Omega \qquad C_1 = C_2 = 1.01\ \text{F}$$

These values represent 1 percent changes from the values given in Example 6.1-3. From Fig. 6.1-3, at the lower band edge we find

$$\begin{aligned}\frac{\Delta|N|}{|N|} &= S_{R_1}^{|N|}\frac{\Delta R_1}{R_1} + S_{R_2}^{|N|}\frac{\Delta R_2}{R_2} + S_{C_1}^{|N|}\frac{\Delta C_1}{C_1} + S_{C_2}^{|N|}\frac{\Delta C_2}{C_2} \\ &\quad + S_{R_3}^{|N|}\frac{\Delta R_3}{R_3} + S_{R_4}^{|N|}\frac{\Delta R_4}{R_4} + S_{R_5}^{|N|}\frac{\Delta R_5}{R_5} + S_{R_6}^{|N|}\frac{\Delta R_6}{R_6} \\ &= (5 \times 0.01) + (5 \times 0.01) + (5 \times 0.01) + (5 \times 0.01) \\ &\quad + (-0.4 \times 0.01) + (0.4 \times 0.01) + (5 \times 0.01) + (-5 \times 0.01) = 0.2\end{aligned}$$

The total effect of the 1 percent changes in the element values is to produce a 20 percent change in the magnitude of the network function at the lower band-edge frequency. A similar calculation at the upper band edge shows a -20 percent change in the magnitude there. Effectively, the magnitude curve has been shifted to a lower resonant frequency. As a check on this, note that the ω_n (resonant frequency) sensitivities give

$$\begin{aligned}\frac{\Delta\omega_n}{\omega_n} &= S_{R_1}^{\omega_n}\frac{\Delta R_1}{R_1} + S_{R_2}^{\omega_n}\frac{\Delta R_2}{R_2} + S_{C_1}^{\omega_n}\frac{\Delta C_1}{C_1} + S_{C_2}^{\omega_n}\frac{\Delta C_2}{C_2} \\ &\quad + S_{R_3}^{\omega_n}\frac{\Delta R_3}{R_3} + S_{R_4}^{\omega_n}\frac{\Delta R_4}{R_4} + S_{R_5}^{\omega_n}\frac{\Delta R_5}{R_5} + S_{R_6}^{\omega_n}\frac{\Delta R_6}{R_6} \\ &= (-0.5 \times 0.01) + (-0.5 \times 0.01) + (-0.5 \times 0.01) + (-0.5 \times 0.01) \\ &\quad + (0 \times 0.01) + (0 \times 0.01) + (-0.5 \times 0.01) + (0.5 \times 0.01) = -0.02\end{aligned}$$

In this case, the total effect of the 1 percent changes in the element values is to produce a -2 percent change in the resonant frequency. These results are readily verified by plotting

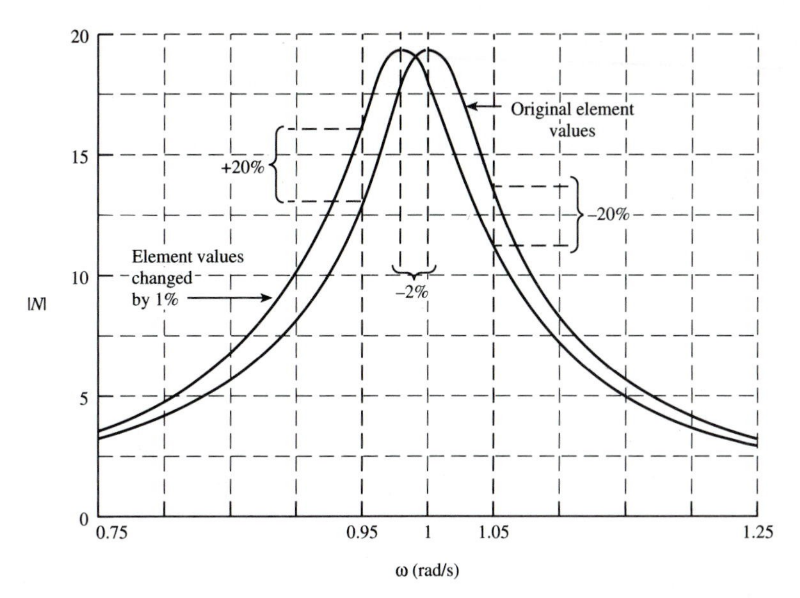

FIGURE 6.1-4
Effect of element value changes in a state-variable filter.

the magnitude characteristic both for the original element values given in Example 6.1-3 and for the modified ones given in this example. This is done in Fig. 6.1-4. The ± 20 percent changes in $|N(j\omega)|$ and the -2 percent change in the resonant frequency are readily apparent. Note that a similar analysis using the resistance values given above but changing the capacitor values to 0.99 (a -1 percent change) will give a magnitude plot virtually indistinguishable from the one shown in Fig. 6.1-4 for the original element values due to sensitivity cancellation.

The state-variable filter described in this section simultaneously realizes low-pass, band-pass, and high-pass functions. This property can be used to extend the application of the circuit to the realization of general biquadratic functions. This will be done in Sec. 6.3. A discussion of commercial implementations of the filter will be given in Sec. 6.4.

6.2 OTHER MULTIPLE-AMPLIFIER FILTERS

In the preceding section we introduced the state-variable multiple-amplifier filter configuration. Many other multiple-amplifier circuits have been described in the literature. In this section we present a discussion of two of the most well-known of these, the Tow-Thomas (or resonator) filter and the closely related Åkerberg-Mossberg filter.

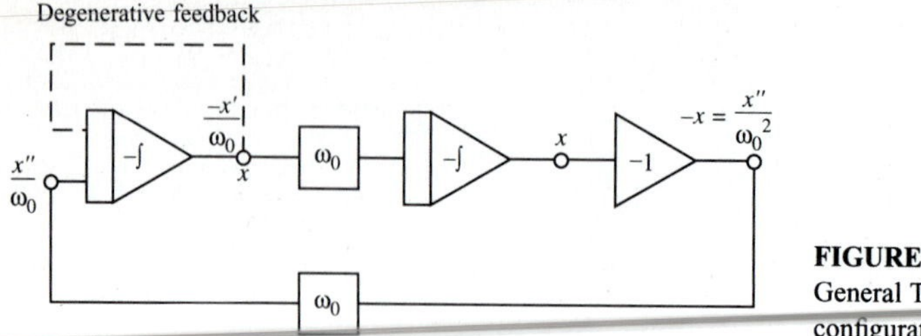

FIGURE 6.2-1
General Tow-Thomas filter configuration.

The Tow-Thomas Filter

The basic approach used in developing these filters is to implement an *RC* oscillator and apply degenerative (negative) feedback to move the natural frequencies off the $j\omega$ axis and into the left-half plane. The development leads to the block diagram flowchart shown in Fig. 6.2-1. Without the degenerate feedback this circuit solves the differential equation

$$x''(t) + \omega_0^2 x(t) = 0 \tag{1}$$

which has the solution

$$x(t) = A \sin \omega_0 t \tag{2}$$

Implementing Fig. 6.2-1 with operational amplifiers results in the circuit shown in Fig. 6.2-2. This is called the Tow-Thomas filter.[1] The resistor R_1 serves as the degenerate feedback path. The input signal V_1 is injected into the circuit through R_4. To analyze this filter, we begin by finding $V_{\mathrm{BP}}(s)/V_1(s)$. The signal $V_{\mathrm{BP}}(s)$ can be expressed as

$$V_{\mathrm{BP}}(s) = \frac{-1/R_4C_1}{s + 1/R_1C_1} V_1(s) - \frac{1/R_3C_1}{s + 1/R_1C_1} V_x(s) \tag{3}$$

However, we see that

$$V_x(s) = -V_{\mathrm{LP}}(s) = \frac{V_{\mathrm{BP}}(s)}{sR_2C_2} \tag{4}$$

so that substituting (4) into (3) results in the transfer functions of the Tow-Thomas active filter:

$$\frac{V_{\mathrm{BP}}(s)}{V_1(s)} = \frac{-s/R_4C_1}{s^2 + s/R_1C_1 + 1/R_2R_3C_1C_2} \tag{5a}$$

$$\frac{V_{\mathrm{LP}}(s)}{V_1(s)} = \frac{1/R_2R_4C_1C_2}{s^2 + s/R_1C_1 + 1/R_2R_3C_1C_2} \tag{5b}$$

[1] This filter is discussed in J. Tow, "Design Formulas for Active *RC* Filters Using Operational-Amplifier Biquad," *Electron. Lett.*, vol. 5, July 24, 1969, pp. 339–341, and L. C. Thomas, "The Biquad: Part I—Some Practical Design Considerations," *IEEE Trans. Circuit Theory*, vol. CT-18, 1971, pp. 350–357.

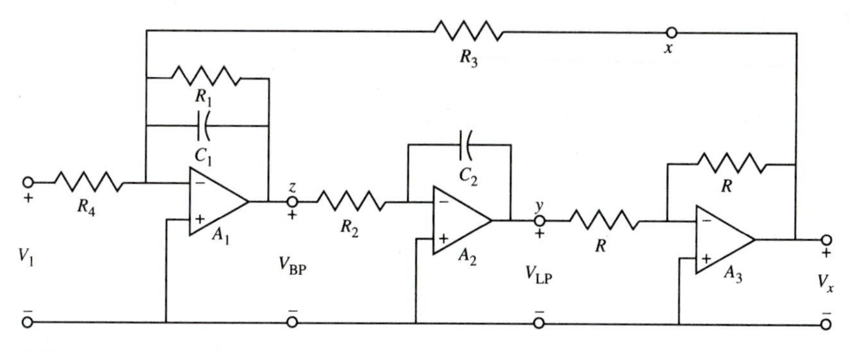

FIGURE 6.2-2
The Tow-Thomas filter.

If the low-pass output is taken from the output of the inverter, then an inverting low-pass realization may also be obtained. The ideal integrator between z and y may be interchanged with the unity gain inverter between y and x to provide a noninverting band-pass transfer function at y. The circuit cannot be directly used to provide a high-pass transfer function.

Design

If we equate the denominator of (5) to that of the standard second-order band-pass function of (1) of Sec. 5.3, we obtain

$$\omega_n = \frac{1}{\sqrt{R_2 R_3 C_1 C_2}} \tag{6a}$$

$$\frac{1}{Q} = \frac{1}{R_1}\sqrt{\frac{R_2 R_3 C_2}{C_1}} \tag{6b}$$

The expressions for H_0 are

Low pass:

$$H_0 = \frac{R_3}{R_4} \tag{7a}$$

Band pass:

$$|H_0| = \frac{R_1}{R_4} \tag{7b}$$

A design procedure can now be developed by assuming that $R_2 = R_3 = R$ and $C_1 = C_2 = C$ and proceeding as follows:

1. Assume ω_n, Q, and H_0 are specified.
2. Let $R_2 = R_3 = R$ and $C_1 = C_2 = C$.
3. Select either R or C and solve for the other using

$$\omega_n = \frac{1}{RC} \tag{8a}$$

4. Calculate

$$R_1 = QR \tag{8b}$$

Low pass:

$$R_4 = \frac{R}{H_0} \tag{8c}$$

Band pass:

$$R_4 = \frac{R_1}{|H_0|} \tag{8d}$$

For a frequency- and impedance-normalized realization, in (8) we set $R = C = 1$ to obtain

$$\omega_n = 1 \qquad Q = R_1 \qquad H_0(\text{LP}) = \frac{1}{R_4} \qquad H_0(\text{BP}) = -\frac{R_1}{R_4} \tag{9}$$

Example 6.2-1 A Tow-Thomas filter. It is desired to design a low-pass and a band-pass second-order Tow-Thomas active filter with $Q = 20, f_n = 1$ kHz, and $|H_0| = 1$. Let us choose $C_1 = C_2 = C = 0.01$ μF. Thus $R = 15.915$ kΩ and $R_1 = 318.31$ kΩ. For the low-pass realization $R_4 = 15.915$ kΩ, and for the band-pass realization $R_4 = 318.31$ kΩ.

The Q and ω_n sensitivities of the Tow-Thomas filter are easily determined from (6) using the relations of Sec. 3.5. We obtain

$$S_{R_1}^{Q} = 1 \qquad S_{R_2,R_3,C_2}^{Q} = -S_{C_1}^{Q} = -\tfrac{1}{2} \qquad S_{R_4}^{Q} = 0 \tag{10a}$$

$$S_{R_2,R_3,C_1,C_2}^{\omega_n} = -\tfrac{1}{2} \qquad S_{R_1,R_4}^{\omega_n} = 0 \tag{10b}$$

These relations illustrate the excellent sensitivity properties of the filter.

Similarly, the function sensitivities for the Tow-Thomas filter have low values. An example follows.

Example 6.2-2 Function sensitivities for a Tow-Thomas filter. The voltage transfer function given for a Tow-Thomas band-pass filter in (5) can be evaluated to obtain the function sensitivities for a specific design of the filter. As an example, consider the realization of a frequency- and impedance-normalized function in which $\omega_n = 1$ rad/s, $Q = 10$, and $|H_0| = 19$. The element values are $C_1 = C_2 = 1$ F, $R_2 = R_3 = 1$ Ω, $R_4 = 1/1.9 = 0.5263$ Ω, and $R_1 = 10$ Ω. The resulting voltage transfer function is

$$\frac{V_{\text{BP}}(s)}{V_1(s)} = \frac{-1.9s}{s^2 + 0.1s + 1}$$

Some of the results obtained from determining the function sensitivities and evaluating them over a range $0 \le \omega \le 2$ rad/s are shown in Fig. 6.2-3. The R_2 and R_3 sensitivity curves are identical with the one for C_2. Note that the curve for R_4 is a straight line, independent of frequency. This occurs since this resistor determines the multiplicative constant. Observations similar to those made for the state-variable filter in Examples 6.1-3 and 6.1-4 can be made using these curves. The details are left as an exercise (see the Problems).

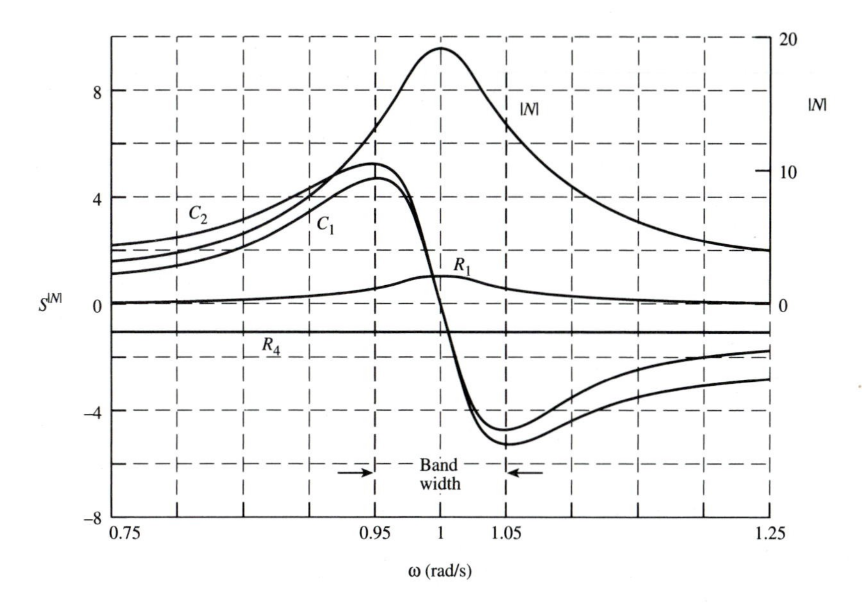

FIGURE 6.2-3
Function sensitivities for a Tow-Thomas filter.

The Åkerberg-Mossberg Filter

The Tow-Thomas filter described above may be considered as a cascade of an inverting damped integrator and a *non*inverting integrator. The noninverting integrator is realized by an inverting integrator and an inverter as shown in Fig. 6.2-4(*a*). An interesting and useful variation of the Tow-Thomas filter occurs if this circuit is replaced by an alternate realization in which one of the operational amplifiers is used as a feedback element rather than a feedforward one. The configuration is shown in Fig. 6.2-4(*b*). It is easily analyzed using the method given in Summary 5.4-1. The details are left as a problem.

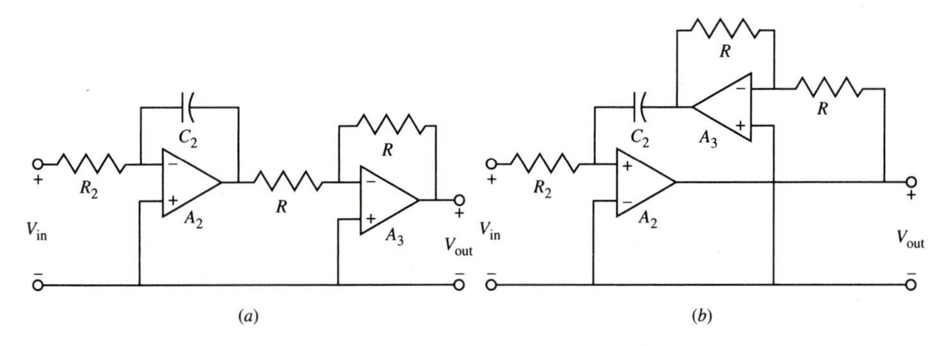

FIGURE 6.2-4
Noninverting integrators.

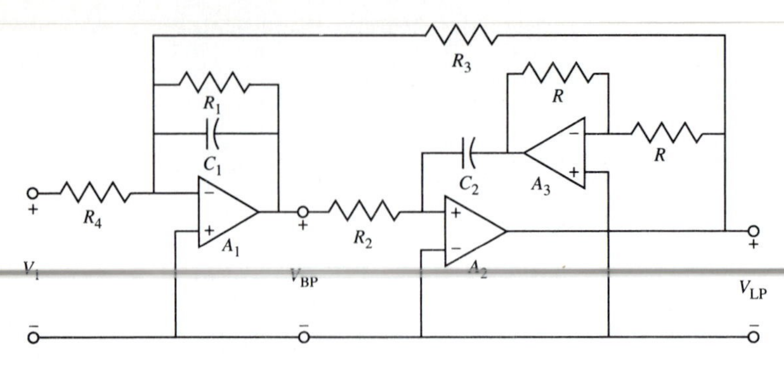

FIGURE 6.2-5
The Åkerberg-Mossberg filter.

For both of the circuits in Fig. 6.2-4 the voltage transfer function is

$$\frac{V_{\text{out}}(s)}{V_{\text{in}}(s)} = \frac{1}{sR_2C_2} \tag{11}$$

This configuration provides better phase compensation for the filter when high-frequency effects, such as the gain bandwidth of the operational amplifier, are considered. The complete form of the Åkerberg-Mossberg filter is shown in Fig. 6.2-5. Its voltage transfer function, design equations, and sensitivities are the same as those given for the Tow-Thomas filter in Eqs. (5) to (10).

6.3 BIQUADRATIC MULTIPLE-AMPLIFIER FILTERS

In the preceding sections of this chapter we introduced multiple-amplifier filters. A common characteristic of these filters is their capability of simultaneously realizing two or more filter functions. For example, the state-variable filter provides low-pass, band-pass, and high-pass outputs; and the Tow-Thomas and Åkerberg-Mossberg filters provide low-pass and band-pass ones. In this section we show how the basic configurations of these filters can be extended to realize general biquadratic network functions.

General State-Variable Biquadratic Filter

The general biquadratic network function can be expressed in the following forms:

$$\frac{V_2(s)}{V_1(s)} = H\frac{s^2 + b_1 s + b_0}{s^2 + a_1 s + a_0} = H\frac{s^2 + (\omega_z/Q_z)s + \omega_z^2}{s^2 + (\omega_p/Q_p) + \omega_p^2} \tag{1}$$

The second form is sometimes referred to as the *standard form* of a biquadratic function. To realize such a function, the low-pass, band-pass, and high-pass outputs of the state-variable filter can be summed using a fourth operational amplifier and some additional

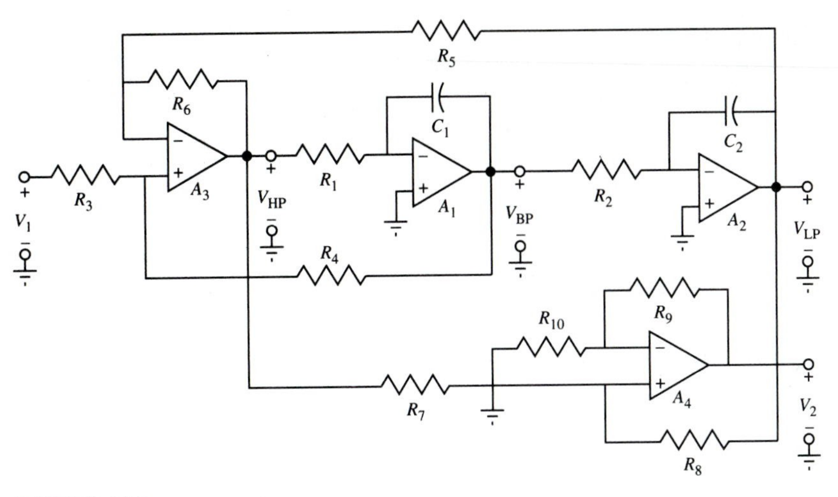

FIGURE 6.3-1
A state-variable filter for biquadratic functions.

resistors. To realize (1) with the coefficients H, b_0, and b_1 all greater than zero, we may modify the basic state-variable configuration shown in Fig. 6.1-2 to the form shown in Fig. 6.3-1 in which an additional operational amplifier designated as A_4 has been added. Since the low-pass and high-pass outputs of the filter are already noninverting, they are summed at the positive input of A_4. The inverting output of the band-pass function is changed to a noninverting one by summing it at the negative input of A_4. The output $V_2(s)$ may be found as a superposition of the three state-variable outputs. Thus we obtain

$$V_2(s) = \frac{1 + R_9/R_{10}}{1 + R_7/R_8} V_{\text{HP}}(s) - \frac{R_9}{R_{10}} V_{\text{BP}}(s) + \frac{1 + R_9/R_{10}}{1 + R_8/R_7} V_{\text{LP}}(s) \tag{2}$$

From (8) of Sec. 6.1 and (2) we obtain

$$\frac{V_2(s)}{V_1(s)} = \frac{1 + R_9/R_{10}}{1 + R_7/R_8} \frac{1 + R_6/R_5}{1 + R_3/R_4} \frac{s^2 + \dfrac{s}{R_1 C_1} \dfrac{1 + R_7/R_8}{1 + R_{10}/R_9} + \dfrac{R_7/R_8}{R_1 R_2 C_1 C_2}}{s^2 + \dfrac{s}{R_1 C_1} \dfrac{1 + R_6/R_5}{1 + R_4 R_3} + \dfrac{R_6/R_5}{R_1 R_2 C_1 C_2}} \tag{3}$$

A useful means of simplifying this equation is to let $R_1 = R_2 = R_3 = R_5 = R_8 = R_9 = 1\ \Omega$ and $C_1 = C_2 = 1$ F. For these values we obtain

$$\frac{V_2(s)}{V_1(s)} = \frac{R_4}{R_{10}} \frac{(1 + R_6)(1 + R_{10})}{(1 + R_4)(1 + R_7)} \frac{s^2 + s\dfrac{1 + R_7}{1 + R_{10}} + R_7}{s^2 + s\dfrac{1 + R_6}{1 + R_4} + R_6} \tag{4}$$

Equating (1) to (4) results in

$$\omega_z = \sqrt{R_7} \qquad Q_z = \sqrt{R_7}\,\frac{1+R_{10}}{1+R_7} \tag{5a}$$

$$\omega_p = \sqrt{R_6} \qquad Q_p = \sqrt{R_6}\,\frac{1+R_4}{1+R_6} \tag{5b}$$

From these equations we see that the state-variable biquad may be designed by using R_6 and R_7 to control ω_p and ω_z and R_4 and R_{10} to control Q_p and Q_z.

State-Variable Filter with $j\omega$-Axis Zeros

A modification of the circuit shown in Fig. 6.3-1 may be used for the frequently encountered case where the zeros of the network function are on the $j\omega$ axis. This occurs in the inverse-Chebyshev and elliptic approximations and requires $b_1 = 0$ and $Q_z = \infty$ in (1). To accomplish this, the left end of the resistor R_{10} in Fig. 6.3-1 is removed from the $V_{BP}(s)$ output and instead connected to ground. The resulting circuit is shown in Fig. 6.3-2. For this connection (2) becomes

$$V_2(s) = \frac{1+R_9/R_{10}}{1+R_7/R_8}\,V_{HP}(s) + \frac{1+R_9/R_{10}}{1+R_8/R_7}\,V_{LP}(s) \tag{6}$$

From (8) of Sec. 6.1 and (6) we obtain

$$\frac{V_2(s)}{V_1(s)} = \frac{1+R_9/R_{10}}{1+R_7/R_8}\,\frac{1+R_6/R_5}{1+R_3/R_4}\,\frac{s^2 + \dfrac{R_7/R_8}{R_1R_2C_1C_2}}{s^2 + \dfrac{s}{R_1C_1}\dfrac{1+R_6/R_5}{1+R_4/R_3} + \dfrac{R_6/R_5}{R_1R_2C_1C_2}} \tag{7}$$

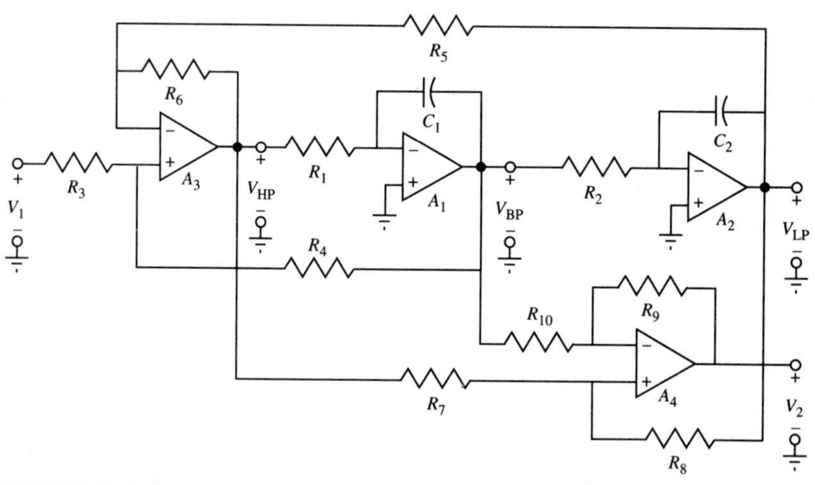

FIGURE 6.3-2
A state-variable filter for $j\omega$-axis zeros.

To simplify these equations, we let $R_1 = R_2 = R_3 = R_5 = R_8 = R_9 = 1\ \Omega$ and $C_1 = C_2 = 1$ F. For these values

$$\frac{V_2(s)}{V_1(s)} = \frac{R_4}{R_{10}} \frac{(1+R_6)(1+R_{10})}{(1+R_4)(1+R_7)} \frac{s^2 + R_7}{s^2 + s\dfrac{1+R_6}{1+R_4} + R_6} \tag{8}$$

Equating (1) to (8) results in

$$\omega_z = \sqrt{R_7} \qquad Q_z = \infty \tag{9a}$$

$$\omega_p = \sqrt{R_6} \qquad Q_p = \sqrt{R_6}\frac{1+R_4}{1+R_6} \tag{9b}$$

From these equations we see that the $j\omega$-axis-zeros version of the state-variable biquadratic filter may be designed by using R_6 and R_7 to control ω_p and ω_z and R_4 to control Q_p. The term R_{10} may be chosen so as to provide additional gain in the realization.

Example 6.3-1 State-variable filter with $j\omega$-axis zeros. As an example of the use of the state-variable filter to realize $j\omega$-axis zeros, consider the realization of a normalized second-order elliptic function with 1-dB ripple in the passband, $0 \le \omega \le 1$ rad/s, and a minimum of 17 dB attenuation in the stopband, $\omega > 2$ rad/s. From Table 2.4-1a, the network function is

$$\frac{V_2(s)}{V_1(s)} = H\frac{s^2 + 7.464102}{s^2 + 0.998942s + 1.170077}$$

From this we find

$$\omega_z = 2.732051 \qquad \omega_p = 1.081701 \qquad Q_p = 1.082847$$

From which, using (9), we obtain

$$R_4 = 1.1724\ \Omega \qquad R_6 = 1.1701\ \Omega \qquad R_7 = 7.4641\ \Omega$$

From (8) we find $H = 0.13836(1 + R_{10})/R_{10}$. As an example of using R_{10} to adjust the gain, let us assume that the gain at direct current is to be 2.0. From the defining equation for $V_2(s)/V_1(s)$ we find

$$2 = 0.13836\frac{1+R_{10}}{R_{10}}\frac{7.464102}{1.170077}$$

Solving, we find $R_{10} = 0.7900\ \Omega$.

State-Variable Filter with Right-Half-Plane Zeros

The state-variable configuration shown in Fig. 6.3-1 can be modified to realize zeros in the right half of the complex-frequency plane. In this case, in (1), the coefficient b_1 and the quality factor Q_z are both negative. The modification consists of moving the right end of the resistor R_{10} from the inverting summing terminal of operational amplifier A_4 to the noninverting one. Another resistor R_{11} must then be connected from the inverting

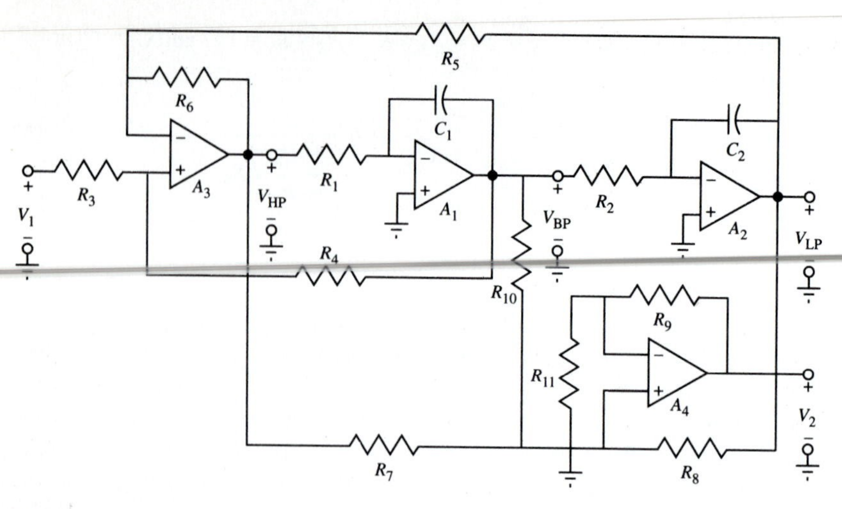

FIGURE 6.3-3
A state-variable filter for right-half-plane zeros.

terminal to ground. The resulting circuit is shown in Fig. 6.3-3. For this configuration (2) becomes

$$V_2(s) = \frac{1 + R_9/R_{11}}{1 + R_7/R_8||R_{10}} V_{\text{HP}}(s) + \frac{1 + R_9/R_{11}}{1 + R_{10}/R_7||R_8} V_{\text{BP}}(s) + \frac{1 + R_9/R_{11}}{1 + R_8/R_7||R_{10}} V_{\text{LP}}(s) \quad (10)$$

where the symbol $R_i||R_j$ stands for *the parallel connection of R_i and R_j*. From (8) of Sec. 6.1 and (10) we obtain

$$\frac{V_2(s)}{V_1(s)} = \frac{(1 + R_9/R_{11})(1 + R_6/R_5)}{1 + R_3/R_4} \times \frac{\dfrac{s^2}{1 + R_7/R_8||R_{10}} - \dfrac{s}{R_1C_1}\dfrac{1}{1 + R_{10}/R_7||R_8} + \dfrac{1}{1 + R_8/R_7||R_{10}}\dfrac{1}{R_1R_2C_1C_2}}{s^2 + \dfrac{s}{R_1C_1}\dfrac{1 + R_6/R_5}{1 + R_4/R_3} + \dfrac{R_6/R_5}{R_1R_2C_1C_2}} \quad (11)$$

One approach to simplifying this equation is to let $R_1 = R_2 = R_3 = R_5 = R_9 = R_{10} = R_{11} = 1\ \Omega$ and $C_1 = C_2 = 1$ F. For these values we obtain

$$\frac{V_2(s)}{V_1(s)} = \frac{2R_8(1 + R_6)}{\left(1 + \dfrac{1}{R_4}\right)(R_8 + R_7R_8 + R_7)} \frac{s^2 - R_7 s + R_7/R_8}{s^2 + s\dfrac{1 + R_6}{1 + R_4} + R_6} \quad (12)$$

Equating (1) to (12) results in

$$\omega_z = \sqrt{\frac{R_7}{R_8}} \qquad Q_z = -\sqrt{\frac{1}{R_7 R_8}} \tag{13a}$$

$$\omega_p = \sqrt{R_6} \qquad Q_p = \sqrt{R_6}\,\frac{1+R_4}{1+R_6} \tag{13b}$$

An example of the use of these equations follows.

Example 6.3-2 All-pass state-variable filter. As an example of the use of the state-variable filter to realize a biquadratic function with zeros in the right half of the complex-frequency plane, consider the realization of an all-pass function with a Thomson phase characteristic normalized for a delay of 2 s. For this we have

$$\frac{V_2(s)}{V_1(s)} = H\frac{s^2 - 3s + 3}{s^2 + 3s + 3}$$

From this equation, using (1), we find

$$\omega_z = \sqrt{3} \qquad Q_z = -\frac{1}{\sqrt{3}} \qquad \omega_p = \sqrt{3} \qquad Q_p = \frac{1}{\sqrt{3}}$$

From the relations of (13) we obtain

$$R_4 = \tfrac{1}{3} \qquad R_6 = 3 \qquad R_7 = 3 \qquad R_8 = 1\ \Omega$$

Tow-Thomas Biquadratic Filter

The Tow-Thomas filter described in Sec. 6.2 can also be modified to realize biquadratic filter functions. To do this, we first rotate the circuit until the output of operational amplifier A_1 is at the far right. An additional capacitor C_3 is added, and the input signal is then applied simultaneously to each amplifier as shown in Fig. 6.3-4. The transfer

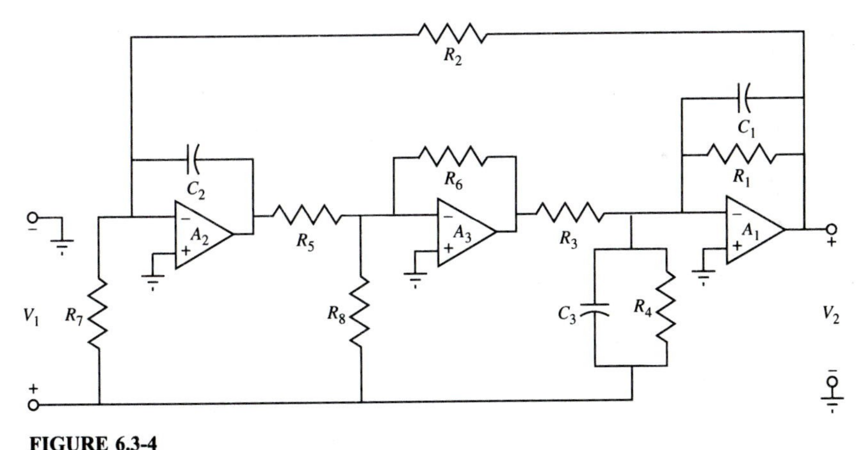

FIGURE 6.3-4
A Tow-Thomas filter for biquadratic functions.

function of this circuit can be shown to be

$$\frac{V_2(s)}{V_1(s)} = \frac{-C_3}{C_1} \frac{s^2 + s\left(\dfrac{1}{R_4} - \dfrac{R_6}{R_8R_3}\right)\dfrac{1}{C_3} + \dfrac{R_6}{R_3R_5R_7C_2C_3}}{s^2 + \dfrac{s}{R_1C_1} + \dfrac{R_6}{R_2R_3R_5C_1C_2}} \tag{14}$$

Equating (1) and (14), we obtain

$$\omega_z = \sqrt{\frac{R_6}{R_3R_5R_7C_2C_3}} \qquad \frac{1}{Q_z} = \left(\frac{1}{R_4} - \frac{R_6}{R_3R_8}\right)\sqrt{\frac{R_3R_5R_7C_2}{R_6C_3}} \tag{15a}$$

$$\omega_p = \sqrt{\frac{R_6}{R_2R_3R_5C_1C_2}} \qquad \frac{1}{Q_p} = \frac{1}{R_1}\sqrt{\frac{R_2R_3R_5C_2}{R_6C_1}} \tag{15b}$$

Simplified normalized design procedures may be developed to permit this circuit to realize a high-pass function (which was not available in the basic configuration shown in Fig. 6.2-2) or any of the usual biquadratic functions. The details follow.

Summary 6.3-1 Biquadratic realizations using the Tow-Thomas filter

High-pass realization. For this case we choose $R_4 = R_7 = R_8 = \infty$ (open circuits), $R_3 = R_5 = R_6 = 1\ \Omega$, and $C_1 = C_2 = 1$ F. From (14), the resulting function is

$$\frac{V_2(s)}{V_1(s)} = -C_3 \frac{s^2}{s^2 + s/R_1 + 1/R_2} \tag{16}$$

The design equations are

$$\omega_p = \frac{1}{\sqrt{R_2}} \qquad Q_p = \frac{R_1}{\sqrt{R_2}} \qquad |H| = C_3 \tag{17}$$

Biquadratic realization—left-half-plane zeros. For this case we choose $R_8 = \infty$, $R_3 = R_5 = R_6 = 1\ \Omega$, and $C_1 = C_2 = C_3 = 1$ F. From (14), the resulting function is

$$\frac{V_2(s)}{V_1(s)} = -\frac{s^2 + s/R_4 + 1/R_7}{s^2 + s/R_1 + 1/R_2} \tag{18}$$

The design equations are

$$\omega_z = \frac{1}{\sqrt{R_7}} \qquad Q_z = \frac{R_4}{\sqrt{R_7}} \qquad \omega_p = \frac{1}{\sqrt{R_2}} \qquad Q_p = \frac{R_1}{\sqrt{R_2}} \qquad |H| = 1 \tag{19}$$

Biquadratic realization—right-half-plane zeros. For this case we choose $R_4 = \infty$, $R_3 = R_5 = R_6 = 1\ \Omega$, and $C_1 = C_2 = C_3 = 1$ F. From (14), the resulting function is

$$\frac{V_2(s)}{V_1(s)} = -\frac{s^2 - s/R_8 + 1/R_7}{s^2 + s/R_1 + 1/R_2} \tag{20}$$

The design equations are

$$\omega_z = \frac{1}{\sqrt{R_7}} \qquad Q_z = -\frac{R_4}{\sqrt{R_7}} \qquad \omega_p = \frac{1}{\sqrt{R_2}} \qquad Q_p = \frac{R_1}{\sqrt{R_2}} \qquad |H| = 1 \tag{21}$$

The jω-axis-zeros realization. For this case we choose $R_4 = R_8 = \infty$, $R_3 = R_5 = R_6 = 1\ \Omega$, and $C_1 = C_2 = C_3 = 1$ F. From (14), the resulting function is

$$\frac{V_2(s)}{V_1(s)} = -\frac{s^2 + 1/R_7}{s^2 + s/R_1 + 1/R_2} \tag{22}$$

The design equations are

$$\omega_z = \frac{1}{\sqrt{R_7}} \qquad \omega_p = \frac{1}{\sqrt{R_2}} \qquad Q_p = \frac{R_1}{\sqrt{R_2}} \qquad |H| = 1 \tag{23}$$

Example 6.3-3 *jω*-Axis-zeros Tow-Thomas filter. It is desired to design a second-order realization using the circuit shown in Fig. 6.3-4. The specifications are $f_z = 1.6$ kHz, $f_p = 1.5$ kHz, $Q_z = \infty$, and $Q_p = 10$. From Summary 6.3-1 we see that $R_4 = R_8 = \infty$, $R_3 = R_5 = R_6 = 1\ \Omega$, and $C_1 = C_2 = C_3 = 1$ F. For a normalized realization with $\omega_{pn} = 1$, $\omega_{zn} = 1.0667$ rad/s, from (23) we obtain $R_2 = 1$, $R_7 = 1/\omega_{zn}^2 = 0.8789$, and $R_1 = 10\ \Omega$. The final filter is obtained by applying a frequency denormalization of $\Omega_n = 2 \times \pi \times 1500$ and a convenient frequency denormalization.

6.4 THE UNIVERSAL ACTIVE FILTER

One of the limitations of the state-variable filter described in Sec. 6.1 is that the band-pass network function can only be realized in an inverting form, while the low-pass and high-pass functions can only be realized as noninverting ones. In addition, for the specified design procedure, the value of the multiplicative constant H_0 is not free to be chosen. In this section we present a *modified state-variable filter* realization that overcomes these difficulties. This circuit is commonly called the *universal active filter*, and in packaged form it is commercially available from several sources.[1]

The Noninverting Configuration ($V_{\text{in}B} = 0$)

The modified state-variable filter is shown in Fig. 6.4-1. It differs from the original circuit given in Fig. 6.1-2 in that two resistors, R_7 and R_8, have been added, thus forming a new input $V_{\text{in}B}$. The original input has been relabeled as $V_{\text{in}A}$. Analysis of Fig. 6.4-1 begins by assuming $V_{\text{in}B} = 0$ and expressing $V_{\text{HP}}(s)$ as a function of all the inputs to the amplifier A_3. We get

$$V_{\text{HP}}(s) = \left[V_{\text{in}A}(s) \frac{R_4||R_7}{R_3 + R_4||R_7} + V_{\text{BP}}(s) \frac{R_3||R_7}{R_4 + R_3||R_7} \right] \left(1 + \frac{R_6}{R_5||R_8} \right) - \frac{R_6}{R_5} V_{\text{LP}}(s) \tag{1}$$

[1] Examples are the UAF41 produced by the Burr-Brown Corporation and the AF-100 produced by National Semiconductor.

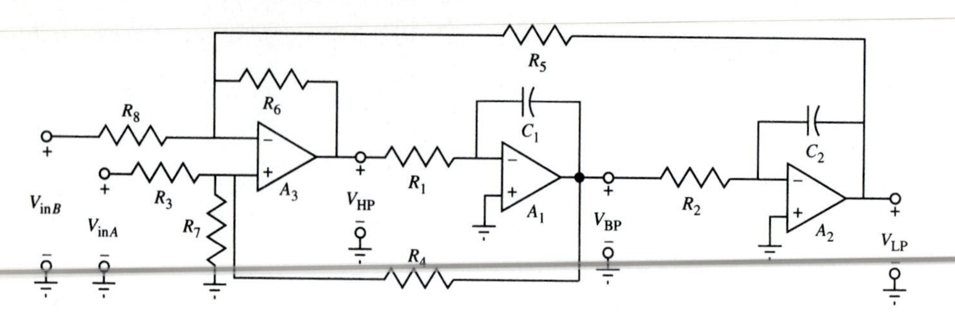

FIGURE 6.4-1
The modified (universal) state-variable filter.

where the symbol || stands for *the parallel connection of*. To simplify the algebra, let us define

$$K_1 = \frac{R_4||R_7}{R_3 + R_4||R_7} = \frac{R_4R_7}{R_3R_7 + R_3R_4 + R_4R_7} = \frac{1}{1 + R_3/R_4 + R_3/R_7} \tag{2a}$$

$$K_2 = \frac{R_3||R_7}{R_4 + R_3||R_7} = \frac{R_3R_7}{R_3R_4 + R_3R_7 + R_4R_7} = \frac{1}{1 + R_4/R_3 + R_4/R_7} \tag{2b}$$

$$K_3 = \frac{R_6}{R_5} \tag{2c}$$

$$K_4 = \frac{R_6}{R_8} \tag{2d}$$

In terms of these quantities, (1) may be rewritten as

$$V_{HP}(s) = [K_1V_{inA}(s) + K_2V_{BP}(s)](1 + K_3 + K_4) - K_3V_{LP}(s) \tag{3}$$

If we now define

$$\omega_1 = \frac{1}{R_1C_1} \quad \text{and} \quad \omega_2 = \frac{1}{R_2C_2} \tag{4}$$

and observe that

$$V_{BP}(s) = -\frac{\omega_1}{s}V_{HP}(s) \tag{5}$$

and

$$V_{LP}(s) = -\frac{\omega_2}{s}V_{BP}(s) = \frac{\omega_1\omega_2}{s^2}V_{HP}(s) \tag{6}$$

then (3) can be written as follows:

$$V_{HP}(s) = -(1 + K_3 + K_4)\frac{\omega_1K_2}{s}V_{HP}(s) - \frac{K_3\omega_1\omega_2}{s^2}V_{HP}(s) + (1 + K_3 + K_4)K_1V_{inA}(s) \tag{7}$$

Solving this, the high-pass transfer function for the circuit in Fig. 6.4-1 with $V_{\text{in}B} = 0$ is

$$\frac{V_{\text{HP}}(s)}{V_{\text{in}A}(s)} = \frac{(1 + K_3 + K_4)K_1 s^2}{s^2 + (1 + K_3 + K_4)K_2 \omega_1 s + K_3 \omega_1 \omega_2} \tag{8}$$

If $V_{\text{in}B} = 0$, then we may set $R_8 = \infty$ so that $K_4 = 0$. In this case (8) can be written as

$$\frac{V_{\text{HP}}(s)}{V_{\text{in}A}(s)} = \frac{(1 + K_3)K_1 s^2}{s^2 + (1 + K_3)K_2 \omega_1 s + \omega_1 \omega_2 K_3} \tag{9}$$

The band-pass and low-pass transfer functions for Fig. 6.4-1 with $V_{\text{in}B} = 0$ and $R_8 = \infty$ may be found from the above using (5) and (6). Thus we obtain

$$\frac{V_{\text{BP}}(s)}{V_{\text{in}A}(s)} = \frac{-(1 + K_3)K_1 \omega_1 s}{s^2 + (1 + K_3)K_2 \omega_1 s + \omega_1 \omega_2 K_3} \tag{10}$$

and

$$\frac{V_{\text{LP}}(s)}{V_{\text{in}A}(s)} = \frac{(1 + K_3)K_1 \omega_1 \omega_2}{s^2 + (1 + K_3)K_2 \omega_1 s + \omega_1 \omega_2 K_3} \tag{11}$$

Equating the denominator of (9), (10), or (11) to the standard second-order polynomial gives the equations

$$\omega_n = \sqrt{\omega_1 \omega_2 K_3} = \sqrt{\frac{R_6/R_5}{R_1 R_2 C_1 C_2}} \tag{12a}$$

$$\frac{1}{Q} = (1 + K_3)K_2 \sqrt{\frac{\omega_1}{\omega_2 K_3}} = \frac{1 + R_6/R_5}{1 + R_4/R_3 + R_4/R_7} \sqrt{\frac{R_2 R_5 C_2}{R_1 R_6 C_1}} \tag{12b}$$

Comparing (9), (10), and (11) with the respective low-, band-, and high-pass general second-order transfer functions yields the expressions for H_0 that are given below:

Low-pass: $H_0 = (1 + K_3)\dfrac{K_1}{K_3} = \dfrac{1 + R_5/R_6}{1 + R_3/R_4 + R_3/R_7}$ [see (1) of Sec. 5.2] (13*a*)

Band pass: $H_0 = -\dfrac{K_1}{K_2} = -\dfrac{R_4}{R_3}$ [see (1) of Sec. 5.3] (13*b*)

High pass: $H_0 = (1 + K_3)K_1 = \dfrac{1 + R_6/R_5}{1 + R_3/R_4 + R_3/R_7}$ [see (3) of Sec 5.3] (13*c*)

Typically $K_3 = \frac{1}{10}$, $C_1 = C_2 = 1000$ pF, $R_4 = R_5 = 100\ \text{k}\Omega$, and $R_6 = 10\ \text{k}\Omega$. For these values (12) and (13) reduce to the following expressions:

$$\omega_n = \sqrt{\frac{\omega_1 \omega_2}{10}} = \frac{3.162 \times 10^8}{\sqrt{R_1 R_2}} \tag{14a}$$

$$Q = 28{,}748\left(\frac{1}{100\ \text{k}\Omega} + \frac{1}{R_3} + \frac{1}{R_7}\right)\sqrt{\frac{R_1}{R_2}} \tag{14b}$$

Low pass:
$$H_0 = \frac{11}{1 + R_3\left(\dfrac{1}{100\ \text{k}\Omega} + \dfrac{1}{R_7}\right)} \tag{14c}$$

Band pass:
$$H_0 = -\frac{100\ \text{k}\Omega}{R_3} \tag{14d}$$

High pass:
$$H_0 = \frac{1.1}{1 + R_3\left(\dfrac{1}{100\ \text{k}\Omega} + \dfrac{1}{R_7}\right)} \tag{14e}$$

A design procedure that assumes these values, namely, $C_1 = C_2 = 1000$ pF, $R_4 = R_5 = 100$ kΩ, $R_6 = 10$ kΩ, and $R_8 = \infty$, and in addition uses $R_1 = R_2$ can be developed for the circuit of Fig. 6.4-1 with $V_{\text{in}B} = 0$. From (14*a*) we have

$$R_1 = R_2 = \frac{5.0329 \times 10^4}{f_n}\ \text{k}\Omega \tag{15}$$

To complete the design, it remains to find the values of R_3 and R_7 from (14*b*) through (14*e*). If we assume that $|H_0| = 1$, then we find

Low pass:
$$R_3 = \frac{316.2\ \text{k}\Omega}{Q} \tag{16a}$$

Band pass:
$$R_3 = 100\ \text{k}\Omega \tag{16b}$$

High pass:
$$R_3 = \frac{31.62\ \text{k}\Omega}{Q} \tag{16c}$$

Solving for R_7 in (14*b*) gives

$$R_7 = \frac{100\ \text{k}\Omega}{3.4785Q - 1 - \dfrac{100\ \text{k}\Omega}{R_3}} \tag{17}$$

Substitution of (16) into (17) gives

Low pass:
$$R_7 = \frac{100\ \text{k}\Omega}{3.162Q - 1} \tag{18a}$$

Band pass:
$$R_7 = \frac{100\ \text{k}\Omega}{3.4785Q - 2} \tag{18b}$$

High pass:
$$R_7 = \frac{100\ \text{k}\Omega}{0.3162Q - 1} \tag{18c}$$

We note that Q must be greater than $\sqrt{10}$ for the high-pass case. If this lower limit of Q is unacceptable, then the value of R_2/R_1 may be adjusted to lower this limitation. If the product of R_1 and R_2 is equal to the original product when R_1 and R_2 were equal, then R_2/R_1 modifies only Q. This concludes the design procedure for network functions in which $|H_0|$ is unity.

Example 6.4-1 A unity-gain inverting state-variable band-pass filter. It is desired to design a band-pass filter with $Q = 20, f_n = 1$ kHz, and $|H_0| = 1$. From the design procedure given above ($R_8 = \infty$, $V_{\text{in}B} = 0$) we have $R_4 = R_5 = 100$ kΩ, $R_6 = 10$ kΩ, and $C_1 = C_2 = 1000$ pF. From (15) we get $R_1 = R_2 = 50.329$ kΩ; from (16*b*), $R_3 = 100$ kΩ; and from (18*b*), $R_7 = 1480$ Ω.

It is of interest to examine the relative magnitudes of $V_{\text{LP}}(j\omega)$, $V_{\text{BP}}(j\omega)$, and $V_{\text{HP}}(j\omega)$ for the filter of Fig. 6.4-1. If we assume that $\omega_1 = \omega_2$ and $R_6/R_5 = 0.1$, then (12*a*) gives $\omega_1 = \omega_2 = \sqrt{10}\omega_n = 3.162\omega_n$. Equation (6) shows that at a constant frequency ω_x

$$|V_{\text{LP}}(j\omega_x)| = 3.162 \left|\frac{\omega_n}{\omega_x}\right| |V_{\text{BP}}(j\omega_x)| = 10 \left|\frac{\omega_n^2}{\omega_x^2}\right| |V_{\text{HP}}(j\omega_x)| \tag{19}$$

Thus the signal level at the various output points in the circuit can differ by a level of as much as 10. Note that this could cause saturation even though it was not indicated by calculations for the magnitude of the output voltage at the particular output point used.

The Inverting Configuration ($V_{\text{in}A} = 0$)

To complete the introduction of the modified state-variable realization, we need to consider the use of the $V_{\text{in}B}$ input. In this configuration we have $V_{\text{in}A} = 0$ and $R_3 = \infty$ in Fig. 6.4-1. Following a development similar to that used for the $V_{\text{in}A}$ configuration, we find

$$V_{\text{HP}}(s) = \frac{R_7}{R_4 + R_7}\left(1 + \frac{R_6}{R_5 || R_8}\right)V_{\text{BP}}(s) - \frac{R_6}{R_8} V_{\text{in}B}(s) - \frac{R_6}{R_5} V_{\text{LP}}(s) \tag{20}$$

This can be simplified to

$$V_{\text{HP}}(s) = -(1 + K_3 + K_4)K_2 \frac{\omega_1 V_{\text{HP}}(s)}{s} - \frac{\omega_1\omega_2 K_3}{s^2} V_{\text{HP}}(s) - K_4 V_{\text{in}B}(s) \tag{21}$$

through the use of (2) and (6) and by noting that since $R_3 = \infty$, $K_2 = 1/(1 + R_4/R_7)$. Solving for the high-pass transfer function, we get

$$\frac{V_{\text{HP}}(s)}{V_{\text{in}B}(s)} = \frac{-K_4 s^2}{s^2 + (1 + K_3 + K_4)K_2\omega_1 s + \omega_1\omega_2 K_3} \tag{22}$$

Equation (6) may be used to obtain the band-pass and low-pass transfer functions for this case. These are

$$\frac{V_{\text{BP}}(s)}{V_{\text{in}B}(s)} = \frac{\omega_1 K_4 s}{s^2 + (1 + K_3 + K_4)K_2\omega_1 s + \omega_1\omega_2 K_3} \tag{23}$$

and

$$\frac{V_{\text{LP}}(s)}{V_{\text{in}B}(s)} = \frac{-\omega_1\omega_2 K_4}{s^2 + (1 + K_3 + K_4)K_2\omega_1 s + \omega_1\omega_2 K_3} \tag{24}$$

Comparing (9), (10), and (11) with the above three equations shows that the universal active filter has the capability to realize second-order low-pass, band-pass, and high-pass filters with either 0 or 180° phase shift at the frequency where H_0 is specified. Equating (22), (23), and (24) to the standard second-order denominator results in

$$\omega_n = \sqrt{\omega_1 \omega_2 K_3} = \sqrt{\frac{R_6/R_5}{R_1 R_2 C_1 C_2}} \tag{25a}$$

$$\frac{1}{Q} = (1 + K_3 + K_4)K_2 \sqrt{\frac{\omega_1}{\omega_2 K_3}} = \frac{1 + R_6/R_5 + R_6/R_8}{1 + R_4/R_7} \sqrt{\frac{R_2 R_5 C_2}{R_1 R_6 C_1}} \tag{25b}$$

The values of H_0 for the three cases are found to be

Low pass:
$$H_0 = -\frac{K_4}{K_3} = -\frac{R_5}{R_8} \tag{26a}$$

Band pass:
$$H_0 = \frac{K_4}{K_2(1 + K_3 + K_4)} = \frac{1 + R_4/R_7}{1 + R_8/R_6 + R_8/R_5} \tag{26b}$$

High pass:
$$H_0 = -K_4 = -\frac{R_6}{R_8} \tag{26c}$$

Assuming that $R_4 = R_5 = 100\ \text{k}\Omega$, $R_6 = 10\ \text{k}\Omega$, and $C_1 = C_2 = 1000$ pF, as we did for the $V_{\text{in}A}(s)$ configuration, we obtain

$$\omega_n = \frac{3.162 \times 10^8}{\sqrt{R_1 R_2}} \tag{27a}$$

$$Q = 0.3162 \sqrt{\frac{R_1}{R_2}} \frac{1 + \dfrac{100\ \text{k}\Omega}{R_7}}{1.1 + \dfrac{10\ \text{k}\Omega}{R_8}} \tag{27b}$$

Low pass:
$$H_0 = -\frac{100\ \text{k}\Omega}{R_8} \tag{27c}$$

Band pass:
$$H_0 = \frac{1 + \dfrac{100\ \text{k}\Omega}{R_7}}{1 + \dfrac{R_8}{10\ \text{k}\Omega} + \dfrac{R_8}{100\ \text{k}\Omega}} = \frac{1 + \dfrac{100\ \text{k}\Omega}{R_7}}{1 + \dfrac{R_8}{9.091\ \text{k}\Omega}} \tag{27d}$$

High pass:
$$H_0 = -\frac{10\ \text{k}\Omega}{R_8} \tag{27e}$$

A design procedure for the $V_{\text{in}B}$ input can be developed using the values given and also letting $R_1 = R_2$. The values of R_1 and R_2 are then found from

$$R_1 = R_2 = \frac{5.0329 \times 10^4}{f_n}\ \text{k}\Omega \tag{28}$$

TABLE 6.4-1
Low-frequency design equations for the state-variable active filter with $H_0 = 1$*

	Noninverting configuration ($R_8 = \infty$, $V_{inB} = 0$)			Inverting configuration ($R_3 = \infty$, $V_{inA} = 0$)		
	Low pass, phase shift 0°	**Band pass, phase shift ±180°**	**High pass, phase shift 0°**	**Low pass, phase shift ±180°**	**Band pass, phase shift 0°**	**High pass, phase shift ±180°**
R_1, R_2 (kΩ)	$\frac{5.0329 \times 10^4}{f_n}$	$\frac{5.0329 \times 10^4}{f_n}$	$\frac{5.0329 \times 10^4}{f_n}$	$\frac{5.0329 \times 10^4}{f_n}$	$\frac{5.0329 \times 10^4}{f_n}$	$\frac{5.0329 \times 10^4}{f_n}$
R_3 (kΩ)	$\frac{316.2}{Q}$	100	$\frac{31.26}{Q}$	—	—	—
R_8 (kΩ)	—	—	—	100	$31.62Q$	10
R_7 (kΩ)	$\frac{100}{3.162Q - 1}$	$\frac{100}{3.4785Q - 2}$	$\frac{100}{0.3162Q - 1}$	$\frac{100}{3.4785Q - 1}$	$\frac{100}{3.4785Q}$	$\frac{100}{6.6402Q - 1}$

* This design procedure assumes that $R_1 = R_2$, $R_4 = R_5 = 100$ kΩ, $R_6 = 10$ kΩ, $C_1 = C_2 = 1000$ pF. Phase shift of transfer function is at frequency where H_0 is specified.

Assuming $|H_0| = 1$ in (27), we get the following design formulas for R_8:

Low pass: $R_8 = 100 \text{ k}\Omega$ (29a)

Band pass: $R_8 = 31.62Q \text{ k}\Omega$ (29b)

High pass: $R_8 = 10 \text{ k}\Omega$ (29c)

Solving for R_7 in (27b) and noting that $R_1 = R_2$ gives

$$R_7 = \frac{100 \text{ k}\Omega}{3.162Q(1.1 + 10 \text{ k}\Omega / R_8) - 1} \tag{30}$$

Substitution of (29) into (30) results in

Low pass: $$R_7 = \frac{100 \text{ k}\Omega}{3.7947Q - 1} \tag{31a}$$

Band pass: $$R_7 = \frac{100 \text{ k}\Omega}{3.4785Q} \tag{31b}$$

High pass: $$R_7 = \frac{100 \text{ k}\Omega}{6.6402Q - 1} \tag{31c}$$

Example 6.4-2 A unity-gain noninverting state-variable band-pass filter. It is desired to design a noninverting band-pass filter with $Q = 20$ and $f_n = 1$ kHz using the V_{inB} input in Fig. 6.4-1. From the design procedure we have initially assumed that $R_4 = R_5 = 100 \text{ k}\Omega$, $R_6 = 10 \text{ k}\Omega$, and $C_1 = C_2 = 1000$ pF. Therefore from (15) we find that $R_1 = R_2 = 50.329 \text{ k}\Omega$. Equations (29b) and (31b) give $R_8 = 632.40 \text{ k}\Omega$ and $R_7 = 1.437 \text{ k}\Omega$. From (27d) we determine that $H_0 = 1$.

Table 6.4-1 summarizes the design procedures for both the V_{inA} and the V_{inB} low-pass, band-pass, and high-pass configurations of the modified state-variable active filter. The design equations of this table are satisfactory as long as f_n is much less than the gain bandwidths of the operational amplifiers. The versatility of this "universal" active filter realization is well illustrated by the table.

6.5 OPERATIONAL AMPLIFIER PARASITICS—GAIN BANDWIDTH

The effects of operational amplifier gain bandwidth limitations on filter performance were introduced in Sec. 5.7 for the single-amplifier filter. Gain bandwidth also limits the performance of multiple-amplifier filters. In this section we present an introductory study of these effects.

State-Variable Filter with Gain Bandwidth

In Sec. 5.7 we introduced a model for the open-loop gain $A(s)$ of an operational amplifier in terms of the gain bandwidth $\text{GB} = A_0\omega_a$, where A_0 is the open-loop dc gain and $-\omega_a$

is the location of the dominant pole. The expression is

$$A(s) = \frac{\text{GB}}{s} \tag{1}$$

If we now analyze the state-variable filter using (1), after considerable algebra we obtain the denominator polynomial $D(s)$ that determines the poles of the network function for the filter:

$$\begin{aligned} D(s) = {} & s^5 \frac{1+m}{\text{GB}^3} + s^4\left[\frac{(1+m)(\omega_1+\omega_2)}{\text{GB}^3} + \frac{3+2m}{\text{GB}^2}\right] \\ & + s^3\left[\frac{(1+m)\omega_1\omega_2}{\text{GB}^3} + \frac{(2+m)(\omega_1+\omega_2)}{\text{GB}^2} + \frac{3+m}{\text{GB}}\right] \\ & + s^2\left[\frac{\omega_1\omega_2}{\text{GB}^2} + \frac{\omega_1+\omega_2+\omega_n/Q}{\text{GB}} + 1\right] + s\left[\frac{\omega_n}{Q}\left(\frac{\omega_2}{\text{GB}}+1\right)\right] + \omega_n^2 \end{aligned} \tag{2}$$

where we have assumed that all the operational amplifiers have the same gain bandwidth and where we have defined

$$m = \frac{R_6}{R_5} \qquad \omega_1 = \frac{1}{R_1C_1} \qquad \omega_2 = \frac{1}{R_2C_2} \qquad \omega_n^2 = m\omega_1\omega_2 \qquad \frac{\omega_n}{Q} = \frac{1+m}{1+R_4/R_3}\omega_1 \tag{3}$$

The fact that (2) is of fifth order represents the effects of the two passive capacitors and the three operational amplifiers. Note that in the limit as GB approaches infinity, (2) reduces to the standard second-order denominator polynomial form

$$D(s) = s^2 + \frac{\omega_n}{Q}s + \omega_n^2 \tag{4}$$

Gain Bandwidth Effects in the State-Variable Filter (Design 1)

Now consider a normalized design in which the specifications for the filter are given as

$$\omega_{n(\text{design})} = \omega_{n0} = 1 \qquad Q_{(\text{design})} = Q_0 \tag{5}$$

To realize this, let us choose design 1 of Sec. 6.1, in which

$$R_1 = R_2 = R_3 = R_5 = R_6 = 1\ \Omega \qquad C_1 = C_2 = 1\ \text{F} \qquad Q_0 = \tfrac{1}{2}(1+R_4) \tag{6}$$

The quantities defined in (3) now become $m = 1$ and $\omega_1 = \omega_2 = 1$ rad/s. Substituting these values in (2) and replacing GB by the frequency-normalized gain bandwidth GB_n, we obtain

$$\begin{aligned} D(s) = {} & s^5\left[\frac{2}{\text{GB}_n^3}\right] + s^4\left[\frac{4}{\text{GB}_n^3} + \frac{5}{\text{GB}_n^2}\right] + s^3\left[\frac{2}{\text{GB}_n^3} + \frac{6}{\text{GB}_n^2} + \frac{4}{\text{GB}_n}\right] \\ & + s^2\left[\frac{1}{\text{GB}_n^2} + \frac{2+1/Q_0}{\text{GB}_n} + 1\right] + s\left[\frac{1}{Q_0}\left(\frac{1}{\text{GB}_n}+1\right)\right] + 1 \end{aligned} \tag{7}$$

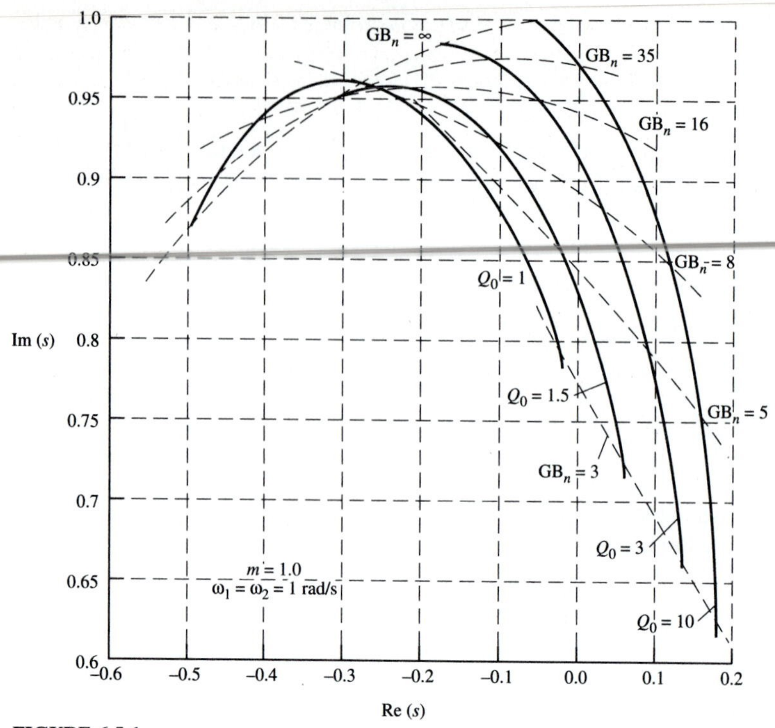

FIGURE 6.5-1
GB effects on the dominant poles of the state-variable design 1 filter.

Factoring this polynomial gives a result that has the form

$$D(s) = \frac{1}{\mathrm{GB}_n^3 X(s)} \left[(s+g_1)(s+g_2)(s+g_3)\left(s^2 + s\frac{\omega_n}{Q} + \omega_n^2 \right) \right] \tag{8}$$

where Q and ω_n are the parameters characterizing the dominant poles actually realized by the filter. The poles located at $s = -g_1, -g_2, -g_3$ are usually located far out on the negative real axis and, as a result, have little effect on the performance of the filter. The term $X(s)$ represents the common factor that has been extracted in order to make the highest degree term of s unity in the bracketed expression. The dominant complex-conjugate poles of (8), as specified by the values of ω_n and Q, may be considerably different than the design values. For given values of Q_0 and GB_n, the complex-conjugate roots of (8) are usually found by applying root-solving techniques to the fifth-order polynomial. A graph showing the resulting pole positions as a function of a range of values of Q_0 and GB_n is given by the plot of the upper half of the complex plane shown in Fig. 6.5-1.[1] Note that the original normalized design values appear on the locus for $\mathrm{GB}_n = \infty$.

[1] A. Oksasoglu and L. P. Huelsman, "Effects and Compensation of Gain Bandwidth in the State-Variable Filter," *Proceedings of the Midwest Symposium on Circuits and Systems*, 1991.

Example 6.5-1 The Q and ω_n of a state-variable filter (design 1). It is desired to find the actual values of Q and ω_n of a state-variable filter that has design specifications $\omega_n = 100 \times 2\pi$ krad/s (100 kHz) and $Q = 3$. The filter is to be realized using operational amplifiers that have a gain bandwidth of 1.6 MHz. The first step is to find the normalized specifications. For these we have $\omega_{n0} = 1$, $Q_0 = 3$, and $\text{GB}_n = 16$. From Fig. 6.5-1, we find the actual normalized pole locations are at $-0.05 \pm j0.95$. The corresponding denominator quadratic factor is $s^2 + 0.1s + 0.905$ for which $\omega_n = 0.951$ and $Q = 9.51$. The denormalized values are $\omega_n = 95.1 \times 2\pi$ krad/s (95.1 kHz) and $Q = 9.51$. These values represent (approximately) a -4.9 percent change in ω_n and a $+217$ percent change in Q. Obviously, the operational amplifier gain bandwidth limitations have produced a serious error in the performance of the filter.

Exercise 6.5-1. Find the design 1 pole locations and the values of Q and ω_n for a normalized ($\omega_{n0} = 1$ rad/s) state-variable filter for each of the following sets of values for Q_0 and GB_n:

		Answers		
Q_0	GB_n	Poles	Q	ω_n
3	35	$-0.11 \pm j0.975$	4.46	0.98
1.5	8	$-0.11 \pm j0.925$	4.23	0.93
10	50	$-0.02 \pm j0.99$	24.8	0.99

Gain Bandwidth Effects in the State-Variable Filter (Design 2)

The choice of design procedure for the state variable filter has a strong influence on the effects produced by the operational amplifier gain bandwidth. As an example of this, consider a normalized design using the specifications given in (5) but choosing the parameters of (3) as $m = 0.1$ and $\omega_1 = \omega_2 = \sqrt{10}$ rad/s (design 2 of Sec. 6.1). Substituting these values in (2) and representing GB by the frequency-normalized gain bandwidth GB_n, we obtain

$$D(s) = s^5\left[\frac{1.1}{\text{GB}_n^3}\right] + s^4\left[\frac{2.2\sqrt{10}}{\text{GB}_n^3} + \frac{3.2}{\text{GB}_n^2}\right] + s^3\left[\frac{11}{\text{GB}_n^3} + \frac{4.2\sqrt{10}}{\text{GB}_n^2} + \frac{3.1}{\text{GB}_n}\right]$$
$$+ s^2\left[\frac{10}{\text{GB}_n^2} + \frac{2\sqrt{10} + 1/Q_0}{\text{GB}_n} + 1\right] + s\left[\frac{1}{Q_0}\left(\frac{\sqrt{10}}{\text{GB}_n} + 1\right)\right] + 1 \qquad (9)$$

The upper half-plane dominant complex-conjugate roots of (9), as found by applying root-solving techniques, are given for a range of values of Q_0 and GB_n in Fig. 6.5-2. The original normalized design values appear on the locus for $\text{GB}_n = \infty$.

Example 6.5-2 The Q and ω_n of a state-variable filter (design 2). It is desired to find the actual values of Q and ω_n of a state-variable filter that has the design specifications $\omega_n = 100 \times 2\pi$ krad/s (100 kHz) and $Q = 3$. The filter is to be realized using operational amplifiers that have a gain bandwidth of 1.6 MHz. The normalized specifications are

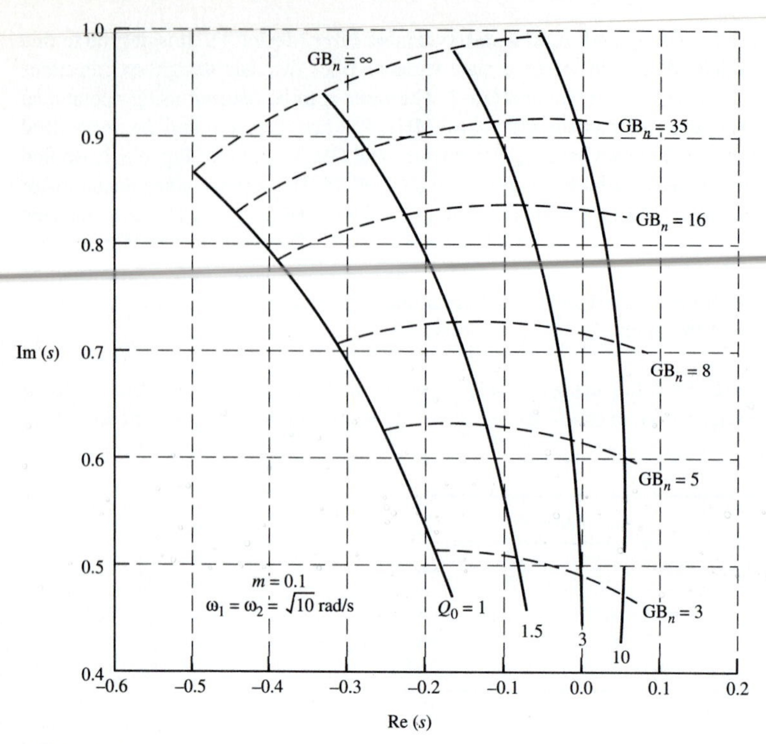

FIGURE 6.5-2
GB effects on the dominant poles of the state-variable design 2 filter.

$\omega_{n0} = 1$, $Q_0 = 3$, and $\mathrm{GB}_n = 16$. From Fig. 6.5-2 we find the actual normalized pole locations are at $-0.08 \pm j0.84$. The corresponding denominator quadratic factor is $s^2 + 0.16s + 0.712$ for which $\omega_n = 0.844$ and $Q = 5.27$. The denormalized values are $\omega_n = 84.4 \times 2\pi$ krad/s (84.4 kHz) and $Q = 5.27$. These values represent (approximately) a -15.6 percent change in ω_n and a $+76$ percent change in Q. If these results are compared with the ones given for design 1 for the same filter specifications in Example 6.5-1, we see that the sensitivity of Q to the gain bandwidth has been reduced. A price for this reduction has been paid, however, since the ω_n sensitivity is considerably larger.

Exercise 6.5-2. Find the design 2 pole locations and the values of Q and ω_n for a normalized ($\omega_{n0} = 1$ rad/s) state-variable filter for each of the following sets of values for Q_0 and GB_n:

		Answers		
Q_0	GB$_n$	Poles	Q	ω_n
3	35	$-0.12 \pm j0.92$	3.87	0.93
1.5	8	$-0.17 \pm j0.725$	2.19	0.74
10	50	$-0.03 \pm j0.96$	16.0	0.96

TABLE 6.5-1
Compensation of the state-variable filter

GB_n	Parameter	Before compensation	After compensation
		Low-Q realization: $\omega_{n0} = 1$, $Q_0 = 0.707$	
5	R_4	0.414	0.822
	ω_2	1.0	0.357
	Poles	$-0.376 \pm j1.061$	$-0.707 \pm j0.707$
	Q	1.495	0.707
	ω_n	1.125	1.0
100	R_4	0.414	0.420
	ω_2	1.0	0.977
	Poles	$-0.710 \pm j0.721$	$-0.707 \pm j0.707$
	Q	0.7124	0.707
	ω_n	1.011	1.0
		High-Q realization: $\omega_{n0} = 1$, $Q_0 = 100$	
5	R_4	199	1.319
	ω_2	1.0	1.470
	Poles	$0.187 \pm j0.737$	$-0.005 \pm j1.0$
	Q	Unstable	99.8
	ω_n	Unstable	1.0
100	R_4	199	38.77
	ω_2	1.0	1.02
	Poles	$0.014 \pm j0.989$	$-0.005 \pm j1.0$
	Q	Unstable	100
	ω_n	Unstable	1.0

Compensation

The effect of operational amplifier gain bandwidth on the state-variable filter illustrated in the preceding discussion can be minimized by compensation. The process is similar to that described in connection with the single-amplifier filter in Sec. 5.7. Although the method is numerically more complex in the state-variable case due to the higher degree of the denominator polynomial of the network function, useful results are obtained in most cases. The process is best illustrated by an example, for which we choose design 1 of the state-variable filter. For this case, to compensate for the effects of gain bandwidth, it has been found[2] that compensation is obtained by varying the ratio ω_2/ω_1 and the ratio R_4/R_3. In terms of the design 1 specifications given in (5) and the quantities defined in (3), we choose $m = 1$, $\omega_1 = 1$ rad/s, and $R_3 = 1\ \Omega$ and vary ω_2 and R_4 to obtain the necessary compensation. The process gives useful results over a wide range of values of Q_0 and GB_n. A specific low-Q and high-Q case are illustrated in Table 6.5-1. The results obtained from applying the process over a range of values of Q_0 and GB_n are given by the plot shown in Fig. 6.5-3. The appropriate intersection of the lines of constant Q_0 and GB_n determine the values of ω_2 and R_4, as determined by the values of the ordinate and abscissa scales, to use in the compensated filter circuit.

[2] Oksasoglu and Huelsman, *op. cit.*

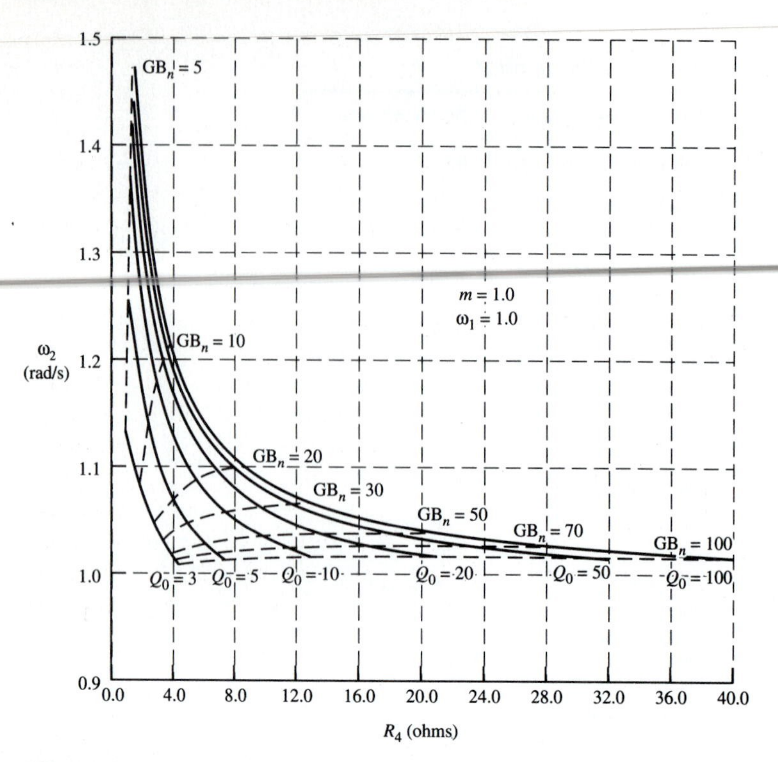

FIGURE 6.5-3
Compensation chart for GB effects in the state-variable design 1 filter.

Example 6.5-3 Compensation of a state-variable filter. It is desired to find the values of R_4 and ω_2 that will compensate a design 1 state-variable filter with a normalized design of $\omega_{n0} = 1$ rad/s and $Q_0 = 3$ if the filter is to be constructed with operational amplifiers for which $GB_n = 16$. The element values corresponding with the design specifications (without compensation) are given in (6) with $R_4 = 5\ \Omega$. From Fig. 6.5-3 we find the values $R_4 = 2.44\ \Omega$ and $\omega_2 = 1.06$ rad/s will provide the necessary compensation. From (3), the desired value of ω_2 can be realized by setting $R_2 = 1/1.06 = 0.943\ \Omega$ (with $C_2 = 1$ F).

Example 6.5-3. Find the normalized ($\omega_{n0} = 1$ rad/s) compensating values for R_4 and ω_2 for each of the following sets of values for Q_0 and GB_n.

		Answers	
Q_0	GB_n	R_4	ω_2
3	35	3.3	1.03
5	10	2.1	1.14
10	50	9.9	1.035

6.6 OPERATIONAL TRANSCONDUCTANCE AMPLIFIERS

The operational amplifier has emerged as the dominant active device used in active filter realizations. Its main application has been in discrete component configurations. The increased demand for the realization of analog filter functions that can be integrated on a single silicon chip with other signal processing components has led to the consideration and use of other types of gain devices. One of the most popular of these is the operational transconductance amplifier (OTA). This device acts as a high-gain VCIS and produces an output current from a voltage input. Among its advantages are that it is easy to implement in monolithic form, has significantly higher bandwidth than the commoner VCVS type of operational amplifier, can be electronically tuned, and leads to very simple filter configurations. The practicality of filters using OTAs has been considerably strengthened by the availability of implementations using complementary metal-oxide-semiconductor (CMOS) technology. This has made possible the use of higher input voltage swings, resulting in increased applications for the circuits using them. Discrete OTAs are available in a wide range of models from many different manufacturers. Among these are the Harris CA3080, the National Semiconductor LM13700, the Signetics NE5517, and the Exar XR-13600. In this section we present an introduction to the properties and applications of OTA filters.

Models for the OTA

The circuit symbol for an ideal OTA is shown in Fig. 6.6-1(a). As shown in Fig. 6.6-1(b), it is modeled by an ideal VCIS characterized by the relation

$$I_o = g_m(V_+ - V_-) \tag{1}$$

with infinite input impedance and infinite output impedance. Typically, g_m is a very small number. In many OTA designs the transconductance is variable and, as indicated by the dashed line in Fig. 6.6-1(a), can be controlled by an input current I_c. In this case we can write $g_m = KI_c$. In this expression, for g_m in microsiemens and I_c in microamperes, a typical value of K might be 15. The linearity of the dependence of g_m on I_c will ordinarily be valid over a several-decade range of I_c, for example, from 0.001 to 1000 μA. The gain bandwidth is also proportional to I_c and may go as high as hundreds of megahertz. Other

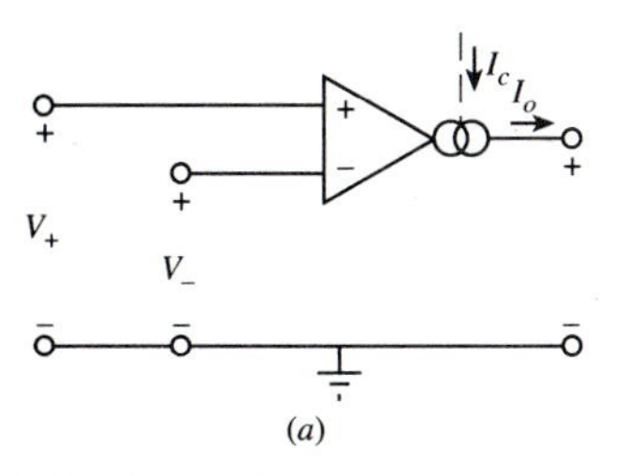

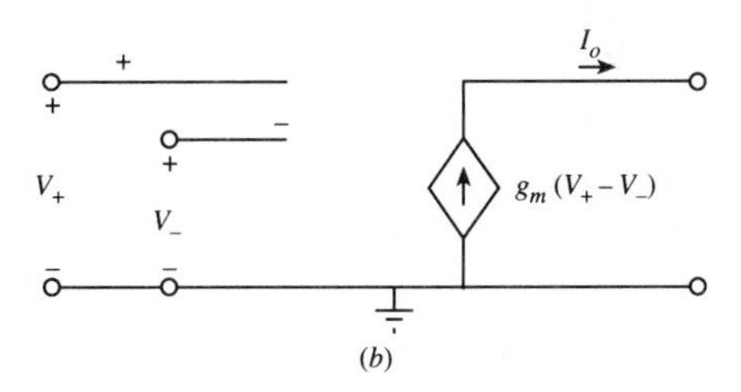

FIGURE 6.6-1
An OTA.

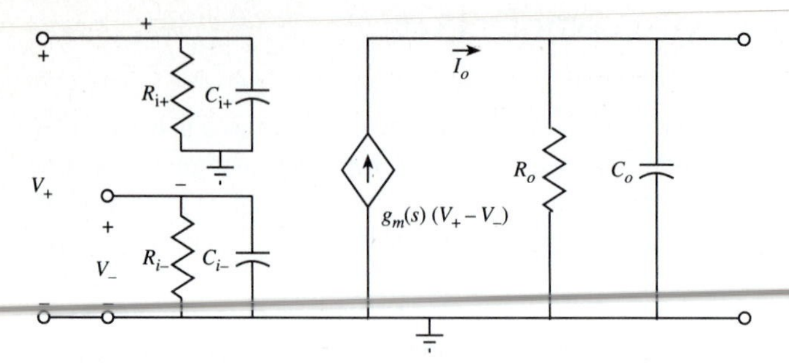

FIGURE 6.6-2
An OTA model with input and output impedances.

bandwidth limiting parasitics, which may be of even greater significance than the intrinsic frequency dependence of g_m, are the input and output impedances. These effects are shown in the more complex model of Fig. 6.6-2. Representative values of the resistors are $R_i = 0.5\ \text{M}\Omega$ and $R_o = 50\ \text{M}\Omega$. The capacitors C_i and C_o typically have values of a few picofarads. In the physical design of an OTA filter, due consideration must also be given to the intrinsic properties of the device. One of these is the limitation on the magnitude of the input signal for linear operation. Another effect that must be included in the design is a provision for dc paths for bias currents at the input nodes.

OTA Resistor Realizations

An important application of OTAs is in the simulation of grounded and floating resistors. Such circuits make possible the realization of large monolithic resistance values in a small chip area. Direct fabrication of such elements is impractical due to the large chip area that would be required. A circuit for the realization of a *grounded resistor* is shown in Fig. 6.6-3(a). For this, from Fig. 6.6-1(a) and (1) we obtain

$$-I_1 = I_o = g_m(0 - V_-) = g_m(-V_1) \tag{2}$$

The input resistance for the circuit is

$$R_{\text{eq}} = \frac{V_1}{I_1} = \frac{1}{g_m} \tag{3}$$

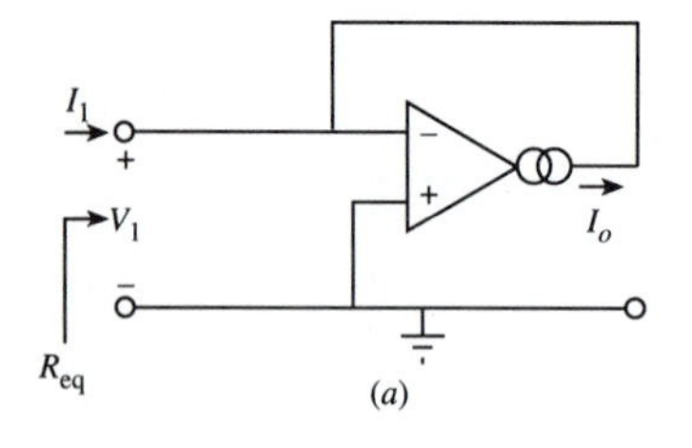

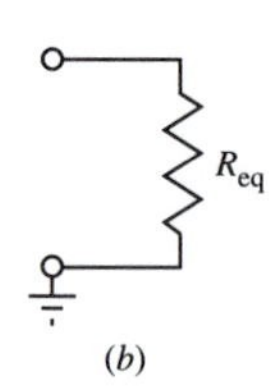

FIGURE 6.6-3
Grounded resistor.

The equivalent resistor is shown in Fig. 6.6-3(b). For example, for $g_m = 10^{-5}$ S, the input resistance is 10^5 Ω. Note that this circuit makes possible the implementation in silicon of large resistance values by the use of a small value of transconductance and thus a desirably small chip area.

Example 6.6-1 The 3-dB frequency of a grounded resistor. If we use the OTA model shown in Fig. 6.6-2 in the grounded resistor circuit shown in Fig. 6.6-3(a), we obtain

$$Z_{\text{in}}(s) = \frac{V_1(s)}{I_1(s)} = \frac{1}{g_m + (sC_i + G_i) + (sC_o + G_o)}$$

The 3-dB frequency for the equivalent resistor is

$$\omega_{3\text{dB}}(s) = \frac{g_m + G_i + G_o}{C_i + C_o} \approx \frac{g_m}{C_i + C_o}$$

For the values $g_m = 20$ μS and $C_i = C_o = 10$ pF, $\omega_{3\text{dB}} = 100$ krad/s, about 160 kHz. This analysis assumes that g_m is not a function of frequency. Such an equivalent resistor would be practical for audiofrequency applications for frequencies up to 20 kHz.

The grounded resistor realization shown in Fig. 6.6-3 can be modified to obtain a *floating resistor* by using the circuit shown in Fig. 6.6-4(a). For this we find

$$I_1 = -g_{m1}(V_2 - V_1) \qquad I_2 = -g_{m2}(V_1 - V_2) \tag{4}$$

From these equations, the y parameters for the two-port network are defined by

$$\begin{bmatrix} I_1 \\ I_2 \end{bmatrix} = \begin{bmatrix} g_{m1} & -g_{m1} \\ -g_{m2} & g_{m2} \end{bmatrix} \begin{bmatrix} V_1 \\ V_2 \end{bmatrix} \tag{5}$$

For the matched case in which $g_m = g_{m1} = g_{m2}$ we obtain the same y parameters as characterize the floating resistor shown in Fig. 6.6-4(b) for which

$$R_{\text{eq}} = \frac{1}{g_m} \tag{6}$$

Another useful OTA equivalent resistor circuit is an implementation of a *summer*. This requires one OTA for each input to be summed. The output currents of these OTAs

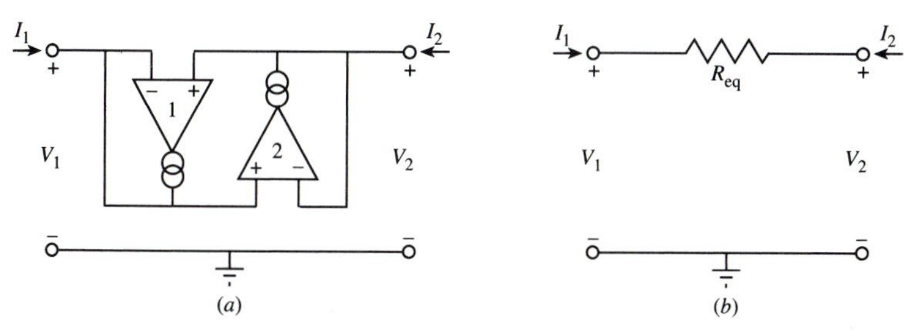

FIGURE 6.6-4
Floating resistor.

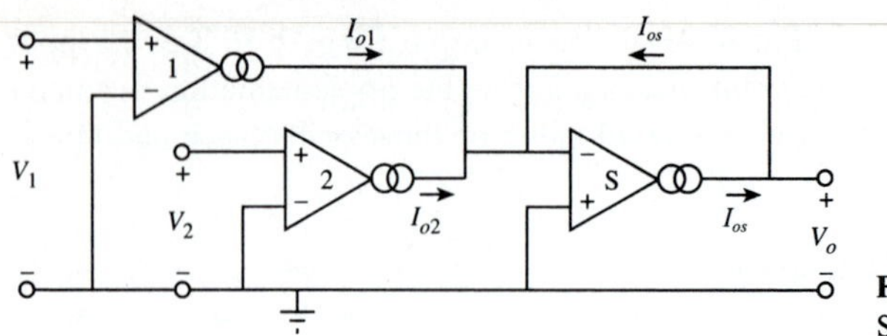

FIGURE 6.6-5
Summer.

are conveniently summed with the OTA realization of a grounded resistor, as shown in Fig. 6.6-3. The configuration for summing two inputs V_1 and V_2 to produce an output V_o is shown in Fig. 6.6-5. For this we obtain

$$V_o = -I_{os}\frac{1}{g_{ms}} = (I_{o1} + I_{o2})\frac{1}{g_{ms}} = (V_1 g_{m1} + V_2 g_{m2})\frac{1}{g_{ms}} = \frac{g_{m1}}{g_{ms}}V_1 + \frac{g_{m2}}{g_{ms}}V_2 \tag{7}$$

Thus V_o is the scaled sum of V_1 and V_2.

OTA Building Blocks

Now let us consider some simple OTA circuits involving a capacitor. These may be used as building blocks in the design of OTA filters. The first circuit to be considered is an *integrator*. The configuration is shown in Fig. 6.6-6. For this we obtain

$$V_o(s) = I_o(s)\frac{1}{sC} = \frac{g_m}{sC}[V_+(s) - V_-(s)] \tag{8}$$

This equation defines a differential-input integrator with time constant C/g_m. The OTA shown in Fig. 6.6-6 may be replaced by two or more OTAs as indicated in Fig. 6.6-5 to provide a summing function.

The second circuit to be considered is the *damped integrator*. This can be realized with a differential input by the circuit shown in Fig. 6.6-7. For this we obtain

$$V_o(s) = -I_{o2}(s)\frac{1}{g_{m2}} = \frac{I_{o1}(s) - V_o(s)sC}{g_{m2}} = \frac{g_{m1}[V_+(s) - V_-(s)] - V_o(s)sC}{g_{m2}} \tag{9}$$

Rearranging terms, we obtain

$$V_o(s) = \frac{g_{m1}}{sC + g_{m2}}[V_+(s) - V_-(s)] \tag{10}$$

This equation defines a differential-input damped integrator. Note that OTA 1 in Fig. 6.6-7 may be replaced by two OTAs to provide a *summing* inverting or noninverting

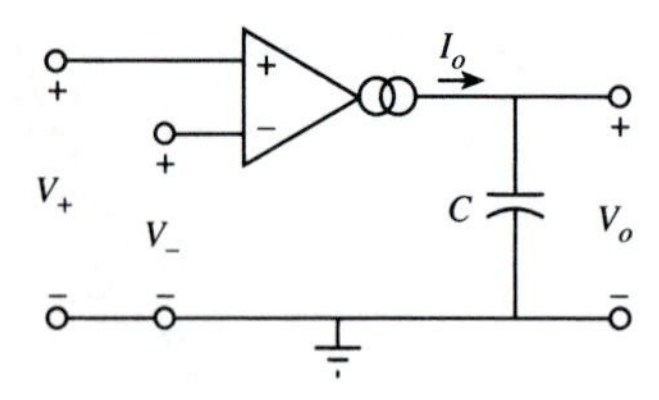

FIGURE 6.6-6
Differential-input integrator.

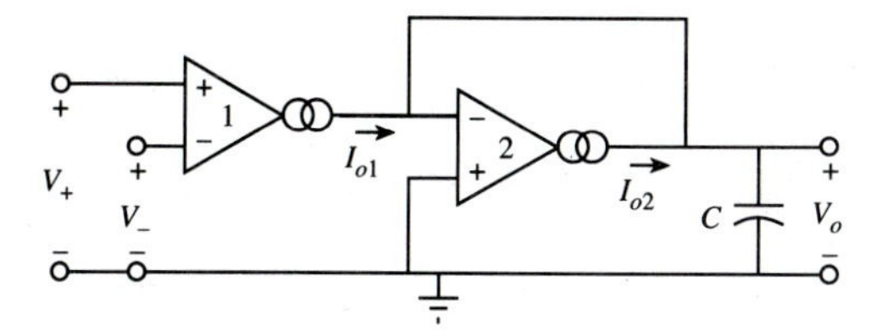

FIGURE 6.6-7
Differential-input damped integrator.

damped integrator. A second damped integrator circuit can be designed using the summing circuit of Fig. 6.6-5 (with a single input) as a feedback element around the integrator shown in Fig. 6.6-6. The result is shown in Fig. 6.6-8. For this circuit we find

$$\frac{V_o(s)}{V_1(s)} = \frac{g_{m1}}{sC + g_{m1}g_{m2}/g_{m3}} \tag{11}$$

As another example of a useful building block, note that we may use OTAs to construct an *impedance inverter.* Such a circuit has an input impedance that is inversely proportional to the impedance of some load. Thus, $Z_{\text{in}}(s) \propto 1/Z_L(s)$. Note that if $Z_L(s) = 1/sC$, $Z_{\text{in}}(s)$ becomes the impedance of an inductor. In this manner we can use OTAs and a capacitor to simulate an inductor. The impedance inversion circuit is shown in Fig. 6.6-9. For this we obtain

$$I_1(s) = -I_{o2}(s) = -(g_{m2})[-V_C(s)] = g_{m2}I_{o1}(s)\frac{1}{sC} = g_{m2}g_{m1}V_1(s)\frac{1}{sC} \tag{12}$$

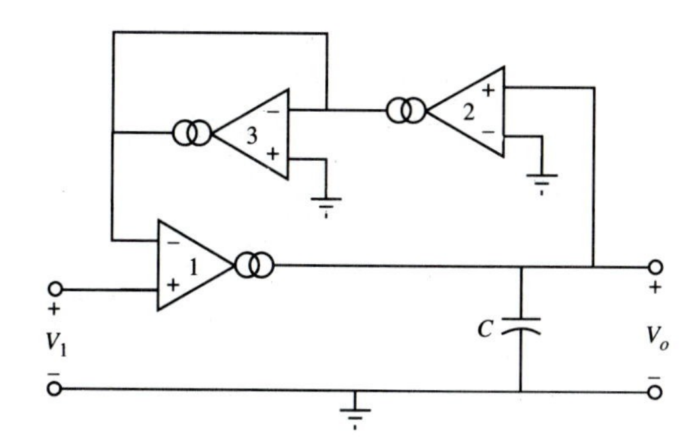

FIGURE 6.6-8
Damped integrator.

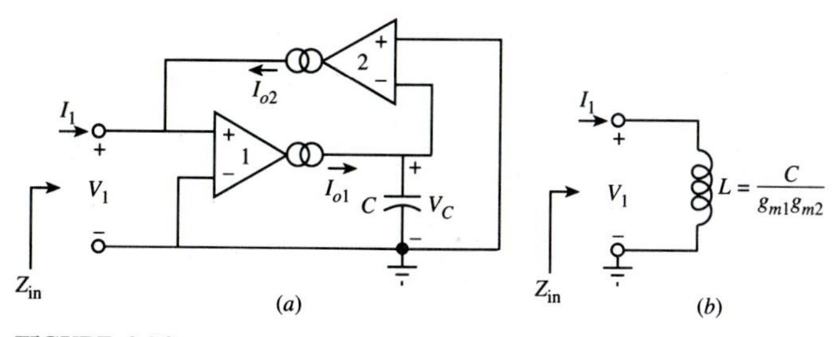

FIGURE 6.6-9
Impedance inverter used to simulate a grounded inductor.

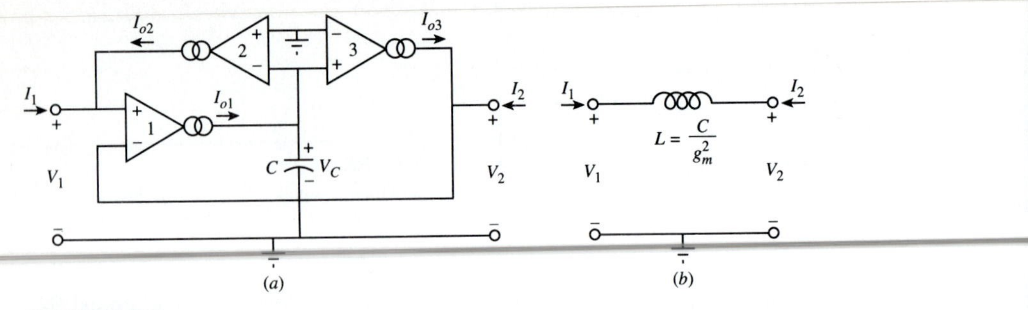

FIGURE 6.6-10
Simulation of a floating inductor.

This may be solved to find the input impedance

$$Z_{\text{in}}(s) = \frac{V_1(s)}{I_1(s)} = \frac{sC}{g_{m1}g_{m2}} \tag{13}$$

We see that the circuit acts like an inductor of value $L = C/(g_{m1}g_{m2})$ henrys.

The circuit described above may be modified to provide a realization of a *floating inductor*. The result is shown in Fig. 6.6-10(*a*). For this circuit we obtain

$$I_1(s) = -I_{o2}(s) = -g_{m2}[-V_C(s)] = g_{m2}I_{o1}(s)\frac{1}{sC} = \frac{g_{m1}g_{m2}}{sC}[V_1(s) - V_2(s)] \tag{14a}$$

$$I_2(s) = -I_{o3}(s) = -g_{m3}V_C(s) = -g_{m3}I_{o1}(s)\frac{1}{sC} = \frac{g_{m1}g_{m3}}{sC}[-V_1(s) + V_2(s)] \tag{14b}$$

From these equations the y parameters for the two-port network are defined by

$$\begin{bmatrix} I_1(s) \\ I_2(s) \end{bmatrix} = \frac{g_{m1}}{sC}\begin{bmatrix} g_{m2} & -g_{m2} \\ -g_{m3} & g_{m3} \end{bmatrix}\begin{bmatrix} V_1(s) \\ V_2(s) \end{bmatrix} \tag{15}$$

For the matched case in which $g_m = g_{m1} = g_{m2} = g_{m3}$, these parameters are identical with those of a floating inductor of value $L = C/g_m^2$ as shown in Fig. 6.6-10(*b*).

Two-Integrator Loop Structures

Operational transductance amplifiers can be used to realize second-order filter functions by using a pair of integrators of the types shown in Figs. 6.6-6 to 6.6-8. A typical two-integrator loop structure as used in the Tow-Thomas filter described in Sec. 6.2 consists of a regular integrator, a damped integrator, and a feedback element. If we use Fig. 6.6-6 for the integrator, Fig. 6.6-7 for the damped integrator, and a straight connection for the feedback, we obtain the circuit shown in Fig. 6.6-11. The circuit realizes the low-pass voltage transfer function

$$\frac{V_o(s)}{V_i(s)} = \frac{g_{m1}g_{m2}/C_1C_2}{s^2 + s(g_{m3}/C_2) + (g_{m1}g_{m2}/C_1C_2)} \tag{16}$$

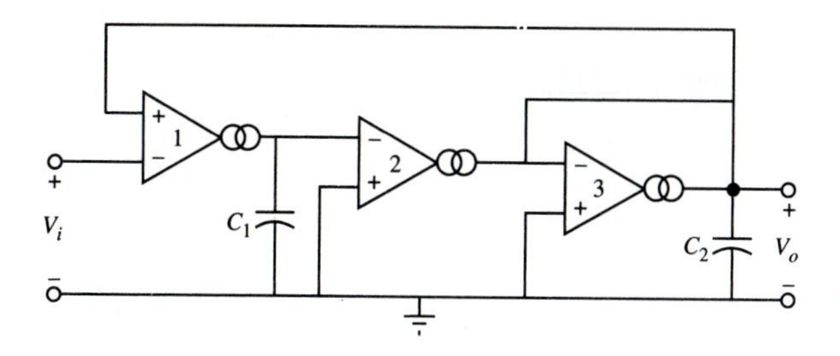

FIGURE 6.6-11
Second-order low-pass filter.

Example 6.6-2 OTA low-pass filter. It is desired to realize a second-order low-pass Butterworth function with a cutoff frequency of 10 kHz using the circuit shown in Fig. 6.6-11. The network function (for unity dc gain) is

$$\frac{V_o(s)}{V_i(s)} = \frac{(2\pi \times 10^4)^2}{s^2 + s\sqrt{2}(2\pi \times 10^4) + (2\pi \times 10^4)^2}$$

Comparing this with (16), we can choose $C_1 = C_2 = 318$ pF, $g_{m1} = g_{m2} = 20\ \mu\text{S}$, and $g_{m3} = 20\sqrt{2}\ \mu\text{S}$.

The circuit shown in Fig. 6.6-11 can be redrawn and generalized to obtain a wide variety of second-order functions. The generalization is achieved in part by allowing the capacitors C_1 and C_2 to be ungrounded so that they may be used as series input elements. The circuit is shown in Fig. 6.6-12.[1] In the figure, certain nodes have been numbered 1 to 7. As shown, switches are used so that nodes 1 to 5 may be either connected to the input $V_i(s)$ or grounded. Another switch is used to select nodes 6 or 7 as output. The

[1] Henrique S. Malavar, "Electronically Controlled Active-C Filters and Equalizers with Operational Transconductance Amplifiers," *IEEE Trans. Circuits Systems*, vol. CAS-31, no. 7, July 1984, pp. 645–649.

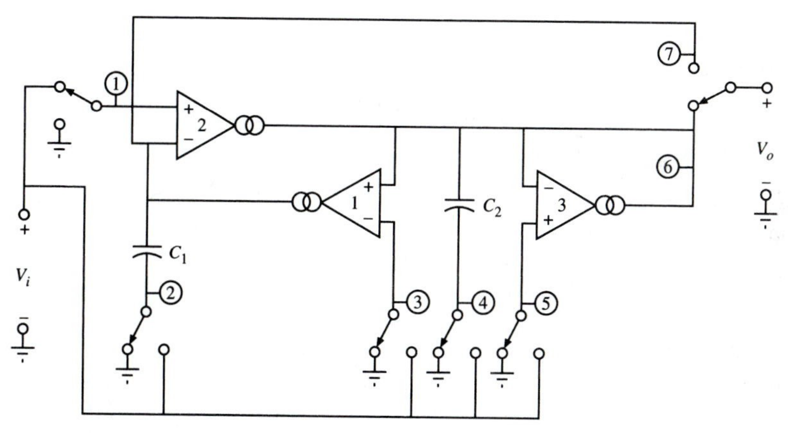

FIGURE 6.6-12
General-purpose second-order filter.

TABLE 6.6-1
Connections and numerator terms for the filter shown in Fig. 6.6-12

Filter type	Numerator polynomial	Input nodes	Output node	Grounded nodes
Low pass I	$\frac{g_{m1}g_{m2}}{C_1C_2}$	3 or 1	6 7	1, 2, 4, 5 2, 3, 4, 5
Low pass II	$\frac{g_{m1}g_{m3}}{C_1C_2}$	5	7	1, 2, 3, 4
Band pass I	$\frac{sg_{m2}}{C_2}$	1 or 4	6 7	2, 3, 4, 5 1, 2, 3, 5
Band pass II	$\frac{sg_{m3}}{C_2}$	5	6	1, 2, 3, 4
High pass	s^2	4	6	1, 2, 3, 5
Notch	$s^2 + \frac{g_{m1}g_{m2}}{C_1C_2}$	3, 4 or 1, 2, 3, 4	6 6	1, 2, 5 5

specific connection information and the resulting form of the numerator of the network function is given in Table 6.6-1. The denominator for all network functions is

$$D(s) = s^2 + s\frac{g_{m3}}{C_2} + \frac{g_{m1}g_{m2}}{C_1C_2} \tag{17}$$

As an example of the use of the table, the case where node 3 is specified as input in the "lowpass I" category corresponds with the circuit shown in Fig. 6.6-11.

Equalizers

The circuit shown in Fig. 6.6-12 may be further modified by adding a fourth OTA to permit the realization of second-order biquadratic network functions. The circuit is shown in Fig. 6.6-13. It realizes

$$\frac{V_o(s)}{V_i(s)} = \frac{s^2 + s(g_{m4}/C_2) + (g_{m1}g_{m2}/C_1C_2)}{s^2 + s(g_{m3}/C_2) + (g_{m1}g_{m2}/C_1C_2)} \tag{18}$$

For $g_{m4} > g_{m3}$, a response peak at resonance is obtained, while for $g_{m4} < g_{m3}$, a notch is obtained. This is sometimes referred to as a "bump equalizer." To obtain a delay equalizer, the input terminals of OTA 4 are reversed. This reverses the sign of the first-degree term in the numerator of (18) and, for $g_{m3} = g_{m4}$, produces an all-pass function that may be used for phase correction.

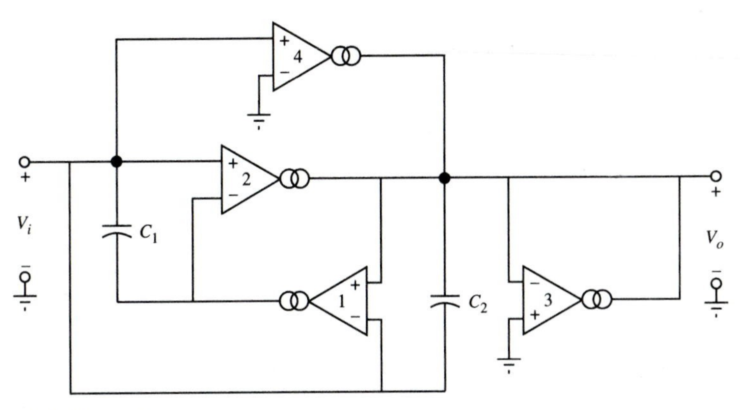

FIGURE 6.6-13
Biquadratic filter.

Programmed State-Variable OTA Filter

The state-variable filter introduced in Sec. 6.1 can be implemented using OTAs to determine the time constant of the integrators. The OTA control currents can then be used to program the frequency characteristic of the filter. A circuit for accomplishing this is shown in Fig. 6.6-14. In the circuit OTA 1 is also used to provide a summation of the feedback signals, thus eliminating the need for the third conventional operational amplifier normally used as a summer. Operational transductance amplifier 2 is used as a signal inverter, thus allowing both feedback signals to be summed at the same terminal of OTA 1. Assuming $Q \gg 1$, the transfer function can be written as

$$\frac{V_{BP}(s)}{V_i(s)} = \frac{sg_m/C}{s^2 + s\dfrac{g_m}{CQ} + \left(\dfrac{g_m}{C}\right)^2} \tag{19a}$$

$$\frac{V_{LP}(s)}{V_i(s)} = \frac{(g_m/C)^2}{s^2 + s\dfrac{g_m}{CQ} + \left(\dfrac{g_m}{C}\right)^2} \tag{19b}$$

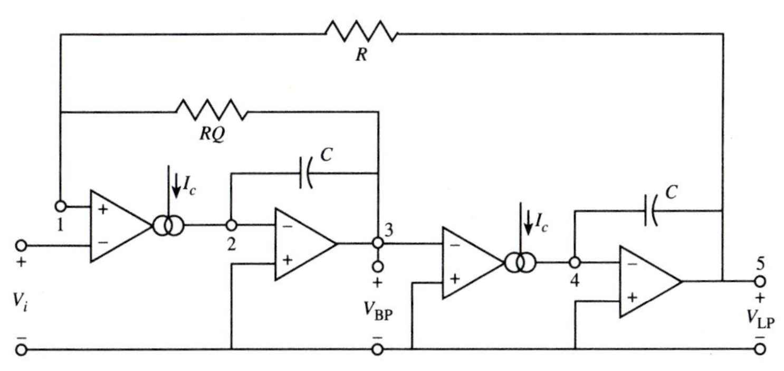

FIGURE 6.6-14
Programmable state-variable filter.

By using a dual OTA to implement the circuit, the inherent similar properties of the two amplifiers makes it possible to use a single current source of value $2I_c$ to control their g_m and thus to control the resonant frequency of the circuit. Modifications of the basic circuit are easily made to obtain a high-pass output and/or to make the Q programmable.

As a practical observation, it should be noted that actual implementations of many of the circuits presented in this section will have their input and feedback signals attenuated in order to maintain the OTA input voltages within the linear range of the device. The details as well as example circuits may be found in manufacturers' data books.[2]

6.7 USE OF COMPUTER PROGRAMS IN CASCADE MULTIPLE-AMPLIFIER SYNTHESIS

The computations required in designing the second-order active filter building blocks used as cascaded elements in this chapter are readily implemented on a PC. Many shareware and commercial PC programs are available that provide a wide range of design capabilities. A useful program is the Filter Master program originally introduced in Sec. 2.8. In addition to the wide variety of approximations provided, the program also designs a multiple-amplifier configuration to meet the specifications. An example follows.

---

[2] See, for example, *General Purpose Linear Devices Databook*, National Semiconductor Corporation, Santa Clara, CA, 1989.

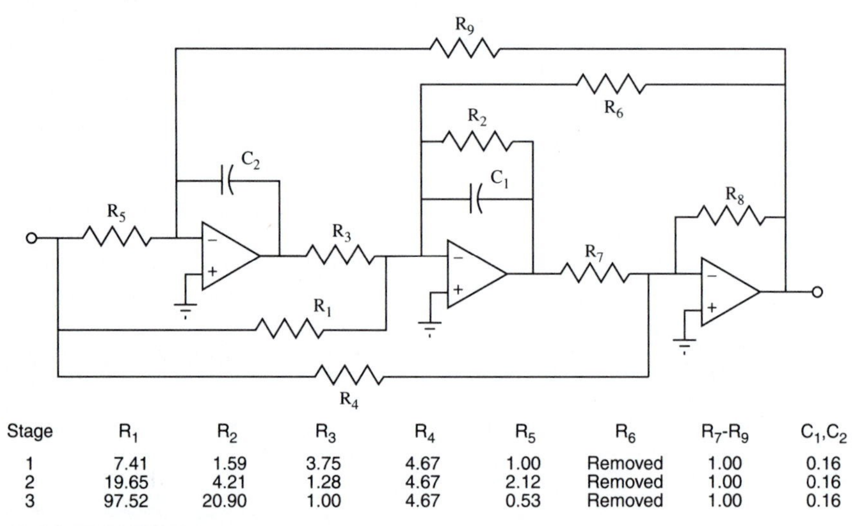

Stage	R_1	R_2	R_3	R_4	R_5	R_6	R_7-R_9	C_1,C_2
1	7.41	1.59	3.75	4.67	1.00	Removed	1.00	0.16
2	19.65	4.21	1.28	4.67	2.12	Removed	1.00	0.16
3	97.52	20.90	1.00	4.67	0.53	Removed	1.00	0.16

All resistor values in kilohms
All capacitor values in microfarads

Press ESC to exit

FIGURE 6.7-1
Sixth-order elliptic filter.

Example 6.7-1 A sixth-order elliptic multiple-amplifier filter. As an example of the use of Filter Master to synthesize a cascaded multiple-amplifier filter, consider the design of an elliptic low-pass filter with a 1-dB ripple passband from 0 to 1 kHz and a stopband with a minimum of 40 dB attenuation starting at 1.1 kHz. When the equivalent normalized specifications of 0 to 1 rad/s passband and stopband starting at 1.1 rad/s are entered into the program, a set of poles and zeros are generated that match the ones given for the same approximation in Table B-1 of Appendix B. If the actual specifications are entered into the program, the resulting design is shown in Fig. 6.7-1.

Information on the availability of Filter Master is given in Sec. 2.8.

Another useful computational tool for the design of multiple-amplifier filters is the Active RC program described in Sec. 5.8. In addition to providing realizations of single-amplifier filters, it also includes low-pass Butterworth and Chebyshev state-variable and Tow-Thomas configurations for use in cascade synthesis.

PROBLEMS

Section 6.1

1. (*a*) Design a state-variable band-pass filter with $Q = 10$ and $f_n = 10$ kHz using design 1. Determine the voltage transfer function that is (ideally) realized by the filter. Choose $C = 10^{-9}$ F.

(*b*) Repeat part (*a*) using design 2.

2. (*a*) Design a state-variable low-pass filter with $Q = 10$ and $f_n = 10$ kHz using design 1. Determine the voltage transfer function that is (ideally) realized by the filter. Choose $C = 10^{-9}$ F.

(*b*) Repeat part (*a*) using design 2.

3. (*a*) Design a state-variable high-pass filter with $Q = 10$ and $f_n = 10$ kHz using design 1. Determine the voltage transfer function that is (ideally) realized by the filter. Choose $C = 10^{-9}$ F.

(*b*) Repeat part (*a*) using design 2.

4. (*a*) Use the state-variable filter configuration given in Fig. 6.1-2 and design 1 to realize a band-pass function with $Q = 100$ and resonant frequency $f_n = 100$ Hz. Let $C_1 = C_2 = 0.1$ μF and $R_3 = R_5 = R_6 = 10$ kΩ. Find the resulting value of H_0 in (1) of Sec. 5.3.

(*b*) If the same circuit is used to realize a low-pass function, what is the value of the constant H_0 in (1) of Sec. 5.2?

(*c*) If the same circuit is used to realize a high-pass function, what is the value of the constant H_0 in (3) of Sec. 5.3?

5. In the state-variable filter configuration shown in Fig. 6.1-2, the elements have the following values:

$$R_1 = R_2 = R_3 = R_5 = R_6 = 10\ \text{k}\Omega \qquad R_4 = 190\ \text{k}\Omega$$
$$C_1 = C_2 = 0.01\ \mu\text{F}$$

Find the voltage transfer function $V_{LP}(s)/V_1(s)$.

6. A normalized state-variable filter is designed for $Q = 10$ and $\omega_n = 1$ rad/s using design 1. Find the percentage change in the magnitude of the voltage transfer function at the frequency 0.975 rad/s if all components have a +1 percent change from their design values.

Section 6.2

1. (*a*) Use the Tow-Thomas circuit shown in Fig. 6.2-2 to realize a band-pass second-order function with $Q = 100$, resonant frequency $f_n = 100$ Hz, and a value of $|H_0|$ in (1) of Sec. 5.3 of unity. Let $C_1 = C_2 = 1\ \mu\text{F}$ and $R_2 = R_3$.
   (*b*) Repeat the design for a low-pass function with the same specifications but with H_0 in (1) of Sec. 5.2 having a magnitude of unity.
2. A Tow-Thomas band-pass filter is designed with $Q = 10$, $\omega_n = 1$ rad/s, and $|H_0| = 19$ using the element values given in Example 6.2-2. Use sensitivity methods to find the percentage change in the magnitude of the voltage transfer function at the lower band-edge frequency (0.95 rad/s) if all element values are changed by +1 percent.
3. Use the method given in Summary 5.4-1 to analyze the noninverting integrator shown in Fig. 6.2-4(*b*).
4. Use the method given in Summary 5.4-1 to analyze the Åkerberg-Mossberg filter shown in Fig. 6.2-5.
5. (*a*) Use the Åkerberg-Mossberg circuit shown in Fig. 6.2-5 to realize a band-pass second-order function with $Q = 100$, resonant frequency $f_n = 100$ Hz, and a value of $|H_0|$ in (1) of Sec. 5.3 of unity. Let $C_1 = C_2 = 1\ \mu\text{F}$ and $R_2 = R_3$.
   (*b*) Repeat the design for a low-pass function with the same specifications.

Section 6.3

1. Use the state-variable filter to design a normalized high-pass function with a 1-dB ripple in the passband, $1 \le \omega \le \infty$ rad/s, and a minimum of 17 dB attenuation in the stopband, $\omega \le 0.5$ rad/s. Find the range over which the gain at infinity may be adjusted by varying the resistor R_{10}.
2. Use the state-variable filter to design a fourth-order all-pass function with a Thomson characteristic and a normalized delay of 2 s.
3. Use the Tow-Thomas filter to realize a high-pass filter with a Butterworth magnitude characteristic, a 3-dB cutoff frequency of 10 kHz, and an infinite frequency gain of 10. Let $C_1 = C_2 = 0.001\ \mu\text{F}$.
4. In the circuit shown in Fig. P6.3-4, assume that all the resistors and capacitors have unity value. Find the voltage transfer function $V_2(s)/V_1(s)$.

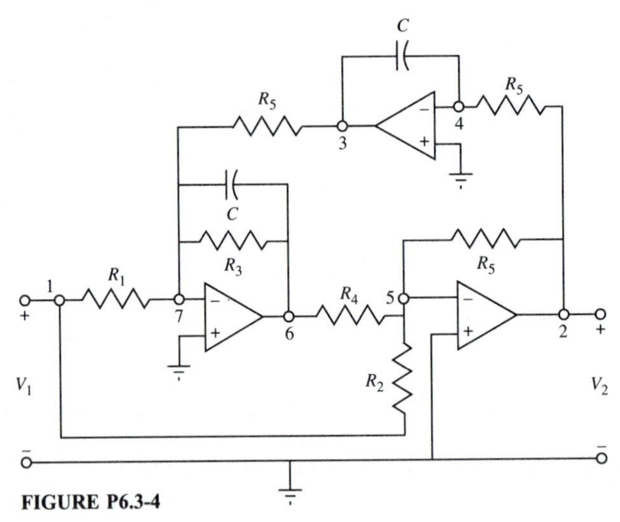

FIGURE P6.3-4

5. Use the Tow-Thomas filter to realize a second-order all-pass function with a Thomson characteristic and a normalized delay of 1 ms. Let $C_1 = C_2 = C_3 = 0.01\ \mu F$.
6. Use the Tow-Thomas filter to realize a second-order low-pass elliptic function with a 1-dB ripple in the passband, $0 \leq \omega \leq 1$ rad/s, and a minimum attenuation of 17 dB in the stopband, $\omega \geq 2$ rad/s.
7. Use the Tow-Thomas filter to design a normalized second-order high-pass function with a 1-dB ripple in the passband, $1 \leq \omega \leq \infty$ rad/s, and a minimum of 17 dB attenuation in the stopband, $\omega \leq 0.5$ rad/s. The gain at infinite frequency should be unity.
8. For the filter circuit shown in Fig. P6.3-8, find the voltage transfer function $V_2(s)/V_1(s)$ for the case where all the resistors except R_3 have the value R and $C_1 = C_2 = C$.

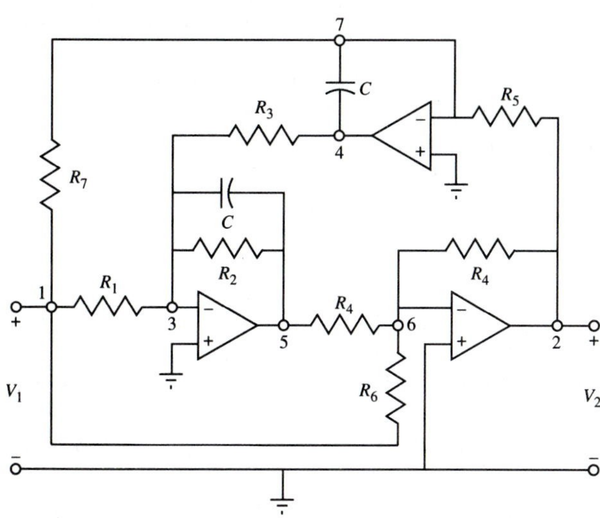

FIGURE P6.3-8

Section 6.4

1. (*a*) Use the design procedure given in Table 6.4-1 to design a band-pass filter in which $H_0 = 1$, $f_n = 1000$ Hz, and $Q = 50$ for the case where the passband gain has a phase of $\pm 180°$.
   (*b*) Repeat part (*a*) for the case where the phase is 0°.
2. (*a*) For the filters designed in Prob. 1, if the magnitude of the voltage at the high-pass output is 1 V peak to peak, what are the magnitudes of the voltages at the low-pass and band-pass outputs when the excitation frequency $\omega = \omega_n/10$?
   (*b*) Repeat part (*a*) for the condition $\omega = \omega_n$.
   (*c*) Repeat part (*a*) for the condition $\omega = 10\omega_n$.

Section 6.5

1. (*a*) A state-variable filter is to have design values (assuming infinite values of GB) of $Q = 10$ and $f_n = 10$ kHz. Use design 1 to find the resistor values for the case where $C_1 = C_2 = 0.001\ \mu F$.
   (*b*) Repeat part (*a*) for design 2.
   (*c*) What is the smallest value of gain bandwidth in the design 1 realization for which the filter will be stable?
   (*d*) What is the smallest value of gain bandwidth in the design 2 realization for which the filter will be stable?

2. (*a*) Design a (normalized) low-pass state-variable filter in which (for infinite operational amplifier GB), $\omega_n = 1$ rad/s and $Q = 4$. The filter must also have the property that it will be stable if realized with operational amplifiers for which $\text{GB}_n \geq 10$. Use the values $C_1 = C_2 = 1$ F and find the values of the other network elements.
   (*b*) Find the network function that is actually realized by the design of part (*a*) (with infinite GB operational amplifiers).
   (*c*) What is the network function that is realized for *R* and *C* values found in part (*a*) if the operational amplifiers have $\text{GB}_n = 10$.
3. (*a*) A state-variable filter is to have $s^2 + 0.1s + 1$ as a factor of the denominator polynomial of its network function. Find the coefficients of (7) for the case where $\text{GB}_n = 100$, and evaluate the resulting polynomial for $s = -0.05 \pm j0.99875$ to determine how closely the desired poles are realized (ideally, the evaluated polynomial should have a value of zero).
   (*b*) Repeat part (*a*) for $\text{GB}_n = 8$.
   (*c*) Evaluate the denominator polynomial found in part (*b*) at the *s*-plane location predicted by the chart shown in Fig. 6.5-1 to verify the results given in the chart (ideally, the evaluated polynomial should have a value of zero).
4. (*a*) A state-variable filter is to have $s^2 + 0.1s + 1$ as a factor of the denominator polynomial of its network function. Find the coefficients of (9) for the case where $\text{GB}_n = 100$, and evaluate the resulting polynomial for $s = -0.05 \pm j0.99875$ to determine how closely the desired poles are realized (ideally, the evaluated polynomial should have a value of zero).
   (*b*) Repeat part (*a*) for $\text{GB}_n = 8$.
   (*c*) Evaluate the denominator polynomial found in part (*b*) at the *s*-plane location predicted by the chart shown in Fig. 6.5-2 to verify the results given in the chart (ideally, the evaluated polynomial should have a value of zero).

Section 6.6

1. Analyze the circuit shown in Fig. P6.6-1 to find the voltage transfer function $V_o(s)/V_i(s)$.

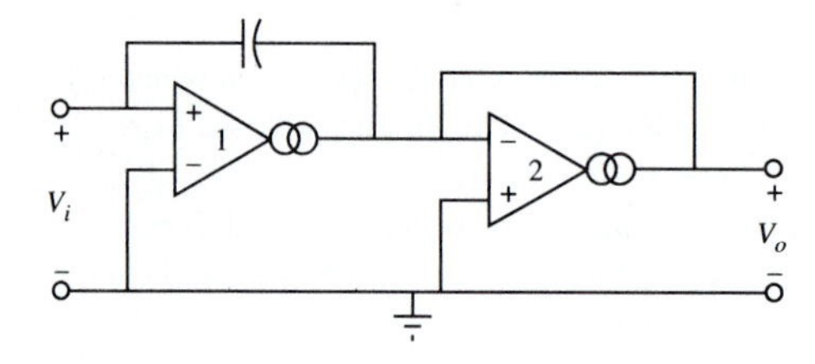

FIGURE P6.6-1

2. Analyze the circuit shown in Fig. P6.6-2 to find the voltage transfer function $V_o(s)/V_i(s)$.

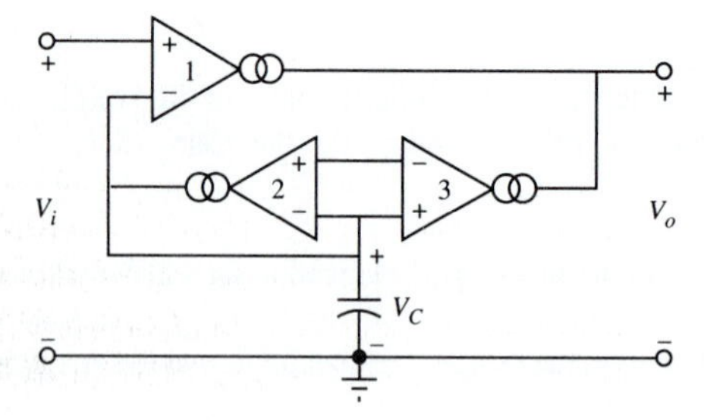

FIGURE P6.6-2

3. Derive the voltage transfer function given in (16) for the circuit shown in Fig. 6.6-11.

4. In Table 6.6-1, verify the "band pass I" results for the case where input node 1 is used.

5. (*a*) Identify the circuit shown in Fig. P6.6-5 as being one of the cases listed in Table 6.6-1.

(*b*) Derive the voltage transfer function for the circuit and show that it agrees with the results given in the table.

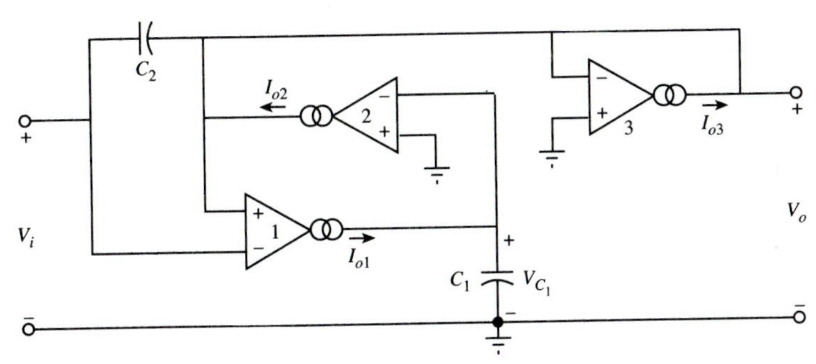

FIGURE P6.6-5

6. Derive the voltage transfer function given in (18) for the circuit shown in Fig. 6.6-13.

7. (*a*) Derive the voltage transfer functions $V_{BP}(s)/V_i(s)$ and $V_{LP}(s)/V_i(s)$ for the circuit shown in Fig. 6.6-14.

(*b*) Show that, for $Q \gg 1$, the results of part (*a*) may be written as shown in (19).

CHAPTER 7

DIRECT REALIZATION METHODS

In the first section of Chap. 5, we pointed out that there are two general methods for using active *RC* filters to realize network functions. The first such method was the subject of the last two chapters. It was the *cascade method* in which *RC* amplifier filters were used to realize second-order functions, and a cascade of such realizations was used to obtain higher order functions. In this chapter we discuss the second general method for using active *RC* filters to realize network functions. It is called the *direct method* and has the characteristic that filter functions of any order are realized by a single circuit configuration.

7.1 ACTIVE NETWORK ELEMENTS FOR DIRECT REALIZATIONS

The first direct method that we shall consider starts with a passive prototype network and uses active *RC* circuits to simulate certain elements of the passive realization. Such a method is called a *passive network simulation method*. It has the advantage of low sensitivities, basically the same ones that characterize the prototype passive network. In addition, since tables of element values for passive network realizations are readily available, the synthesis procedures for such simulation methods become very straightforward. We shall encounter other advantages and disadvantages of these methods in connection with our discussion of individual filter types.

Transmission (*ABCD*) Parameters

The characteristics of active *RC* networks that simulate the elements of passive realizations are best defined by using the transmission (*ABCD*) two-port parameters. They are defined by the equation

$$\begin{bmatrix} V_1(s) \\ I_1(s) \end{bmatrix} = \begin{bmatrix} A(s) & B(s) \\ C(s) & D(s) \end{bmatrix} \begin{bmatrix} V_2(s) \\ -I_2(s) \end{bmatrix} \tag{1}$$

where the voltage and current variables for the two-port system are as shown in Fig. 7.1-1. From (1), the individual parameters are readily seen to be

$$A(s) = \left. \frac{V_1(s)}{V_2(s)} \right|_{I_2(s)=0} \qquad B(s) = \left. \frac{V_1(s)}{-I_2(s)} \right|_{V_2(s)=0} \tag{2a}$$

$$C(s) = \left. \frac{I_1(s)}{V_2(s)} \right|_{I_2(s)=0} \qquad D(s) = \left. \frac{I_1(s)}{-I_2(s)} \right|_{V_2(s)=0} \tag{2b}$$

Note that $V_1(s)$ and $I_1(s)$ are the excitation variables and $V_2(s)$ and $-I_2(s)$ are the resulting responses. These equations specify the necessary testing conditions that must be placed on a two-port network to find the transmission parameters. For example, to find $A(s)$, from the first equation of (2*a*) we must apply a voltage $V_1(s)$ at port 1 and determine the resulting response voltage $V_2(s)$ at port 2 under the condition that the port is open-circuited $[I_2(s) = 0]$. This test and the ones for the other parameters are illustrated in the following example.

Example 7.1-1 The *ABCD* parameters of a simple network. As an example of the determination of the *ABCD* parameters, consider a two-port network consisting of a series impedance $Z_1(s)$ and a shunt impedance $Z_2(s)$. The network and the tests for the four parameters are shown in Fig. 7.1-2. The results are summarized as

$$\begin{bmatrix} V_1(s) \\ \\ I_1(s) \end{bmatrix} = \begin{bmatrix} \dfrac{Z_1(s) + Z_2(s)}{Z_2(s)} & Z_1(s) \\ \dfrac{1}{Z_2(s)} & 1 \end{bmatrix} \begin{bmatrix} V_2(s) \\ \\ -I_2(s) \end{bmatrix}$$

When finding the transmission parameters of some circuit configurations, especially those that contain VCVSs, situations may be encountered in which the response variables have a (theoretically) infinite value. From (2), however, the two-port parameters are proportional to the inverse of these response variables. Thus they will have a value of zero. This situation occurs most frequently in two cases: (1) when the testing

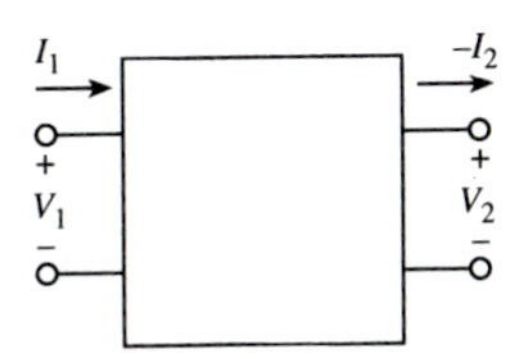

FIGURE 7.1-1
Two-port network.

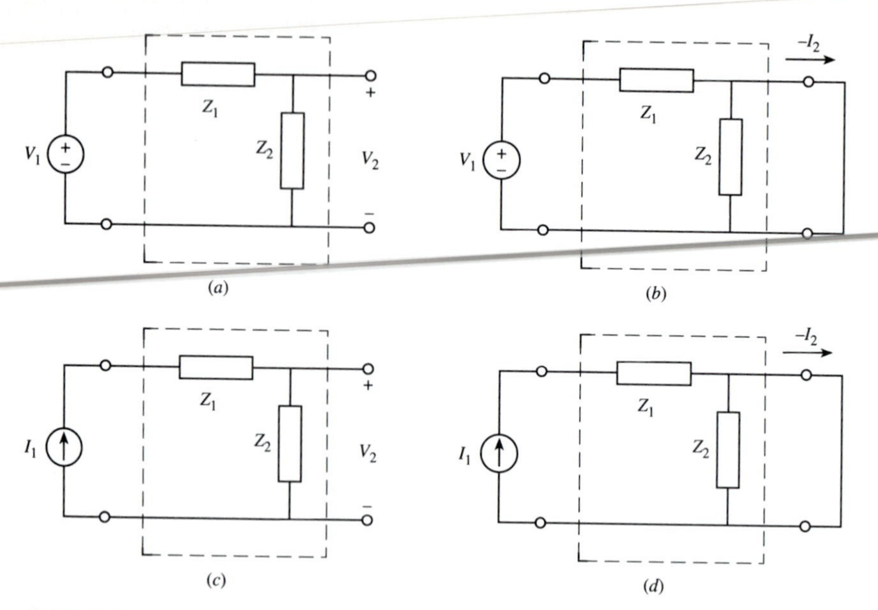

FIGURE 7.1-2
Tests for *ABCD* parameters: (*a*) $A(s)$; (*b*) $B(s)$; (*c*) $C(s)$; (*d*) $D(s)$.

condition requires that a short circuit be placed across a voltage source (and the resulting current from the source is infinite) and (2) when a current source is connected to an open circuit (and the resulting voltage across the source is infinite). Both of these cases are illustrated in the following example.

Example 7.1-2 The *ABCD* parameters of a noninverting VCVS. As an example of the situation where the response variables of a network have infinite values, consider the determination of the *ABCD* parameters for a noninverting VCVS as realized by an operational amplifier. The tests are shown in Fig. 7.1-3. Note that in Fig. 7.1-3(*b*), for the determination of $B(s)$, the voltage source that is the output element of the operational amplifier is effectively short circuited by the connection $[V_2(s) = 0]$ where $-I_2(s)$ is defined. As a result, $-I_2(s) = \infty$ and the parameter $B(s) = 0$. Similarly, in the test for $C(s)$ shown in Fig. 7.1-3(*c*), the output of the current source $I_1(s)$ is connected to an open circuit (the input impedance of the operational amplifier), and the input and output voltages of the operational amplifier are infinite. Since $V_2(s) = \infty$, $C(s) = 0$. A similar situation results in Fig. 7.1-3(*d*) and produces $D(s) = 0$. The resulting transmission parameters for the noninverting VCVS are

$$\begin{bmatrix} V_1(s) \\ \\ I_1(s) \end{bmatrix} = \begin{bmatrix} \dfrac{R_1}{R_1 + R_2} & 0 \\ \\ 0 & 0 \end{bmatrix} \begin{bmatrix} V_2(s) \\ \\ -I_2(s) \end{bmatrix}$$

Exercise 7.1-2. (*a*) Find the transmission parameters for the two-port network consisting of the series impedance $Z_1(s)$ shown in Fig. 7.1-4(*a*).

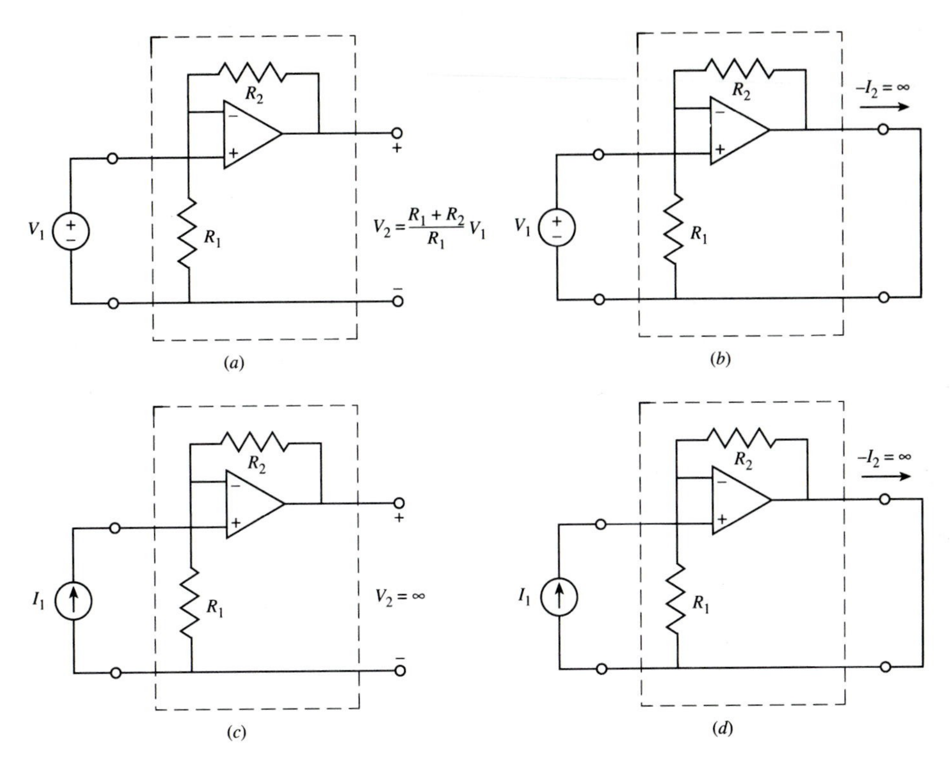

FIGURE 7.1-3
Tests for *ABCD* parameters: (*a*) $A(s)$; (*b*) $B(s)$; (*c*) $C(s)$; (*d*) $D(s)$.

(*b*) Find the transmission parameters for the two-port network consisting of the shunt admittance $Y_2(s) = 1/Z_2(s)$ shown in Fig. 7.1-4(*b*).

Answers

$$(a)\quad \begin{bmatrix} 1 & Z_1(s) \\ 0 & 1 \end{bmatrix} \qquad (b)\quad \begin{bmatrix} 1 & 0 \\ Y_2(s) & 1 \end{bmatrix}$$

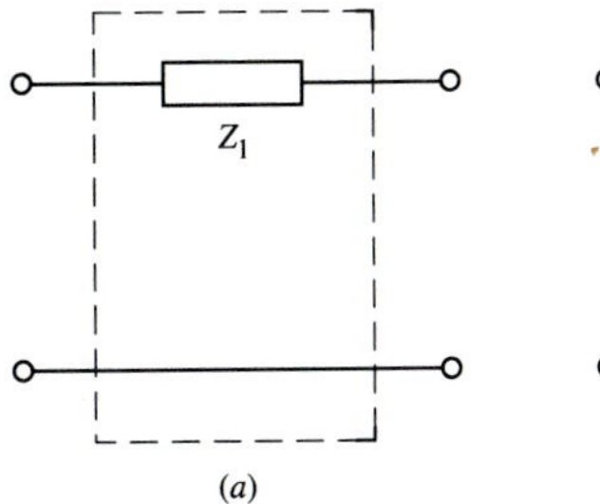

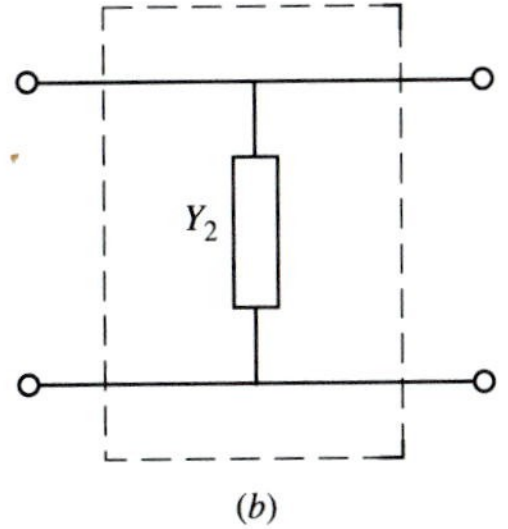

FIGURE 7.1-4
Example two-port networks.

For some two-port networks, a simplified analysis procedure may be used to determine some or all of the transmission parameters. For example, if a circuit has the property that $V_1(s) = V_2(s)$ no matter what is the value of $-I_2(s)$, then from (1) we can write $A(s) = 1$ and $B(s) = 0$. An example of this is given for the two-port network shown in Fig. 7.1-4(*b*) [see the transmission parameters given in Exercise 7.1-2(*b*)]. A second simplification occurs if a circuit has the property that $I_1(s) = -I_2(s)$ no matter what is the value of $V_2(s)$. In this case, from (1) we can write $C(s) = 0$ and $D(s) = 1$. An example of this is given for the two-port system shown in Fig. 7.1-4(*a*) [see the transmission parameters given in Exercise 7.1-2(*a*)]. A further use of these simplifications is given in connection with the circuit that follows.

Generalized Impedance Converter

We may now define a two-port active network device that may be used to develop several types of active *RC* passive network simulation realizations. It is called a *generalized impedance converter* (GIC). The circuit configuration is shown in Fig. 7.1-5. To analyze this circuit, we first recall that the voltage across the input terminals of an operational amplifier is (ideally) zero. In the figure this requires

$$V_{Z1}(s) - V_{Z2}(s) = 0 \quad \rightarrow \quad V_{Z1}(s) = V_{Z2}(s) \tag{3a}$$
$$V_{Z3}(s) - V_{Z4}(s) = 0 \quad \rightarrow \quad V_{Z3}(s) = V_{Z4}(s) \tag{3b}$$

As a result, $V_1(s) = V_2(s)$, and we find $A(s) = 1$ and $B(s) = 0$. To determine the values of $C(s)$ and $D(s)$, we start at the right end of the network and proceed with the steps in the following summary.

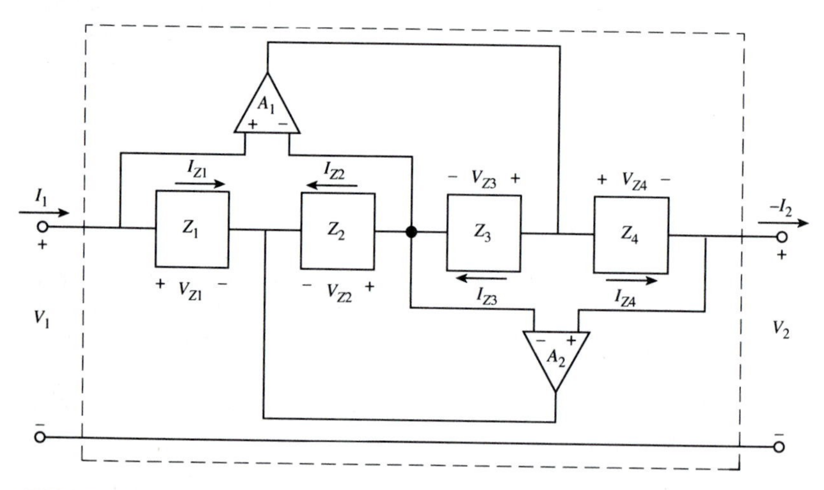

FIGURE 7.1-5
Realization of a GIC.

Summary 7.1-1 Analysis of a GIC realization. To analyze the two-operational-amplifier GIC configuration shown in Fig. 7.1-5, we may proceed as follows [for simplicity the (s) notation has been deleted]:

Step 1. $V_{Z4} = Z_4 I_{Z4} = Z_4(-I_2)$

Step 2. $V_{Z3} = Z_3 I_{Z3} = V_{Z4} = Z_4(-I_2)$ [from (3b) and step 1]

Step 3. $I_{Z3} = \dfrac{Z_4}{Z_3}(-I_2)$ [from step 2]

Step 4. $I_{Z2} = I_{Z3} = \dfrac{Z_4}{Z_3}(-I_2)$ [no current flows in the input terminals of the operational amplifiers]

Step 5. $V_{Z2} = Z_2 I_{Z2} = \dfrac{Z_2 Z_4}{Z_3}(-I_2)$ [from step 4]

Step 6. $V_{Z1} = Z_1 I_{Z1} = V_{Z2} = \dfrac{Z_2 Z_4}{Z_3}(-I_2)$ [from (3a) and step 5]

Step 7. $I_1 = I_{Z1} = \dfrac{Z_2 Z_4}{Z_1 Z_3}(-I_2)$ [from step 6]

In step 7 of the summary given above, we see that $I_1(s)$ is a function of the variable $-I_2(s)$ and not a function of $V_2(s)$. From this we conclude that $C(s) = 0$ and $D(s) = Z_2(s)Z_4(s)/Z_1(s)Z_3(s)$. In summary, the transmission parameters for the circuit shown in Fig. 7.1-5 are

$$\begin{bmatrix} 1 & 0 \\ 0 & \dfrac{Z_2(s)Z_4(s)}{Z_1(s)Z_3(s)} \end{bmatrix} \tag{4}$$

Note that in the GIC, the VCVSs that comprise the outputs of the operational amplifiers (these are not shown explicitly in Fig. 7.1-5) have one terminal connected to ground, that is, to the power supply ground. As a result, the negative reference polarities of the voltages $V_1(s)$ and $V_2(s)$ are also at ground potential. This means that the GIC cannot be used as a floating (ungrounded) element.

Generalized Impedance Network

If a terminating impedance $Z_5(s)$ is connected at the right port (port 2) of a two-port network as shown in Fig. 7.1-6, we define a one-port or two-terminal network that, as shown in the figure, has an input impedance $Z_{\text{in}}(s) = V_1(s)/I_1(s)$. An expression for this impedance in terms of the transmission parameters of the two-port network is easily found by taking the ratio of the separate equations of (1) and using the relation $V_2(s) = Z_5(s)[-I_2(s)]$. From this we obtain

$$Z_{\text{in}}(s) = \frac{V_1(s)}{I_1(s)} = \frac{A(s)V_2(s) - B(s)I_2(s)}{C(s)V_2(s) - D(s)I_2(s)} = \frac{A(s)Z_5(s) + B(s)}{C(s)Z_5(s) + D(s)} \tag{5}$$

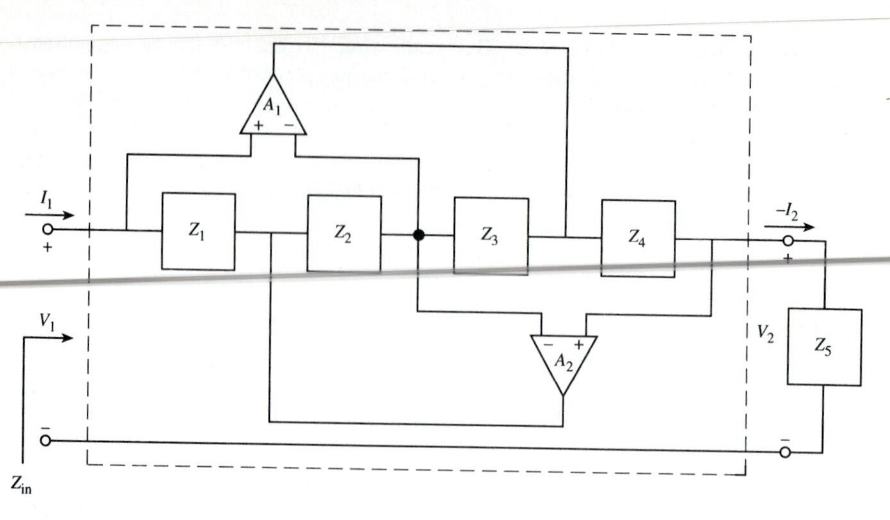

FIGURE 7.1-6
Realization of a GIN.

If, as shown in Fig. 7.1-6, the two-port network is the GIC shown in Fig. 7.1-5, then, using the *ABCD* parameters given in (4) in the expression for $Z_{in}(s)$ of (5), we obtain

$$Z_{in}(s) = \frac{Z_1(s)Z_3(s)Z_5(s)}{Z_2(s)Z_4(s)} \tag{6}$$

Similarly, by defining $Y_{in}(s) = 1/Z_{in}(s)$ and $Y_i(s) = 1/Z_i(s)$ $(i = 1, 2, \ldots 5)$, (6) may be written in the form

$$Y_{in}(s) = \frac{Y_1(s)Y_3(s)Y_5(s)}{Y_2(s)Y_4(s)} \tag{7}$$

The device defined by (6) and (7) is called a *generalized impedance network* (GIN). Note that since the GIC is grounded, one end of $Z_{in}(s)$ [the end where the negative reference polarity of $V_1(s)$ is defined] is also grounded. Note that this device can be used to realize a wide range of network elements, some of which are not otherwise physically realizable.

Example 7.1-3 Forms of $Z_{in}(s)$ realizable by a GIN. As examples of some of the types of impedances a GIN can realize, consider the case where one of the elements $Z_i(s)$ $(i = 1, 2, \ldots 5)$ is a capacitor (impedance $1/sC$) and the other impedances are resistors. For this choice for $Z_2(s)$ or $Z_4(s)$, $Z_{in}(s)$ acts like an inductor (impedance Ks). For the case where both $Z_2(s)$ and $Z_4(s)$ are capacitors, $Z_{in}(s)$ acts like an element with impedance Ks^2 that has no passive equivalent. Similarly, for any two of impedances $Z_1(s)$, $Z_3(s)$, and $Z_5(s)$ capacitors, $Z_{in}(s)$ acts like an element with impedance K/s^2, which also has no passive equivalent. We shall find applications for many of these GIN impedances and also for the GIC itself in the following sections of this chapter.

7.2 INDUCTANCE SIMULATION

One of the simplest ways of obtaining an active *RC* direct realization is the passive simulation method. In this approach, the starting point is a passive *RLC* network (usually normalized) that satisfies the specific filter requirements. The inductors in the network are then replaced by active *RC* equivalents. The resulting elements are called synthetic inductors.

Synthetic Inductors

The GIN introduced in Sec.7.1 is ideally suited for the realization of a synthetic inductor. If $Z_4(s)$ is a capacitor C_4 and the remaining elements Z_i $(i = 1, 2, 3, 5)$ are resistors R_i, then, from (6) of Sec. 7.1 we obtain

$$Z_{\text{in}}(s) = \frac{sR_1R_3C_4R_5}{R_2} \qquad L_{\text{eq}} = \frac{R_1R_3C_4R_5}{R_2} \tag{1}$$

A synthetic inductor realized by a GIN turns out to have excellent characteristics. If the amplifiers A_1 and A_2 of Fig. 7.1-6 are matched, then the effect of the nonideal parameters of the amplifiers on the synthetic inductance are minimized. In addition, since $Z_1(s)$ is always a resistance, the dc bias currents at the inputs of the amplifiers do not present difficulties. The only disadvantage of the synthetic inductance approach to network synthesis is that since the realization of the synthetic inductor is an active device, it must have one terminal grounded. Thus the *RLC* network must also be of a type that has all inductors grounded. This restricts the synthetic inductance approach primarily to high-pass realizations. The FDNR technique, to be discussed in Sec. 7.3, will be seen to apply to low-pass and band-pass realizations. Floating synthetic inductors can be made, but their performance and simplicity are not as good as those of the grounded synthetic inductors.

In order to test the properties (and idealness) of a synthetic inductor, we may use a test circuit in which a GIN is driven by a voltage source with an output resistance R_s as shown in Fig. 7.2-1. Such a circuit allows us to observe the voltages $V_{o1}(s)$ and $V_{o2}(s)$ at the output terminals of the operational amplifiers with respect to the input GIN voltage $V_1(s)$. For the general case, an analysis technique similar to that outlined in Summary 7.1-1 may be used (see the Problems) to determine

$$\frac{V_{o1}(s)}{V_1(s)} = 1 + \frac{Z_4(s)}{Z_5(s)} \qquad \frac{V_{o2}(s)}{V_1(s)} = 1 - \frac{Z_2(s)Z_4(s)}{Z_3(s)Z_5(s)} \tag{2}$$

For the resistor and capacitor assignments used for (1), these equations become

$$\frac{V_{o1}(s)}{V_1(s)} = 1 + \frac{1}{sR_5C_4} = \frac{s + 1/R_5C_4}{s} \tag{3a}$$

$$\frac{V_{o2}(s)}{V_1(s)} = 1 - \frac{R_2}{sR_3R_5C_4} = \frac{s - R_2/(R_3R_5C_4)}{s} \tag{3b}$$

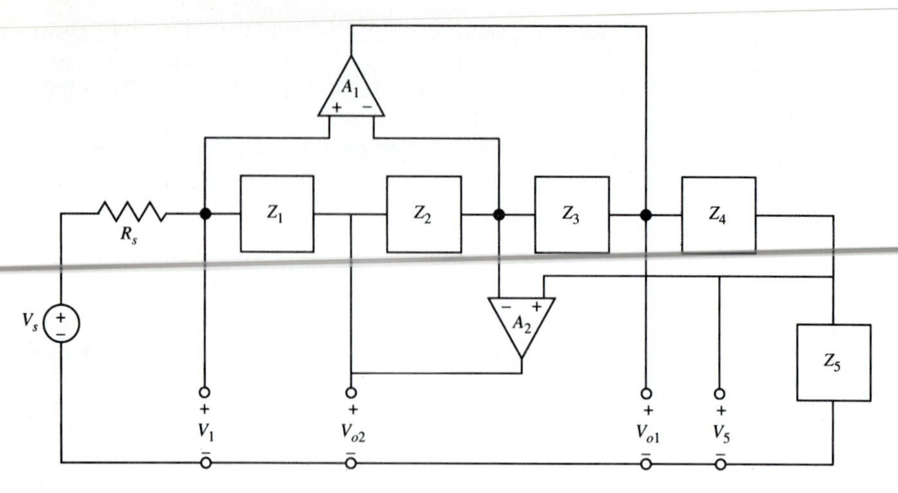

FIGURE 7.2-1
A GIN excited by a nonideal voltage source.

Bode plots of these functions are shown in Fig. 7.2-2. These functions have two applications. First, they illustrate the very desirable condition that no saturation of the operational amplifier outputs occurs as frequency increases. Second, they demonstrate a breakpoint that may be used as a tuning criteria to adjust the element values of the realization to more closely match its performance to that of the desired inductance value.

The steps used in applying the synthetic inductance technique are given in the following summary.

Summary 7.2-1 Synthesis using synthetic inductors

1. Design a passive *RLC* high-pass normalized network using the tables in App. A and the low-pass to high-pass transformation given in Sec. 2.5.

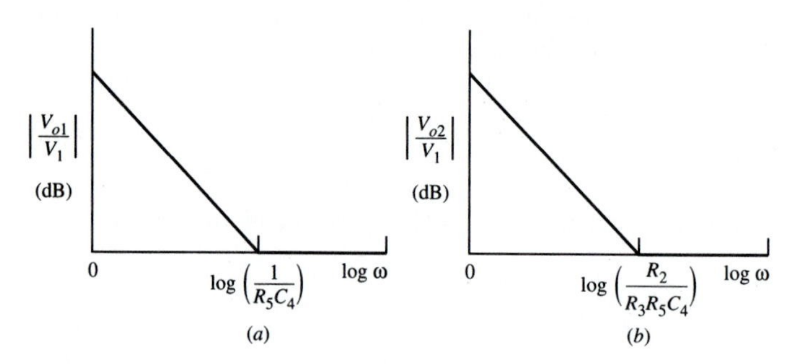

FIGURE 7.2-2
Bode plots: (*a*) Eq. (3*a*); (*b*) Eq. (3*b*).

2. Design the synthetic inductors that will replace the inductors in the passive prototype by using $Z_4 = sC_4$ in the circuit of Fig. 7.1-6. The other impedances are resistors $[Z_i(s) = R_i\ (i = 1, 2, 3, 5)]$. Choose normalized values of the resistors and the capacitor that satisfy (1).
3. Denormalize the resistors and capacitors (including those in the synthetic inductor) to meet the frequency specifications and to provide a proper impedance level.

Example 7.2-1 A fifth-order high-pass Butterworth synthetic inductor filter. A high-pass filter is to be realized using the synthetic inductor approach. The cutoff frequency is to be 10^4 rad/s, and the source and load resistors are 1000 Ω each. Using App. A and applying the low-pass to high-pass transformation of Sec. 2.5, we obtain the normalized high-pass prototype network shown in Fig. 7.2-3(a). The synthetic inductors are easily designed from (1) by letting $R_1 = R_2 = R_3 = 1.0\ \Omega$, $C = 1$ F, and $R_5 = 0.618\ \Omega$. The resulting normalized realization is shown in Fig. 7.2-3(b). In order to obtain the denormalized realization, all capacitors should be multiplied by 10^{-7} and all resistors by 10^3. The synthetic inductors should be tuned as discussed above.

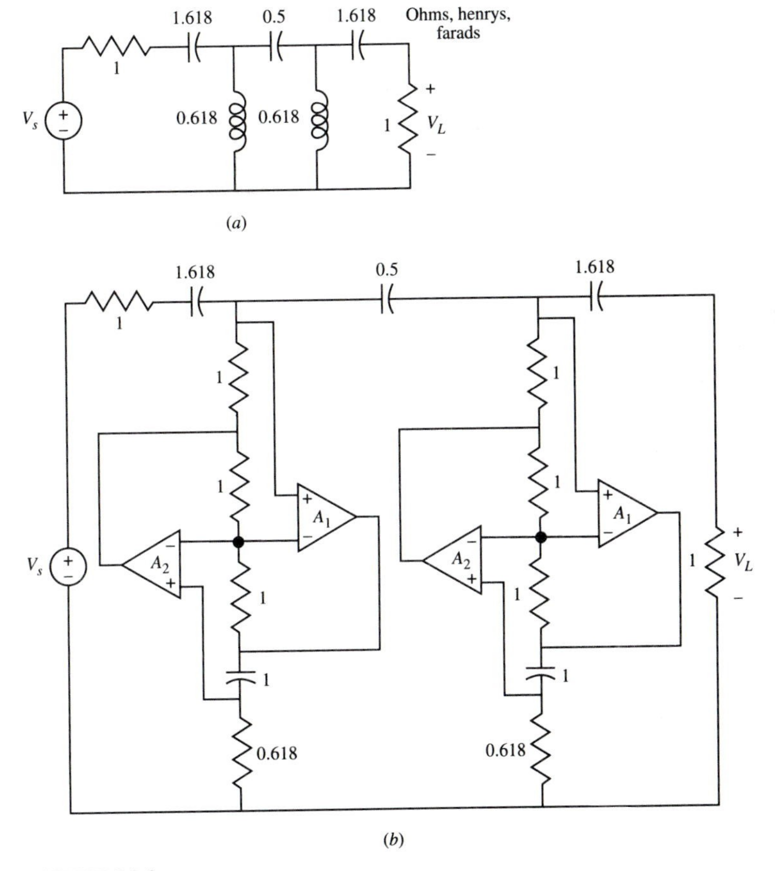

FIGURE 7.2-3
The fifth-order high-pass filter realized in Example 7.2-1.

Sensitivity

One of the major advantages of the passive simulation method of active *RC* synthesis is that the excellent sensitivity characteristics of the high-pass passive prototype are retained in the active realization. This is true not only for the prototype resistors and capacitors that appear directly in the active realization but also for the elements of the GINs that realize the synthetic inductors. To see this, note from (1) that

$$S^{L_{eq}}_{R_1,R_3,C_4,R_5} = 1 \qquad S^{L_{eq}}_{R_2} = -1 \tag{4}$$

If we now consider an arbitrary sensitivity S^y, where y is any usual network characteristic such as the network function $N(s)$, the magnitude function $|N(j\omega)|$, the Q, a numerator denominator coefficient a_i or b_i, and so forth, then we have

$$S^y_{R_1,R_3,C_4,R_5} = S^y_{L_{eq}} S^{L_{eq}}_{R_1,R_3,C_4,R_5} = S^y_{L_{eq}} \tag{5a}$$

$$S^y_{R_2} = S^y_{L_{eq}} S^{L_{eq}}_{R_2} = -S^y_{L_{eq}} \tag{5b}$$

From the above we see that the sensitivities to the elements of the GIN are ± 1 times the sensitivities to the inductor that is simulated.

Example 7.2-2 Sensitivity of a synthetic inductor filter. As an example of the determination of the sensitivities of a synthetic inductor filter, consider the third-order high-pass filter shown in Fig. 7.2-4(*a*). The elements have been numbered so as to correspond with the element designations given for the analysis method described in Summary 4.9-1. Applying this method, for the "four element—output *Z*" figure shown in Fig. 4.9-15(*b*), we may identify

$$Z_1(s) = R_1 \qquad Y_2(s) = sC_2 \qquad Z_3(s) = sL_3 \qquad Y_4(s) = \frac{sC_4}{sC_4R_4 + 1}$$

From Table 4.9-5 we obtain

$$\frac{V_2(s)}{V_1(s)} = \frac{s^3R_1C_2L_3C_4/(sC_4R_4+1)}{1 + s^2L_3C_4/(sC_4R_4+1) + sR_1C_2 + s^2C_2L_3 + s^3R_1C_2L_3C_4/(sC_4R_4+1)}$$

This can be put in the form

$$\frac{V_2(s)}{V_1(s)} = \frac{s^3R_1C_2L_3C_4}{s^3(R_1+R_4)C_2L_3C_4 + s^2(R_1C_2R_4C_4 + C_2L_3 + L_3C_4) + s(R_1C_2 + R_4C_4) + 1}$$

Comparing this with the general form

$$\frac{V_2(s)}{V_1(s)} = \frac{a_3s^3}{b_3s^3 + b_2s^2 + b_1s + b_0}$$

we find

$$S^{a_3}_{L_3} = S^{b_3}_{L_3} = 1 \qquad S^{b_2}_{L_3} = \frac{L_3(C_2+C_4)}{b_2} < 1 \qquad S^{b_1}_{L_3} = S^{b_0}_{L_3} = 0$$

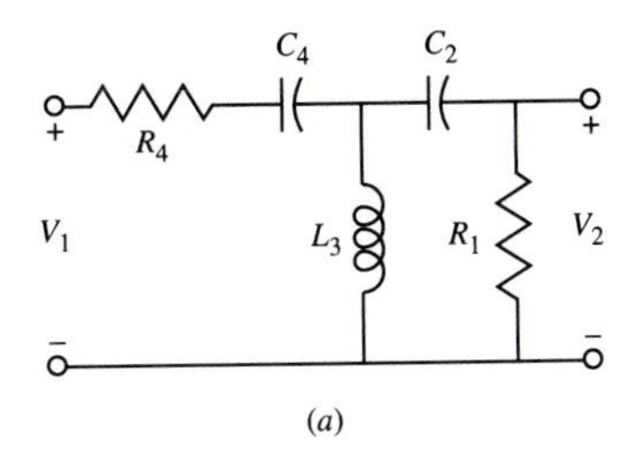

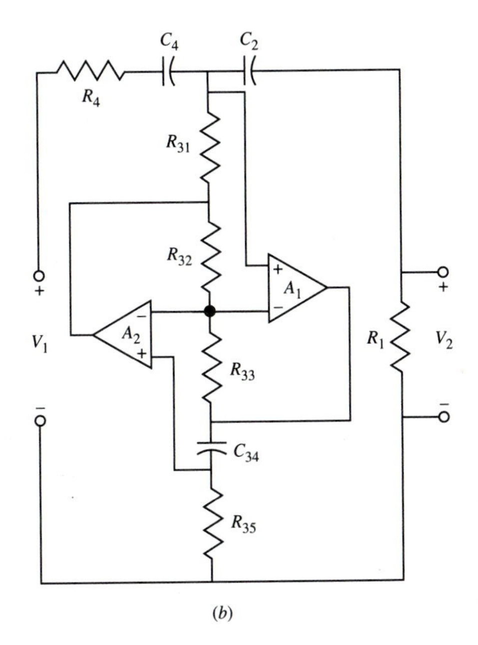

FIGURE 7.2-4
Third-order high-pass filter.

A synthetic inductor realization of the prototype passive filter is shown in Fig. 7.2-4(b). From (5) we see that

$$S^{a_3}_{R_{31},R_{33},C_{34},R_{35}} = S^{b_3}_{R_{31},R_{33},C_{34},R_{35}} = 1 \qquad S^{a_3}_{R_{32}} = S^{b_3}_{R_{32}} = -1$$

$$S^{b_2}_{R_{31},R_{33},C_{34},R_{35}} = \frac{L_3(C_2 + C_4)}{b_2} \qquad S^{b_2}_{R_{32}} = \frac{-L_3(C_2 + C_4)}{b_2}$$

$$S^{b_1}_{R_{31},R_{32},R_{33},C_{34},R_{35}} = S^{b_0}_{R_{31},R_{32},R_{33},C_{34},R_{35}} = 0$$

The results obtained above can be applied to determine the function sensitivity for a particular approximation. For example, consider a normalized high-pass Butterworth characteristic with a 3-dB frequency of 1 rad/s. For this we have

$$R_1 = 1\ \Omega \qquad C_2 = 1\ \text{F} \qquad L_3 = \tfrac{1}{2}\ \text{H} \qquad C_4 = 1\ \text{F} \qquad R_4 = 1\Omega$$

$$a_3 = \tfrac{1}{2} \qquad b_3 = 1 \qquad b_2 = 2 \qquad b_1 = 2 \qquad b_0 = 1$$

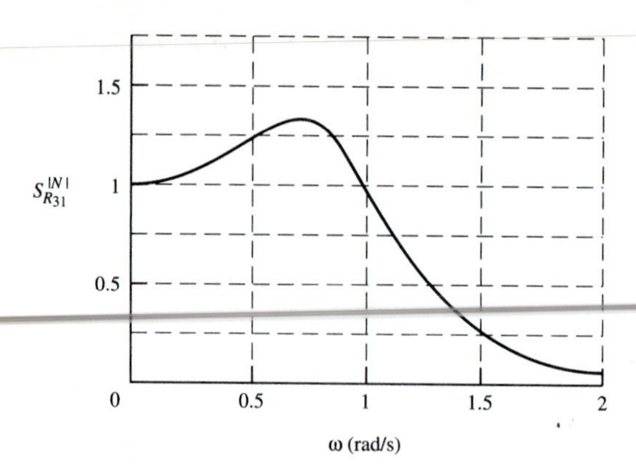

FIGURE 7.2-5
Sensitivity of $|N(j\omega)|$ for Example 7.2-2.

The function sensitivity for the GIN element R_{31} is found as follows:

$$S_{R_{31}}^{N(s)} = \frac{a_3 S_{R_{31}}^{a_3} s^3}{a_3 s^3} - \frac{b_3 S_{R_{31}}^{b_3} s^3 + b_2 S_{R_{31}}^{b_2} s^2}{b_3 s^3 + b_2 s^2 + b_1 s + b_0}$$

$$= 1 - \frac{s^3 + s^2}{s^3 + 2s^2 + 2s + 1} = \frac{s^2 + 2s + 1}{s^3 + 2s^2 + 2s + 1} = \frac{s+1}{s^2 + s + 1}$$

where $N(s) = V_2(s)/V_1(s)$. The function sensitivity for the elements R_{33}, C_{34}, and R_{35} is the same. The function sensitivity for the element R_{32} is the negative of this expression. A plot of the real part (for $s = j\omega$) of this expression is given in Fig. 7.2-5. As pointed out in Sec. 3.2, this defines the sensitivity of $|N(j\omega)|$.

Synthesis of Coupled Resonator Band-Pass Filters

The use of synthetic inductors as a passive simulation method is not only applicable to high-pass functions, but it may also be used to realize narrow-band band-pass functions. To do this, the passive prototype band-pass filter must have been designed using the coupled resonator method given in Sec. 4.10 in which all the inductors are grounded. An example of the procedure follows.

Example 7.2-3 Fourth-order synthetic inductor band-pass filter. A normalized fourth-order coupled resonator band-pass filter with a center frequency of 1 rad/s, a bandwidth of 0.1 rad/s, and a maximally flat (Butterworth) magnitude characteristic is shown in Fig. 7.2-6. The synthesis procedure used to produce the network is discussed in

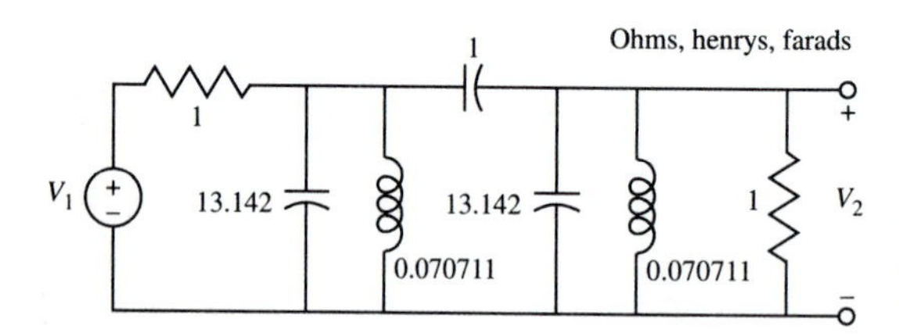

FIGURE 7.2-6
Fourth-order coupled resonator band-pass filter.

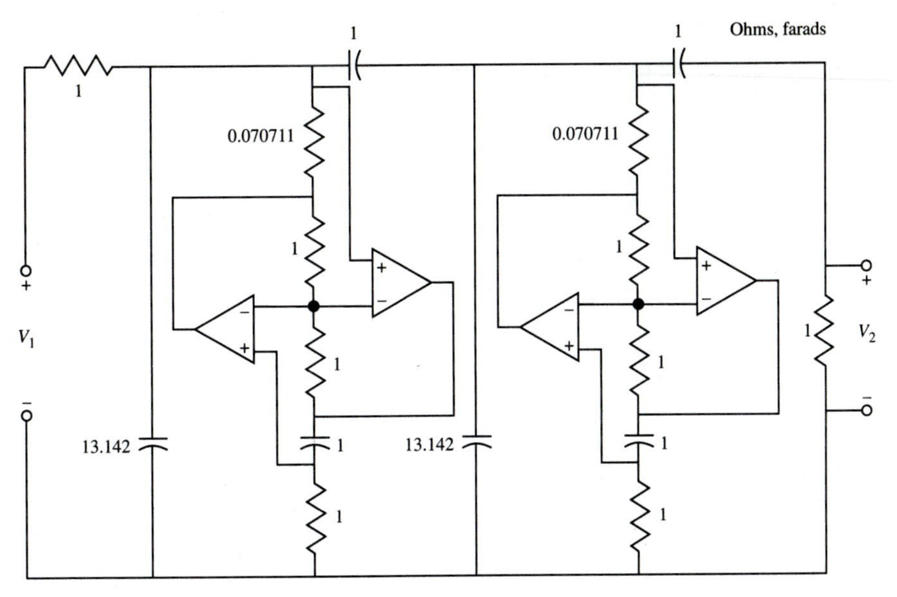

FIGURE 7.2-7
Synthetic inductor realization of coupled resonator band-pass filter.

Example 4.10-1. An active *RC* synthetic inductor realization of this filter is shown in Fig. 7.2-7. The realization has excellent sensitivity properties and provides a useful and easily implemented active filter.

7.3 FREQUENCY-DEPENDENT NEGATIVE RESISTORS

One of the most useful devices for implementing passive simulation methods is the *frequency-dependent negative resistor* (FDNR). This is a one-port active network element with the symbol shown in Fig. 7.3-1. Its input admittance is

$$Y(s) = \frac{I(s)}{V(s)} = s^2 D \tag{1}$$

where D is a positive real constant whose units are farad-seconds (Fs). If in the relation of (1) we let $s = j\omega$, we find $Y(j\omega) = -\omega^2 D$, which is a negative-valued quantity, is also

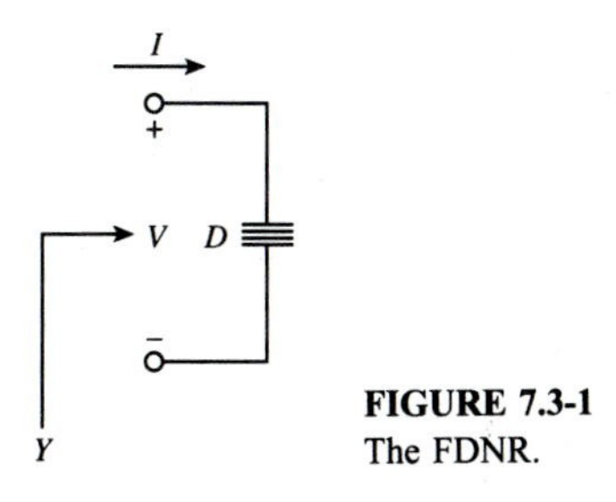

FIGURE 7.3-1
The FDNR.

a function of frequency and is real, that is, resistive in nature. Thus, its characteristics are aptly described by the term frequency-dependent negative resistor. An FDNR can be realized using the GIN of Fig. 7.1-6. For the general circuit the input admittance is

$$Y_{\text{in}}(s) = \frac{Y_1(s)Y_3(s)Y_5(s)}{Y_2(s)Y_4(s)} \tag{2}$$

where $Y_i(s) = 1/Z_i(s)$. To realize an FDNR, any two of the quantities $Y_1(s)$, $Y_3(s)$, and $Y_5(s)$ can be chosen as capacitors with the other elements being resistors. The preferred choice is $Y_1(s)$ and $Y_5(s)$. Note that if we let $Y_1(s) = Y_5(s) = sC$ and $Y_2(s) = Y_3(s) = Y_4(s) = G$, (2) reduces to

$$Y_{\text{in}}(s) = \left(\frac{C^2}{G}\right)s^2 = Ds^2 \tag{3}$$

This is the input of an FDNR in which $D = C^2/G$. The circuit is shown in Fig. 7.3-2.

Properties of an FDNR

Let us now determine some properties of an FDNR. Since it has a real negative value for $s = j\omega$, it is a potentially unstable one-port network. To see this, consider the circuit shown in 7.2-1 in which a GIN is driven by a voltage source $V_s(s)$ having a source impedance of R_s. Since the input voltages across the operational amplifier input terminals are zero, $V_1(s) = V_5(s)$. The voltage transfer function from V_s to V_1 or V_5 is

$$\frac{V_1(s)}{V_s(s)} = \frac{V_5(s)}{V_s(s)} = \frac{1}{1 + \dfrac{R_s Z_2(s) Z_4(s)}{Z_1(s) Z_3(s) Z_5(s)}} \tag{4}$$

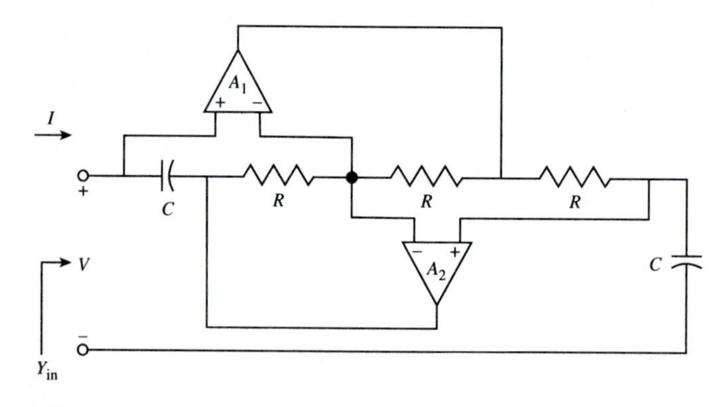

FIGURE 7.3-2
Using the GIN to realize an FDNR.

For a realization in which $Z_1(s) = 1/sC_1$, $Z_2(s) = R_2$, $Z_3(s) = R_3$, $Z_4(s) = R_4$, and $Z_5(s) = 1/sC_5$, (4) becomes

$$\frac{V_1(s)}{V_s(s)} = \frac{R_3/R_sR_2R_4C_1C_5}{s^2 + R_3/R_sR_2R_4C_1C_5} = \frac{\omega_0^2}{s^2+\omega_0^2} \tag{5}$$

where we have defined

$$\omega_0 = \sqrt{\frac{R_3}{R_sR_2R_4C_1C_5}} \tag{6}$$

The frequency ω_0 is the radian frequency at which the FDNR "resonates" with the source resistor R_s, that is, the frequency of potential instability. In practice, the instability predicted in (5) may not occur, since the gains of the operational amplifiers A_1 and A_2, being finite, will cause the poles to shift slightly to the left of the $j\omega$ axis. The voltage transfer function of (5), however, will still have a high Q and, due to amplifier saturation, be difficult to measure. As an alternate and more observable characteristic, consider the transfer functions from $V_1(s)$ to $V_{o1}(s)$ and $V_{o2}(s)$ given for the general case in (2) of Sec. 7.2. For the choices of elements used here, we obtain

$$\frac{V_{o1}(s)}{V_1(s)} = 1 + sR_4C_5 = \frac{s + \omega_0\sqrt{R_sR_2C_1/R_3R_4C_5}}{\omega_0\sqrt{R_sR_2C_1/R_3R_4C_5}} \tag{7a}$$

$$\frac{V_{o2}(s)}{V_1(s)} = -\left[\frac{s - \omega_0\sqrt{R_sR_3C_1/R_2R_4C_5}}{\omega_0\sqrt{R_sR_3C_1/R_2R_4C_5}}\right] \tag{7b}$$

Either of these expressions is useful for experimentally verifying the performance of the FDNR realization.

Example 7.3-1 Bode plots of FDNR transfer functions. As an example of the determination of (7), consider the case where $C_1 = C_5 = 0.01\ \mu\text{F}$ and $R_2 = R_3 = R_4 = R_s = 10\ \text{k}\Omega$. For these values we find $D = 10^{-12}$ Fs and $\omega_0 = 10^4$ rad/s. The magnitudes of the transfer functions defined in (7) are identical and have the Bode plot shown in Fig. 7.3-3.

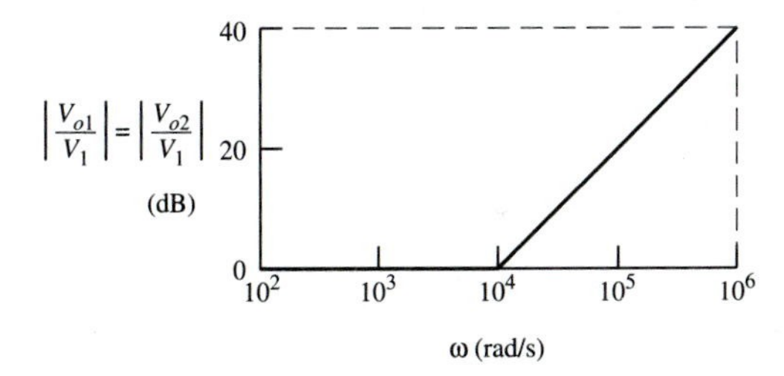

FIGURE 7.3-3
A Bode magnitude plot for (7).

Synthesis of Low-Pass Functions Using FDNRs

To apply the FDNR to the synthesis of filters, it is necessary to first perform a transformation proposed by Bruton.[1] This transformation is called the *RLC-CRD transformation*. It has the form of the basic impedance normalization given in (4) of Sec. 1.4 but for which the impedance normalization constant z_n is a function of s. Specifically, $z_n = 1/s$. Thus we may write

$$Z(s) = z_n Z_n(s) = \frac{Z_n(s)}{s} \qquad Y(s) = \frac{1}{z_n} Y_n(s) = sY_n(s) \tag{8}$$

As an example, consider the admittance of a general passive *RLC* network branch expressed as

$$Y_{RLC}(s) = G + \frac{1}{sL} + sC \tag{9}$$

Such a branch obviously consists of a paralleled resistor, inductor, and capacitor with values of $1/G$ ohms, L henrys, and C farads, respectively. Applying the *RLC-CRD* transformation to this branch, we obtain

$$Y_{CRD}(s) = sG + s^2C + \frac{1}{L} \tag{10}$$

which consists of a paralleled capacitor, FDNR, and resistor, with values of G farads, C $(= D)$ farad-seconds, and L ohms, respectively. The gain of a VCVS, since it is dimensionless, will be unaffected by such a transformation. Similarly, if the *RLC-CRD* transformation is applied to a two-port network, characterized by the transmission parameters defined in (2) of Sec. 7.1, then since the $A(s)$ and $D(s)$ parameters are dimensionless, they will not be affected by the transformation. The $B(s)$ and $C(s)$ parameters, however, since they have the dimensions of impedance and admittance, will be multiplied by $1/s$ and s, respectively. These results are summarized in Table 7.3-1.

Since the FDNR is an active element, it is important from power supply considerations that one end of it always be grounded. This is equivalent to requiring that all the capacitors in the prototype network to which the *RLC-CRD* transformation is to be applied also be grounded. Such a requirement is generally satisfied by low-pass filter realizations. Thus, the use of FDNRs is especially suitable to the realization of such filters. The design procedure may be outlined as follows.

Summary 7.3-1 Synthesis of low-pass functions using FDNRs

1. Design an *RLC* low-pass normalized prototype network using the tables in App. A.
2. Transform the *RLC* network into a *CRD* normalized network using the relationships of Table 7.3-1.

---

[1] L. T. Bruton, "Network Transfer Functions Using the Concepts of Frequency-dependent Negative Resistance," *IEEE Trans. Circuit Theory*, vol. CT-16, August 1969, pp. 406–408.

TABLE 7.3-1

Effect of the *RLC-CRD* transformation on various network elements

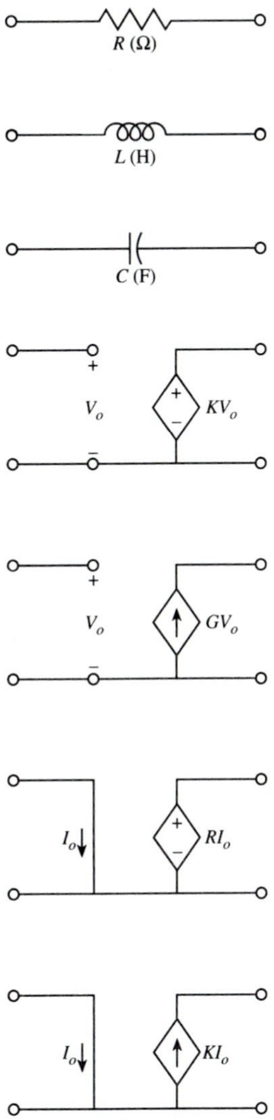

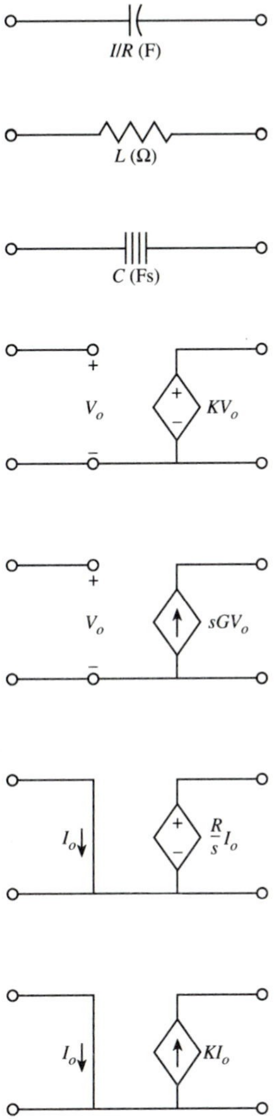

RLC network	*CRD* network
R (Ω)	$1/R$ (F)
L (H)	L (Ω)
C (F)	C (Fs)
V_o (+, −); KV_o	V_o (+, −); KV_o
V_o (+, −); GV_o	V_o (+, −); sGV_o
I_o; RI_o	I_o; $\frac{R}{s}I_o$
I_o; KI_o	I_o; KI_o
$\begin{bmatrix} A & B \\ C & D \end{bmatrix}$	$\begin{bmatrix} A & B/s \\ sC & D \end{bmatrix}$

3. Design the FDNRs using normalized resistor and capacitor values.
4. Provide paths for dc current from all amplifier input terminals to ground if they do not already exist. (Since the source and load resistors have been transformed to capacitors, it will be necessary to shunt one or both with resistors chosen so as to have negligible effect on the realization.)
5. Denormalize all the resistors and capacitors of the realization.

Example 7.3-2 A fifth-order low-pass Butterworth FDNR filter. It is desired to use FDNRs to realize a fifth-order double-terminated low-pass Butterworth filter having a cutoff frequency of 10^4 rad/s. The source and load resistors are required to have a value of 1000 Ω each. Using App. A, we obtain the normalized low-pass prototype network shown in Fig. 7.3-4(*a*). Applying the *RLC-CRD* transformation produces the network of Fig. 7.3-4(*b*). We now select an FDNR realization in which $Z_1(s) = 1/sC$, $Z_2(s) = Z_3(s) = R$, $Z_4(s) = R_4$, and $Z_5(s) = 1/sC$. From (1) and (2), $D = C^2R_4$. Choosing C as 1.0 F provides a realization in which all capacitors are equal valued. Resistance R can be any value since in (2) R_2 and R_3, which are equal to this value, cancel. Thus we may select $R = 1.0\ \Omega$. Resistor R_4 will be the element used to tune the FDNR. Since it is required that $D = 1.618$ Fs, the nominal value of R_4 is 1.618 Ω. The normalized realization of the filter is shown in Fig. 7.3-4(*c*). Here R_A and R_B have been added to permit dc current to flow from the positive terminals of A_{11} and A_{21} to ground. The terms R_A and R_B are determined in the following manner. At direct current we have

$$\frac{V_2(0)}{V_1(0)} = \frac{R_B}{R_A + R_B + 3.236} \tag{11}$$

The 3.236 corresponds to the total series resistance between R_A and R_B. If (11) is equal to $R_L/(R_s + R_L)$, then the circuit in Fig. 7.3-4(*c*) will have the same dc response as the prototype one in Fig. 7.3-4(*a*). This provides one relationship for solving for R_A and R_B. The other relationship is found by letting R_A or $R_B \gg R_L$ or R_s. This is necessary so that at the normalized cutoff frequency of 1 rad/s, R_A and R_B do not load C_s and C_L, respectively. Therefore, by equating (11) to 0.5 and by selecting $R_A = 100\ \Omega$, we get $R_B = 103.24\ \Omega$, as shown in Fig. 7.3-4(*c*). The final network is found by multiplying each capacitor by 10^{-7} and each resistor by 10^3.

In physically constructing the filter described in the preceding example, each of the FDNRs should be tuned as shown in Fig. 7.3-5 by adjusting R_4 so that the break frequency of (7) occurs at 6180 rad/s. This value is found by substituting the ideal values of the passive components of the FDNR into (6) and (7). After each FDNR is tuned, the network of Fig. 7.3-4(*c*) may be constructed. After the filter circuit is assembled, very little large-scale tuning can be accomplished because of the complex way in which the individual components affect the overall circuit behavior. The filter characteristic, however, is sensitive to changes in the values of the R_4 resistors in the FDNRs and the values of C_s and C_L. Small changes in the filter response can thus be made by carefully adjusting these components.

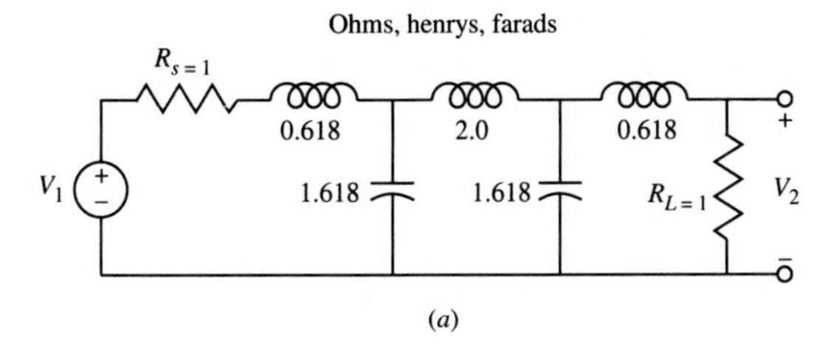

(a)

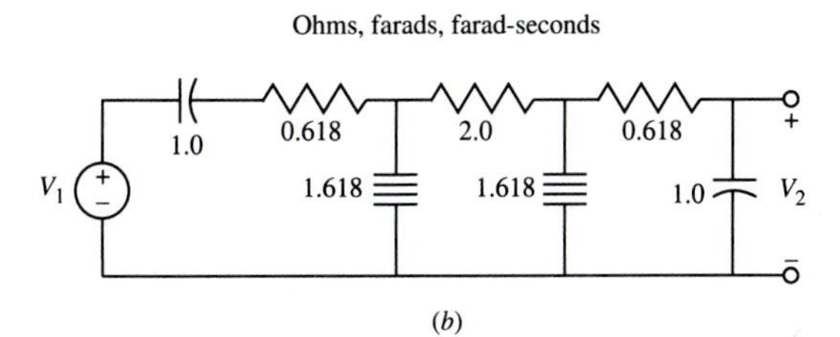

(b)

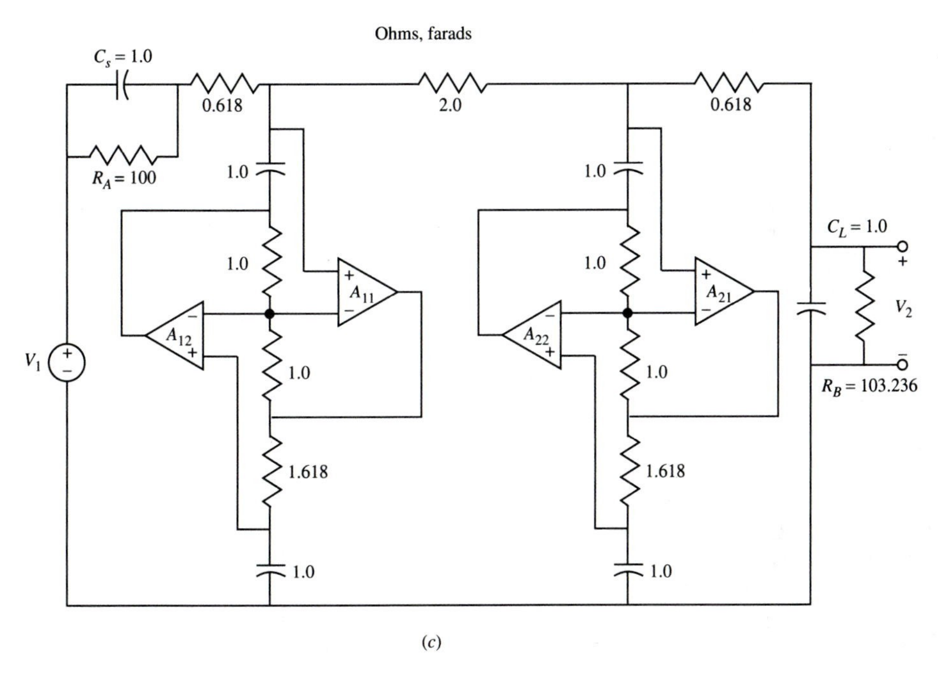

(c)

FIGURE 7.3-4
The fifth-order filter realized in Example 7.3-2.

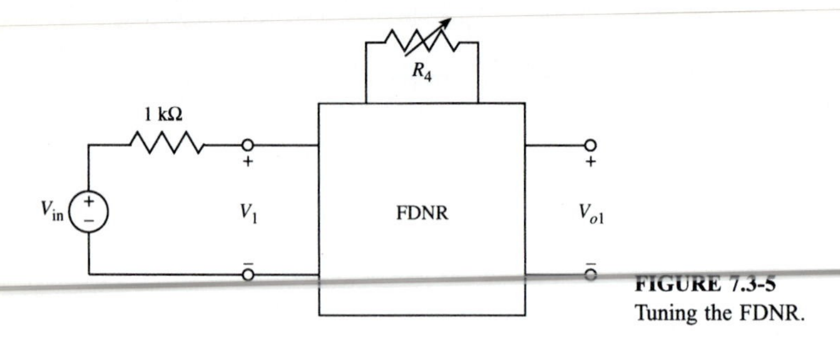

FIGURE 7.3-5
Tuning the FDNR.

Sensitivity

Like the synthetic inductance passive simulation method of active *RC* synthesis, the use of FDNRs retains the excellent sensitivity properties of the low-pass prototype network. This is true not only for the transformed prototype resistors and capacitors that appear directly in the active realization, but also for the elements of the GINs that realize the FDNRs. To see this, note that a prototype capacitor C_{prot} is transformed to an FDNR of value $D_{\text{FDNR}} = C_{\text{prot}}$ that is realized as

$$C_{\text{prot}} = D_{\text{FDNR}} = \frac{C_1 R_2 R_4 C_5}{R_3} \tag{12}$$

As a result

$$S^{C_{\text{prot}}}_{C_1,R_2,R_4,C_5} = 1 \qquad S^{C_{\text{prot}}}_{R_3} = -1 \tag{13}$$

If we now consider an arbitrary sensitivity S^y, where y is any usual network characteristic such as the network function $N(s)$, the magnitude function $|N(j\omega)|$, the Q, a numerator denominator coefficient a_i or b_i, and so forth, then we have

$$S^y_{C_1,R_2,R_4,C_5} = S^y_{C_{\text{prot}}} S^{C_{\text{prot}}}_{C_1,R_2,R_4,C_5} = S^y_{C_{\text{prot}}} \tag{14a}$$

$$S^y_{R_3} = S^y_{C_{\text{prot}}} S^{C_{\text{prot}}}_{R_3} = -S^y_{C_{\text{prot}}} \tag{14b}$$

From the above we see that the sensitivities to the elements of the GIN are ± 1 times the sensitivities to the prototype capacitor simulated.

Example 7.3-3 Sensitivity of a low-pass FDNR filter. As an example of the determination of the sensitivities of an FDNR filter, consider the third-order low-pass filter shown in Fig. 7.3-6(*a*). After the *RLC-CRD* transformation the network appears as shown in Fig. 7.3-6(*b*). The elements of the two figures are related as $C_1 = 1/R_{\text{out}}$, $R_2 = L_2$, $D_3 = C_3$, $R_4 = L_4$, and $C_4 = 1/R_{\text{in}}$. The elements have been numbered so as to correspond with the element designations given for the analysis method described in Summary 4.9-1. Applying this method for the "four element—output *Y*" figure shown in Fig. 4.9-15(*a*), we may identify

$$Y_1(s) = sC_1 \qquad Z_2(s) = R_2 \qquad Y_3(s) = s^2 D_3 \qquad Z_4(s) = \frac{1 + sR_4C_4}{sC_4}$$

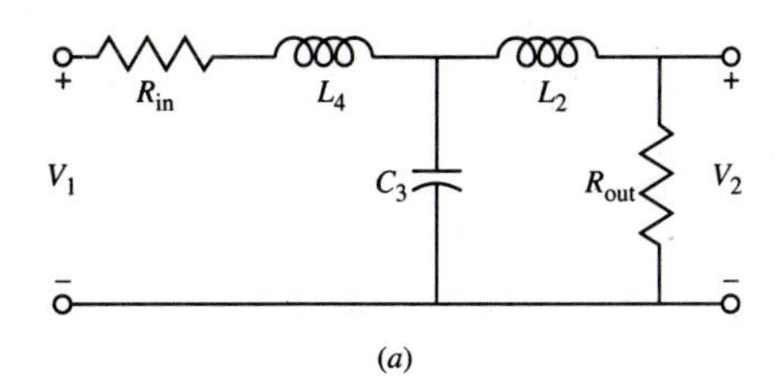

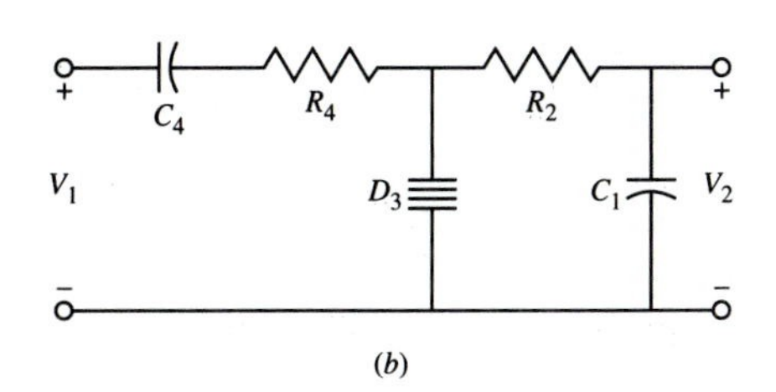

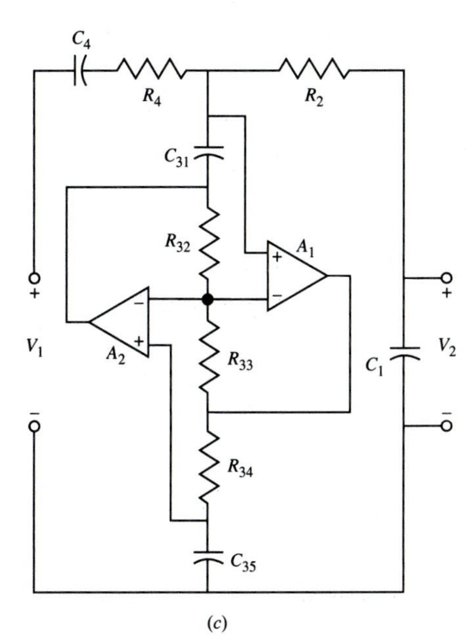

FIGURE 7.3-6
Third-order low-pass FDNR filter.

From Table 4.9-5 we obtain

$$\frac{V_2(s)}{V_1(s)} = \frac{1}{1 + \dfrac{s^2 D_3(1 + sR_4C_4)}{sC_4} + sC_1R_2 + \dfrac{sC_1(1 + sR_4C_4)}{sC_4} + \dfrac{s^3C_1R_2D_3(1 + sR_4C_4)}{sC_4}}$$

This can be put in the form

$$\frac{V_2(s)}{V_1(s)} = \frac{1}{s^3C_1R_2D_3R_4 + s^2(C_1R_2D_3/C_4 + D_3R_4) + s(C_1R_2 + C_1R_4 + D_3/C_4) + 1 + C_1/C_4}$$

Comparing this with the general form

$$\frac{V_2(s)}{V_1(s)} = \frac{a_0}{b_3s^3 + b_2s^2 + b_1s + b_0}$$

we find

$$S_{D_3}^{b_3} = S_{D_3}^{b_2} = 1 \qquad S_{D_3}^{b_1} = \frac{D_3/C_4}{b_1} < 1 \qquad S_{D_3}^{b_0} = S_{D_3}^{a_0} = 0$$

An FDNR realization of the filter is shown in Fig. 7.3-6(c). From (12) we see that

$$S_{C_{31},R_{32},R_{34},C_{35}}^{b_3} = S_{C_{31},R_{32},R_{34},C_{35}}^{b_2} = 1 \qquad S_{R_{33}}^{b_3} = S_{R_{33}}^{b_2} = -1$$

$$S_{C_{31},R_{32},R_{34},C_{35}}^{b_1} = \frac{D_3/C_4}{b_1} \qquad S_{R_{33}}^{b_1} = \frac{-D_3/C_4}{b_1}$$

$$S_{C_{31},R_{32},R_{33},R_{34},C_{35}}^{b_0} = S_{C_{31},R_{32},R_{33},R_{34},C_{35}}^{a_0} = 0$$

The results obtained above can be applied to determine the function sensitivity for a particular approximation. For example, consider a normalized low-pass Butterworth characteristic with a 3-dB frequency of 1 rad/s. For this we have

$$C_4 = 1\text{ F} \qquad R_4 = 1\ \Omega \qquad D_3 = 2\text{ Fs} \qquad R_2 = 1\ \Omega \qquad C_1 = 1\text{ F}$$

$$a_0 = 1 \qquad b_3 = 2 \qquad b_2 = 4 \qquad b_1 = 4 \qquad b_0 = 2$$

The function sensitivity for the element C_{31} is found as follows:

$$S_{C_{31}}^{N(s)} = \frac{-(b_3S_{C_{31}}^{b_3}s^3 + b_2S_{C_{31}}^{b_2}s^2 + b_1S_{C_{31}}^{b_1}s)}{b_3s^3 + b_2s^2 + b_1s + b_0}$$

$$= \frac{-(s^3 + 2s^2 + s)}{s^3 + 2s^2 + 2s + 1} = \frac{-(s^2 + s)}{s^2 + s + 1}$$

where $N(s) = V_2(s)/V_1(s)$. The function sensitivity for the elements R_{32}, R_{34}, and C_{35} is the same. The function sensitivity for the element R_{33} is the negative of this expression. A plot of the real part of the expression for $s = j\omega$ is given in Fig. 7.3-7. From the development given in Sec. 3.2, this is the sensitivity of $|N(j\omega)|$.

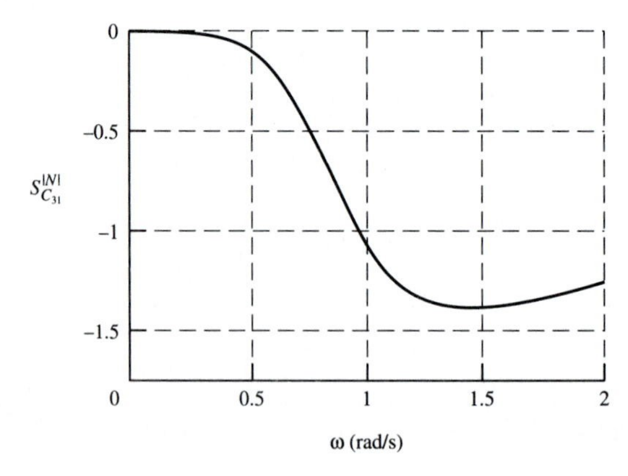

FIGURE 7.3-7
Sensitivity of an FDNR filter.

Synthesis of Band-Pass Functions Using FDNRs

In the preceding paragraphs we discussed the use of FDNRs to realize low-pass functions. We now consider the more difficult problem of the realization of band-pass ones. In Fig. 7.3-8(*a*), an inner section of a passive low-pass structure is shown. If the low-pass to band-pass transformation of (6) of Sec. 2.5 is applied, the circuit of Fig. 7.3-8(*b*) results. If the *RLC-CRD* transformation is applied, we obtain the circuit shown in Fig. 7.3-8(*c*). This circuit contains two floating FDNRs as well as a grounded FDNR and a resistor. To avoid having to directly realize the floating FDNRs (which would require floating power supplies), we instead use the circuit of Fig. 7.3-9, where a cascade of two GICs with a passive network is shown. The transmission parameters of the first GIC are

$$T_{\text{GIC1}} = \begin{bmatrix} 1 & 0 \\ 0 & \dfrac{Z_2(s)Z_4(s)}{Z_1(s)Z_3(s)} \end{bmatrix} \tag{15}$$

The transmission parameters of the passive network will be designated as

$$T_{\text{passive}} = \begin{bmatrix} A(s) & B(s) \\ C(s) & D(s) \end{bmatrix} \tag{16}$$

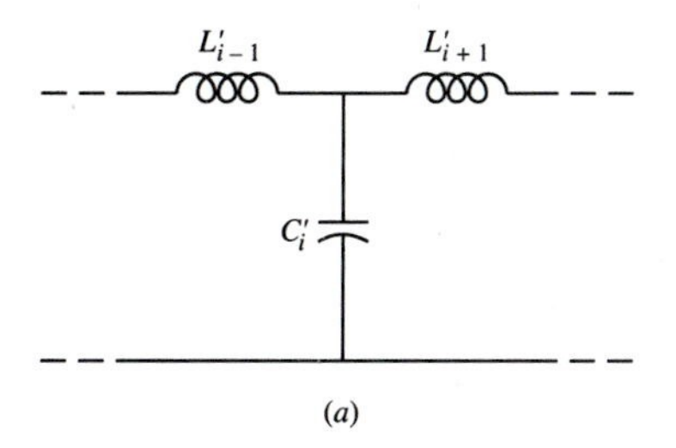

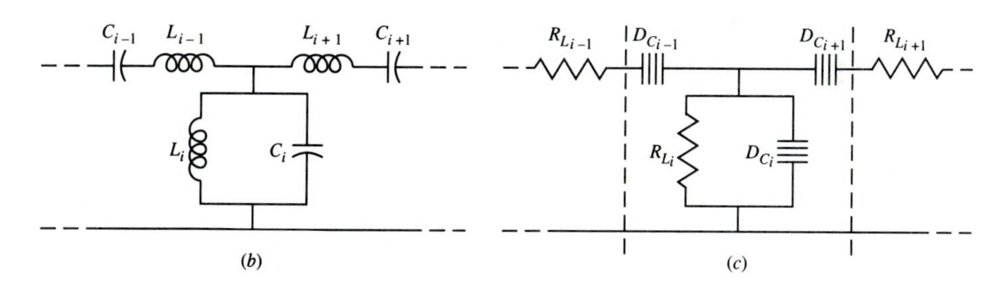

FIGURE 7.3-8
Using FDNRs to realize a band-pass filter.

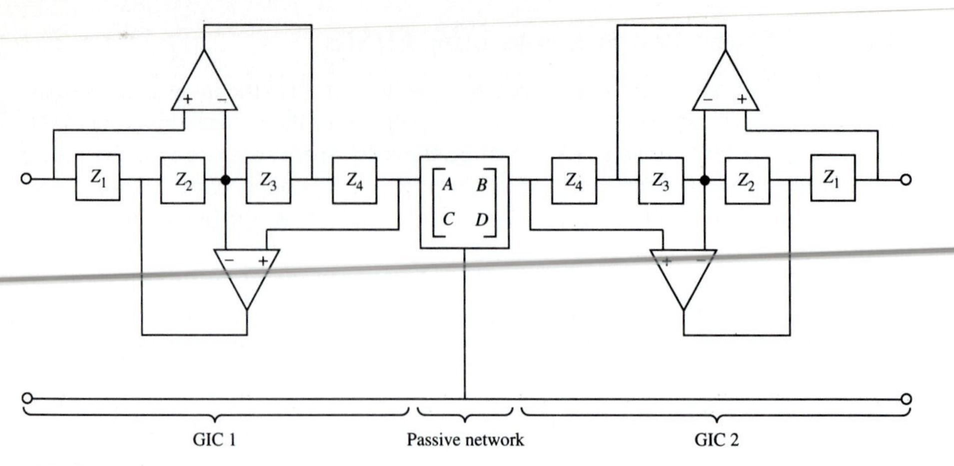

FIGURE 7.3-9
A cascade of two GICs with a passive network.

The transmission parameters of the second GIC are

$$T_{\text{GIC2}} = \begin{bmatrix} 1 & 0 \\ 0 & \dfrac{Z_1(s)Z_3(s)}{Z_2(s)Z_4(s)} \end{bmatrix} \tag{17}$$

Cascading these three networks results in the overall transmission parameters

$$T_{\text{GIC1}} \times T_{\text{passive}} \times T_{\text{GIC2}} = \begin{bmatrix} A(s) & B(s)\dfrac{Z_1(s)Z_3(s)}{Z_2(s)Z_4(s)} \\ C(s)\dfrac{Z_2(s)Z_4(s)}{Z_1(s)Z_3(s)} & D(s) \end{bmatrix} \tag{18}$$

Thus the configuration of Fig. 7.3-9 is equivalent to a related passive network whose elements have been impedance normalized by $Z_1(s)Z_3(s)/Z_2(s)Z_4(s)$. If $Z_1(s)$, $Z_2(s)$, $Z_3(s)$, and $Z_4(s)$ are selected so that

$$\frac{Z_1(s)Z_3(s)}{Z_2(s)Z_4(s)} = \frac{1}{s^2} \tag{19}$$

then the elements of the passive network are subjected to a transformation that is the same as two *RLC-CRD* transformations. Hence, Fig. 7.3-9 can be used to realize Fig. 7.2-8(*c*) by using a "pretransformed" passive network in which the FDNRs are replaced by resistors and the resistors are replaced by elements whose impedance is proportional

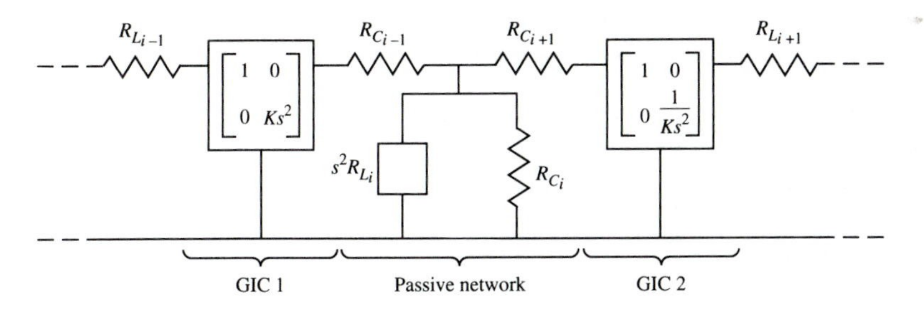

FIGURE 7.3-10
A realization of the filter section of Fig. 7.3-8c.

to s^2. As a result, the structure of Fig. 7.3-8(c) can be realized through the use of two GICs and an element with the input impedance of $s^2R_{L_i}$ as shown in Fig. 7.3-10. This latter element can be realized using the GIN of Fig. 7.3-11. The concepts illustrated in Figs. 7.3-8 through 7.3-11 can also be applied to use GICs and FDNRs to realize high-pass functions.

Example 7.3-4 Sixth-order band-pass FDNR filter. As an example of the techniques described above, consider the realization of a normalized sixth-order broadband band-pass filter with a center frequency of 1 rad/s, a 3-dB bandwidth of 1 rad/s, and a Butterworth magnitude characteristic. The prototype *RLC* band-pass filter, derived by applying the low-pass to band-pass transformation to a third-order low-pass filter, is shown in Fig. 7.3-12(a). After applying the *RLC-CRD* transformation, we obtain the network shown in Fig. 7.3-12(b). The center section of the network, as indicated by the dashed lines, is realized by the two GICs and the transformed elements shown in Fig. 7.3-12(c). The complete circuit realization is shown in Fig. 7.3-12(d). To simplify the latter figure, the operational amplifiers are not shown and the passive elements of the GICs and the GIN have been enclosed in dashed lines.

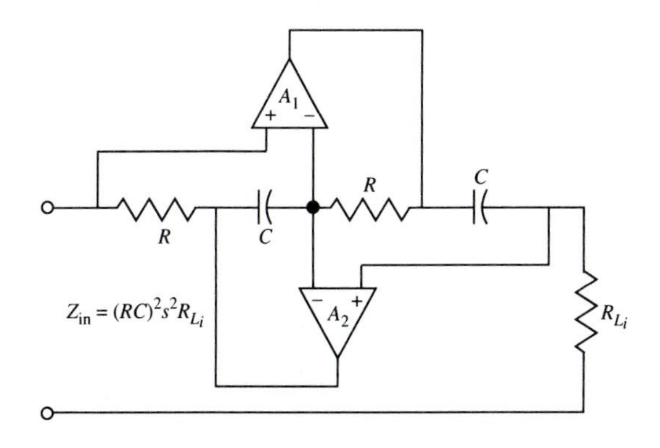

FIGURE 7.3-11
A realization of an element with an impedance proportional to s^2.

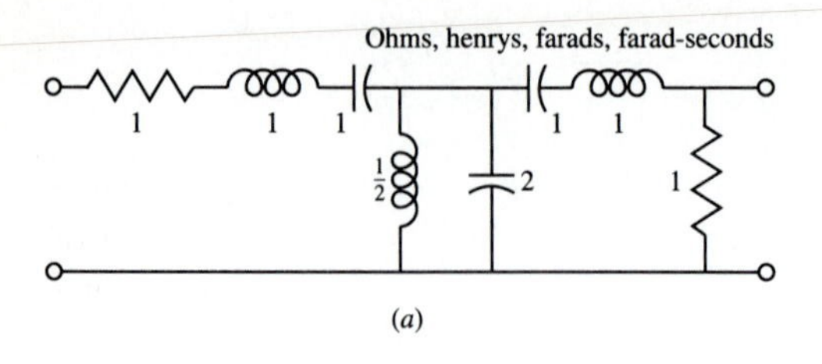

(a)

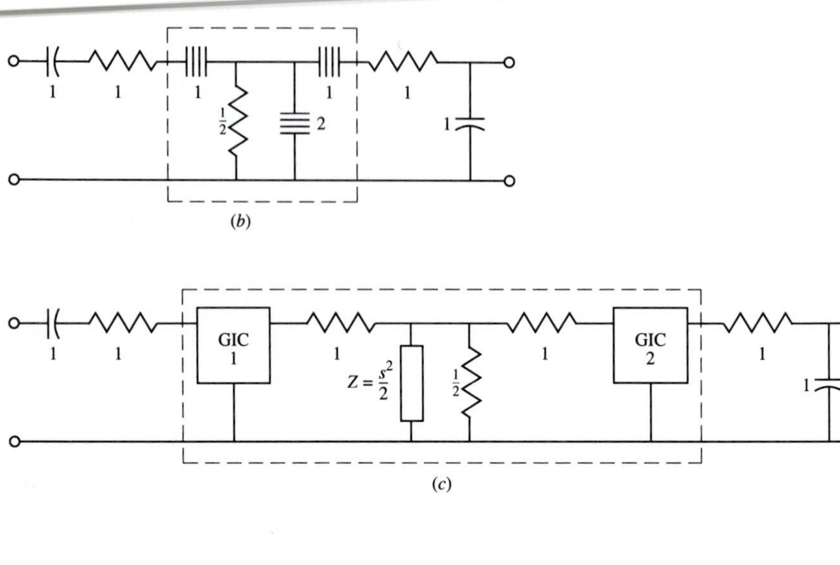

(b)

(c)

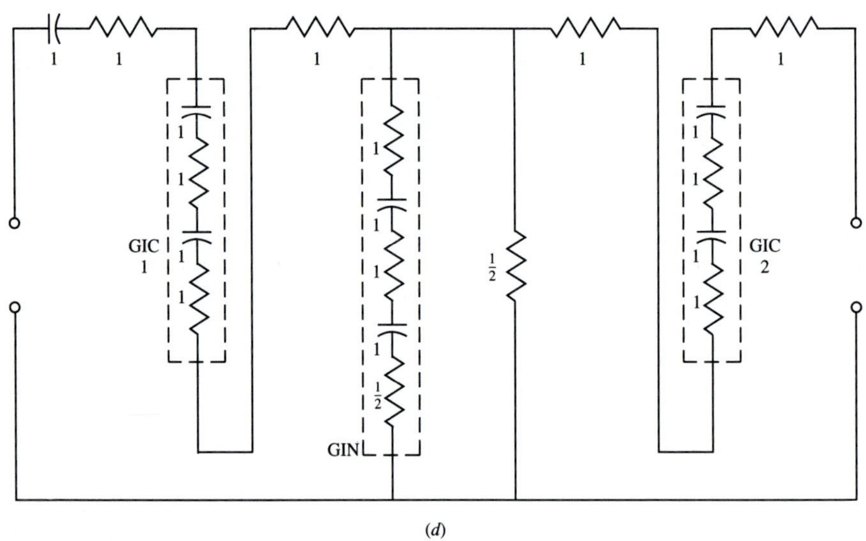

(d)

FIGURE 7.3-12
Sixth-order band-pass FDNR filter.

7.4 LEAPFROG REALIZATION TECHNIQUES

In the preceding sections of this chapter we discussed two approaches to the direct method of filter synthesis, namely, the use of synthetic inductance and the use of FDNRs. Both of these techniques are examples of the simulation of the elements of a passive network through the use of active *RC* circuits. In this section we shall present a discussion of a somewhat different simulation approach to the direct method of synthesis. It is known as the *leapfrog technique*,[1] and it uses negative feedback to simulate the voltage and current relations of a passive *RLC* ladder network. It results in a useful and stable realization, one advantage of which is the repeated use of nearly identical component circuits as building blocks of the overall realization.

Developing the Leapfrog Structure

The leapfrog technique is based on the use of an active *RC* circuit in which the voltages at various portions of the network are analogs of the shunt branch voltages and series branch currents of the prototype *RLC* passive network being simulated. To see this, consider the passive ladder network shown in Fig. 7.4-1. The branch relationships of this network can be written as

$$I_1 = (V_1 - V_2)Y_1 \tag{1a}$$
$$V_2 = (I_1 - I_3)Z_2 \tag{1b}$$
$$I_3 = (V_2 - V_4)Y_3 \tag{1c}$$
$$V_4 = (I_3 - I_5)Z_4 \tag{1d}$$
$$I_5 = (V_4 - V_6)Y_5 \tag{1e}$$
$$\cdots\cdots\cdots\cdots\cdots$$

[1] F. E. J. Girling and E. F. Good, "Active Filters 12: The Leap-Frog or Active-Ladder Synthesis," *Wireless World*, vol. 76, July 1970, pp. 341–345.

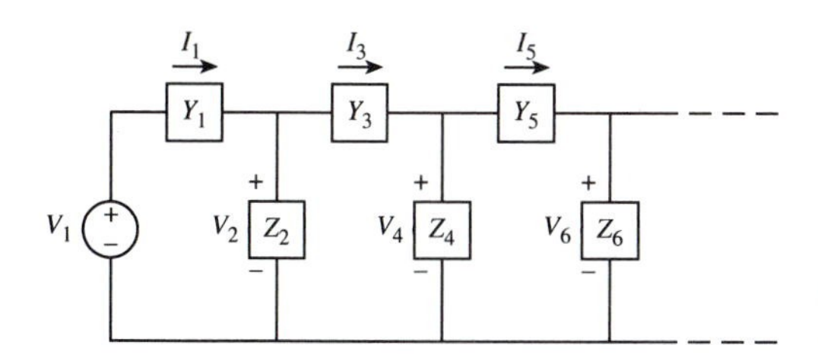

FIGURE 7.4-1
A passive ladder network.

This set of equations may be simulated by representing the currents $I_1, I_3, I_5, \ldots$ of Fig. 7.4-1 by voltages $V_{i1}, V_{i3}, V_{i5}, \ldots$. In this case the equations of (1) become

$$V_{i1} = (V_1 - V_2)Y_1 \tag{2a}$$
$$V_2 = (V_{i1} - V_{i3})Z_2 \tag{2b}$$
$$V_{i3} = (V_2 - V_4)Y_3 \tag{2c}$$
$$V_4 = (V_{i3} - V_{i5})Z_4 \tag{2d}$$
$$V_{i5} = (V_4 - V_6)Y_5 \tag{2e}$$
$$\cdots\cdots\cdots\cdots\cdots$$

Note that in this usage the quantities $Y_1, Y_3, Y_5, \ldots$ and $Z_2, Z_4, \ldots$ become dimensionless voltage transfer functions, although for convenience we have retained the Y and Z notation. To simplify the realization of these equations, we now introduce minus signs into (2*b*), (2*c*), and (2*d*) to obtain

$$V_{i1} = (V_1 - V_2)Y_1 \tag{3a}$$
$$-V_2 = (V_{i1} - V_{i3})(-Z_2) \tag{3b}$$
$$-V_{i3} = (-V_2 + V_4)Y_3 \tag{3c}$$
$$V_4 = (-V_{i3} + V_{i5})(-Z_4) \tag{3d}$$
$$V_{i5} = (V_4 - V_6)Y_5 \tag{3e}$$
$$\cdots\cdots\cdots\cdots\cdots$$

A block diagram of a network realization of these equations is given in Fig. 7.4-2. The "leapfrog" arrangement of the signal paths gives the filter its name. Note that the $Y_1, Y_3, Y_5, \ldots$ voltage transfer functions are realized as noninverting while the $Z_2, Z_4, \ldots$ ones are realized as inverting. It will turn out that the inverting ones are easier to realize than the noninverting ones.

An alternate realization may be developed by introducing minus signs into (2) in a different way to produce

$$-V_{i1} = (V_1 - V_2)(-Y_1) \tag{4a}$$
$$-V_2 = (-V_{i1} + V_{i3})Z_2 \tag{4b}$$
$$V_{i3} = (-V_2 + V_4)(-Y_3) \tag{4c}$$
$$V_4 = (V_{i3} - V_{i5})Z_4 \tag{4d}$$
$$-V_{i5} = (V_4 - V_6)(-Y_5) \tag{4e}$$
$$\cdots\cdots\cdots\cdots\cdots$$

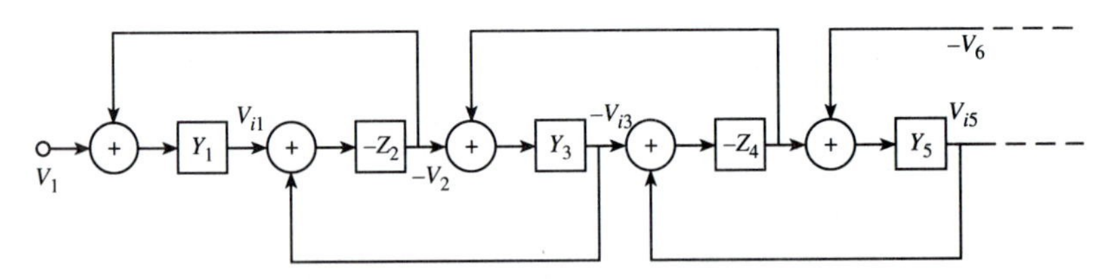

FIGURE 7.4-2
A block diagram of a simulation of the network of Fig. 7.4-1.

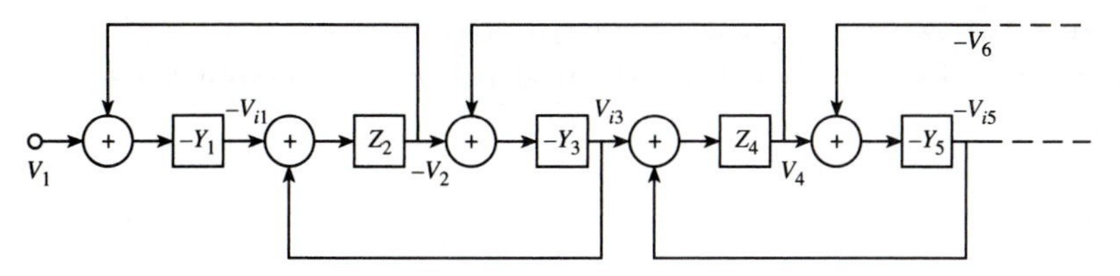

FIGURE 7.4-3
An alternate block diagram of a simulation of the network of Fig. 7.4-1.

These equations are realized by the configuration shown in Fig. 7.4-3. In this realization all the Y_i voltage transfer functions are inverting. Both Figs. 7.4-2 and 7.4-3 show valid realizations of the leapfrog structure.

Leapfrog Low-Pass Filters

Now let us consider the application of the leapfrog concept to the realization of a low-pass filter in which all the transmission zeros are at infinity. In this case, the series elements Y_i of Fig. 7.4-1 will be inductors and the shunt elements Z_i will be capacitors. The initial and final elements will include the source and load resistors, respectively. If the realization is driven by a voltage source and is of odd order, the final element will be as shown in Fig. 7.4-4(a) (for $n = 5$). For the voltage-excitation even-order case, the

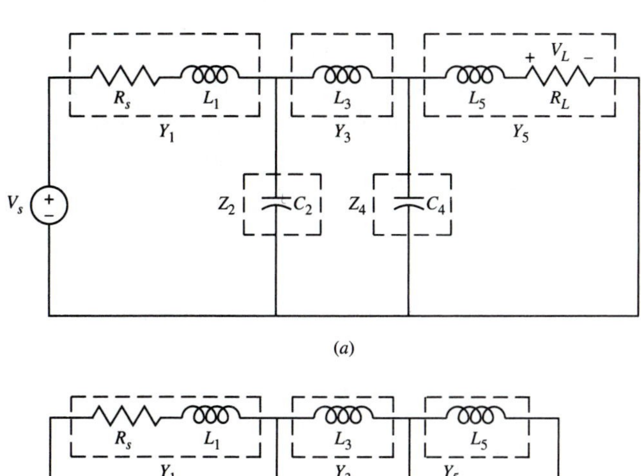

FIGURE 7.4-4
Passive low-pass filter configurations: (a) fifth-order (odd); (b) sixth-order (even).

final element will appear as in Fig. 7.4-4(*b*) (for $n = 6$). Other cases are covered in the Problems. For Fig. 7.4-4(*a*) the voltage transfer functions of the boxes of Fig. 7.4-2 are found as

$$T_1(s) = Y_1(s) = \frac{1/L_1}{s + R_s/L_1} \tag{5a}$$

$$T_2(s) = -Z_2(s) = -\frac{1}{sC_2} \tag{5b}$$

$$T_3(s) = Y_3(s) = \frac{1}{sL_3} \tag{5c}$$

$$T_4(s) = -Z_4(s) = -\frac{1}{sC_4} \tag{5d}$$

$$T_5(s) = Y_5(s) = \frac{1/L_5}{s + R_L/L_5} \tag{5e}$$

Of these voltage transfer functions, $T_2(s)$ and $T_4(s)$ may be realized by conventional operational amplifier inverting integrators, and $T_3(s)$ may be realized by such an integrator in cascade with an inverter, or by a noninverting integrator, using the circuit of Fig. 7.4-5, for which

$$\frac{V_o(s)}{V_i(s)} = \frac{2}{sRC} \tag{6}$$

The terms $T_1(s)$ and $T_5(s)$ may be realized by a damped noninverting integrator. The term *damped* reflects the fact that the pole is displaced from the origin. A realization for a damped inverting integrator is shown in Fig. 7.4-6. Its voltage transfer function is

$$\frac{V_o(s)}{V_i(s)} = \frac{-1/RC_x}{s + 1/R_xC_x} \tag{7}$$

The resulting leapfrog realization of Fig. 7.4-4(*a*), using inverters and using the form of Fig. 7.4-2, is given in Fig. 7.4-7(*a*). All the resistor values are specified as functions of an arbitrary resistance value R, which may be chosen to achieve a desired impedance

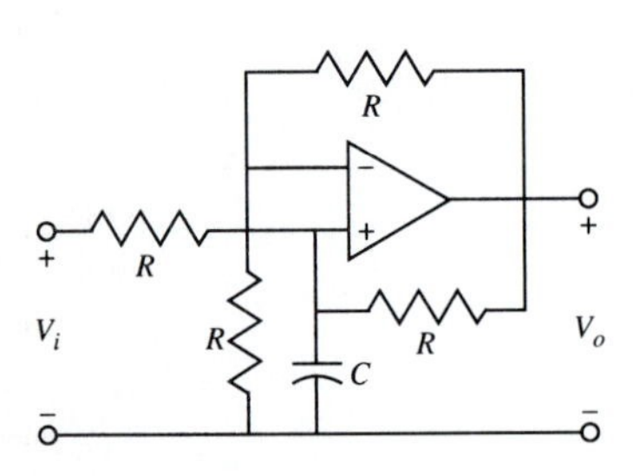

FIGURE 7.4-5
A noninverting integrator.

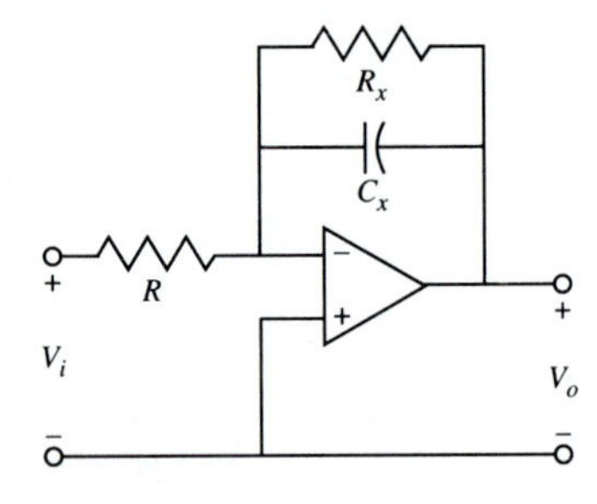

FIGURE 7.4-6
A damped inverting integrator.

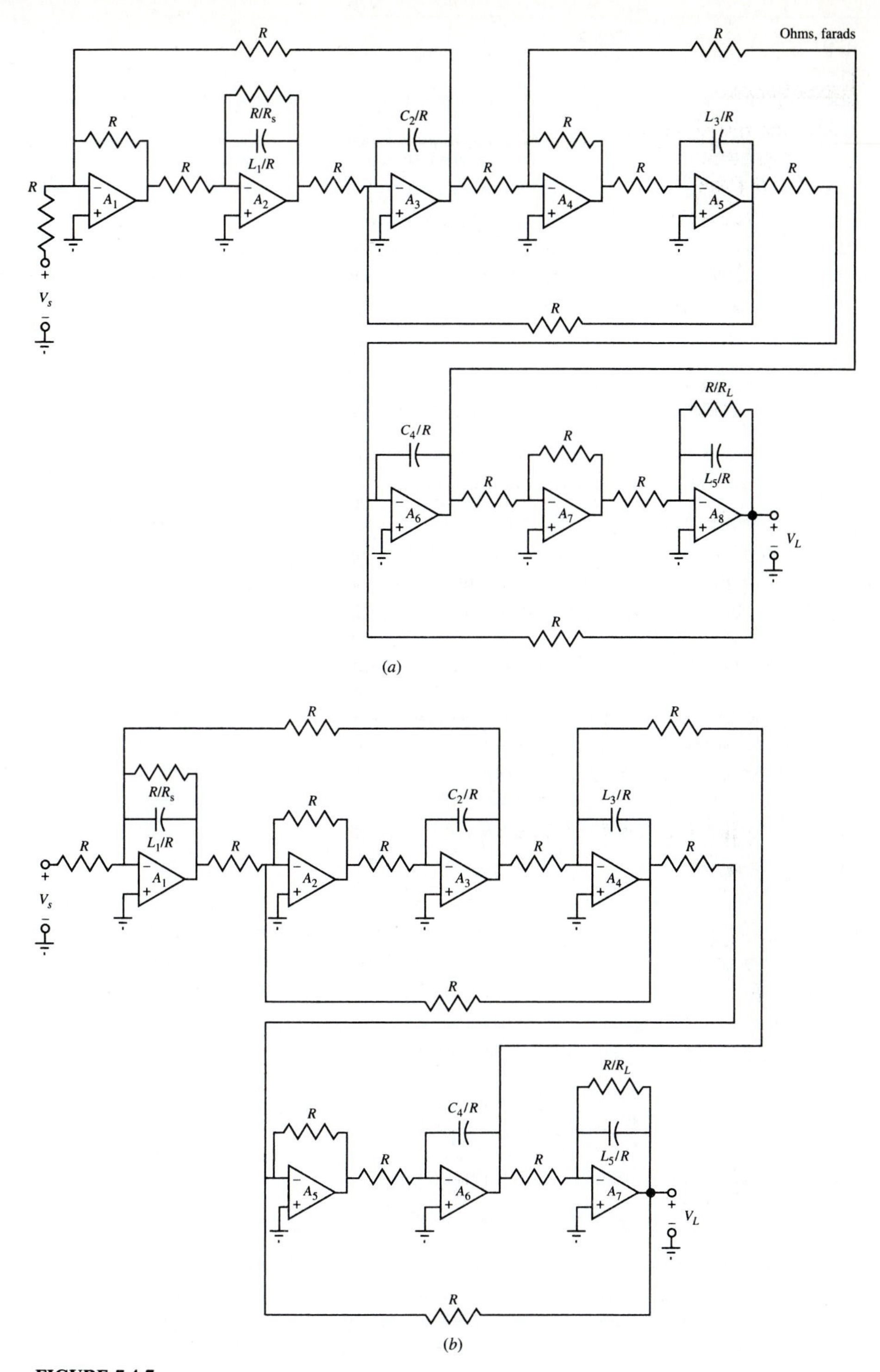

FIGURE 7.4-7
A realization of Fig. 7.4-4*a*: (*a*) using the simulation of Fig. 7.4-2; (*b*) using the simulation of Fig. 7.4-3.

normalization. The values of the capacitors are given in terms of R and of the element values in Fig. 7.4-4(a).

A similar realization of Fig. 7.4-4(a) using the configuration of Fig. 7.4-3 is possible. In this case, only two inverters are required, resulting in one less amplifier. The circuit is shown in Fig. 7.4-7(b). It is of interest to note that in all the above circuits the impedance level of the leapfrog realizations is completely independent of the impedance level in the prototype network. For example, assuming ideal operational amplifiers, the leapfrog realizations will have an output impedance of zero and an input impedance of R.

Summary 7.4-1 Design of low-pass leapfrog filters. The design procedure for low-pass leapfrog filters may be summarized as follows:

1. Design a normalized low-pass prototype using App. A.
2. Select the general configuration form of either Fig. 7.4-2 or Fig. 7.4-3.
3. Design the inner elements of the leapfrog filter using inverting and noninverting integrators and using a normalized value of R of unity.
4. Design the input and output elements of the leapfrog filter using (7) and Fig. 7.4-6.
5. Perform the necessary frequency and impedance denormalizations.

Example 7.4-1 A third-order low-pass Butterworth leapfrog filter. It is desired to realize a third-order Butterworth low-pass filter with a cutoff frequency of 1 krad/s. The excitation voltage source is assumed to have zero internal resistance. From App. A we obtain the circuit shown in Fig. 7.4-8(a). Using the leapfrog configuration of Fig. 7.4-3 and selecting $R = 1$, we obtain the filter circuit of Fig. 7.4-8(b). Using a frequency denormalization of 10^3 and an impedance denormalization of 10^4, all the resistors become equal to 10 kΩ and all the capacitor values are multiplied by 10^{-7}.

Sensitivity

One of the advantages of the leapfrog filter realization is that it retains the excellent low-sensitivity properties of the passive prototype network. As an example of this, consider the realization of the transfer function $T_1(s)$ given in (5a) by the damped inverting integrator shown in Fig. 7.4-6 with transfer function given in (7). Comparing the two equations (and ignoring the minus sign), we obtain

$$L_1 = RC_x \qquad R_s = \frac{R}{R_x} \tag{8}$$

For these expressions, the following relations hold:

$$S_R^{L_1} = 1 \qquad S_{C_x}^{L_1} = 1 \qquad S_R^{R_s} = 1 \qquad S_{R_x}^{R_s} = -1 \tag{9}$$

The sensitivities of the damped integrator components R, R_x, and C_x are seen to be ± 1 times those of the prototype network components R_s and L_1. As another example, if an expression of the form given in (5b) is compared with the voltage transfer function of a conventional inverting integrator ($= 1/sRC$), we obtain

$$C_2 = RC \qquad \text{for which} \qquad S_R^{C_2} = 1 \qquad S_C^{C_2} = 1 \tag{10}$$

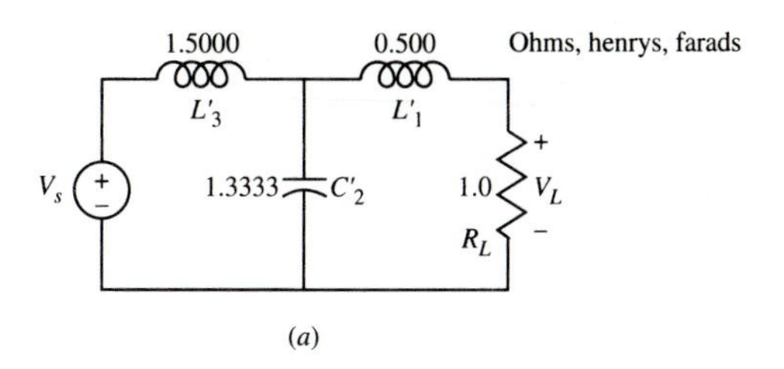

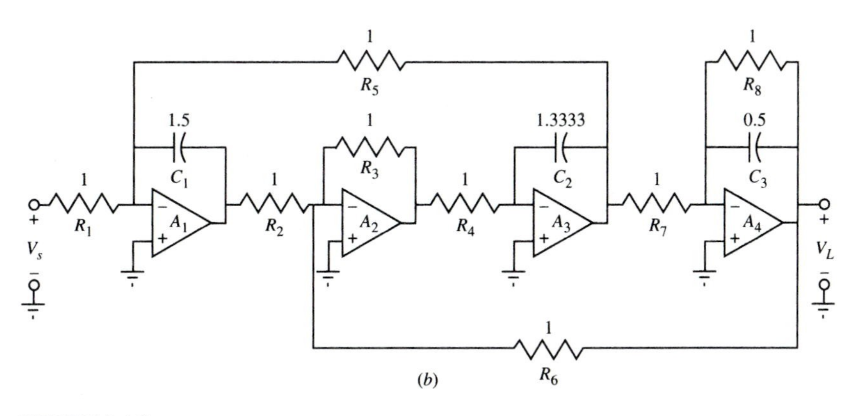

FIGURE 7.4-8
The third-order Butterworth leapfrog filter of Example 7.4-1.

In this case the sensitivities of the integrator components R and C are the same as those of the prototype network component C_2. If we extend these examples to the other sections of a leapfrog filter, we obtain a similar conclusion, namely that the sensitivities of the passive elements associated with the integrators of a leapfrog filter are ± 1 times those of the related components in the passive prototype network. The sensitivities of the other passive components of a leapfrog filter, the ones used for summation and feedback, are readily shown to have similar sensitivities. This is verified in the following example.

Example 7.4-2 Sensitivities of a leapfrog filter. As an example of the determination of the sensitivities of a leapfrog filter, consider the third-order filter with a Butterworth magnitude characteristic shown in Fig. 7.4-8(a). The voltage transfer function is given as

$$N(s) = \frac{V_2(s)}{V_1(s)} = \frac{R_L}{s^3 \mathrm{L}_1' C_2' L_3' + s^2 C_2' L_3' R_L + s(L_1' + L_3') + R_L}$$

A plot of the magnitude sensitivities for the various components of the passive prototype filter with the element values given in Fig. 7.4-8(a) is given in Fig. 7.4-9. For reference, a plot of $|N(j\omega)|$ is also shown in the figure. Note that the sensitivities are all very low, with magnitudes less than 1.5. If we now make a literal analysis of the circuit shown in Fig.

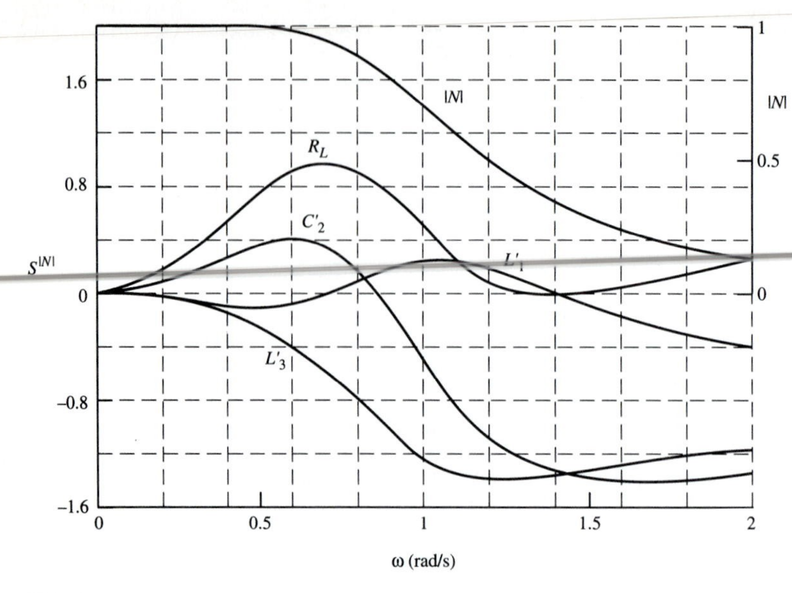

FIGURE 7.4-9
Magnitude sensitivities for prototype filter.

7.4-8(b), we obtain

$$N(s) = \frac{V_2(s)}{V_1(s)}$$

$$= \frac{R_3R_5R_6R_8}{s^3C_1C_2C_3R_1R_2R_4R_5R_6R_7R_8 + s^2C_1C_2R_1R_2R_4R_5R_6R_7 + s(C_1R_1R_2R_3R_5R_8 + C_3R_1R_3R_6R_7R_8) + R_1R_3R_6R_7}$$

The magnitude sensitivities for the various filter components with the element values shown in Fig. 7.4-8(b) are shown in Fig. 7.4-10. Note the following:

1. All the sensitivities are low, with magnitudes less than 1.5.
2. The sensitivities for R_2 and C_1 are the same as the sensitivity for L_3' (shown in Fig. 7.4-9).
3. The sensitivities for R_4 and C_2 are the same as the sensitivity for C_2' (shown in Fig. 7.4-9).
4. The sensitivity for C_3 is the same as the one for L_1' (shown in Fig. 7.4-9).

Leapfrog Band-Pass Filters

The leapfrog technique described above for low-pass filters is also applicable to band-pass filters with zeros at the origin and at infinity. In this case the low-pass to band-pass transformation is applied to each element of a prototype low-pass network. The general form of the resulting series and shunt branches is given in Fig. 7.4-11(a) and (b), respectively. Usually R_i in Fig. 7.4-11(a) will be zero and R_j in Fig. 7.4-11(b) will be

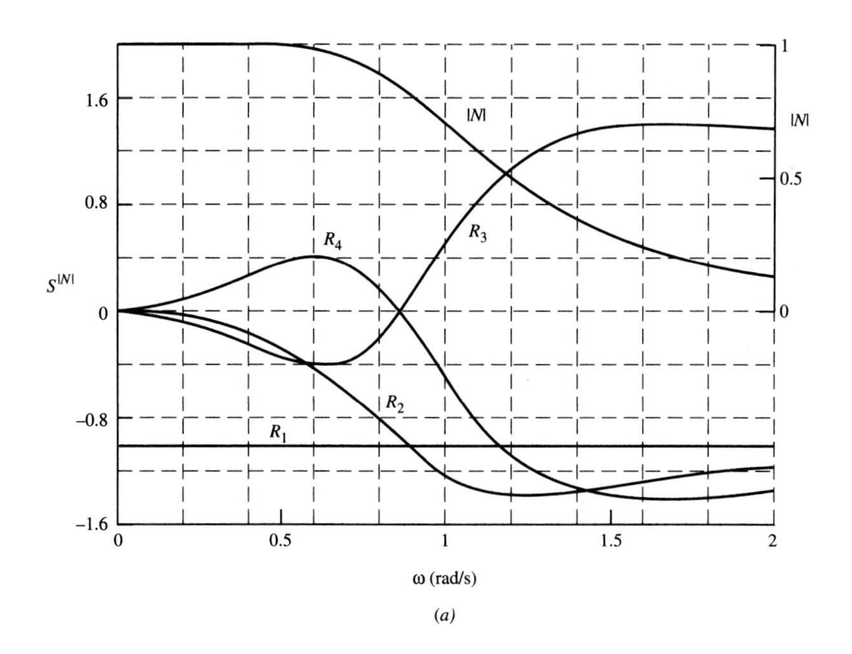

(a)

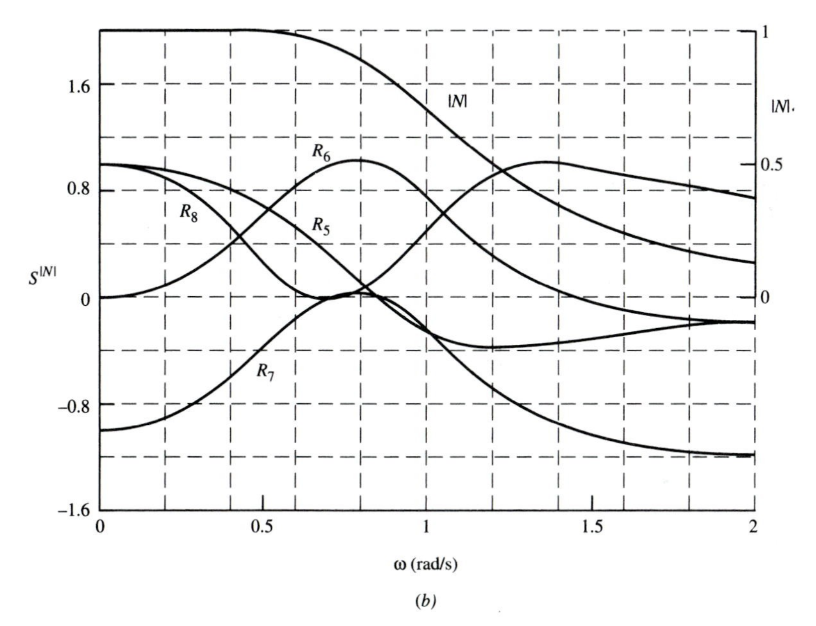

(b)

FIGURE 7.4-10
Magnitude sensitivities for leapfrog filter.

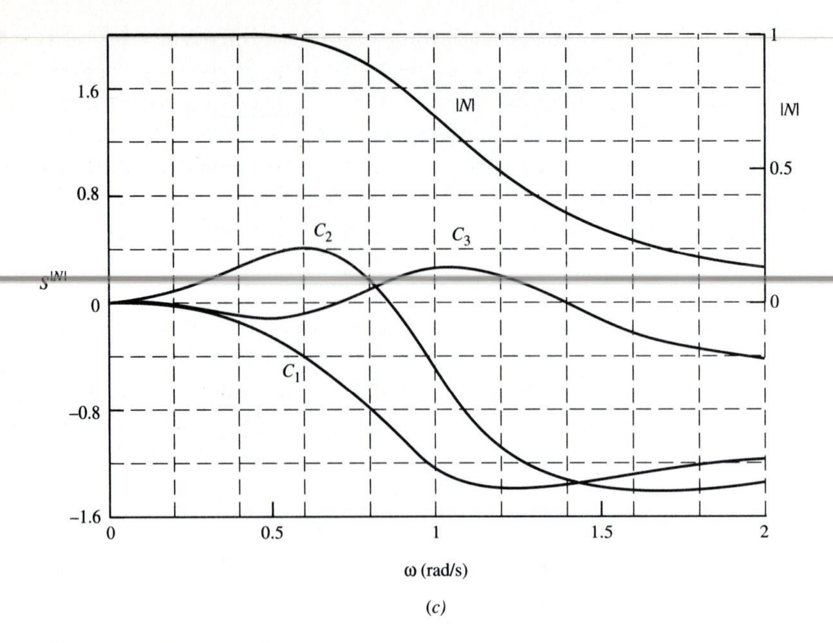

FIGURE 7.4-10 (*Continued*)

infinity. The admittance of Fig. 7.4-11(*a*) is found as

$$Y_i(s) = \frac{(1/L_i)s}{s^2 + (R_i/L_i)s + 1/L_iC_i} \tag{11}$$

The impedance of Fig. 7.4-11(*b*) is found as

$$Z_j(s) = \frac{(1/C_j)s}{s^2 + (1/R_jC_j)s + 1/L_jC_j} \tag{12}$$

Obviously, for the band-pass case the voltage transfer functions required are second-order band-pass ones. Thus the integrators of the low-pass realization will be replaced

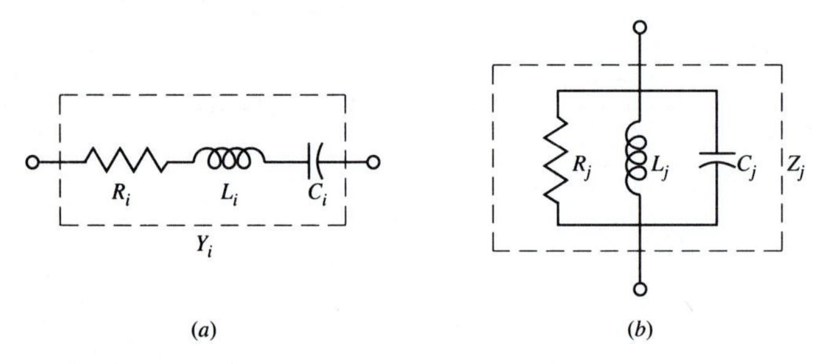

FIGURE 7.4-11
Series and shunt branches of a band-pass network.

by any of the second-order band-pass filter circuits developed in Chaps. 5 and 6. We see that the leapfrog technique provides another method of obtaining higher order realizations from second-order ones. In practice the resulting realizations have excellent characteristics. There are several important requirements that such usage places on the second-order band-pass realization used. First, it must be capable of both inverting and noninverting operation, so that additional inverters are not required. Second, when $R_i = 0$ or $R_j = \infty$, the Q of the second-order realization must become infinite. The third requirement is that the realization be capable of summing two or more signals. In order to meet this last requirement, the Tow-Thomas, or modified state-variable circuits are most appropriate. The infinite-gain realization will require an inverter, whereas the Tow-Thomas and state-variable ones can realize both a noninverting and an inverting band-pass function. In order to get very high values of Q with the infinite-gain realization, one must also employ positive as well as negative feedback. If the Tow-Thomas is to be used, the circuit shown in Fig. 7.4-12, which provides a summing capability, may be used. The transfer function is

$$V_{o2}(s) = -V_{o1}(s) = \left(\frac{s/R_4C_1}{s^2 + s/R_1C_1 + 1/R_2R_3C_1C_2}\right)(V_{i1} + V_{i2}) \tag{13}$$

Example 7.4-3 A sixth-order Butterworth band-pass leapfrog filter. It is desired to use the leapfrog technique to design a six-pole band-pass Butterworth filter with a center frequency of 1 kHz and a bandwidth of one octave. In order to have a stable first stage, we select the doubly terminated structure from App. A with $R = 1$. The low-pass prototype is shown in Fig. 7.4-13(a). From Sec. 2.5 we find that the octave bandwidth for a center frequency of 1 rad/s is $1/\sqrt{2}$ rad/s. Applying the transformation from low-pass prototype to band-pass prototype with a bandwidth of $1/\sqrt{2}$ gives the circuit in Fig. 7.4-13(b). If we use the circuit of Fig. 7.4-12 for the transfer function blocks of Fig. 7.4-13(b), then the choice between Figs. 7.4-2 and 7.4-3 provides no simplification of circuitry. Selecting the form of Fig. 7.4-2, we obtain the realization of Fig. 7.4-13(c). Each of the boxes represents the circuit of Fig. 7.4-12. If C is selected as 1 F, then $R = 1\ \Omega$. The frequency denormalization is accomplished by dividing all capacitors by $2\pi \times 10^3$. The resistors may be scaled

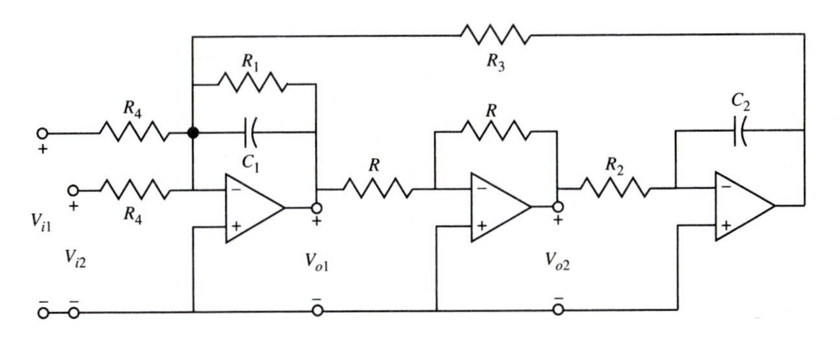

FIGURE 7.4-12
A modification of the Tow-Thomas filter for use in band-pass leapfrog filter realizations.

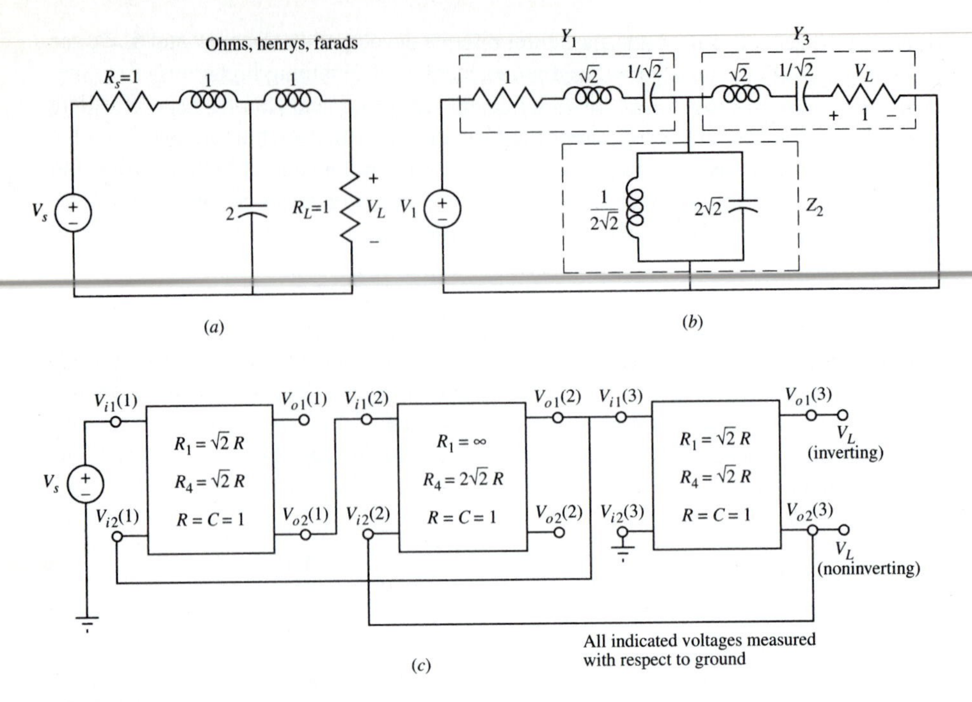

FIGURE 7.4-13
The sixth-order band-pass leapfrog filter realized in Example 7.4-2.

to an arbitrary level to give practical values of both resistors and capacitors. For example, impedance denormalizing by 10^4 gives $R = 10\ \text{k}\Omega$ and $C = 0.0159\ \mu\text{F}$. Resistances R_1 and R_4 of each block are calculated from the relationships given in Fig. 7.4-13(*c*).

The leapfrog method is probably one of the best methods available for realizing band-pass filters, since it makes use of the second-order band-pass blocks developed in Chap. 6 and yet provides sensitivities very close to those of passive networks. In practice, in some cases, the individual second-order blocks can be detuned several percent before a noticeable change occurs in the frequency response. The leapfrog method can also be used to realize network functions with $j\omega$-axis zeros.

7.5 THE PRIMARY RESONATOR BLOCK

In the preceding section we showed how active *RC* circuits, each realizing a second-order band-pass voltage transfer function, could be used as building blocks in a leapfrog structure to obtain a high-order band-pass filter. In this section we present a similar technique called the PRB (*p*rimary *r*esonator *b*lock).[1] It also uses second-order band-

[1] G. Hurtig III, U.S. Patent 3,720,881, March 1973.

pass realizations as building blocks to obtain a high-order band-pass filter. It has an advantage over the leapfrog technique, however, in that it does not require the second-order realizations to have infinite Q. It also has the disadvantage of having design equations somewhat more complicated than those of the leapfrog technique.

General PRB Configuration

A block diagram of the general configuration for a PRB of order $2n$ is shown in Fig. 7.5-1. Each of the boxes represents a circuit realizing a second-order voltage transfer function and having (ideally) zero output impedance. Feedback is provided from the outputs of the n cascaded second-order stages to the input of the operational amplifier by the resistors R_i ($i = 1, 2, \ldots, n$). The overall voltage transfer function is

$$\frac{V_2(s)}{V_1(s)} = \frac{-a_0 T_1(s) T_2(s) \cdots T_n(s)}{1 + a_1 T_1(s) + a_2 T_1(s) T_2(s) + \cdots + a_n T_1(s) T_2(s) \cdots T_n(s)} \tag{1}$$

where the $T_i(s)$ are the individual transfer functions of the second-order band-pass stages and where

$$a_i = \frac{R_f}{R_i} \qquad i = 0, 1, \ldots, n \tag{2}$$

Design of PRB Band-Pass Filters

The PRB technique may be used for band-pass realizations in a manner similar to that developed for the leapfrog technique. As a starting point, consider the inverting nth-order low-pass, normalized (bandwidth of 1 rad/s) transfer function

$$N_{LP}(s) = \frac{-Hb_0}{b_n s^n + b_{n-1} s^{n-1} + \cdots + b_1 s + b_0} = \frac{-Hb_0}{P(s)} \tag{3}$$

The normalized low-pass to band-pass transformation of (6) of Sec. 2.5 may now be modified to the form

$$s = Q\frac{p^2 + 1}{p} \tag{4}$$

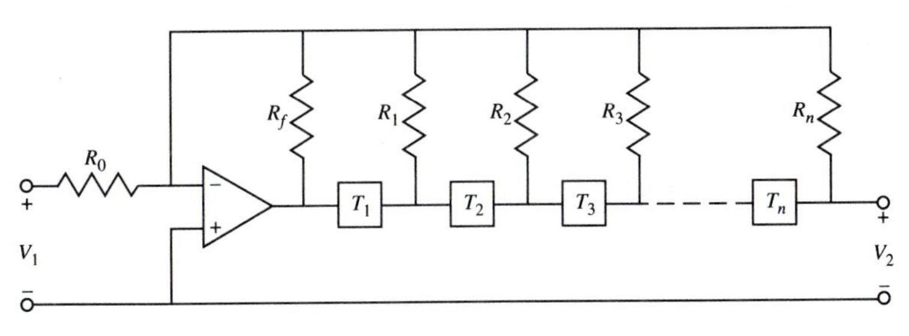

FIGURE 7.5-1
The PRB.

where the band-pass filter resulting from such a change of variable has a center frequency of 1 rad/s and a bandwidth of $1/Q$. Substituting (4) into (3) and redefining the p complex-frequency variable as s, we obtain

$$N_{\mathrm{BP}}(s) = N_{\mathrm{LP}}\left(Q\frac{s^2+1}{s}\right) = \frac{V_2(s)}{V_1(s)} = \frac{-Hb_0 s^n/Q^n}{D_1(s)} \tag{5}$$

where
$$D_1(s) = \sum_{i=0}^{n} \frac{b_{n-i}s^i(s^2+1)^{n-i}}{Q^i} \tag{6}$$

Now let us assume that all the blocks in Fig. 7.5-1 are identical, and realize

$$T(s) = T_i(s) = \frac{H_0 s/Q_p}{s^2 + (1/Q_p)s + 1} \qquad i = 1, 2, \ldots, n \tag{7}$$

For this case, (1) simplifies to

$$\frac{V_2(s)}{V_1(s)} = \frac{-a_0 H_0^n s^n/Q_p^n}{D_2(s)} \tag{8}$$

where

$$D_2(s) = \left(s^2 + \frac{1}{Q_p}s + 1\right)^n + \sum_{i=1}^{n} \frac{a_i H_0^i s^i[s^2 + (1/Q_p)s + 1]^{n-i}}{Q_p^i} \tag{9}$$

Applying the binomial theorem to $D_2(s)$, we obtain

$$D_2(s) = \sum_{i=0}^{n} \binom{n}{i}(s^2+1)^{n-i}(s/Q_p)^i + \sum_{i=1}^{n}\sum_{j=0}^{n-i} a_i H_0^i \binom{n-i}{j}(s^2+1)^{n-i-j}(s/Q_p)^{i+j} \tag{10}$$

where

$$\binom{x}{y} = \frac{x!}{y!(x-y)!} \tag{11}$$

The elements of the feedback circuit of Fig. 7.5-1 are obtained by equating (8) and (10) with (5) and (6), resulting in

$$\frac{a_0 H_0^n}{Q_p^n} = \frac{Hb_0}{Q^n} \tag{12}$$

and
$$\binom{n}{k}\frac{1}{Q_p^k} + \sum_{i=1}^{k} a_i H_0^i \binom{n-i}{k-i}\frac{1}{Q_p^k} = \frac{b_{n-k}}{Q^k} \qquad k = 1, 2, \ldots, n \tag{13}$$

One approach to solving (11) and (12) is to make the assignments

$$H_0 = \frac{1}{c} \qquad Q_p = \frac{Q}{c} \tag{14}$$

where c is arbitrary, being chosen if possible so as to make the a_i nonnegative. Substituting (14) into (12) and (13) yields

$$a_0 = Hb_0 \tag{15}$$

and

$$a_1 = b_{n-1} - cn$$

$$a_k = b_{n-k} - \binom{n}{k}c^k - \sum_{i=1}^{k-1}\binom{n-i}{k-i}a_i c^{k-i} \qquad k = 2, 3, \ldots, n \tag{16}$$

where in (16) we have solved explicitly for a_k in terms of $a_1, a_2, \ldots, a_{k-1}$. Thus for a chosen value of c we may obtain successively $a_1, a_2, \ldots, a_n$.

We may also solve (16) explicitly for a_k in terms c, n, k, and the b_i. The pattern that emerges as each successive a_k is obtained is given by

$$a_k = (-1)^k \sum_{i=0}^{k} (-1)^i \binom{n-i}{k-i} c^{k-i} b_{n-i} \qquad k = 1, 2, \ldots, n \tag{17}$$

where we note that $b_n = 1$. Equation (17) may be readily verified by substitution into (16) and observing the vanishing of the various powers of c.

The range on c for nonnegative a_k depends on the low-pass prototype coefficients b_i. It is interesting to note from (17) that $a_n = P(-c)$, where $P(s)$ is the denominator polynomial defined in (3). Since $P(s)$ is strictly Hurwitz, all its zeros occur in the left half of the s plane. Also $P(0) > 0$. Therefore, denoting the real zero of $P(s)$ that is nearest the origin by $-\sigma$ $(\sigma > 0)$, we note that if $0 < c < \sigma$, the result is $a_n > 0$. Of course, from the first equation of (16), $a_1 \geq 0$ if

$$0 < c < \frac{b_{n-1}}{n} \tag{18}$$

Other bounds on c are obtained, for the various cases, by considering the other equations of (16) or (17).

In applying the PRB technique, we note that if c is chosen so that one of the a_i is zero, then this eliminates one of the feedback resistors. For example, if $c = b_{n-1}/n$, then by (16), $a_1 = 0$ and thus R_1 is infinite (open circuit). Another consideration is that from (17) we see that the feedback resistors are independent of the gain and center frequency of the filter and are determined solely by the b_i of the low-pass prototype filter. The gain H is set by the input resistor R_0, which by (2) and (15) is given by

$$R_0 = \frac{R_f}{Hb_0} \tag{19}$$

In Fig. 7.5-1, we have assumed that the component second-order band-pass realizations are noninverting. If inverting realizations are used, then additional inverters are required. Finally, the constant c was found to be constrained as given in (18). This implies that c is normally less than unity, which by (14) causes Q_p to be large for a moderate value of Q. If we select $c = b_{n-1}/n$, then

$$Q = \frac{b_{n-1}Q_p}{n} \tag{20}$$

which gives the smallest value of Q_p for a specified Q. Two examples of the design of PRB filters follow.

Example 7.5-1 Fourth-order broadband PRB band-pass filter. As an example of the synthesis of a PRB band-pass filter, consider a normalized (center frequency of 1 rad/s) design for a fourth-order filter with a bandwidth of 1 rad/s that has a maximally flat magnitude characteristic. The prototype low-pass function defined in (3) becomes

$$N_{\mathrm{LP}}(s) = \frac{-Hb_0}{b_2 s^2 + b_1 s + b_0} = \frac{-H}{s^2 + \sqrt{2}s + 1}$$

If we apply (4) with $Q = 1$, we obtain (the p variable is redefined as s for convenience)

$$N_{\mathrm{BP}}(s) = \frac{-Hs^2}{s^4 + \sqrt{2}s^3 + 3s^2 + \sqrt{2}s + 1}$$

For $H = 1$, from (15) we find $a_0 = Hb_0 = 1$. The value of the a_1 feedback coefficient is found from (16) as

$$a_1 = b_{n-1} - nc = b_1 - 2c = \sqrt{2} - 2c$$

To simplify the realization, we choose $c = 1/\sqrt{2}$ so that $a_1 = 0$. The a_2 coefficient is now found as

$$a_2 = b_0 - \binom{2}{2} c^2 = 1 - (1)\tfrac{1}{2} = \tfrac{1}{2}$$

From (14) we obtain

$$H_0 = \frac{1}{c} = \sqrt{2} \qquad Q_p = \frac{Q}{c} = \sqrt{2}$$

The network function for the primary resonator blocks is

$$T(s) = \frac{(H_0/Q_p)s}{s^2 + s/Q_p + 1} = \frac{s}{s^2 + s/\sqrt{2} + 1}$$

This function can be realized by the infinite-gain single-amplifier band-pass filter shown in Fig. 5.4-6. From the design equations of (21) of Sec. 5.4 for this circuit we obtain

$$C_2 = C_3 = 1\ \mathrm{F}$$

$$R_1 = \frac{Q_p}{|H_0|} = \frac{\sqrt{2}}{\sqrt{2}} = 1\ \Omega$$

$$R_5 = \frac{Q_p}{2Q_p^2 - |H_0|} = \frac{\sqrt{2}}{2 \times (\sqrt{2})^2 - \sqrt{2}} = 0.54692\ \Omega$$

$$R_6 = 2Q_p = 2 \times \sqrt{2} = 2.8284\ \Omega$$

The final step is to set $R_0 = R_f = 1\ \Omega$ in Fig. 7.5-1. From (2) we find $R_1 = 1/a_1 = \infty$ and $R_2 = 1/a_2 = 2\ \Omega$. The resulting filter is shown in Fig. 7.5-2. A plot of the magnitude of the voltage transfer function is shown in Fig. 7.5-3.

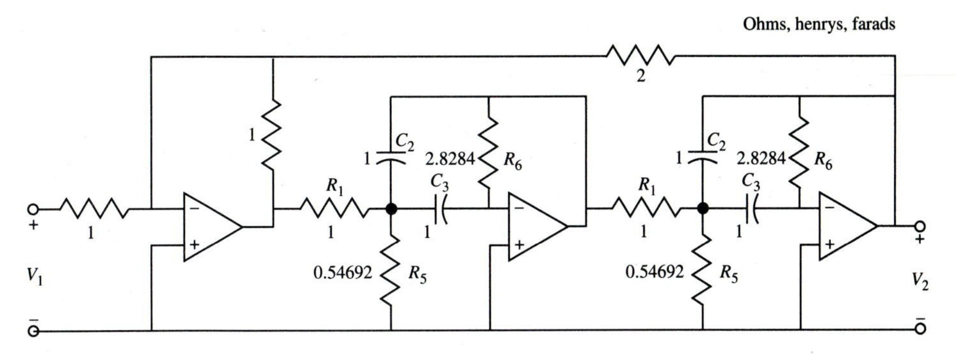

FIGURE 7.5-2
The PRB band-pass filter.

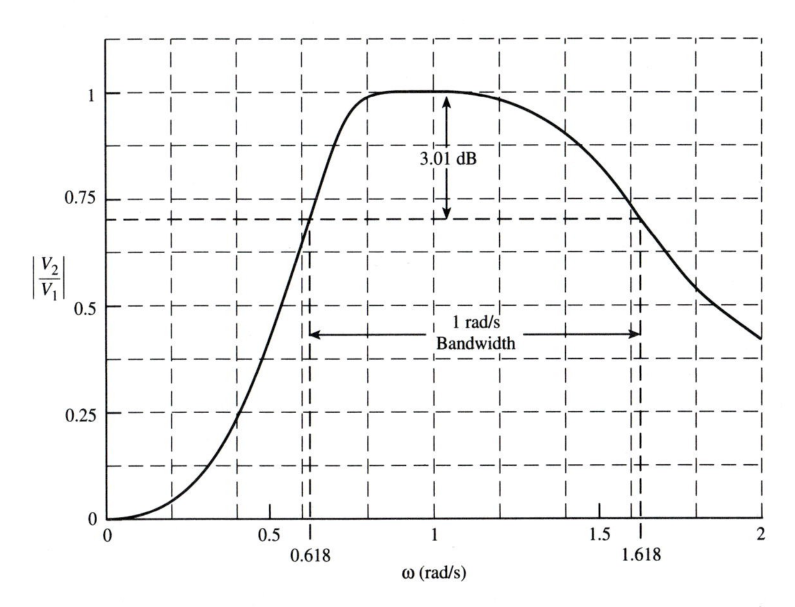

FIGURE 7.5-3
Voltage transfer function magnitude for the filter shown in Fig. 7.5-2.

Example 7.5-2 A maximally flat magnitude eighth-order band-pass PRB filter. It is desired to realize a maximally flat magnitude eighth-order band-pass filter with a center frequency of 3000 Hz, a -3-dB bandwidth of 600 Hz, and a -30-dB bandwidth of not more than 1500 Hz. The center-frequency gain is to be unity. We begin by letting Q_p be defined by (20). From the filter specifications we see that $Q = 5$ and $\omega_0 = 6000\pi$ rad/s. From Table 2.1-3*a* for $n = 4$, we obtain

$$\frac{V_2(s)}{V_1(s)} = \frac{1}{s^4 + 2.613126s^3 + 3.414214s^2 + 2.613126s + 1}$$

Thus from (3) we get $b_0 = 1$, $b_1 = b_3 = 2.613126$, and $b_2 = 3.414214$. If $c = b_{n-1}/n$, then $c = b_3/4 = 0.6532815$. Therefore

$$Q_p = \frac{Q}{c} = \frac{5}{0.6532815} = 7.654$$

$$H_0 = \frac{1}{c} = 1.5307$$

$$\omega_p = 6000\pi \text{ rad/s}$$

where ω_p is the undamped natural frequency of (7). From (16) we get

$$a_1 = 0$$

$$a_2 = b_2 - \binom{4}{2}c^2 - \sum_{i=1}^{1}\binom{4-i}{2-i}a_i c^{2-i} = 0.8536$$

$$a_3 = b_1 - \binom{4}{3}c^3 - \sum_{i=1}^{2}\binom{4-i}{3-i}a_i c^{3-i} = 0.3838$$

$$a_4 = b_0 - c^4 - \sum_{i=1}^{3} a_i c^{4-i} = 0.2036$$

From (2) we can select R_f and solve for the various R_i. If $R_f = 10$ kΩ, then $R_1 = \infty$, $R_2 = 11.718$ kΩ, $R_3 = 26.055$ kΩ, and $R_4 = 49.116$ kΩ. The final design where the second-order band-pass filters are inverting and represented by blocks is given in Fig. 7.5-4.

Sensitivity

A sensitivity analysis of a PRB filter can be made by first determining the sensitivity of the overall network function to the second-order components with transfer function $T_i(s)$

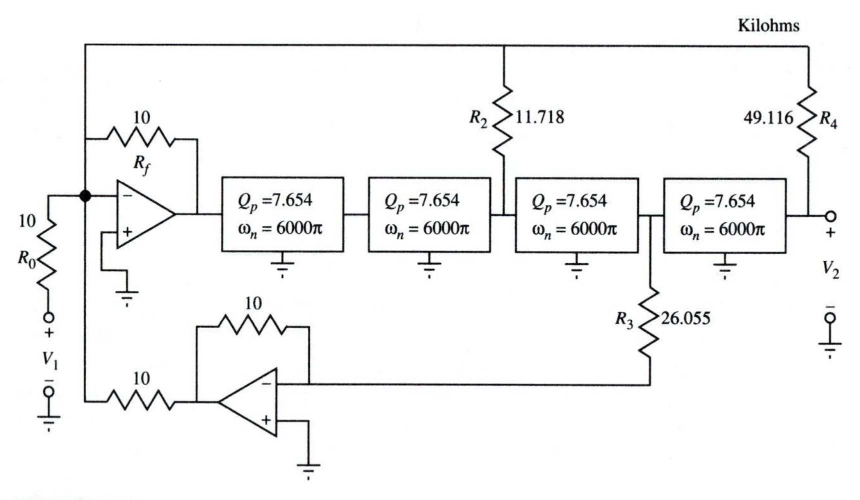

FIGURE 7.5-4
The PRB eighth-order band-pass filter realized in Example 7.5-2.

and then determining the sensitivities of the $T_i(s)$ to the individual resistors and capacitors that produce the functions. From (1), if all the second-order blocks have identical transfer functions $T(s)$, the overall network function is

$$N(s) = \frac{V_2(s)}{V_1(s)} = \frac{A(s)}{B(s)} = \frac{-a_0 T(s)^n}{1 + \sum_{j=1}^{n} a_j T(s)^j} \tag{21}$$

The sensitivity of $N(s)$ to any component x may be determined as

$$S_x^{N(s)} = S_{T(s)}^{N(s)} S_x^{T(s)} = [S_{T(s)}^{A(s)} - S_{T(s)}^{B(s)}] S_x^{T(s)} \tag{22}$$

where

$$S_{T(s)}^{A(s)} = n \qquad S_{T(s)}^{B(s)} = \frac{\sum_{j=1}^{n} j a_j T(s)^j}{1 + \sum_{j=1}^{n} a_j T(s)^j} \tag{23}$$

Thus (22) becomes

$$S_x^{N(s)} = S_{T(s)}^{N(s)} S_x^{T(s)} = \left[\frac{n + \sum_{j=1}^{n} (n-j) a_j T(s)^j}{1 + \sum_{j=1}^{n} a_j T(s)^j} \right] S_x^{T(s)} \tag{24}$$

The sensitivity of $|N(j\omega)|$ with respect to s is

$$S_x^{|N(j\omega)|} = \mathrm{Re}[S_{T(j\omega)}^{N(j\omega)} S_x^{T(j\omega)}] \tag{25}$$

Due to the complexity of the expressions, the computations can be fairly tedious.

Example 7.5-3 Sensitivity of a PRB filter. As an example of the determination of the sensitivity of a PRB filter, consider the normalized fourth-order broadband band-pass filter designed in Example 7.5-1. In this case the order of the low-pass prototype is $n = 2$. From (24), we find

$$S_{T(s)}^{N(s)} = \frac{2 + a_1 T(s)}{1 + a_1 T(s) + a_2 T(s)^2}$$

For the case where $a_1 = 0$ and $T(s)$ is given by (7), this becomes

$$S_{T(s)}^{N(s)} = \frac{2(s^2 + s/Q_p + 1)^2}{(s^2 + s/Q_p + 1)^2 + a_2 (H_0 s/Q_p)^2}$$

For the values $H_0 = \sqrt{2}$, $Q_p = \sqrt{2}$, and $a_2 = \frac{1}{2}$ found in the example we obtain

$$S_{T(s)}^{N(s)} = \frac{2(s^2 + s/\sqrt{2} + 1)^2}{(s^2 + s/\sqrt{2} + 1)^2 + \frac{1}{2}s^2}$$

The sensitivities for the individual elements of the infinite-gain single-amplifier filters used to realize the second-order functions $T(s)$ and identified in Fig. 7.5-2 are found from (19) of Sec. 5.4:

$$S_{R_1}^{T(s)} = \frac{-s^2 - s/\sqrt{2} - 1 + 1/2\sqrt{2}}{D(s)}$$

$$S_{C_2}^{T(s)} = \frac{-s^2 - s/2\sqrt{2}}{D(s)}$$

$$S_{C_3}^{T(s)} = \frac{s/2\sqrt{2} + 1}{D(s)}$$

$$S_{R_5}^{T(s)} = \frac{1 - 1/2\sqrt{2}}{D(s)}$$

$$S_{R_6}^{T(s)} = \frac{s/\sqrt{2} + 1}{D(s)}$$

where
$$D(s) = s^2 + \frac{s}{\sqrt{2}} + 1$$

If we apply (25) to these functions, we obtain the magnitude sensitivity plots shown in Fig. 7.5-5. To illustrate and verify the results shown in these plots, we may choose a specific frequency and determine, by direct analysis, the percentage change in the network function magnitude as each of the components (in both subcircuits) is given a small perturbation.

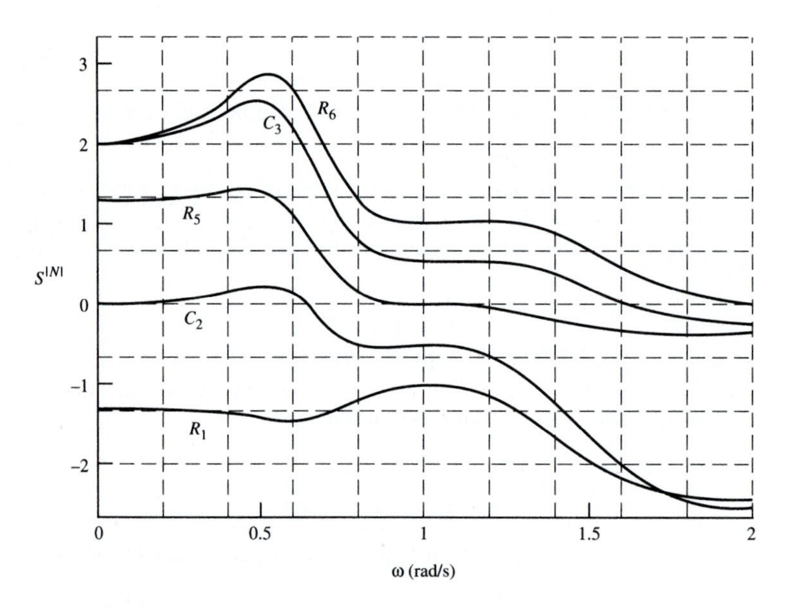

FIGURE 7.5-5
Sensitivities for the magnitude of the voltage transfer function for the filter shown in Fig. 7.5-2.

The results for the choice $\omega = 0.5$ rad/s are

$$\begin{aligned}
1\% \text{ change in } R_1 &\rightarrow -1.4\% \text{ change in } |N(j0.5)| \\
1\% \text{ change in } C_2 &\rightarrow 0.3\% \text{ change in } |N(j0.5)| \\
1\% \text{ change in } C_3 &\rightarrow 2.5\% \text{ change in } |N(j0.5)| \\
1\% \text{ change in } R_5 &\rightarrow 1.4\% \text{ change in } |N(j0.5)| \\
1\% \text{ change in } R_6 &\rightarrow 2.8\% \text{ change in } |N(j0.5)|
\end{aligned}$$

The results are easily seen to agree with the sensitivities shown in Fig. 7.5-5. Note that, from the plots, all of the sensitivities in this realization are low in value.

Many other filter configurations involving second-order building blocks and various combinations of feedback have appeared in the literature. Although some excellent realizations have been developed, many of them are extremely difficult to design, requiring the use of a digital computer in the process. The procedures given in this and the preceding section represent good compromises between filter performance and design simplicity.

7.6 SWITCHED CAPACITOR FILTERS

In the previous sections of this chapter we introduced filter realizations based on the use of a passive *RLC* prototype. In this section we present another such design. It provides an alternate technique for physically realizing the leapfrog filter described in Sec. 7.4. It can also be used for realizing second-order filter sections. It uses a device called a switched capacitor and is especially suited to very large scale integrated circuit (VLSI) implementations.

The Switched Capacitor

The basic premise in the development of switched capacitor filters is that a resistor may be simulated by using periodically operated metal-oxide-semiconductor (MOS) switches and a capacitor. Such filters are especially attractive for realization using VLSI technologies in which the basic elements are MOS transistors and picofarad-range capacitors. Switched capacitor filters have three main advantages:

1. Resistors are eliminated. In VLSI realizations, resistors have the disadvantages of requiring large amounts of die area. In addition, they have tolerance levels and drift parameters that are usually too large to be acceptable.
2. The frequency properties of the filter are determined by the *ratio* of capacitor values. These ratios can usually be controlled to 0.1 percent accuracy and maintain excellent tracking with respect to time and temperature variations. Active-RC filters, on the other hand, have frequency characteristics determined by *RC* products that provide dramatically inferior performance in VLSI implementations.
3. The frequency performance of the filters is proportional to a clock frequency that can be controlled to provide programmable filter characteristics.

Disadvantages of switched capacitor filters include noise and spurious high-frequency responses, both of which are inherent in the circuits due to the nature of the sampling processes. These disadvantages are reduced by using a clock frequency that is much higher than the highest signal frequency component.

A switched capacitor circuit realization of a resistor may be modeled with a SPDT (single-pole, double-throw) switch as shown in Fig. 7.6-1(*a*). With the switch at the left (position 1) as shown in the figure, the capacitor C_R charges to the voltage V_1. Thus, $Q_1 = C_R V_1$. When the switch is moved to the right (position 2), C_R discharges to V_2. Now we have $Q_2 = C_R V_2$. The net charge transfer per cycle of the switch is

$$\Delta Q = Q_1 - Q_2 = C_R(V_1 - V_2) \tag{1}$$

If the switch cycles back and forth at $f_c = 1/T_c$ cycles per second, the average current (for constant values of V_1 and V_2) is

$$I_{\text{avg}} = \frac{\Delta Q}{\Delta T} = \frac{\Delta Q}{T_c} = C_R f_c(V_1 - V_2) \tag{2}$$

If f_c is much higher than the highest frequency component in V_1 and V_2, then the charge transfer process may be treated as a continuous one and the switched capacitor circuit shown in Fig. 7.6-1(*a*) may be modeled as an equivalent resistor as shown in Fig. 7.6-1(*b*), where

$$R_{\text{eq}} = \frac{V_1 - V_2}{I_{\text{avg}}} = \frac{1}{C_R f_c} \tag{3}$$

A modification of the switched capacitor circuit shown in Fig. 7.6-1(*a*) can be made by replacing the SPDT switch with two SPST (single-pole, single-throw) switches as shown in Fig. 7.6-2(*a*). To describe the sequencing of these switches, we use the timing diagram shown in Fig. 7.6-2(*b*). Note that we avoid having switch 1 and switch 2 closed at the same time, as this would place a short between V_1 and V_2. Such a switch sequencing is called *nonoverlapping*.

An implementation of the switched capacitor circuit shown in Fig. 7.6-2(*a*) is shown in Fig. 7.6-3(*a*). It consists of two MOS switches driven by the nonoverlapping clock voltage waveforms of Fig. 7.6-3(*b*). When the gate voltage applied to these switches is high, the typical enhancement nMOS transistor has a channel resistance of the order of $10^3\ \Omega$. When its gate voltage is low, the channel resistance is $10^{12}\ \Omega$ or more. As a result of the large ratio of these resistance values, the transistor functions effectively

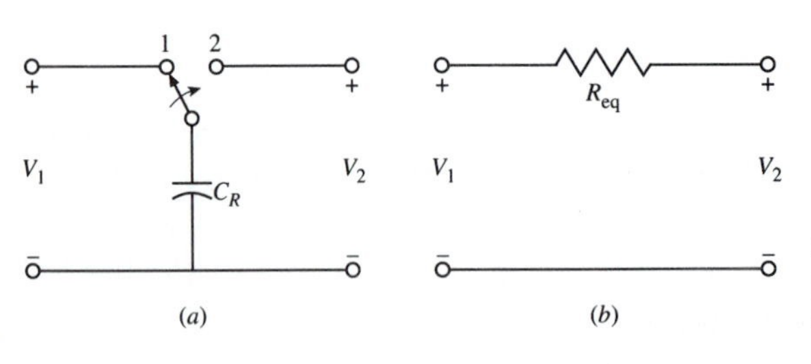

FIGURE 7.6-1
Switched capacitor realization of a resistor.

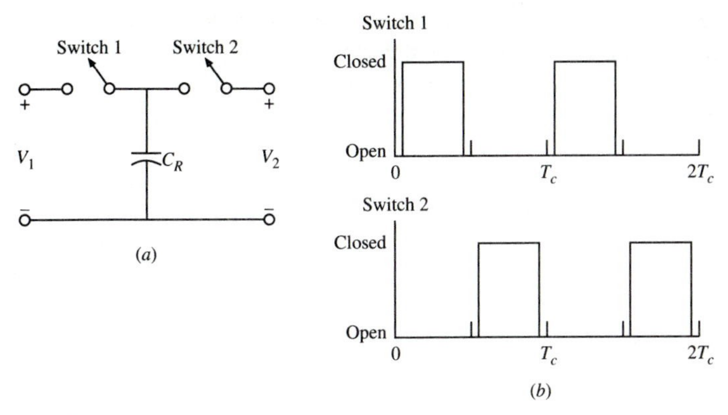

FIGURE 7.6-2
Use of two SPST switches.

as an on-off switch. The two transistors shown in Fig. 7.6-3(a) thus function in the same way as the two SPST switches shown in Fig. 7.6-2(a) [and also the SPDT switch shown in Fig. 7.6-1(a)], and the circuit simulates an equivalent resistor. The implementation of such an equivalent resistor is very attractive in VLSI technology because it provides very effective usage of silicon chip area.

Example 7.6-1 A switched capacitor equivalent resistor. As an example of typical values encountered in the implementation of a switched capacitor equivalent resistor, consider the case where a 1-pF capacitor is switched with $f_c = 100$ kHz. From (3) we obtain

$$R_{\text{eq}} = \frac{1}{C_R f_c} = \frac{1}{10^{-12} \times 10^5} = 10^7\ \Omega$$

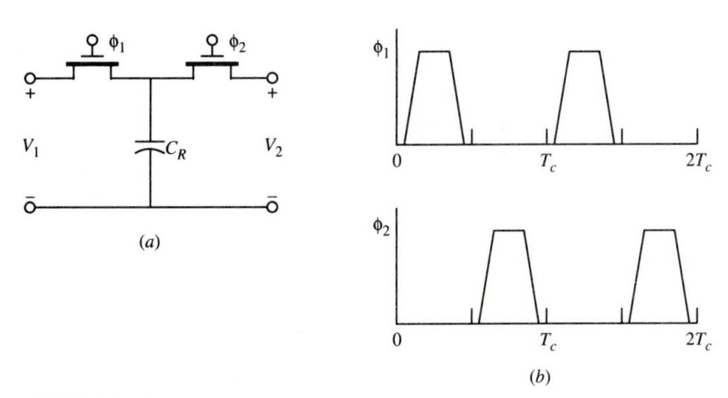

FIGURE 7.6-3
The MOS implementation of a switched capacitor resistor.

The circuit is equivalent to a 10-MΩ resistor. An approximate estimate of the chip area required to implement this is 0.01 mm^2. As a comparison, a 10-MΩ resistor implemented by other techniques, such as a polysilicon line or diffusion, would require at least 100 times more area.

Switched capacitor equivalent resistors can be combined with other capacitors and with operational amplifiers to obtain many of the filter functions normally realized with active *RC* filters. It should be noted that since these circuits involve sampling of the input data, to avoid aliasing problems, the minimum value of the switching frequency f_c is $2f_0$, where f_0 is the highest signal frequency component to be filtered. The frequency $2f_0$ is called the *Nyquist frequency*. In practice, f_c is usually chosen considerably higher. Values of $f_c = 100f_0$ are common.

Effect of Parasitics

In order to minimize the silicon chip area required for a given switched capacitor filter implementation, the capacitor values used in the design are usually quite small, on the order of a few picofarads. For such values, the effects of the parasitic capacitances encountered in the implementation become very significant. The nature of these parasitics can be seen from an examination of the structure of a typical MOS double polysilicon capacitor as shown in Fig. 7.6-4. In the figure, the following quantities are defined:

The basic capacitance between the polysilicon layers, C_R, is used to define R_{eq} in (3).

The capacitance between the bottom polysilicon layer and the substrate, C_B, is typically 10 to 20 percent of the value of C_R.

The accumulated capacitance of all the interconnections (the routing metalization) that are connected to the top polysilicon layer is C_M.

The nonlinear junction capacitance (not shown), C_J, is associated with the source-drain diffusion of the MOS switches.

The total capacitance between the top polysilicon layer and the substrate is C_T. In general, $C_T = C_M + C_J$.

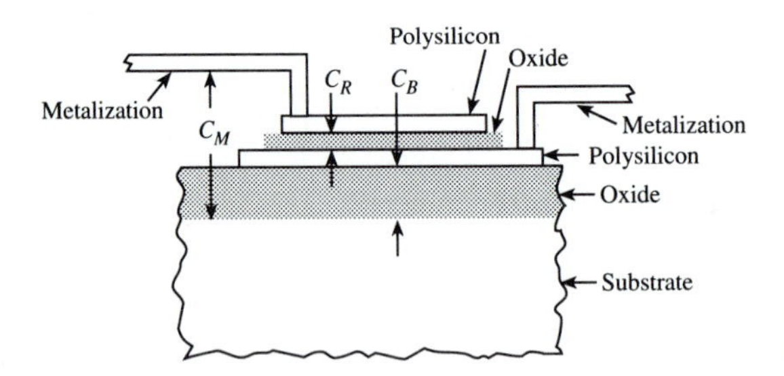

FIGURE 7.6-4
Parasitic capacitances.

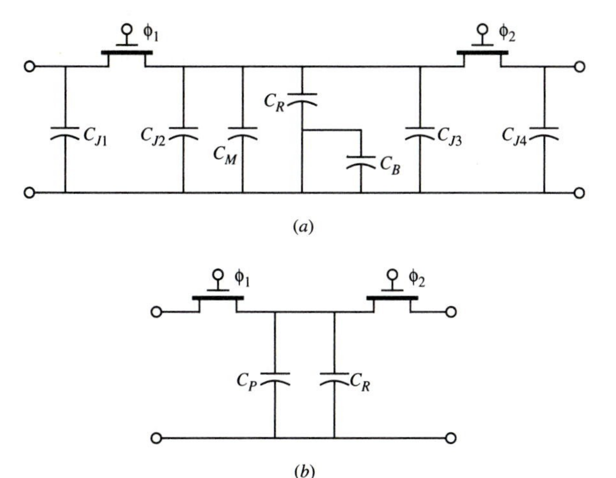

FIGURE 7.6-5
Parasitic capacitances in a switched capacitor equivalent resistor.

The net result of all these parasitics is to convert the MOS switched capacitor circuit shown in Fig. 7.6-3(a) to the more realistic configuration shown in Fig. 7.6-5(a). Some of these parasitic capacitors may be ignored due to other circuit considerations. For example, when a voltage source (with zero internal impedance) is used as an excitation at the left end of the circuit, the element C_{J1} has no effect. Similarly, since the output of the circuit is usually connected to an operational amplifier input terminal (which has zero voltage between it and ground), the element C_{J4} may be ignored. Finally, C_B is effectively grounded and thus has little effect. The remaining parasitics can be modeled by a single capacitor C_P as shown in Fig. 7.6-5(b). Since C_P may be as much as 20 percent of the value of C_R, considerable error can be introduced into the filter realization by ignoring this parasitic. Note that it is not feasible to compensate for the effect of C_P by simply making the sum $C_R + C_P$ equal to the desired value as determined in (3), since the value of C_P is not easily determined and may vary considerably due to its nonlinear components.

To provide compensation for the effects of parasitics in a switched capacitor realization, the single SPDT switch shown in the model in Fig. 7.6-1(a) is replaced by a pair of SPDT switches as shown in Fig. 7.6-6(a). An implementation of the dual-switch

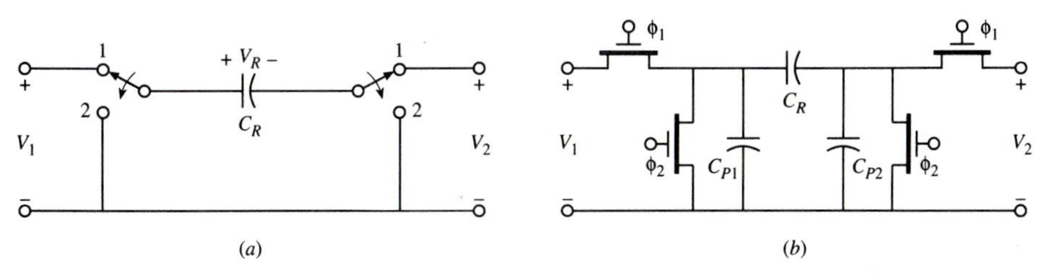

FIGURE 7.6-6
Parasitic-insensitive switched capacitor equivalent resistor.

circuit showing the resulting significant parasitic capacitors C_{P1} and C_{P2} is shown in Fig. 7.6-6(*b*). Note that in this configuration, these parasitics may also include a component due to the capacitance C_B. When the two switches in Fig. 7.6-6(*a*) are in position 1 (as shown), corresponding with ϕ_1 high in Fig. 7.6-6(*b*), the charge on C_R is given by $Q_1 = C_R V_R = C_R(V_1 - V_2)$. When the two switches are in position 2, corresponding with ϕ_2 high, C_R is discharged and $Q_2 = 0$. In addition, the parasitic capacitors C_{P1} and C_{P2} are discharged and have no effect on the voltages V_1 and V_2. The net charge transferred per cycle of the switch is

$$\Delta Q = C_R(V_1 - V_2) \tag{4}$$

If this cycle is repeated at a frequency of f_c, (2) and (3) apply and the circuit provides a parasitic-insensitive realization for a resistor of value $R_{\text{eq}} = 1/C_R f_c$.

A modification of the two-switch design of Fig. 7.6-6 is shown in Fig. 7.6-7. When the switches are in the positions shown in Fig. 7.6-7(*a*) (the left switch in position 1 and the right switch in position 2), corresponding with ϕ_1 high in Fig. 7.6-7(*b*), $V_R = V_1$ and C_R is charged to $Q_1 = C_R V_1$. In the next half-cycle, with the switches reversed (the left switch in position 1 and the right switch in position 2), corresponding with ϕ_2 high in Fig. 7.6-7(*b*), $V_R = -V_2$ and C_R is charged to $Q_2 = -C_R V_2$. The net charge input to C_R is $Q_{\text{in}} = Q_1 - Q_2 = C_R(V_1 + V_2)$. Due to the phase reversal provided by the switches, the net charge out of the capacitor is $Q_{\text{out}} = -Q_{\text{in}} = -C_R(V_1 + V_2)$. The net charge transferred per cycle is

$$\Delta Q = -C_R(V_1 + V_2) \tag{5}$$

In the most frequently used application of this circuit, the right terminal is connected to the inverting input of an operational amplifier that is at zero potential. In this case $V_2 = 0$, and if the switching cycle is repeated at a frequency of f_c, the circuit provides a parasitic-insensitive realization for a negative-valued resistor $R_{\text{eq}} = -1/C_R f_c$. Note that the same result can be obtained by interchanging ϕ_1 and ϕ_2 throughout Fig. 7.6-7(*b*). As in the previous circuit, the parasitic capacitors shown in Fig. 7.6-7(*b*) are discharged through the switches and the circuit provides a parasitic-insensitive realization. We shall shortly see that this modified circuit makes it easy to realize a *non*inverting switched capacitor integrator.

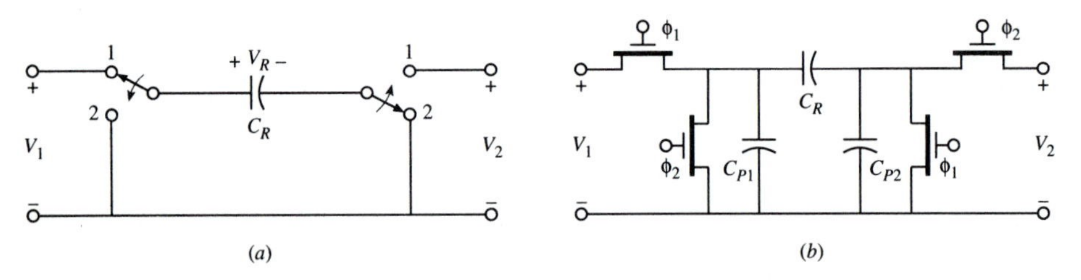

FIGURE 7.6-7
Parasitic-insensitive switched capacitor equivalent negative resistor.

Basic Switched Capacitor Integrators

As an example of the use of switched capacitor circuits, consider the inverting integrator shown in Fig. 7.6-8(a). The voltage transfer function is given as

$$\frac{V_{\text{out}}(s)}{V_{\text{in}}(s)} = \frac{-1}{sR_1C_2} \tag{6}$$

If we represent R_1 by the switched capacitor resistor implementation shown in Fig. 7.6-6(b) (with $C_R = C_1$), we obtain the parasitic-insensitive switched capacitor inverting integrator circuit shown in Fig. 7.6-8(b). For simplicity, the parasitic capacitors are not shown. Assuming that the highest signal frequency component is much lower than f_c, the voltage transfer function for this circuit is

$$\frac{V_{\text{out}}(s)}{V_{\text{in}}(s)} = \frac{-C_1 f_c}{sC_2} \tag{7}$$

This circuit provides a parasitic-insensitive realization of an *inverting* integrator.

A realization of a *noninverting* integrator is shown in Fig. 7.6-9(a) (another realization of this function is given in Fig. 7.4-5). Its voltage transfer function is

$$\frac{V_{\text{out}}(s)}{V_{\text{in}}(s)} = \frac{1}{sR_1C_2} \tag{8}$$

To realize this function with an *SC* circuit, we need only reverse the phases of the clock signals on the left pair of switches in the inverting integrator realization shown in Fig. 7.6-8(b). The result is shown in Fig. 7.6-9(b). From the preceding discussion we conclude that such a phase reversal introduces a negative sign in the transfer function. Thus we obtain

$$\frac{V_{\text{out}}(s)}{V_{\text{in}}(s)} = \frac{C_1 f_c}{sC_2} \tag{9}$$

The circuit provides a parasitic-insensitive switched capacitor realization of a noninverting integrator.

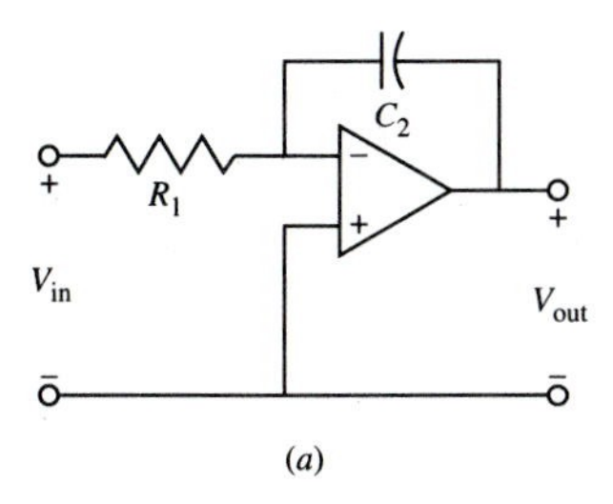

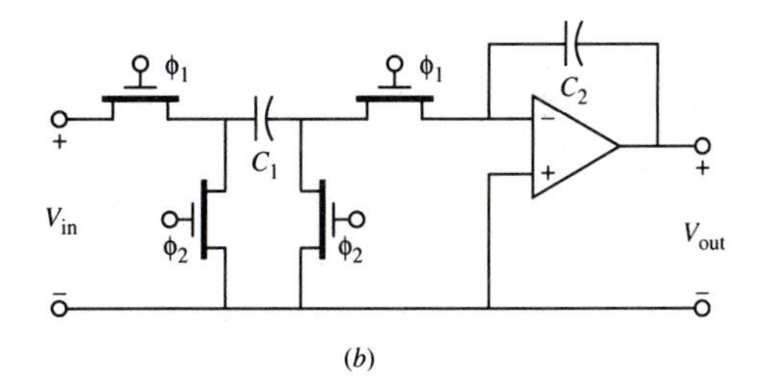

FIGURE 7.6-8
Inverting integrator.

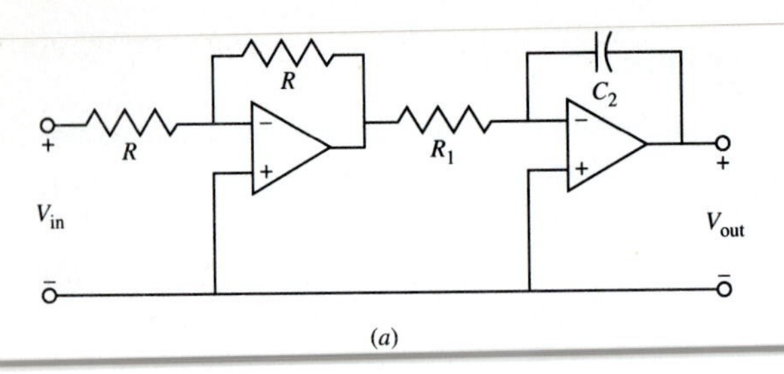

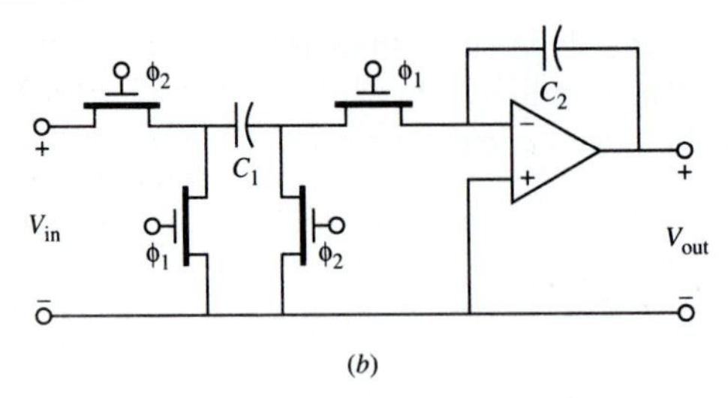

FIGURE 7.6-9
Noninverting integrator.

Other Switched Capacitor Circuits

Another useful active *RC* building block is the lossy integrator. It is used in the Tow-Thomas filter (Sec. 6.2) as well as in the leapfrog simulation method (Sec. 7.4). A realization of a lossy integrator is shown in Fig. 7.6-10(*a*). The voltage transfer function is

$$\frac{V_{out}(s)}{V_{in}(s)} = \frac{-1/R_1C_2}{s + 1/R_3C_2} \tag{10}$$

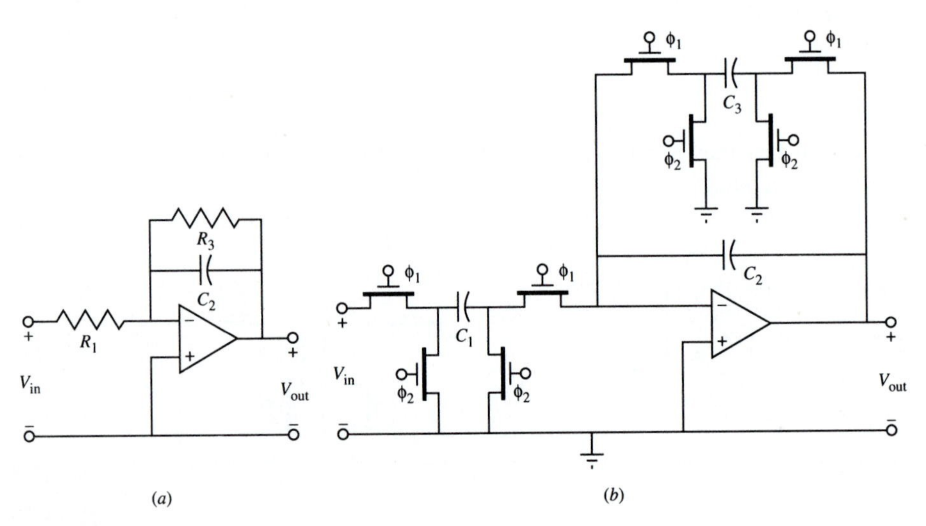

FIGURE 7.6-10
Lossy (damped) integrator.

A parasitic-insensitive switched capacitor realization of the lossy integrator is shown in Fig. 7.6-10(*b*). Its voltage transfer function is

$$\frac{V_{out}(s)}{V_{in}(s)} = \frac{-C_1 f_c/C_2}{s + C_3 f_c/C_2} \tag{11}$$

Note that in this realization, there are two switches wired to the (negative) input terminal of the operational amplifier. These are used to connect the right plate of C_1 and the left plate of C_3 to the amplifier. As shown in the figure, the phases of the switches are chosen so that both inputs are connected at the same time (ϕ_1 high). In general, this results in simpler analysis and improved performance and should always be done. Another advantage that results from this choice of phasing can be seen by noting that during the phase when the capacitor plates referred to above are *not* connected to the amplifier, they are both connected to ground (ϕ_2 high). Since the connections made to these plates are the same in both phases, the plates may be connected together and switched with a single pair of switches. This connection eliminates two switches and is referred to as *switch sharing*. The resulting circuit is shown in Fig. 7.6-11.

The inverting and noninverting integrator circuits are easily modified to provide other useful integrator configurations. Three such circuits are the summing integrator, the differential integrator, and the integrator/summer. A schematic of a *summing integrator* is shown in Fig. 7.6-12. The output voltage is given as

$$V_{out}(s) = -\frac{C_1 f_c}{sC_3} V_1(s) - \frac{C_2 f_c}{sC_3} V_2(s) \tag{12}$$

To simplify the figure, single SPDT switches are shown in the switched capacitor equivalent resistor implementations in Fig. 7.6-12 and in many of the figures throughout the remainder of this section. In most applications, these switches would need to be implemented by dual-switch configurations [as shown in Fig. 7.6-6(*b*)] to obtain stray

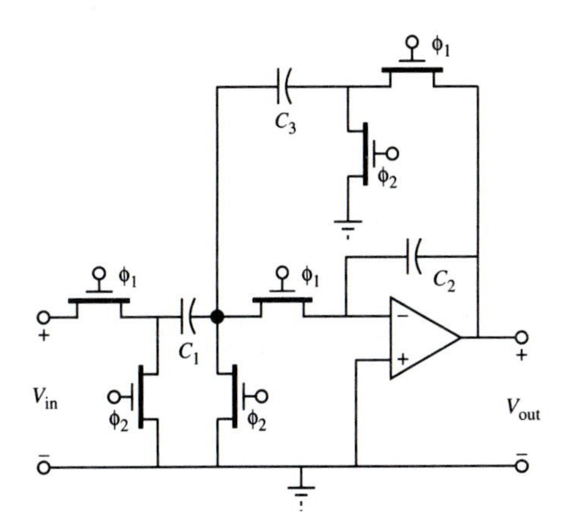

FIGURE 7.6-11
Lossy integrator with switch sharing.

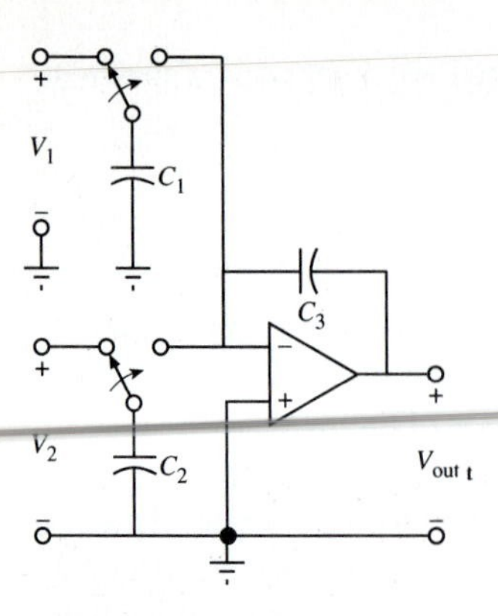

FIGURE 7.6-12
Summing integrator.

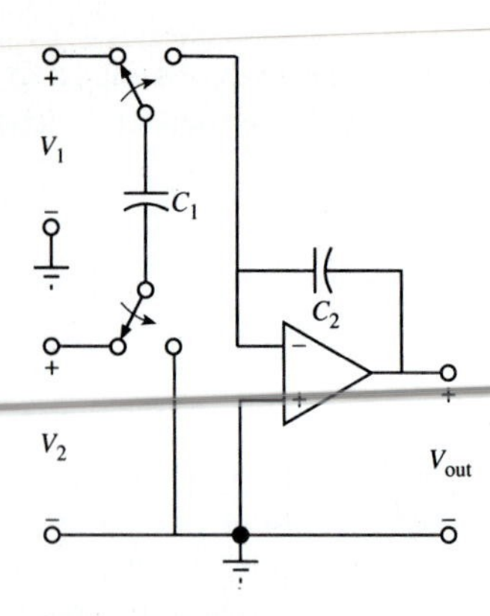

FIGURE 7.6-13
Differential integrator.

insensitive performance. With such configurations, either or both of the inputs are easily changed from inverting to noninverting by using the circuit shown in Fig. 7.6-7(*b*). A schematic of *differential-input integrator* is shown in Fig. 7.6-13. The output voltage is given as

$$V_{out}(s) = -\frac{C_1 f_c}{sC_2}[V_1(s) - V_2(s)] \tag{13}$$

This circuit is a stray-insensitive one. It is easily modified to obtain a *differential-input lossy integrator*. Such a circuit is shown in Fig. 7.6-14. The output voltage is given as

$$V_{out}(s) = \frac{-C_1 f_c/C_2}{s + C_3 f_c/C_2}[V_1(s) - V_2(s)] \tag{14}$$

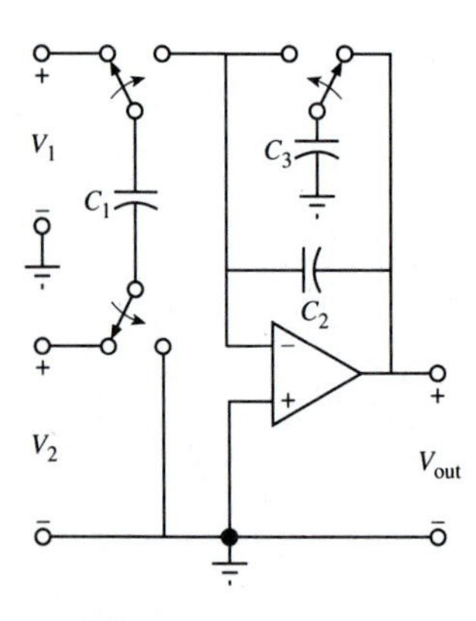

FIGURE 7.6-14
Differential lossy integrator.

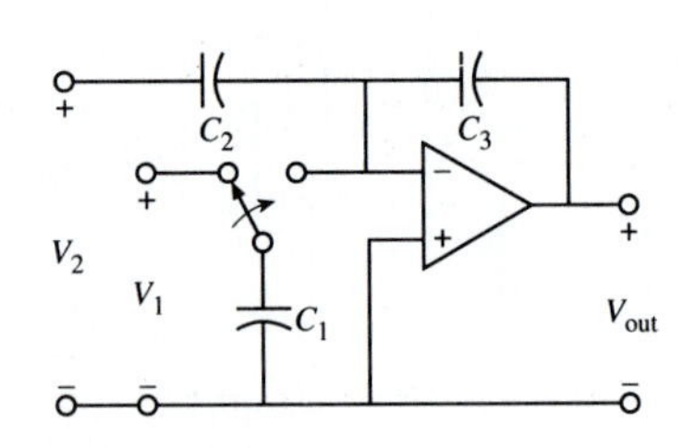

FIGURE 7.6-15
Integrator/summer.

A schematic of an *integrator/summer* is shown in Fig. 7.6-15. The output voltage is

$$V_{\text{out}}(s) = -\frac{C_1 f_c}{sC_3} V_1(s) - \frac{C_2}{C_3} V_2(s) \tag{15}$$

Dual-switch configurations would usually be required for this circuit.

Switched Capacitor Tow-Thomas Circuit

The circuits described above can be used as building blocks to realize many filter functions. As an example, consider the Tow-Thomas filter described in Sec. 6.2 and shown in Fig. 6.2-2. It consists of a lossy integrator, an inverting integrator, and an inverter. In a switched capacitor realization, the functions of inverting integrator and inverter can be combined. Thus only two circuits are required, one to provide lossy integration and summation and the other for noninverting integration. A switched capacitor realization of the Tow-Thomas filter is shown in Fig. 7.6-16. For this the voltage transfer functions are

$$\begin{aligned} \frac{V_{\text{BP}}(s)}{V_{\text{in}}(s)} &= -\frac{sC_1 f_c/C_2}{s^2 + sC_3 f_c/C_2 + (C_1 f_c/C_2)^2} \\ \frac{V_{\text{LP}}(s)}{V_{\text{in}}(s)} &= \frac{(C_1 f_c/C_2)^2}{s^2 + sC_3 f_c/C_2 + (C_1 f_c/C_2)^2} \end{aligned} \tag{16}$$

Switched Capacitor Leapfrog Filter

The leapfrog filter introduced in Sec. 7.4 is readily implemented using switched capacitor techniques. As an example, consider the fifth-order low-pass filter shown in Figure 7.4-4(*a*). The switched capacitor circuits of choice for use in the realization are the differential-input regular and lossy integrators. To implement the realization, the

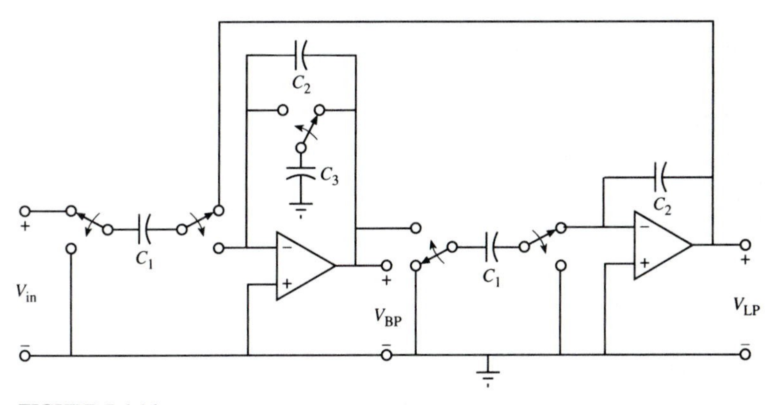

FIGURE 7.6-16
Swiched capacitor Tow-Thomas filter.

equations of (4) of Sec. 7.4 may be modified so that all the transfer functions are inverting. The new equations (for the fifth-order case) become

$$\begin{aligned} -V_{i1} &= [V_1 - (+V_2)](-Y_1) \\ V_2 &= [-V_{i1} - (-V_{i3})](-Z_2) \\ -V_{i3} &= [V_2 - (+V_4)](-Y_3) \\ V_4 &= [-V_{i3} - (-V_{i5})](-Z_4) \\ -V_{i5} &= V_4(-Y_5) \end{aligned} \tag{17}$$

A block diagram realization of these equations is shown in Fig. 7.6-17. The voltage transfer functions realized as $Y_1(s)$, $Z_2(s)$, $Y_3(s)$, $Z_4(s)$, and $Y_5(s)$ are defined by the relations given in (5) of Sec. 7.4. A switched capacitor realization of the block diagram is shown in Fig. 7.6-18. The values of the capacitors used to replace resistors are

$$C_0 = \frac{1}{Rf_c} \qquad C_{R_s} = \frac{R_s}{Rf_c} \qquad C_{R_L} = \frac{R_L}{Rf_c} \tag{18}$$

where R is an impedance scaling constant. The values of the capacitors used in the integrators are expressed in terms of the element values given in the prototype filter shown in Fig. 7.4-4(a) as follows:

$$C_{L_1} = \frac{L_1}{R} \qquad C_{C_2} = \frac{C_2}{R} \qquad C_{L_3} = \frac{L_3}{R} \qquad C_{C_4} = \frac{C_4}{R} \qquad C_{L_5} = \frac{L_5}{R} \tag{19}$$

Example 7.6-2 A fifth-order low-pass filter. As an example, consider the use of switched capacitor techniques to realize a fifth-order Butterworth low-pass filter with a bandwidth of 1 kHz using a clock frequency $f_c = 100$ kHz. From App. A we find $L'_1 = L'_5 = 0.618$ H, $C'_2 = C'_4 = 1.618$ F, and $L'_3 = 2$ H for the normalized filter. For these values the voltage transfer function is

$$\frac{V_L(s)}{V_s(s)} = \frac{0.5}{s^5 + 3.236068s^4 + 5.236068s^3 + 5.236068s^2 + 3.236068s + 1}$$

If we choose $C_0 = 1/Rf_c = 1$ pF, we obtain $R = 10^7\ \Omega$ and $C_{R_s} = C_{R_L} = 1$ pF. Using these values and frequency denormalizing by $2\pi \times 10^3$, we obtain $C_{L'_1} = C_{L'_5} = 9.836$ pF, $C_{C'_2} = C_{C'_4} = 25.751$ pF, and $C_{L'_3} = 31.831$ pF. The realization is shown in Fig. 7.6-18.

Switched capacitor techniques may also be used to realize band-pass leapfrog filters by using the switched capacitor Tow-Thomas filter shown in Fig. 7.6-16 as a

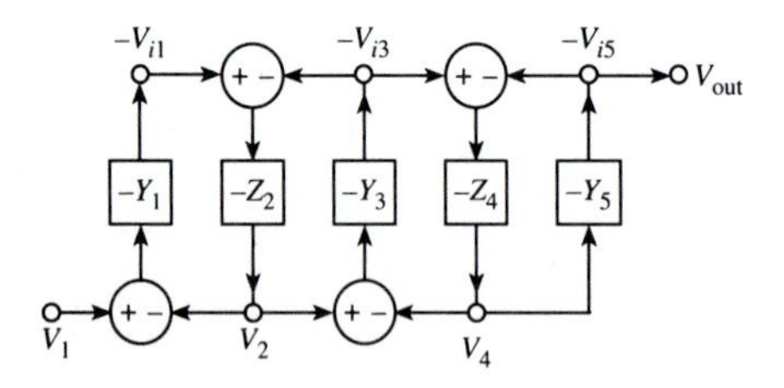

FIGURE 7.6-17
Block diagram of Eqs. (17).

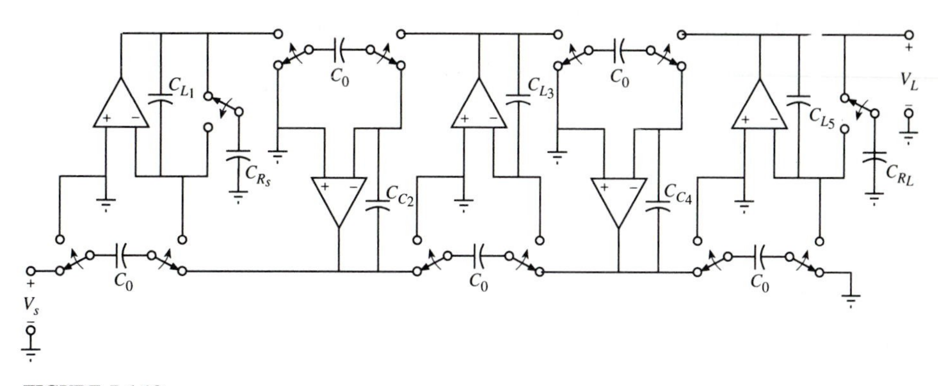

FIGURE 7.6-18
Fifth-order switched capacitor low-pass filter.

building block and following the leapfrog construction for band-pass filters given in Sec. 7.4.

Commercial implementations of switched capacitor filters are available from several manufacturers. Like the Universal Active Filter described in Sec. 6.1, these implementations allow the user to obtain a wide range of filter characteristics. An example of such a product is the Universal Switched Capacitor Filter, which provides two dual-integrator-loop sections, each of which independently realizes the state-variable filter described in Sec. 6.1. It is available as the MF10 (National Semiconductor) and the LTC1060 (Linear Technology). Examples of fixed-type high-performance switched capacitor filters are the Gould S3528 low-pass and S3529 high-pass seven-pole elliptic configurations.

PROBLEMS

Section 7.1

1. (*a*) Find the transmission parameters for the operational amplifier realization of the current-controlled voltage source shown in Fig. P7.1-1(*a*). Assume that the operational amplifier is ideal with infinite gain, zero output impedance, and infinite input impedance.

(*b*) Repeat part (*a*) for the voltage-controlled current source realization shown in Fig. P7.1-1(*b*).

(*c*) Repeat part (*a*) for the current-controlled current source realization shown in Fig. P7.1-1(*c*).

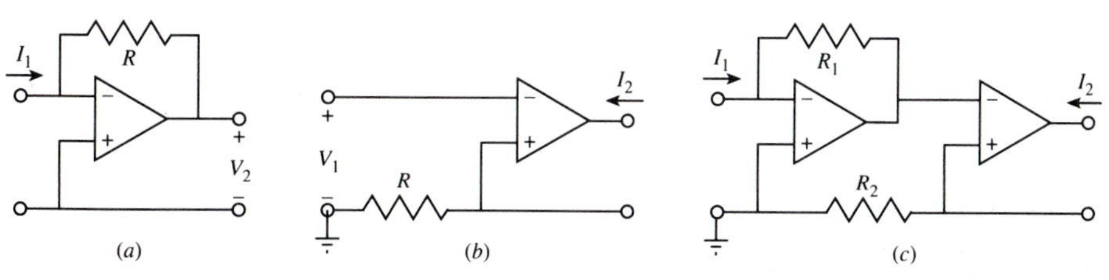

FIGURE P7.1-1

2. (*a*) Realize the impedance $Z_{in}(s) = Ks$ using the GIN shown in Fig. 7.1-6 with single resistors or capacitors for the impedances $Z_1(s)$ through $Z_5(s)$. If the value of the resistors is 1 kΩ and that of the capacitors is 1 μF, find the value of the constant K and give its units.
   (*b*) Repeat part (*a*) for the impedance $Z_{in}(s) = K/s$.
   (*c*) Repeat part (*a*) for the impedance $Z_{in}(s) = Ks^2$.
   (*d*) Repeat part (*a*) for the impedance $Z_{in}(s) = K/s^2$.
   (*e*) Repeat part (*a*) for the impedance $Z_{in}(s) = K/s^3$.
3. The GIN network shown in Fig. 7.1-6 is to be converted into a two-port GIC by the following steps: (*a*) Redefine $V_2(s)$ at the terminals where the impedance $Z_4(s)$ is connected and put the positive reference terminal at the left; (*b*) remove the impedance $Z_4(s)$ from the circuit to create the second port; and (*c*) redefine the current $I_2(s)$ as the current flowing into the positive terminal of $V_2(s)$. Now find the transmission parameters of the network.
4. A GIN circuit is shown in Fig. P7.1-4. If a second port is created by removing the impedance $Z_5(s)$, find the transmission parameters for the resulting two-port network.

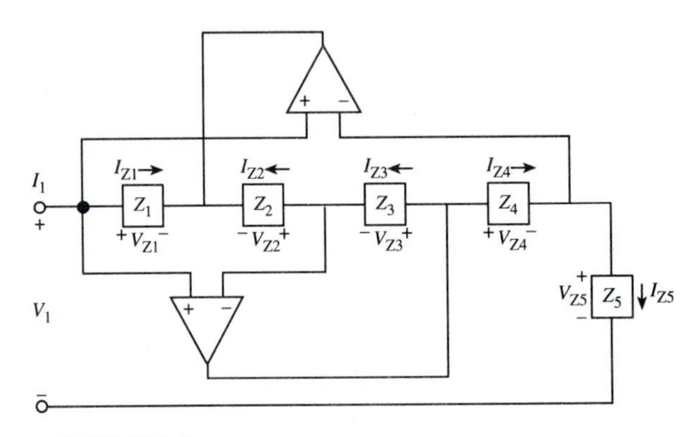

FIGURE P7.1-4

5. A GIN circuit is shown in Fig. P7.1-4. If a second port is created by removing the impedance $Z_4(s)$, find the transmission parameters for the resulting two-port network.
6. A GIN circuit is shown in Fig. P7.1-4. If a second port is created by removing the impedance $Z_3(s)$, find the transmission parameters for the resulting two-port network.

Section 7.2

1. If the values of the resistors and capacitors in the circuit shown in Fig. 7.3-2 are constrained to the ranges 1 to 100 kΩ and 100 pF to 1 μF, respectively, what range of values for a synthetic inductor can be obtained?
2. Find the network functions $V_{o1}(s)/V_1(s)$ and $V_{o2}(s)/V_1(s)$ for the GIN shown in Fig. 7.2-1 for the case where $Z_4(s) = 1/sC$ and $Z_1(s) = Z_2(s) = Z_3(s) = Z_5(s) = R$. Let $C = 0.1$ μF and $R = 10$ kΩ. Draw Bode plots of the magnitude of each of the transfer functions.
3. Design a synthetic inductor third-order high-pass filter with a Butterworth magnitude characteristic and a 3-dB frequency of 10 kHz. Use a double-resistance-terminated prototype realization. In the final filter, all the capacitors should have a value of 0.001 μF.

4. Use the synthetic inductor method to design a fourth-order single-resistance-terminated high-pass filter with a Butterworth voltage transfer function. The terminating resistor should have a value of 1000 Ω. The 3-dB frequency should be 500 Hz. The resistors in the synthetic inductor should all be equal and should have a value of 10 kΩ.
5. Find the function sensitivity of the voltage transfer function of a second-order normalized Butterworth synthetic inductor high-pass filter to the elements R_1, R_2, R_3, C_4, and R_5 of the synthetic inductor. The prototype filter should be a double-resistance-terminated one.
6. Design a sixth-order band-pass coupled resonator filter with a bandwidth of 0.1 rad/s and a center frequency of 1 rad/s. Use a Butterworth magnitude characteristic in the low-pass double-resistance-terminated prototype. Let the prototype resonator inductors have a value of 0.07071 H.

Section 7.3

1. If the values of the resistors and capacitors in the circuit shown in Fig. 7.3-2 are constrained to the ranges 1 to 100 kΩ and 100 pF to 1 μF, respectively, what range of values for the constant D defined in (1) can be obtained?
2. Develop the relationships showing how the unnormalized value of the constant D defined in (1) is related to a normalized constant D_n by the frequency and impedance denormalization constants Ω_n and z_n defined in Sec. 1.4.
3. Design an FDNR realization for a third-order Butterworth low-pass voltage transfer function. The prototype normalized filter should have equal 1-Ω resistance terminations. In the final realization, the bandwidth should be 10^3 rad/s, and all the capacitors should have the value 0.1 μF.
4. Using FDNRs, design a filter that has a fourth-order low-pass Butterworth voltage transfer function. The cutoff frequency should be 5 kHz, and the prototype source and load resistors should have a value of 500 Ω. All the FDNR resistors should have a value of 7500 Ω.
5. Use FDNR techniques to find a realization of a fourth-order low-pass Butterworth filter based on an equal double-resistance-terminated prototype. The denormalized filter should have a bandwidth of 10^4 rad/s. All the capacitors used in the realization should have a value of 0.01 μF.
6. (*a*) Find a realization for a second-order Butterworth single-resistance-terminated low-pass filter with a 1-rad/s bandwidth using an FDNR in which $Z_1(s) = R_1$, $Z_2(s) = R_2$, $Z_4(s) = R_4$, and $Z_3(s) = sC_3$, and $Z_5(s) = sC_5$.
   (*b*) Find the following sensitivities for the realization: $S_{R_1}^Q$, $S_{R_2}^Q$, $S_{C_3}^Q$, $S_{R_4}^Q$, and $S_{C_5}^Q$.
7. (*a*) For an FDNR realization of a second-order equal double-resistance-terminated Butterworth low-pass filter with a normalized bandwidth of 1 rad/s, find the function sensitivity $S^{N(s)}$, where $N(s)$ is the voltage transfer function, to all the resistor values, both those of the network and those of the FDNR. Let the FDNR resistors be R_2, R_4, and R_5.
   (*b*) Find the percentage change in the magnitude of the network function at a frequency of 1 rad/s if all the resistor values are changed by +1 percent.
8. (*a*) For a single FDNR realization of a third-order equal double-resistance-terminated Butterworth low-pass filter with a normalized bandwidth of 1 rad/s, use sensitivity methods to find the percentage change in the magnitude of the voltage transfer function at 1 rad/s if all resistor values are changed by +10 percent.
   (*b*) Find the worst-case percentage change in the magnitude of the voltage transfer function caused by ±10 percent changes in each of the resistors.

Section 7.4

1. Draw a block diagram similar to the one shown in Fig. 7.4-2 for the circuit shown in Fig. P7.4-1. Use the voltages V_{i1}, V_{i2}, and V_{i4} to simulate the currents I_1, I_2, and I_4. Label the variables simulated at each of the points in the block diagram.

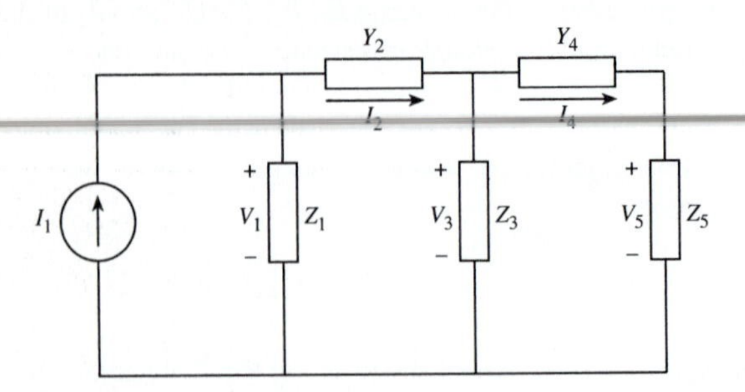

FIGURE P7.4-1

2. (*a*) Make a leapfrog realization for a normalized third-order low-pass Butterworth double-resistance-terminated lossless ladder filter. Use the block diagram shown in Fig. 7.4-2.
   (*b*) Repeat part (*a*) using the block diagram shown in Fig. 7.4-3.
3. (*a*) Design a leapfrog filter realization of a normalized fourth-order low-pass Butterworth double-resistance-terminated lossless ladder filter. Use the block diagram shown in Fig. 7.4-2.
   (*b*) Repeat part (*a*) using the block diagram shown in Fig. 7.4-3.
4. Use the leapfrog method to design a filter having a fifth-order low-pass Butterworth characteristic. The cutoff frequency should be 1 krad/s. All resistors should have a value of 10 kΩ.
5. A leapfrog filter is shown in Fig. P7.4-5. Find the voltage transfer function for the filter. Assume $C = 10^{-8}$ F and $R = 10^4$ Ω.

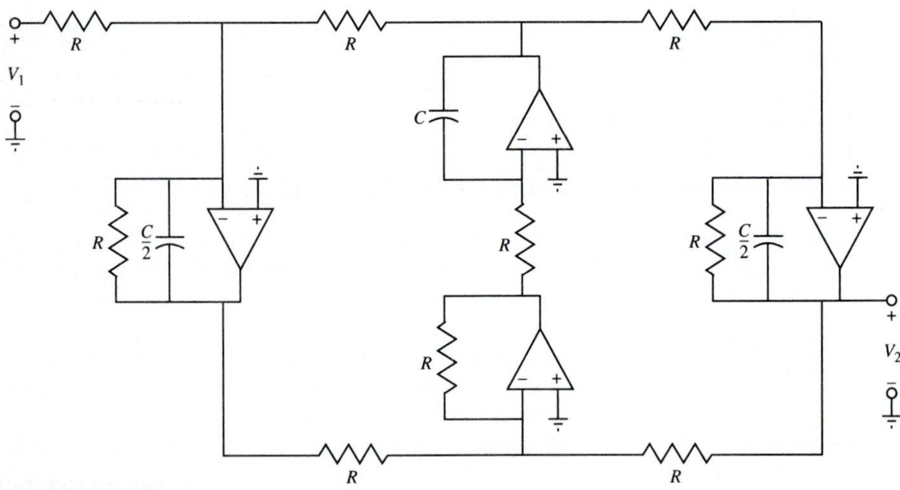

FIGURE P7.4-5

6. (*a*) A leapfrog filter is designed to realize a third-order low-pass function with a Butterworth magnitude characteristic. Find the percentage change in $|N(j1)|$ if all the capacitors are changed by +1 percent.
   (*b*) Find the percentage change in $|N(j1)|$ if all the resistors are changed by +1 percent.
7. Find a leapfrog realization of a normalized fourth-order band-pass function with a maximally flat magnitude characteristic, a center frequency of 1 rad/s, and a bandwidth of 1 rad/s. Use a double-resistance-terminated (1-Ω resistor) lossless ladder filter as a prototype and use the modified Tow-Thomas (resonator) circuit shown in Fig. 7.4-12 to realize the individual second-order stages. Let all the passive elements have unity value except R_1 and R_4. Use the square symbols defined in Fig. 7.4-13(*c*) to represent the resonators.
8. Design a sixth-order Chebyshev band-pass filter with a voltage transfer function that has a center frequency of 1000 Hz and an octave bandwidth with a 0.1-dB ripple. Use the modified Tow-Thomas (resonator) circuit shown in Fig. 7.4-12 to realize the individual second-order stages. All the capacitors should have a value of 0.01 μF. Use an impedance denormalization $z_n = 10^3$ on the elements of the prototype normalized filter.
9. Design a normalized fourth-order Butterworth band-pass filter. Use a double-resistance-terminated (1-Ω resistor) prototype. The filter should have a center frequency of 1 rad/s and a bandwidth of 1 rad/s. Use no more than three operational amplifiers in the realization. Assume that the specifications allow the overall realization to be either inverting or noninverting.

Section 7.5

1. Use PRB techniques to design a normalized fourth-order band-pass filter with a center frequency of 1 rad/s, a bandwidth of 0.1 rad/s, and a Butterworth magnitude characteristic. The gain at resonance in the fourth-order band-pass structure should be unity. Use the (inverting) second-order band-pass infinite-gain single-amplifier filter configuration of Sec. 5.4 for the component blocks. In these circuits let $C_2 = C_3 = 1$ F. Choose the c parameter in the PRB method so that $a_1 = 0$.
2. Design a sixth-order band-pass filter using the PRB configuration. Start with a third-order Butterworth low-pass network function. The resulting band-pass structure should have a gain of 2 at the normalized center frequency of 1 rad/s. Its bandwidth should be 0.1 rad/s. Choose the c parameter so that $a_1 = 0$. *Do not* realize the individual stages; just indicate their parameters, namely, Q_p and H_0. Let $R_f = 1$ and draw the overall configuration showing the values of the resistors.
3. The design shown in Fig. P7.5-3 is proposed to realize a fourth-order band-pass filter with a Butterworth characteristic using two second-order band-pass circuits and based on the PRB technique. In the design equations, H_{BP4} is the gain at resonance of the overall filter, and Q_{BP4}

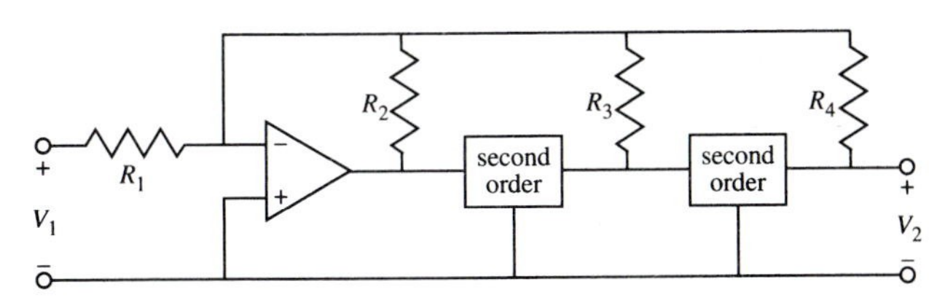

FIGURE P7.5-3

is the Q. In addition, H_{BP2} and Q_{BP2} are the gain at resonance and the Q of the second-order circuits. The design equations are

$$R_1 = \frac{10}{H_{BP4}} \qquad R_2 = 10 \qquad R_3 = 10\sqrt{2} \qquad R_4 = 16$$

$$H_{BP2} = 1\sqrt{2} \qquad Q_{BP2} = 2\sqrt{2}\,Q_{BP4}$$

Determine whether or not these design relations are correct.

4. In a given PRB configuration, the coefficients have the values

$$a_0 = 1 \qquad a_1 = 1 \qquad a_2 = 2 \qquad a_3 = 3 \qquad a_4 = 4 \qquad \text{etc.}$$

In addition, H of the overall filter is required to be positive. The second-order blocks, however, are required to have $H_0 < 0$; that is, they are inverting. Modify the basic PRB circuit shown in Fig. 7.5-1 to accommodate the inverting second-order blocks. Do this by adding another operational amplifier and (if necessary) additional resistors.

5. Determine the percentage change in the magnitude of the voltage transfer function for the PRB filter discussed in Example 7.5-3 if all the elements of both subcircuits are given a 1 percent change in value. Make the evaluation at a frequency of 1 rad/s.

Section 7.6

1. Design a second-order band-pass filter using the switched capacitor Tow-Thomas configuration shown in Fig. 7.6-16. Let the clock frequency be 100 kHz and the undamped natural frequency $\omega_n = 2\pi \times 10^3$ rad/s. Let $Q = 10$ and choose $C_3 = 1$ pF.
2. Repeat the design given in Example 7.6-2 for the case where the filter is to have a 1-dB-ripple Chebyshev characteristic rather than a Butterworth one.
3. Design a fourth-order switched capacitor low-pass filter with a Butterworth magnitude characteristic, a cutoff frequency of 1 kHz, and a clock frequency of 100 kHz. Choose $C_0 = 1$ pF.
4. (*a*) Find the voltage transfer function $V_{out}(s)/V_{in}(s)$ for the switched capacitor filter circuit shown in Fig. P7.6-4.

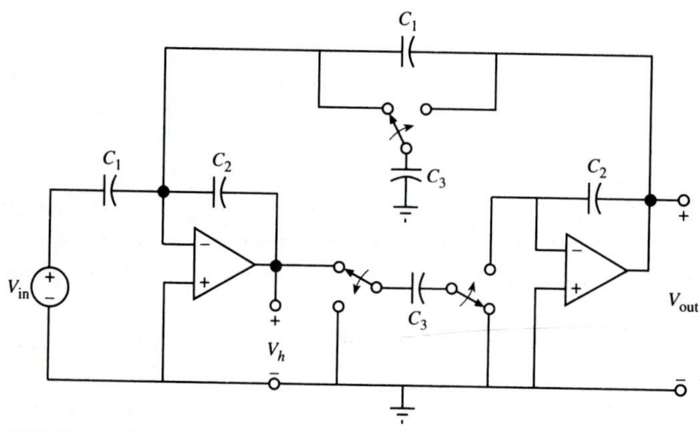

FIGURE P7.6-4

(*b*) Find the voltage transfer function $V_h(s)/V_{in}(s)$ for the same circuit.

(*c*) For the normalized values $\omega_c = 100$, $\omega_n = 1$, and $Q = 10$, find the values of the circuit components.

(*d*) Repeat part (*c*) for $Q = 100$.

APPENDIX A

PASSIVE LOW-PASS FILTER REALIZATIONS

In this appendix we present tables for passive filter realizations of various types of low-pass network functions. The first type of filter is the single-resistance-terminated lossless ladder network. The form of the realization for a voltage source excitation is shown in Figs. A-1(*a*) (for even-order functions) and A-1(*b*) (for odd-order functions). For a current source excitation the realizations have the form shown in Figs. A-2(*a*) (for even-order functions) and A-2(*b*) (for odd-order functions). The element values for (normalized) Butterworth, Chebyshev ($\frac{1}{2}$- and 1-dB ripple), and Thomson filters are given in Table A-1.

Element values for some single-resistance-terminated realizations for inverse-Chebyshev filters are given in Table A-2. The form of the filters is shown in Fig. A-3

A second type of filter realization is the double-resistance-terminated lossless ladder one. The form of the realization is shown in Figs. A-4(*a*) (for even-order functions) and A-4(*b*) (for odd-order functions). An alternate realization form is shown in Figs. A-5(*a*) (for even-order functions) and A-5(*b*) (for odd-order functions). The element values for (normalized) Butterworth, Chebyshev ($\frac{1}{2}$- and 1-dB ripple), and

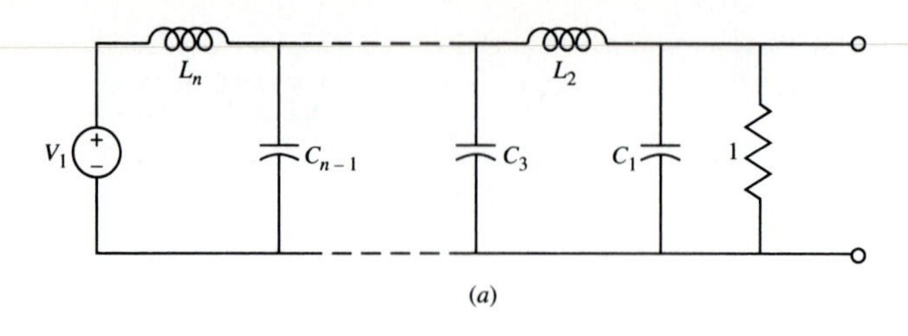

(*a*)

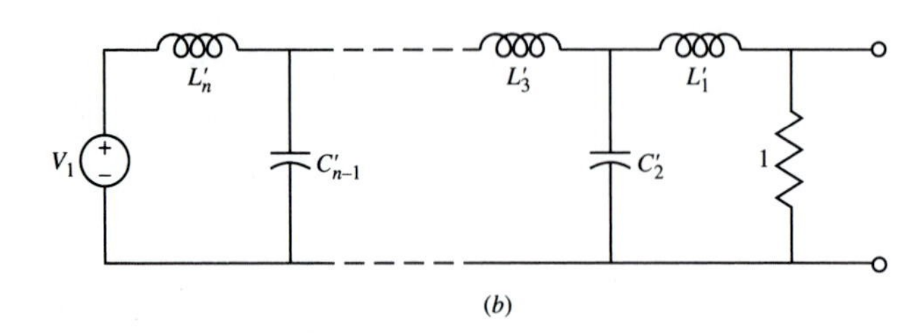

(*b*)

FIGURE A-1
Network configurations for Table A-1 (voltage source excitation): (*a*) even; (*b*) odd.

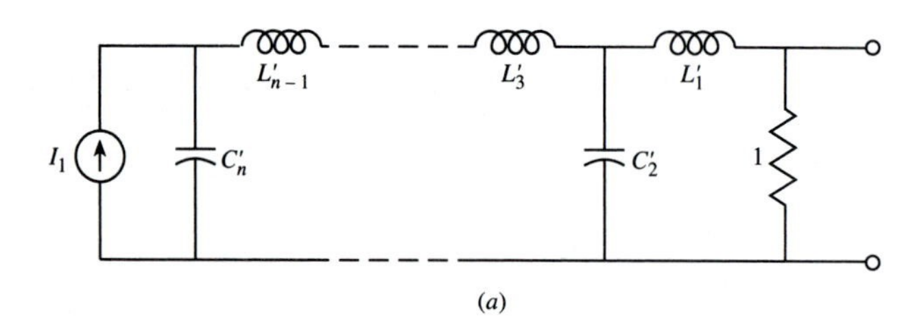

(*a*)

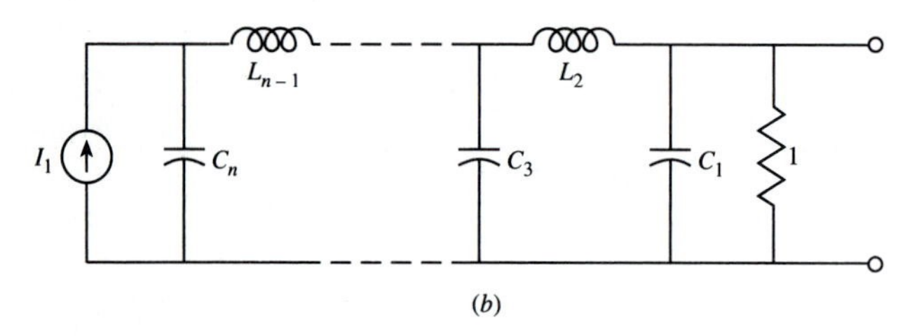

(*b*)

FIGURE A-2
Network configurations for Table A-1 (current source excitation): (*a*) even; (*b*) odd.

TABLE A-1

Element values for low-pass single-resistance-terminated lossless ladder realizations[†]

Elements in Figs. A-1(a) (even) and A-2(b) (odd)

n	C_1	L_2	C_3	L_4	C_5	L_6	C_7	L_8	C_9	L_{10}
					Butterworth (1 rad/s bandwidth)					
2	0.7071	1.4142								
3	0.5000	1.3333	1.5000							
4	0.3827	1.0824	1.5772	1.5307						
5	0.3090	0.8944	1.3820	1.6944	1.5451					
6	0.2588	0.7579	1.2016	1.5529	1.7593	1.5529				
7	0.2225	0.6560	1.0550	1.3972	1.6588	1.7988	1.5576			
8	0.1951	0.5776	0.9370	1.2588	1.5283	1.7287	1.8246	1.5607		
9	0.1736	0.5155	0.8414	1.1408	1.4037	1.6202	1.7772	1.8424	1.5628	
10	0.1564	0.4654	0.7626	1.0406	1.2921	1.5100	1.6869	1.8121	1.8552	1.5643
					½-dB ripple Chebyshev (1 rad/s bandwidth)					
2	0.7014	0.9403								
3	0.7981	1.3001	1.3465							
4	0.8352	1.3916	1.7279	1.3138						
5	0.8529	1.4291	1.8142	1.6426	1.5388					
6	0.8627	1.4483	1.8494	1.7101	1.9018	1.4042				
7	0.8686	1.4596	1.8675	1.7371	1.9712	1.7254	1.5982			
8	0.8725	1.4666	1.8750	1.7508	1.9980	1.7838	1.9571	1.4379		
9	0.8752	1.4714	1.8856	1.7591	2.0116	1.8055	2.0203	1.7571	1.6238	
10	0.8771	1.4748	1.8905	1.7645	2.0197	1.8165	2.0432	1.8119	1.9816	1.4539
					1-dB ripple Chebyshev (1 rad/s bandwidth)					
2	0.9110	0.9957								
3	1.0118	1.3332	1.5088							
4	1.0495	1.4126	1.9093	1.2817						
5	1.0674	1.4441	1.9938	1.5908	1.6652					
6	1.0773	1.4601	2.0270	1.6507	2.0491	1.3457				
7	1.0832	1.4694	2.0437	1.6736	2.1192	1.6489	1.7118			
8	1.0872	1.4751	2.0537	1.6850	2.1453	1.7021	2.0922	1.3691		
9	1.0899	1.4790	2.0601	1.6918	2.1583	1.7213	2.1574	1.6707	1.7317	
10	1.0918	1.4817	2.0645	1.6961	2.1658	1.7306	2.1803	1.7215	2.1111	1.3801
					Thomson (1s delay at dc)					
2	0.3333	1.0000								
3	0.1667	0.4800	0.8333							
4	0.1000	0.2899	0.4627	0.7101						
5	0.0667	0.1948	0.3103	0.4215	0.6231					
6	0.0476	0.1400	0.2246	0.3005	0.3821	0.5595				
7	0.0357	0.1055	0.1704	0.2288	0.2827	0.3487	0.5111			
8	0.0278	0.0823	0.1338	0.1806	0.2227	0.2639	0.3212	0.4732		
9	0.0222	0.0660	0.1077	0.1463	0.1811	0.2129	0.2465	0.2986	0.4424	
10	0.0182	0.0541	0.0886	0.1209	0.1549	0.1880	0.2057	0.2209	0.2712	0.4161
n	L'_1	C'_2	L'_3	C'_4	L'_5	C'_6	L'_7	C'_8	L'_9	C'_{10}

Elements in Figs. A-1(b) (odd) and A-2(a) (even)

[†] Reprinted by permission from L. Weinberg, *Network Analysis and Synthesis*, McGraw-Hill Book Company, New York, 1962; republished by R. E. Krieger Publishing Co., Melbourne, Fla., 1975.

Thomson network functions are given in Table A-3 for the equal-resistance case. In this case, no solution exists for even-order Chebyshev functions. Table A-4 gives the element values for the case where the load and source resistances are related by a factor of 2. In this case no solution exists for even-order 1-db-ripple Chebyshev functions.

The form of the filter for double-resistance-terminated lossless ladder realizations of inverse-Chebyshev functions is given in Figs. A-6 and A-7. The element values for

TABLE A-2
Element values for low-pass single-resistance-terminated lossless ladder inverse-Chebyshev realizations

Elements in Fig. A-3(a) (with voltage source excitation for $n = 2, 4$ and current source excitation for $n = 3$)

n	K_s	C_1	L_2	C_2	C_3	L_4	C_4
2	20	1.5000	3.0000	0.1667			
	30	2.7669	5.5337	0.0904			
	40	4.9750	9.9499	0.0503			
3	20	0.2658	1.3216	0.5675	2.0776		
	30	0.7323	2.3220	0.3230	3.0006		
	40	1.2871	3.6708	0.2043	4.3898		
4	20	0.3138	1.0781	0.1358	1.5247	1.0650	0.8015
	30	0.4717	1.4867	0.0985	2.1150	1.7344	0.4922
	40	0.6687	2.0084	0.0729	2.9708	2.6507	0.3320
n	K_s	L'_1	C'_2	L'_2	L'_3	C'_4	L'_4

Elements in Fig. A-3(b) (with current source excitation for $n = 2, 4$ and voltage source excitation for $n = 3$)

various stopband attenuation specifications and for various orders are given in Table A-5.

The form of the filter for double-resistance-terminated lossless ladder realizations of elliptic low-pass functions is given in Fig. A-6. An alternate realization form is shown in Fig. A-7. The element values for various values of passband ripple and for various orders are given in Tables A-6 and A-7. Table A-6 is for equal-resistance-terminated realizations for even-order (case C) and odd-order filters (0.1- and 1.0-dB passband ripple). Table A-7 is for case B (unequal-resistance terminations) for even-order filters (0.1- and 1.0-dB passband ripple).

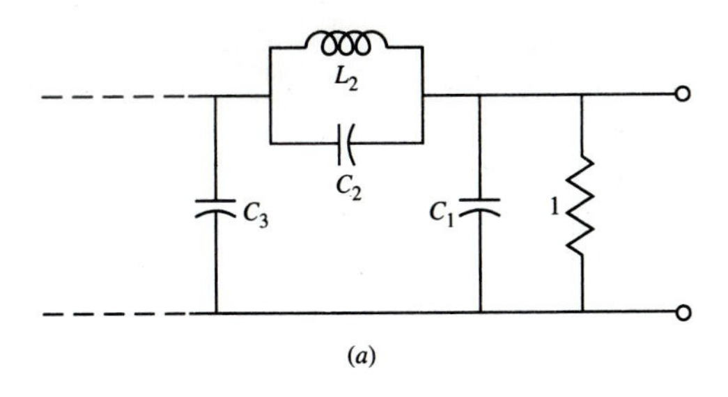

(*a*)

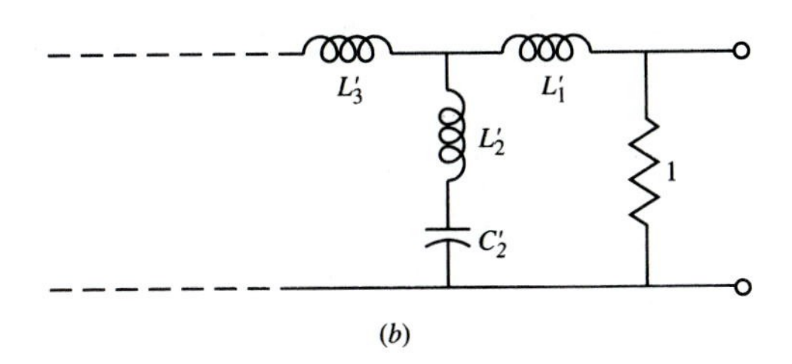

(*b*)

FIGURE A-3
Network configurations for Table A-2.

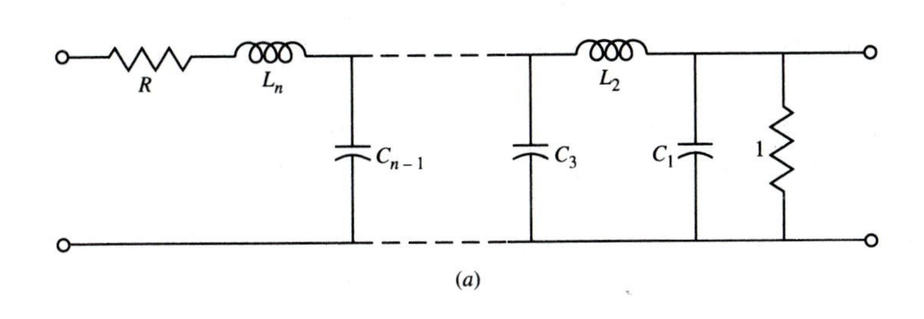

(*a*)

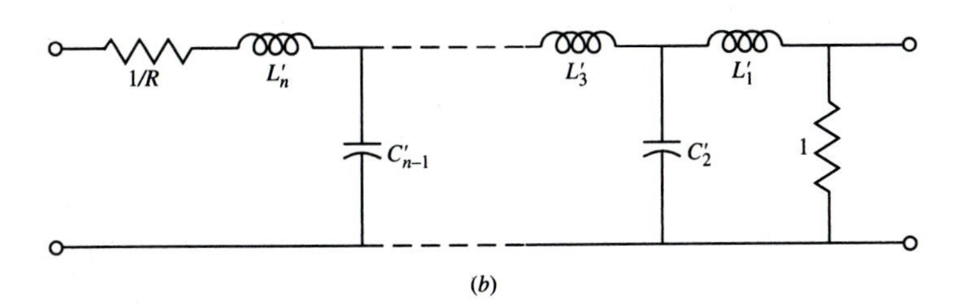

(*b*)

FIGURE A-4
Network configurations for Tables A-3 and A-4: (*a*) even; (*b*) odd.

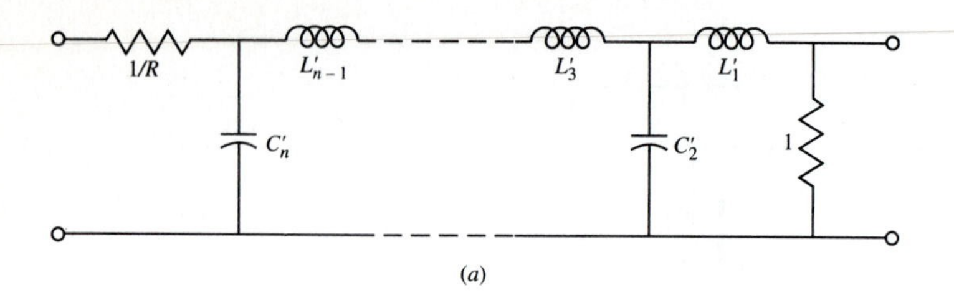

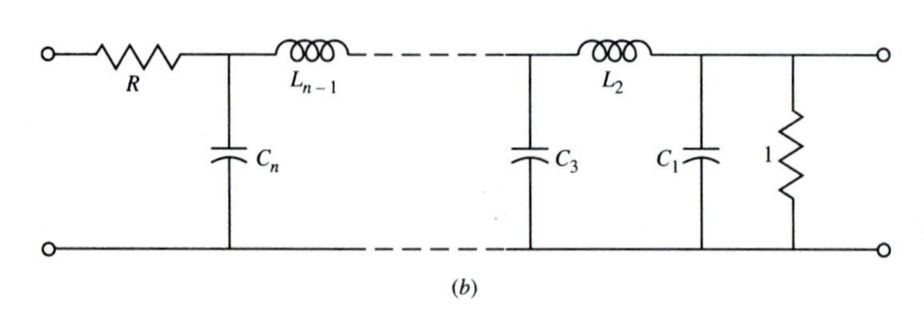

FIGURE A-5
Network configurations for Tables A-3 and A-4: (*a*) even; (*b*) odd.

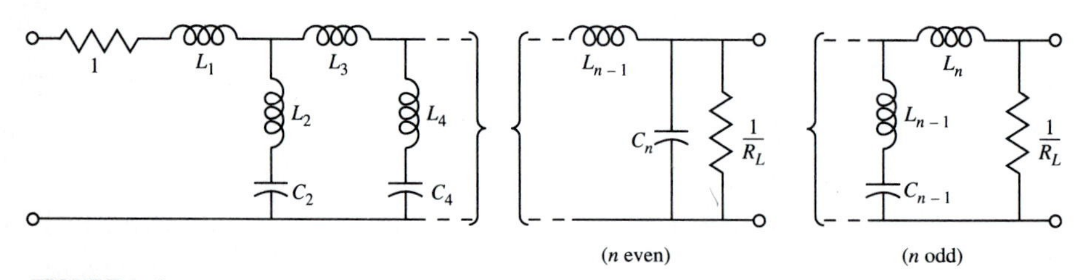

FIGURE A-6
Network configurations for Tables A-5, A-6, and A-7.

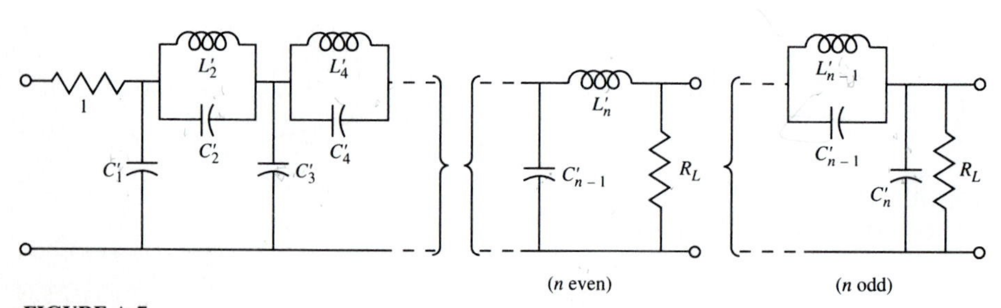

FIGURE A-7
Network configurations for Tables A-5, A-6, and A-7.

TABLE A-3

Element values for low-pass double-resistance-terminated lossless ladder realizations with $R = 1\ \Omega$†

Elements in Figs. A-4(a) (even) and A-5(b) (odd)

n	C_1	L_2	C_3	L_4	C_5	L_6	C_7	L_8	C_9	L_{10}
						Butterworth (1 rad/s bandwidth)				
2	1.4142	1.4142								
3	1.0000	2.0000	1.0000							
4	0.7654	1.8478	1.8478	0.7654						
5	0.6180	1.6180	2.0000	1.6180	0.6180					
6	0.5176	1.4142	1.9319	1.9319	1.4142	0.5176				
7	0.4450	1.2470	1.8019	2.0000	1.8019	1.2470	0.4450			
8	0.3902	1.1111	1.6629	1.9616	1.9616	1.6629	1.1111	0.3902		
9	0.3473	1.0000	1.5321	1.8794	2.0000	1.8794	1.5321	1.0000	0.3473	
0	0.3129	0.9080	1.4142	1.7820	1.9754	1.9754	1.7820	1.4142	0.9090	0.3129
						$\frac{1}{2}$-dB ripple Chebyshev (1 rad/s bandwidth)				
3	1.5963	1.0967	1.5963							
5	1.7058	1.2296	2.5408	1.2296	1.7058					
7	1.7373	1.2582	2.6383	1.3443	2.6383	1.2582	1.7373			
9	1.7504	1.2690	2.6678	1.3673	2.7239	1.3673	2.6678	1.2690	1.7504	
						1-dB ripple Chebyshev (1 rad/s bandwidth)				
3	2.0236	0.9941	2.0236							
5	2.1349	1.0911	3.0009	1.0911	2.1349					
7	2.1666	1.1115	3.0936	1.1735	3.0936	1.1115	2.1666			
9	2.1797	1.1192	3.1214	1.1897	3.1746	1.1897	3.1214	1.1192	2.1797	
						Thomson (1s delay at dc)				
2	1.5774	0.4226								
3	1.2550	0.5528	0.1922							
4	1.0598	0.5116	0.3181	0.1104						
5	0.9303	0.4577	0.3312	0.2090	0.0718					
6	0.8377	0.4116	0.3158	0.2364	0.1480	0.0505				
7	0.7677	0.3744	0.2944	0.2378	0.1778	0.1104	0.0375			
8	0.7125	0.3446	0.2735	0.2297	0.1867	0.1387	0.0855	0.0289		
9	0.6678	0.3203	0.2547	0.2184	0.1859	0.1506	0.1111	0.0682	0.0230	
0	0.6305	0.3002	0.2384	0.2066	0.1808	0.1539	0.1240	0.0911	0.0557	0.0187
n	L'_1	C'_2	L'_3	C'_4	L'_5	C'_6	L'_7	C'_8	L'_9	C'_{10}

Elements in Figs. A-4(b) (odd) and A-5(a) (even)

† Reprinted by permission from L. Weinberg, *Network Analysis and Synthesis*, McGraw-Hill Book Company, New York, 1962; republished by R. E. Krieger Publishing Co., Melbourne, Fla., 1975.

TABLE A-4

Element values for low-pass double-resistance-terminated lossless ladder realizations with $R = \frac{1}{2}\ \Omega$†

Elements in Figs. A-4(*a*) (even) and A-5(*b*) (odd)

n	C_1	L_2	C_3	L_4	C_5	L_6	C_7	L_8	C_9	L_{10}
						Butterworth (1 rad/s bandwidth)				
2	3.3461	0.4483								
3	3.2612	0.7789	1.1811							
4	3.1868	0.8826	2.4524	0.2175						
5	3.1331	0.9237	3.0510	0.4955	0.6857					
6	3.0938	0.9423	3.3687	0.6542	1.6531	0.1412				
7	3.0640	0.9513	3.5532	0.7512	2.2726	0.3536	0.4799			
8	3.0408	0.9558	3.6678	0.8139	2.6863	0.5003	1.2341	0.1042		
9	3.0223	0.9579	3.7426	0.8565	2.9734	0.6046	1.7846	0.2735	0.3685	
10	3.0072	0.9588	3.7934	0.8864	3.1795	0.6808	2.1943	0.4021	0.9818	0.0825
						$\frac{1}{2}$-dB ripple Chebyshev (1 rad/s bandwidth)				
2	1.5132	0.6538								
3	2.9431	0.6503	2.1903							
4	1.8158	1.1328	2.4881	0.7732						
5	3.2228	0.7645	4.1228	0.7116	2.3197					
6	1.8786	1.1884	2.7589	1.2403	2.5976	0.7976				
7	3.3055	0.7899	4.3575	0.8132	4.2419	0.7252	2.3566			
8	1.9012	1.2053	2.8152	1.2864	2.8479	1.2628	2.6310	0.8063		
9	3.3403	0.7995	4.4283	0.8341	4.4546	0.8235	4.2795	0.7304	2.3719	
10	1.9117	1.2127	2.8366	1.2999	2.8964	1.3054	2.8744	1.2714	2.6456	0.8104
						1-dB ripple Chebyshev (1 rad/s bandwidth)				
3	3.4774	0.6153	2.8540							
5	3.7211	0.6949	4.7448	0.6650	2.9936					
7	3.7916	0.7118	4.9425	0.7348	4.8636	0.6757	3.0331			
9	3.8210	0.7182	5.0013	0.7485	5.0412	0.7429	4.9004	0.6797	3.0495	
						Thomson (1 s delay at dc)				
2	2.6180	0.1910								
3	2.1156	0.2613	0.3618							
4	1.7893	0.2461	0.6127	0.0530						
5	1.5686	0.2217	0.6456	0.1015	0.1393					
6	1.4102	0.1999	0.6196	0.1158	0.2894	0.0246				
7	1.2904	0.1821	0.5797	0.1171	0.3497	0.0542	0.0735			
8	1.1964	0.1676	0.5395	0.1135	0.3685	0.0683	0.1684	0.0142		
9	1.1202	0.1558	0.5030	0.1081	0.3580	0.0744	0.2195	0.0336	0.0453	
10	1.0569	0.1460	0.4710	0.1024	0.3586	0.0763	0.2456	0.0450	0.1100	0.0925
n	L'_1	C'_2	L'_3	C'_4	L'_5	C'_6	L'_7	C'_8	L'_9	C'_{10}

Elements in Figs. A-4(*b*) (odd) and A-5(*a*) (even)

† Reprinted by permission from L. Weinberg, *Network Analysis and Synthesis*, McGraw-Hill Book Company, New York, 1962; republished by R. E. Krieger Publishing Co., Melbourne, Fla., 1975.

TABLE A-5

Element values for inverse-Chebyshev low-pass double-resistance-terminated realizations with $R_L = 1\ \Omega$†

Elements in Fig. A-6

n	L_1	C_2	L_2	L_3	C_4	L_4	L_5	C_6	L_6	L_7	C_8	L_8	L_9	C_{10}
(a) $K_s = 20$ dB														
3	1.17172	2.34344	0.32004	1.17172										
4	0.51237	1.57309	0.52663	1.98949	0.92877									
6	0.06700	0.85352	0.54375	1.44096	0.88341	1.05071	0.71706	0.48813						
(b) $K_s = 30$ dB														
3	1.86644	3.73287	0.20092	1.86644										
4	0.90568	2.41716	0.34273	2.72155	1.21008									
5	0.57338	1.75205	0.19719	2.22791	1.25023	0.72347	0.02010							
6	0.23756	1.16423	0.39863	1.78558	1.37482	0.67515	1.09099	0.57512						
8	0.04821	0.70576	0.39858	1.05109	0.99731	0.96302	1.32734	0.86607	0.78415	0.51202	0.36953			
(c) $K_s = 40$ dB														
3	2.83849	5.67699	0.13211	2.83849										
4	1.36485	3.46005	0.23943	3.68487	1.58967									
5	0.51237	1.85507	0.48759	2.81098	2.25281	0.15336	0.78454							
6	0.05487	1.11490	0.83254	2.36029	2.31724	0.20028	1.69836	0.68229						
8	0.14485	0.88070	0.31941	1.32607	1.35961	0.70641	1.60294	1.15706	0.58695	0.74046	0.41718			
10	0.03923	0.58981	0.31568	0.85233	0.96909	0.81388	1.26133	1.05164	0.92704	1.11780	0.76829	0.63447	0.40554	0.29722
n	C_1'	L_2'	C_2'	C_3'	L_4'	C_4'	C_5'	L_6'	C_6'	C_7'	L_8'	C_8'	C_9'	L_{10}'

Elements in Fig. A-7

† Element values provided by David Báez-López, Head and Professor of the Electrical Engineering Department, Universidad de las Americas, Puebla, Mexico.

TABLE A-6

Element values for elliptic low-pass double-resistance-terminated lossless ladder realizations with $R_L = 1\ \Omega$ (case C for even cases)†

n	ω_s	K_s	L_1	C_2	L_2	L_3	C_4	L_4	L_5	C_6	L_6	L_7	C_8	L_8	L_9	C_{10}
			Elements in Fig. A-6													
3	1.05	1.748	.35550	.15374	5.39596	.35550										
	1.10	3.374	.44626	.26993	2.70353	.44626										
	1.20	6.691	.57336	.44980	1.30805	.57336										
	1.50	14.848	.77031	.74561	.47797	.77031										
	2.00	24.010	.89544	.93759	.20697	.89544					0.1-dB passband ripple					
4	1.05	3.284	.00442	.17221	4.93764	1.01224	.84445									
	1.10	6.478	.17279	.32758	2.30986	1.04894	.89415									
	1.20	12.085	.37139	.56638	1.09294	1.11938	.92440									
	1.50	23.736	.62815	.94009	.40730	1.24711	.93518									
	2.00	36.023	.77554	1.17646	.17957	1.33473	.93382									
5	1.05	13.841	.70813	.76630	.73572	1.12761	.20138	4.38116	.04985							
	1.10	20.050	.81296	.92418	.49338	1.22445	.37193	2.13500	.29125							
	1.20	28.303	.91441	1.06516	.31628	1.38201	.60131	1.09329	.52974							
	1.50	43.415	1.02789	1.21517	.15134	1.63179	.93525	.44083	.81549							
	2.00	58.901	1.08758	1.29322	.07317	1.79387	1.14330	.20038	.97720							
6	1.05	18.727	.44177	.71651	.90905	.83142	.36274	2.44680	.80463	.99857						
	1.10	26.230	.57630	.88798	.61282	.97304	.59060	1.35666	.94305	1.01381						
	1.20	36.113	.70984	1.06266	.39136	1.15974	.87407	.76185	1.09176	1.02462						
	1.50	54.202	.86595	1.27403	.18554	1.43106	1.27235	.33007	1.28253	1.03317						
	2.00	72.761	.95131	1.39297	.08926	1.60132	1.51866	.15421	1.39521	1.03621						

			C'_1	L'_2	C'_2	C'_3	L'_4	C'_4	C'_5	L'_6	C'_6	C'_7	L'_8	C'_8	C'_9	L'_{10}
7	1.05	30.470	.91937	1.07659	.34220	1.09623	.40518	2.20850	.84335	.50342	1.51827	.41098				
	1.10	39.357	.98821	1.16726	.24374	1.27743	.59720	1.35681	1.04029	.67881	.96669	.58282				
	1.20	50.963	1.05029	1.24872	.16124	1.48377	.82869	.81542	1.28723	.87428	.58918	.75395				
	1.50	72.129	1.11593	1.33554	.07857	1.75687	1.15174	.37160	1.63827	1.12502	.26822	.95588				
	2.00	93.809	1.14910	1.37979	.03822	1.92026	1.35221	.17692	1.85664	1.27023	.12694	1.06720				
8	1.05	36.268	.68105	1.01040	.47466	.91103	.48838	1.83667	.67613	.70502	1.12616	.98462	1.04901			
	1.10	46.399	.77921	1.13273	.33839	1.08668	.70667	1.15081	.86562	.92698	.74468	1.09627	1.06140			
	1.20	59.639	.87290	1.25079	.22404	1.28897	.97154	.69915	1.09601	1.17713	.46698	1.21321	1.07163			
	1.50	83.807	.97779	1.38443	.10930	1.56051	1.34331	.32099	1.41679	1.50420	.21807	1.35777	1.08093			
	2.00	108.575	1.03311	1.45551	.05321	1.72464	1.57463	.15328	1.61419	1.69718	.10439	1.44022	1.08484			
9	1.05	47.276	1.02597	1.21654	.20583	1.29803	.60675	1.36728	.76114	.44746	2.01083	.94136	.74312	.84407	.63916	
	1.10	58.707	1.07226	1.27741	.14773	1.46403	.79046	.92320	1.00154	.63575	1.28473	1.14956	.89544	.57633	.77015	
	1.20	73.629	1.11295	1.33139	.09815	1.64257	.99964	.58858	1.29050	.86538	.78945	1.39229	1.05143	.36877	.89697	
	1.50	100.842	1.15493	1.38761	.04800	1.86765	1.27611	.27927	1.69055	1.18960	.36523	1.72037	1.23694	.17437	1.04130	
	2.00	128.717	1.17576	1.41568	.02339	1.99761	1.44055	.13471	1.93638	1.39225	.17487	1.91900	1.33887	.08371	1.11842	
10	1.05	53.576	.82096	1.17319	.30389	1.09174	.66563	1.26609	.64760	.54466	1.65398	.76756	.97842	.69517	1.11421	1.08154
	1.10	66.262	.89544	1.26213	.21933	1.25856	.87195	.85540	.86007	.76733	1.06638	.95500	1.18228	.48206	1.20554	1.09096
	1.20	82.830	.96461	1.34441	.14648	1.44189	1.10918	.54607	1.11622	1.03951	.65893	1.17011	1.39629	.31255	1.29689	1.10032
	1.50	113.056	1.03982	1.43372	.07199	1.67758	1.42540	.25986	1.47272	1.42505	.30603	1.45794	1.65959	.14962	1.40489	1.10919
	2.00	144.023	1.07857	1.47977	.03516	1.81552	1.61465	.12563	1.69286	1.66661	.14673	1.63110	1.80819	.07224	1.46428	1.11311

Elements in Fig. A-7

TABLE A-6 (*Continued*)

1.0-dB passband ripple

n	ω_s	K_s	L_1	C_2	L_2	L_3	C_4	L_4	L_5	C_6	L_6	L_7	C_8	L_8	L_9	C_{10}
			Elements in Fig. A-6													
3	1.05	8.134	1.05507	.25223	3.28904	1.05507										
	1.10	11.480	1.22525	.37471	1.94752	1.22525										
	1.20	16.209	1.42450	.52544	1.11977	1.42450										
	1.50	25.176	1.69200	.73340	.48592	1.69200										
	2.00	34.454	1.85199	.85903	.22590	1.85199										
4	1.05	11.322	.63708	.35277	2.41039	1.11522	1.39953									
	1.10	15.942	.80935	.54042	1.40015	1.18107	1.45001									
	1.20	22.293	1.00329	.77733	.79634	1.26621	1.49217									
	1.50	34.179	1.25675	1.11431	.34362	1.38981	1.53225									
	2.00	46.481	1.40677	1.32367	.15960	1.46762	1.55071									
5	1.05	24.135	1.56191	.67560	.83449	1.55460	.26584	3.31881	.88528							
	1.10	30.471	1.69691	.77511	.58827	1.79892	.39922	1.98907	1.12109							
	1.20	38.757	1.82812	.87005	.38720	2.09095	.56347	1.16672	1.38094							
	1.50	53.875	1.97687	.97694	.18824	2.49161	.79362	.51950	1.71889							
	2.00	69.360	2.05594	1.03392	.09152	2.73567	.93561	.24486	1.91939							
6	1.05	29.133	1.07458	.80116	.81300	.92735	.51753	1.71498	.92186	1.60511						
	1.10	36.680	1.22059	.94235	.57746	1.10900	.75718	1.05819	1.01676	1.64682						
	1.20	46.571	1.37146	1.08633	.38284	1.32610	1.05110	.63354	1.12484	1.68498						
	1.50	64.661	1.55425	1.25876	.18779	1.62529	1.46557	.28655	1.26961	1.72482						
	2.00	83.221	1.65661	1.35450	.09179	1.80860	1.72376	.13586	1.35729	1.74424						

			C'_1	L'_2	C'_2	C'_3	L'_4	C'_4	C'_5	L'_6	C'_6	C'_7	L'_8	C'_8	C'_9	L'_{10}
7	1.05	40.926	1.82156	.86343	.42668	1.67632	.34381	2.60271	1.23696	.46779	1.63392	1.22362				
	1.10	49.816	1.91040	.92662	.30705	1.93579	.48016	1.68753	1.55276	.59277	1.10699	1.41994				
	1.20	61.422	1.99168	.98474	.20446	2.22804	.64444	1.04856	1.92724	.73012	.70551	1.62539				
	1.50	82.588	2.07882	1.04761	.10016	2.61372	.87393	.48973	2.44021	.90483	.33349	1.87717				
	2.00	104.268	2.12329	1.07993	.04884	2.84446	1.01638	.23538	2.75306	1.00567	.16034	2.01924				
8	1.05	46.727	1.34673	1.00922	.47521	1.08540	.54692	1.64007	.70773	.86533	.91752	1.00315	1.72154			
	1.10	56.858	1.46597	1.10092	.34817	1.28842	.75883	1.07170	.90030	1.11345	.61997	1.07811	1.75961			
	1.20	70.098	1.58346	1.18748	.23599	1.52224	1.01386	.66997	1.12753	1.39664	.39359	1.15904	1.79429			
	1.50	94.266	1.71869	1.28337	.11791	1.83710	1.36932	.31489	1.43764	1.77189	.18513	1.26147	1.83033			
	2.00	119.034	1.79131	1.33358	.05807	2.02800	1.58946	.15185	1.62645	1.99561	.08878	1.32075	1.84787			
9	1.05	57.736	1.95471	.95672	.26172	1.94887	.46951	1.76694	1.12605	.35392	2.54224	1.40979	.63099	.99407	1.47897	
	1.10	69.167	2.01503	.99976	.18875	2.18048	.60062	1.21501	1.47339	.48929	1.66927	1.71691	.73764	.69962	1.63779	
	1.20	84.089	2.06867	1.03833	.12585	2.43022	.74985	.78464	1.88531	.65376	1.04499	2.06810	.84652	.45803	1.79580	
	1.50	111.302	2.12470	1.07895	.06173	2.74601	.94696	.37634	2.45079	.88531	.49076	2.53625	.97584	.22103	1.97957	
	2.00	139.176	2.15275	1.09942	.03012	2.92871	1.06415	.18236	2.79680	1.02979	.23642	2.81748	1.04688	.10706	2.07916	
10	1.05	64.036	1.52461	1.11270	.32041	1.31260	.68239	1.23499	.69671	.59634	1.51065	.77127	1.18327	.57482	1.06945	1.79994
	1.10	76.722	1.62236	1.17191	.23621	1.51411	.87139	.85595	.92029	.82679	.98969	.94303	1.42284	.40056	1.12735	1.83423
	1.20	93.289	1.71564	1.22500	.16075	1.73787	1.08551	.55798	1.18808	1.10906	.61761	1.13757	1.68092	.25963	1.18655	1.86522
	1.50	123.515	1.81947	1.28114	.08056	2.02843	1.36690	.27098	1.55972	1.50976	.28886	1.39429	2.00434	.12388	1.25749	1.89727
	2.00	154.482	1.87378	1.30964	.03972	2.19972	1.53360	.13227	1.78908	1.76138	.13884	1.54729	2.18935	.05966	1.29688	1.91286

Elements in Fig. A-7

† Computed using a program written by David Báez-López, "Sensitivity and Synthesis of Elliptic Functions," Ph.D. dissertation, University of Arizona, Tucson, 1978.

TABLE A-7

Element values for elliptic low-pass double-resistance-terminated lossless ladder realizations for case B†

0.1-dB passband ripple
$R_L = 0.73781\ \Omega$

			Elements in Fig. A-6													
n	ω_s	K_s	L_1	C_2	L_2	L_3	C_4	L_4	L_5	C_6	L_6	L_7	C_8	L_8	L_9	C_{10}
4	1.05	4.485	.15780	.18091	4.73822	1.20743	.82637									
	1.10	8.308	.33411	.33438	2.28333	1.26881	.84827									
	1.20	14.387	.53773	.55478	1.12558	1.36980	.85261									
	1.50	26.320	.79962	.88310	.43628	1.53672	.84068									
	2.00	38.697	.95051	1.08631	.19517	1.64684	.83004									
6	1.05	20.307	.57153	.65752	1.01346	.92972	.32584	2.72744	1.03524	.88809						
	1.10	27.889	.70783	.81703	.67992	1.10484	.51890	1.54640	1.19779	.88523						
	1.20	37.827	.84244	.98082	.43111	1.32791	.75659	.88144	1.37708	.87992						
	1.50	55.966	.99836	1.17887	.20248	1.64500	1.08849	.38623	1.61158	.87198						
	2.00	74.548	1.08280	1.28970	.09690	1.84134	1.29301	.18123	1.75160	.86710						
8	1.05	37.529	.79899	.92996	.52853	1.02243	.42917	2.09084	.78906	.59418	1.34317	1.24821	.89367			
	1.10	47.686	.89538	1.04944	.37264	1.22051	.61878	1.31484	1.01703	.77697	.89307	1.38901	.89115			
	1.20	60.949	.98606	1.16509	.24409	1.44606	.84957	.79988	1.29323	.98124	.56277	1.53864	.88795			
	1.50	85.138	1.08612	1.29596	.11768	1.74572	1.17448	.36725	1.67684	1.24564	.26411	1.72539	.88347			
	2.00	109.915	1.13830	1.36543	.05694	1.92553	1.37705	.17531	1.91249	1.40030	.12672	1.83237	.88075			
10	1.05	54.608	.92833	1.09111	.33387	1.21271	.58502	1.44313	.74643	.46770	1.92642	.91319	.80993	.84618	1.41635	.89491
	1.10	67.307	.99839	1.18097	.23833	1.39449	.76963	.97094	.99183	.65631	1.24697	1.14180	.97201	.59039	1.53570	.89383
	1.20	83.887	1.06237	1.26448	.15755	1.59212	.98322	.61709	1.28687	.88676	.77257	1.40608	1.14191	.38429	1.65727	.89178
	1.50	114.123	1.13092	1.35538	.07660	1.84367	1.26943	.29213	1.69639	1.21247	.35973	1.76038	1.34833	.18475	1.80184	.88893
	2.00	145.095	1.16590	1.40227	.03721	1.98983	1.44131	.14083	1.94870	1.41617	.17269	1.97377	1.46362	.08940	1.88151	.88719
			C'_1	L'_2	C'_2	C'_3	L'_4	C'_4	C'_5	L'_6	C'_6	C'_7	L'_8	C'_8	C'_9	L'_{10}
			Elements in Fig. A-7													

TABLE A-7 (*Continued*)

			Elements in Fig. A-6													
n	ω_s	K_s	L_1	C_2	L_2	L_3	C_4	L_4	L_5	C_6	L_6	L_7	C_8	L_8	L_9	C_{10}
4	1.05	13.243	.95111	.26779	3.20104	1.90749	.80699									
	1.10	18.140	1.16239	.39958	1.91077	2.05228	.80907									
	1.20	24.700	1.40135	.56068	1.11374	2.23453	.80633									
	1.50	36.771	1.71483	.78307	.49201	2.49368	.79924									
	2.00	49.156	1.90048	.91820	.23091	2.65459	.79441									
6	1.05	30.730	1.40432	.58067	1.14761	1.37588	.31837	2.79144	1.79883	.82259						
	1.10	38.342	1.56906	.69149	.80335	1.66832	.45609	1.75937	1.99786	.82076						
	1.20	48.285	1.73631	.80659	.52424	2.01190	.62218	1.07185	2.22816	.81822						
	1.50	66.425	1.93461	.94611	.25229	2.47740	.85305	.49283	2.53990	.81461						
	2.00	85.008	2.04359	1.02402	.12205	2.75884	.99547	.23540	2.72966	.81243						
8	1.05	47.987	1.67197	.76069	.64615	1.55530	.34969	2.56605	1.17579	.48610	1.64181	2.03669	.82473			
	1.10	58.146	1.79301	.84446	.46309	1.83957	.48544	1.67598	1.51273	.61510	1.12807	2.21830	.82355			
	1.20	71.408	1.90884	.92585	.30716	2.16030	.65000	1.04547	1.91110	.75902	.72753	2.41639	.82209			
	1.50	95.597	2.03847	1.01816	.14980	2.58395	.88106	.48955	2.45568	.94522	.34805	2.66846	.82006			
	2.00	120.374	2.10671	1.06721	.07286	2.83733	1.02492	.23554	2.78753	1.05416	.16833	2.81453	.81883			
10	1.05	65.067	1.83096	.86998	.41873	1.81829	.45061	1.87358	1.10022	.35991	2.50341	1.35196	.63152	1.08524	2.24376	.82561
	1.10	77.767	1.92046	.93288	.30171	2.07454	.58273	1.28236	1.45156	.49556	1.65147	1.67847	.74584	.76942	2.40166	.82484
	1.20	94.346	2.00313	.99155	.20092	2.35262	.73518	.82529	1.86993	.66055	1.03714	2.05233	.86549	.50703	2.56480	.82390
	1.50	124.583	2.09263	1.05564	.09835	2.70617	.93910	.39489	2.44690	.89307	.48839	2.55089	1.01109	.24638	2.76130	.82261
	2.00	155.554	2.13863	1.08880	.04792	2.91150	1.06145	.19123	2.80119	1.03824	.23556	2.85037	1.09249	.11976	2.87052	.82181
			C'_1	L'_2	C'_2	C'_3	L'_4	C'_4	C'_5	L'_6	C'_6	C'_7	L'_8	C'_8	C'_9	L'_{10}
			Elements in Fig. A-7													

1.0-dB passband ripple
$R_L = 0.37598\ \Omega$

† Computed using a program written by David Báez-López, "Sensitivity and Synthesis of Elliptic Functions," Ph.D. dissertation, University of Arizona, Tucson, 1978.

APPENDIX B

NETWORK FUNCTIONS WITH AN ELLIPTIC MAGNITUDE CHARACTERISTIC

In this appendix we present tables of network functions that have an elliptic magnitude characteristic. The tables are an extension of the cases presented in Table 2.4-1.

TABLE B-1

Elliptic functions with poles at p_i and having the form†

$$N(s) = H\prod_i \frac{s^2 + c_i}{s^2 + a_i s + b_i}$$

Odd and case *A* even: 0.1-dB passband ripple

n	ω_s	K_s (dB)	c_i	p_i	a_i	b_i
2	1.05	.343	1.438664	$-.075407 \pm j1.180400$	.150814	1.399030
	1.10	.559	1.714083	$-.129483 \pm j1.268507$	.258966	1.625877
	1.20	1.075	2.235990	$-.236268 \pm j1.393844$	.472537	1.998624
	1.50	3.210	3.927051	$-.534107 \pm j1.568367$	1.068213	2.745046
	2.00	7.418	7.464102	$-.843443 \pm j1.581991$	1.686887	3.214092

(*continued*)

TABLE B-1 (*Continued*)

Odd and case A even: 0.1-dB passband ripple

n	ω_s	K_s (dB)	c_i	p_i	a_i	b_i
3	1.05	1.748	1.205410	$-.044853 \pm j1.079332$	.089707	1.166969
				-2.812966		
	1.10	3.374	1.370314	$-.085421 \pm j1.121848$	.170843	1.265840
				-2.240832		
	1.20	6.691	1.699617	$-.156766 \pm j1.170259$	.313532	1.394082
				-1.744102		
	1.50	14.848	2.806014	$-.289646 \pm j1.212428$	.579292	1.553876
				-1.298182		
	2.00	24.010	5.153209	$-.381858 \pm j1.217905$	.763717	1.629108
				-1.116765		
4	1.05	6.397	1.153634	$-.618576 \pm j1.143244$	1.237152	1.689644
			3.312518	$-.037598 \pm j1.045948$	.075196	1.095422
	1.10	10.721	1.290925	$-.703816 \pm j\ .976495$	1.407633	1.448899
			4.349930	$-.066734 \pm j1.066126$	.133467	1.141079
	1.20	17.051	1.572430	$-.108448 \pm j1.086869$	.216897	1.193044
			6.224402	$-.726853 \pm j\ .798154$	1.453706	1.165365
	1.50	29.064	2.535553	$-.698734 \pm j\ .616949$	1.397469	.868855
			12.099310	$-.173627 \pm j1.108114$	.347253	1.258062
	2.00	41.447	4.593261	$-.670443 \pm j\ .535639$	1.340886	.736403
			24.227201	$-.216254 \pm j1.116820$	.432509	1.294053
5	1.05	13.841	1.133422	$-.266902 \pm j1.015887$	.533804	1.103263
			1.773739	$-.030115 \pm j1.028040$	.060229	1.057772
				-1.128858		
	1.10	20.050	1.259320	$-.329692 \pm j\ .953299$	.659383	1.017475
			2.193093	$-.049533 \pm j1.039346$	.099067	1.082694
				$-.932112$		
	1.20	28.303	1.521127	$-.379155 \pm j\ .875398$	.758311	.910081
			2.968367	$-.075430 \pm j1.051645$	.150860	1.111647
				$-.782858$		
	1.50	43.415	2.425515	$-.417037 \pm j\ .775766$	.834075	.775733
			5.437645	$-.114129 \pm j1.066151$	.228259	1.149703
				$-.649753$		
	2.00	58.901	4.364951	$-.429092 \pm j\ .721329$	.858183	.704436
			10.567732	$-.138913 \pm j1.073567$	.277825	1.171844
				$-.590933$		
6	1.05	22.088	1.123326	$-.647026 \pm j\ .628506$	1.294052	.813662
			1.438664	$-.151511 \pm j\ .985417$	.303023	.994002
			6.528768	$-.023386 \pm j1.018380$	.046771	1.037644
	1.10	29.686	1.243362	$-.599771 \pm j\ .517581$	1.199542	.627615
			1.714083	$-.194450 \pm j\ .951604$	.388900	.943361
			8.826455	$-.036964 \pm j1.025840$	.073927	1.053714
	1.20	39.630	1.495035	$-.547628 \pm j\ .429686$	1.095257	.484527
			2.235990	$-.235429 \pm j\ .907696$	.470859	.879339
			12.952671	$-.054595 \pm j1.034294$	.109189	1.072744
	1.50	57.772	2.369289	$-.487832 \pm j\ .349732$	.975663	.360292
			3.927051	$-.277388 \pm j\ .846778$	.554775	.793976
			25.827242	$-.080385 \pm j1.044897$	.160770	1.098271

(*continued*)

TABLE B-1 (*Continued*)

Odd and case *A* even: 0.1-dB passband ripple

n	ω_s	K_s (dB)	c_i	p_i	a_i	b_i
	2.00	76.355	4.248155	$-.457235 \pm j\ .314305$	.914470	.307851
			7.464102	$-.296650 \pm j\ .810843$	.593299	.745467
			52.356841	$-.096676 \pm j1.050734$	.193352	1.113387
7	1.05	30.470	1.117521	$-.362386 \pm j\ .791219$	.724773	.757351
			1.308341	$-.097930 \pm j\ .979496$	.195860	.969003
			2.714372	$-.018274 \pm j1.012906$	.036549	1.026313
				$-.697913$		
	1.10	39.357	1.234128	$-.372606 \pm j\ .706869$	.745212	.638500
			1.523943	$-.129118 \pm j\ .957427$	.258235	.933339
			3.514769	$-.028279 \pm j1.018274$	.056558	1.037683
				$-.599630$		
	1.20	50.963	1.479872	$-.371195 \pm j\ .627191$	.742389	.531154
			1.941341	$-.161448 \pm j\ .928552$	.322896	.888274
			4.966697	$-.041080 \pm j1.024498$	.082161	1.051284
				$-.519720$		
	1.50	72.129	2.336522	$-.359474 \pm j\ .543128$	.718947	.424210
			3.313990	$-.198305 \pm j\ .887345$	.396611	.826705
			9.530078	$-.059559 \pm j1.032553$	.119118	1.069713
				$-.443710$		
	2.00	93.809	4.180043	$-.350169 \pm j\ .501993$	.700339	.374616
			6.201776	$-.217125 \pm j\ .862267$	.434251	.790648
			18.961095	$-.071124 \pm j1.037140$	.142248	1.080719
				$-.408602$		
8	1.05	38.872	1.113864	$-.518427 \pm j\ .408220$	1.036854	.435410
			1.243118	$-.223630 \pm j\ .856767$	.447259	.784060
			1.906139	$-.068627 \pm j\ .979963$	.137255	.965037
			11.046606	$-.014531 \pm j1.009559$	.029062	1.019421
	1.10	49.032	1.228286	$-.466763 \pm j\ .344817$	.933526	.336767
			1.427103	$-.246416 \pm j\ .795957$	.492832	.694269
			2.380411	$-.092340 \pm j\ .964023$	.184680	.937868
			15.106224	$-.022205 \pm j1.013628$	.044410	1.027934
	1.20	62.296	1.470253	$-.419007 \pm j\ .293877$	.838014	.261931
			1.789509	$-.261711 \pm j\ .732751$	.523423	.605416
			3.252831	$-.118065 \pm j\ .943252$	.236131	.903664
			22.383599	$-.031932 \pm j1.018411$	.063864	1.038180
	1.50	86.485	2.315697	$-.369042 \pm j\ .246297$	.738083	.196854
			2.995660	$-.270992 \pm j\ .659775$	.541983	.508739
			6.021824	$-.149084 \pm j\ .913365$	.298167	.856462
			45.059978	$-.045849 \pm j1.024717$	.091699	1.052148
	2.00	111.263	4.136734	$-.344561 \pm j\ .224683$	.689123	.169205
			5.545063	$-.272775 \pm j\ .621654$	.545549	.460860
			11.766674	$-.165771 \pm j\ .894945$	.331543	.828407
			91.761533	$-.054503 \pm j1.028377$	.109007	1.060529
9	1.05	47.276	1.111406	$-.355257 \pm j\ .616991$	.710513	.506886
			1.205410	$-.149551 \pm j\ .893074$	.299102	.819946
			1.594271	$-.050863 \pm j\ .982010$	.101725	.966931
			3.993674	$-.011772 \pm j1.007371$	.023544	1.014935
				$-.514179$		

(*continued*)

TABLE B-1 (*Continued*)
Odd and case *A* even: 0.1-dB passband ripple

n	ω_s	K_s (dB)	c_i	p_i	a_i	b_i
	1.10	58.707	1.224347	$-.341731 \pm j\ .544813$	.683461	.413601
			1.370314	$-.173149 \pm j\ .847267$	.346297	.747841
			1.937719	$-.069513 \pm j\ .969793$	.139026	.945331
			5.299248	$-.017844 \pm j1.010567$	.035688	1.021564
				$-.448275$		
	1.20	73.629	1.463756	$-.323598 \pm j\ .480740$	.647197	.335827
			1.699617	$-.192808 \pm j\ .797069$	.385616	.672494
			2.579104	$-.090331 \pm j\ .953984$	.180663	.918245
			7.652393	$-.025490 \pm j1.014360$	.050980	1.029577
				$-.392972$		
	1.50	100.842	2.301616	$-.299692 \pm j\ .415879$	.599383	.262770
			2.806014	$-.210182 \pm j\ .735885$	.420364	.585703
			4.636336	$-.116302 \pm j\ .931223$	.232604	.880703
			15.014341	$-.036362 \pm j1.019421$	.072724	1.040542
				$-.339006$		
	2.00	128.717	4.107442	$-.286338 \pm j\ .384821$	.572676	.230077
			5.153209	$-.217131 \pm j\ .702561$	.434263	.540739
			8.922191	$-.130710 \pm j\ .917130$	.261421	.858212
			30.201059	$-.043092 \pm j1.022392$	.086183	1.047142
				$-.313657$		
10	1.05	55.681	1.109672	$-.422209 \pm j\ .302646$	.844417	.269855
			1.181462	$-.246341 \pm j\ .730422$	.492682	.594200
			1.438664	$-.106317 \pm j\ .916451$	.212634	.851186
			2.533648	$-.039276 \pm j\ .984256$	.078553	.970303
			16.859610	$-.009704 \pm j1.005863$	.019407	1.011854
	1.10	68.382	1.221564	$-.377369 \pm j\ .259863$	.754739	.209936
			1.333834	$-.251250 \pm j\ .665084$	.502500	.505463
			1.714083	$-.127784 \pm j\ .880488$	.255569	.791588
			3.261942	$-.054334 \pm j\ .974525$	.108668	.952652
			23.183803	$-.014626 \pm j1.008442$	.029253	1.017169
	1.20	84.962	1.459158	$-.337237 \pm j\ .224734$	.674475	.164234
			1.641431	$-.250236 \pm j\ .602564$	.500472	.425702
			2.235990	$-.147516 \pm j\ .839832$	.295032	.727079
			4.585492	$-.071471 \pm j\ .962020$	.142943	.930591
			34.512249	$-.020797 \pm j1.011524$	.041594	1.023613
	1.50	115.199	2.291641	$-.296075 \pm j\ .191193$	.592150	.124215
			2.682641	$-.243660 \pm j\ .535025$	.487320	.345621
			3.927051	$-.167330 \pm j\ .788570$	.334660	.649842
			8.750764	$-.093337 \pm j\ .944057$	.186675	.899955
			69.791499	$-.029532 \pm j1.015670$	.059064	1.032457
	2.00	146.171	4.086686	$-.276117 \pm j\ .175698$	.552234	.107110
			4.897971	$-.238417 \pm j\ .501208$	.476833	.308052
			7.464102	$-.176418 \pm j\ .759877$	.352837	.608537
			17.363454	$-.105713 \pm j\ .932921$	.211426	.881517
			142.430967	$-.034920 \pm j1.018121$	.069840	1.037791

† This table and Tables B-2 through B-6 were computed using a program from David Báez-López, "Determination of Elliptic Network Functions," M.S. thesis, University of Arizona, Tucson, 1977.

TABLE B-2

Odd and case *A* even: 1.0-dB passband ripple

n	ω_s	K_s (dB)	c_i	p_i	a_i	b_i
2	1.05	2.816	1.438664	$-.157083 \pm j1.068900$	.314166	1.167222
	1.10	4.025	1.714083	$-.229129 \pm j1.075841$	.458258	1.209934
	1.20	6.150	2.235990	$-.320565 \pm j1.064452$	.641131	1.235820
	1.50	11.194	3.927051	$-.439709 \pm j1.010488$	.879418	1.214431
	2.00	17.095	7.464102	$-.499471 \pm j\ .959482$	.998942	1.170077
3	1.05	8.134	1.205410	$-.065504 \pm j1.017106$	.131007	1.038796
				$-.947805$		
	1.10	11.480	1.370314	$-.097651 \pm j1.016303$	.195302	1.042407
				$-.816161$		
	1.20	16.209	1.699617	$-.136461 \pm j1.010059$	.272923	1.038841
				$-.701999$		
	1.50	25.176	2.806014	$-.187698 \pm j\ .994225$	.375396	1.023714
				$-.591015$		
	2.00	34.454	5.153209	$-.217034 \pm j\ .981575$	.434067	1.010594
				$-.539958$		
4	1.05	15.840	1.153634	$-.400926 \pm j\ .723958$	.801852	.684857
			3.312518	$-.036963 \pm j1.004642$	.073925	1.010671
	1.10	20.832	1.290925	$-.399229 \pm j\ .638481$	.798458	.567042
			4.349930	$-.054484 \pm j1.003351$	.108969	1.009681
	1.20	27.432	1.572430	$-.386971 \pm j\ .560447$	.773942	.463847
			6.224402	$-.075673 \pm j1.000256$	.151346	1.006238
	1.50	39.518	2.535553	$-.364988 \pm j\ .480692$	.729977	.364281
			12.099310	$-.104409 \pm j\ .993937$	.208819	.998811
	2.00	51.906	4.593261	$-.351273 \pm j\ .442498$	.702546	.319197
			24.227201	$-.121478 \pm j\ .989176$	.242957	.993226
5	1.05	24.135	1.133422	$-.181185 \pm j\ .858432$	.362371	.769820
			1.773739	$-.023559 \pm j1.001164$	.047118	1.002885
				$-.511794$		
	1.10	30.471	1.259320	$-.202145 \pm j\ .804785$	.404289	.688541
			2.193093	$-.034621 \pm j1.000221$	.069241	1.001640
				$-.446562$		
	1.20	38.757	1.521127	$-.217568 \pm j\ .748167$	.435136	.607089
			2.968367	$-.048084 \pm j\ .998478$	.096167	.999271
				$-.391579$		
	1.50	53.875	2.425515	$-.228875 \pm j\ .681678$	.457749	.517069
			5.437645	$-.066541 \pm j\ .995254$	.133081	.994957
				$-.337846$		
	2.00	69.360	4.364951	$-.232338 \pm j\ .646440$	.464676	.471866
			10.567732	$-.077625 \pm j\ .992914$	.155249	.991903
				$-.312599$		
6	1.05	32.523	1.123326	$-.340554 \pm j\ .466561$	.681109	.333656
			1.438664	$-.099253 \pm j\ .910440$	.198505	.838752
			6.528768	$-.016283 \pm j1.000095$	.032567	1.000456
	1.10	40.142	1.243362	$-.315089 \pm j\ .409244$	.630179	.266762
			1.714083	$-.118730 \pm j\ .874514$	.237461	.778873
			8.826455	$-.023927 \pm j\ .999416$	.047854	.999404

(continued)

Table B-2 (*Continued*)

Odd and case A even: 1.0-dB passband ripple

n	ω_s	K_s (dB)	c_i	p_i	a_i	b_i
	1.20	50.089	1.495035	$-.289467 \pm j\ .359828$	.578933	.213267
			2.235990	$-.136580 \pm j\ .834256$	.273161	.714637
			12.952671	$-.033261 \pm j\ .998304$	.066522	.997718
	1.50	68.231	2.369289	$-.260804 \pm j\ .310775$	.521608	.164600
			3.927051	$-.154480 \pm j\ .783831$	.308960	.638255
			25.827242	$-.046116 \pm j\ .996374$	.092233	.994887
	2.00	86.814	4.248155	$-.246136 \pm j\ .287533$	.492272	.143258
			7.464102	$-.162691 \pm j\ .755715$	.325383	.597574
			52.356841	$-.053871 \pm j\ .995015$	.107742	.992957
7	1.05	40.926	1.117521	$-.206293 \pm j\ .681553$	.412587	.507071
			1.308341	$-.061953 \pm j\ .937640$	.123905	.883008
			2.714372	$-.011920 \pm j\ .999752$	.023840	.999646
				$-.352248$		
	1.10	49.816	1.234128	$-.206797 \pm j\ .621264$	.413594	.428734
			1.523943	$-.077646 \pm j\ .911762$	.155292	.837339
			3.514769	$-.017524 \pm j\ .999244$	.035048	.998795
				$-.310175$		
	1.20	61.422	1.479872	$-.203316 \pm j\ .564153$	.406632	.359606
			1.941341	$-.093371 \pm j\ .881886$	.186742	.786440
			4.966697	$-.024380 \pm j\ .998471$	.048760	.997539
				$-.274069$		
	1.50	82.588	2.336522	$-.195790 \pm j\ .502754$	.391580	.291095
			3.313990	$-.110878 \pm j\ .843206$	.221757	.723291
			9.530078	$-.033844 \pm j\ .997189$	.067688	.995531
				$-.238188$		
	2.00	104.268	4.180043	$-.190684 \pm j\ .472047$	.381369	.259189
			6.201776	$-.119725 \pm j\ .821044$	.239450	.688447
			18.961095	$-.039566 \pm j\ .996309$	.079132	.994196
				$-.221131$		
8	1.05	49.331	1.113864	$-.274078 \pm j\ .341142$	.548157	.191497
			1.243118	$-.130371 \pm j\ .789532$	.260743	.640357
			1.906139	$-.042262 \pm j\ .953923$	.084524	.911755
			11.046606	$-.009102 \pm j\ .999652$	.018205	.999387
	1.10	59.491	1.228286	$-.248664 \pm j\ .300352$	.497328	.152045
			1.427103	$-.139688 \pm j\ .737745$	.279377	.563780
			2.380411	$-.054736 \pm j\ .934315$	.109472	.875940
			15.106224	$-.013388 \pm j\ .999259$	.026777	.998698
	1.20	72.755	1.470253	$-.225049 \pm j\ .265193$	.450098	.120974
			1.789509	$-.145710 \pm j\ .685429$	.291419	.491044
			3.252831	$-.067891 \pm j\ .911293$	.135782	.835065
			22.383599	$-.018638 \pm j\ .998689$	.037276	.997727
	1.50	96.945	2.315697	$-.200053 \pm j\ .230188$	.400106	.093008
			2.995660	$-.149085 \pm j\ .625789$	.298170	.413838
			6.021824	$-.083400 \pm j\ .880916$	.166800	.782968
			45.059978	$-.025894 \pm j\ .997774$	.051789	.996224
	2.00	121.722	4.136734	$-.187672 \pm j\ .213541$	.375344	.080821
			5.545063	$-.149543 \pm j\ .594686$	.299085	.376015
			11.766674	$-.091635 \pm j\ .863229$	.183270	.753561
			91.761533	$-.030287 \pm j\ .997159$	.060574	.995244

(*continued*)

Table B-2 (*Continued*)

Odd and case *A* even: 1.0-dB passband ripple

n	ω_s	K_s (dB)	c_i	p_i	a_i	b_i
9	1.05	57.736	1.111406	$-.195247 \pm j\ .551838$	.390495	.342647
			1.205410	$-.087514 \pm j\ .850482$	.175028	.730978
			1.594271	$-.030697 \pm j\ .964496$	.061394	.931195
			3.993674	$-.007178 \pm j\ .999640$	.014357	.999331
				$-.269878$		
	1.10	69.167	1.224347	$-.186736 \pm j\ .497328$	.373473	.282206
			1.370314	$-.098792 \pm j\ .807558$	.197584	.661910
			1.937719	$-.040723 \pm j\ .949094$	.081447	.902438
			5.299248	$-.010563 \pm j\ .999327$	.021126	.998766
				$-.238541$		
	1.20	84.089	1.463756	$-.176578 \pm j\ .447596$	.353156	.231522
			1.699617	$-.108069 \pm j\ .762286$	.216138	.592758
			2.579104	$-.051655 \pm j\ .930819$	.103311	.869092
			7.652393	$-.014711 \pm j\ .998888$	.029422	.997993
				$-.211419$		
	1.50	111.302	2.301616	$-.163781 \pm j\ .395695$	.327563	.183399
			2.806014	$-.116231 \pm j\ .708498$	.232461	.515479
			4.636336	$-.065021 \pm j\ .906413$	.130043	.825813
			15.014341	$-.020451 \pm j\ .998201$	.040901	.996824
				$-.184275$		
	2.00	139.176	4.107442	$-.156746 \pm j\ .370225$	.313491	.161635
			5.153209	$-.119505 \pm j\ .679573$	.239009	.476101
			8.922191	$-.072341 \pm j\ .892058$	.144681	.801001
			30.201059	$-.023928 \pm j\ .997747$	.047856	.996072
				$-.171309$		
10	1.05	66.141	1.109672	$-.225977 \pm j\ .269047$	.451954	.123452
			1.181462	$-.138223 \pm j\ .680726$	.276445	.482493
			1.438664	$-.062031 \pm j\ .888118$	.124062	.792601
			2.533648	$-.023347 \pm j\ .971768$	.046694	.944878
			16.859610	$-.005806 \pm j\ .999659$	.011612	.999352
	1.10	78.842	1.221564	$-.203387 \pm j\ .237538$	.406775	.097791
			1.333834	$-.139376 \pm j\ .625733$	.278752	.410967
			1.714083	$-.072952 \pm j\ .852542$	.145905	.732149
			3.261942	$-.031537 \pm j\ .959337$	.063074	.921321
			23.183803	$-.008547 \pm j\ .999404$	.017093	.998881
	1.20	95.422	1.459158	$-.182921 \pm j\ .210279$	.365841	.077677
			1.641431	$-.137940 \pm j\ .572984$	.275879	.347338
			2.235990	$-.082871 \pm j\ .813874$	.165743	.669259
			4.585492	$-.040676 \pm j\ .944481$	.081352	.893698
			34.512249	$-.011907 \pm j\ .999055$	.023815	.998253
	1.50	125.659	2.291641	$-.161666 \pm j\ .183031$	.323333	.059636
			2.682641	$-.133903 \pm j\ .515483$	.267806	.283652
			3.927051	$-.092777 \pm j\ .766550$	.185554	.596207
			8.750764	$-.052134 \pm j\ .924481$	.104268	.857383
			69.791499	$-.016560 \pm j\ .998520$	.033120	.997316
	2.00	156.630	4.086686	$-.151264 \pm j\ .170031$	.302528	.051791
			4.897971	$-.130976 \pm j\ .486399$	.261951	.253739
			7.464102	$-.097323 \pm j\ .740520$	.194647	.557842
			17.363454	$-.058541 \pm j\ .912636$	.117082	.836331
			142.430967	$-.019380 \pm j\ .998171$	.038760	.996720

TABLE B-3
Case *B* even: 0.1-dB passband ripple

n	ω_s	K_s (dB)	c_i	p_i	a_i	b_i
4	1.05	4.485	1.166586	$-.852205 \pm j\ .918084$	1.704411	1.569133
				$-.040627 \pm j1.053933$	.081255	1.112426
	1.10	8.308	1.309737	$-.073101 \pm j1.076009$	.146201	1.163139
				$-.796684 \pm j\ .809361$	1.593368	1.289771
	1.20	14.387	1.601406	$-.118529 \pm j1.095968$	.237059	1.215196
				$-.746825 \pm j\ .700487$	1.493650	1.048429
	1.50	26.320	2.595517	$-.183989 \pm j1.112642$	.367978	1.271824
				$-.693645 \pm j\ .580990$	1.387290	.818692
	2.00	38.697	4.716540	$-.223115 \pm j1.118520$	.446230	1.300868
				$-.665767 \pm j\ .521238$	1.331534	.714935
6	1.05	20.307	1.125244	$-.639897 \pm j\ .558952$	1.279794	.721895
			1.500649	$-.166247 \pm j\ .984394$	.332494	.996669
				$-.024644 \pm j1.019788$	.049288	1.040576
	1.10	27.889	1.246215	$-.206675 \pm j\ .947127$	.413349	.939764
			1.800145	$-.038741 \pm j1.027332$	.077482	1.056912
				$-.588055 \pm j\ .479414$	1.176111	.575647
	1.20	37.827	1.499501	$-.243668 \pm j\ .901847$	.487336	.872702
			2.364951	$-.056665 \pm j1.035618$	.113330	1.075715
				$-.538160 \pm j\ .410727$	1.076320	.458312
	1.50	55.966	2.378628	$-.280868 \pm j\ .842354$	.561736	.788448
			4.189497	$-.082131 \pm j1.045701$	.164262	1.100236
				$-.483133 \pm j\ .343213$	.966267	.351213
	2.00	74.548	4.267406	$-.298091 \pm j\ .808340$	.596182	.742272
			8.001486	$-.097774 \pm j1.051149$	.195549	1.114475
				$-.455028 \pm j\ .311687$	.910055	.304199
8	1.05	37.529	1.114416	$-.506414 \pm j\ .386312$	1.012827	.405692
			1.252994	$-.232472 \pm j\ .847640$	.464944	.772537
			2.034525	$-.015010 \pm j1.009939$	.030021	1.020203
				$-.071703 \pm j\ .979388$	.143406	.964341
	1.10	47.686	1.229114	$-.251503 \pm j\ .787764$	.503006	.683826
			1.441163	$-.022814 \pm j1.014037$	.045627	1.028791
			2.557144	$-.095249 \pm j\ .962910$	.190499	.936269
				$-.458373 \pm j\ .332941$	.916746	.320955
	1.20	60.949	1.471557	$-.264032 \pm j\ .726645$	.528064	.597725
			1.810911	$-.032591 \pm j1.018790$	.065183	1.038996
			3.516389	$-.120409 \pm j\ .941863$	.240819	.901605
				$-.413887 \pm j\ .287903$	.827774	.254190
	1.50	85.138	2.318435	$-.271506 \pm j\ .656662$	.543012	.504921
			3.039648	$-.046374 \pm j1.024969$	.092748	1.052713
			6.556721	$-.150389 \pm j\ .912223$	.300778	.854768
				$-.366858 \pm j\ .244169$	.733715	.194203
	2.00	109.915	4.142383	$-.272874 \pm j\ .620149$	.545748	.459045
			5.635349	$-.054826 \pm j1.028517$	.109652	1.060853
			12.861234	$-.166427 \pm j\ .894253$	.332854	.827387
				$-.343582 \pm j\ .223806$	.687165	.168138

(continued)

TABLE B-3 *(Continued)*

Case B even: 0.1-dB passband ripple

n	ω_s	K_s (dB)	c_i	p_i	a_i	b_i
10	1.05	54.608	1.109888	$-.414731 \pm j\ .293587$	.829462	.258195
			1.184476	$-.249669 \pm j\ .721758$	.499337	.583270
			1.459121	$-.109291 \pm j\ .913996$	.218582	.847333
			2.745052	$-.040314 \pm j\ .983945$	.080628	.969774
				$-.009914 \pm j1.006005$	.019828	1.012145
	1.10	67.307	1.221890	$-.252592 \pm j\ .658710$	.505184	.497701
			1.338203	$-.130002 \pm j\ .877950$	.260004	.787698
			1.742587	$-.014883 \pm j1.008599$	.029767	1.017493
			3.552883	$-.055378 \pm j\ .974066$	.110756	.951871
				$-.372589 \pm j\ .254860$	.745178	.203776
	1.20	83.887	1.459673	$-.250516 \pm j\ .598450$	.501033	.420901
			1.648160	$-.148915 \pm j\ .837636$	.297830	.723809
			2.278803	$-.072379 \pm j\ .961490$	.144758	.929701
			5.019460	$-.021068 \pm j1.011673$	.042135	1.023927
				$-.334463 \pm j\ .222157$	.668925	.161219
	1.50	114.123	2.292725	$-.243524 \pm j\ .533167$	.487049	.343572
			2.696572	$-.167901 \pm j\ .787240$	.335803	.647937
			4.014312	$-.093900 \pm j\ .943626$	.187801	.899247
			9.631781	$-.029742 \pm j1.015773$	.059483	1.032679
				$-.294931 \pm j\ .190246$	.589862	.123178
	2.00	145.095	4.088922	$-.238296 \pm j\ .500354$	.476591	.307139
			4.926617	$-.176655 \pm j\ .759173$	.353311	.607551
			7.642845	$-.106015 \pm j\ .932657$	.212031	.881089
			19.166456	$-.035048 \pm j1.018180$	.070095	1.037920
				$-.275610 \pm j\ .175300$	.551220	.106691

TABLE B-4

Case B even: 1.0-dB passband ripple

n	ω_s	K_s (dB)	c_i	p_i	a_i	b_i
4	1.05	13.243	1.166586	$-.422751 \pm j\ .627439$	.845502	.572399
				$-.043154 \pm j1.006305$	.086309	1.014512
	1.10	18.140	1.309737	$-.403039 \pm j\ .574970$	.806078	.493031
				$-.061668 \pm j1.004223$	.123336	1.012268
	1.20	24.700	1.601406	$-.383555 \pm j\ .523216$	.767109	.420870
				$-.082727 \pm j1.000117$	.165455	1.007078
	1.50	36.771	2.595517	$-.361063 \pm j\ .465487$	.722126	.347045
				$-.109356 \pm j\ .993062$	.218712	.998131
	2.00	49.156	4.716540	$-.348975 \pm j\ .435853$	.697949	.311751
				$-.124305 \pm j\ .988442$	.248611	.992470
6	1.05	30.730	1.125244	$-.332474 \pm j\ .436556$	.664948	.301121
			1.500649	$-.107214 \pm j\ .902829$	.214429	.826595
				$-.017380 \pm j1.000183$	.034760	1.000669
	1.10	38.342	1.246215	$-.308336 \pm j\ .391152$	.616673	.248071
			1.800145	$-.124563 \pm j\ .866975$	.249126	.767162
				$-.025185 \pm j\ .999425$	.050371	.999485
	1.20	48.285	1.499501	$-.284782 \pm j\ .349826$	.569565	.203480
			2.364951	$-.140216 \pm j\ .827910$	.280431	.705096
				$-.034509 \pm j\ .998231$	.069019	.997656

(*continued*)

TABLE B-4 (*Continued*)

Case *B* even: 1.0-dB passband ripple

n	ω_s	K_s (dB)	c_i	p_i	a_i	b_i
	1.50	66.425	2.378628	$-.258566 \pm j\ .306851$	.517132	.161014
			4.189497	$-.155955 \pm j\ .780065$	.311909	.632824
				$-.047022 \pm j\ .996256$	.094044	.994738
	2.00	85.008	4.267406	$-.245081 \pm j\ .285837$	.490161	.141767
			8.001486	$-.163303 \pm j\ .753736$	.326605	.594786
				$-.054402 \pm j\ .994927$	.108805	.992839
8	1.05	47.987	1.114416	$-.268114 \pm j\ .329175$	.536228	.180241
			1.252994	$-.134219 \pm j\ .780319$	.268437	.626912
			2.034525	$-.044114 \pm j\ .952060$	.088228	.908365
				$-.009435 \pm j\ .999653$	.018870	.999396
	1.10	58.146	1.229114	$-.244589 \pm j\ .293222$	.489178	.145803
			1.441163	$-.141813 \pm j\ .730432$	.283626	.553642
			2.557144	$-.056360 \pm j\ .932329$	.112720	.872413
				$-.013771 \pm j\ .999245$	.027542	.998679
	1.20	71.408	1.471557	$-.222550 \pm j\ .261268$	.445099	.117789
			1.810911	$-.146649 \pm j\ .680312$	.293298	.484330
			3.516389	$-.069126 \pm j\ .909468$	.138252	.831911
				$-.019020 \pm j\ .998660$	.038040	.997684
	1.50	95.597	2.318435	$-.198966 \pm j\ .228644$	.397931	.091866
			3.039648	$-.149279 \pm j\ .623251$	.298559	.410726
			6.556721	$-.084056 \pm j\ .879706$	.168112	.780949
				$-.026174 \pm j\ .997742$	.052349	.996174
	2.00	120.374	4.142383	$-.187178 \pm j\ .212872$	.374357	.080350
			5.635349	$-.149573 \pm j\ .593460$	.299147	.374567
			12.861234	$-.091959 \pm j\ .862553$	.183918	.752454
				$-.030453 \pm j\ .997137$	.060905	.995209
10	1.05	65.067	1.109888	$-.222316 \pm j\ .263196$	.444633	.118697
			1.184476	$-.139553 \pm j\ .673297$	.279106	.472804
			1.459121	$-.063624 \pm j\ .885077$	.127249	.787410
			2.745052	$-.023970 \pm j\ .971078$	.047940	.943567
				$-.005940 \pm j\ .999655$	.011880	.999346
	1.10	77.767	1.221890	$-.201018 \pm j\ .234049$	.402036	.095187
			1.338203	$-.139872 \pm j\ .620398$	.279743	.404458
			1.742587	$-.074098 \pm j\ .849790$	.148196	.727633
			3.552883	$-.032133 \pm j\ .958580$	.064266	.919909
				$-.008701 \pm j\ .999395$	.017402	.998866
	1.20	94.346	1.459673	$-.181522 \pm j\ .208352$	.363045	.076361
			1.648160	$-.138003 \pm j\ .569550$	.276006	.343432
			2.278803	$-.083577 \pm j.\ 811692$	.167155	.665829
			5.019460	$-.041174 \pm j\ .943762$	.082348	.892382
				$-.012062 \pm j\ .999042$	.024123	.998231
	1.50	124.583	2.292725	$-.161078 \pm j\ .182269$	.322155	.059168
			2.696572	$-.133813 \pm j\ .513911$	.267627	.282010
			4.014312	$-.093062 \pm j\ .765311$	.186124	.594362
			9.631781	$-.052431 \pm j\ .923983$	.104862	.856494
				$-.016673 \pm j\ .998508$	.033347	.997296
	2.00	155.554	4.088922	$-.151000 \pm j\ .169700$	.302000	.051599
			4.926617	$-.130907 \pm j\ .485667$	.261815	.253009
			7.642845	$-.097442 \pm j\ .739879$	.194883	.556916
			19.166456	$-.058698 \pm j\ .912351$	.117396	.835830
				$-.019447 \pm j\ .998163$	.038895	.996707

TABLE B-5

Case C even: 0.1-dB passband ripple

n	ω_s	K_s (dB)	c_i	p_i	a_i	b_i
4	1.05	3.284	1.176045	$-.041450 \pm j1.062080$	.082900	1.129732
				$-1.142752 \pm j1.056906$	2.285503	2.422931
	1.10	6.478	1.321589	$-.076408 \pm j1.089646$	.152816	1.193166
				$-1.041973 \pm j\ .905418$	2.083946	1.905490
	1.20	12.085	1.615455	$-.128382 \pm j1.115527$	.256764	1.260882
				$-.953405 \pm j\ .756606$	1.906811	1.481435
	1.50	23.736	2.611679	$-.206296 \pm j1.136431$	.412592	1.334033
				$-.863022 \pm j\ .597833$	1.726044	1.102212
	2.00	36.023	4.733595	$-.817435 \pm j\ .520713$	1.634871	.939343
				$-.253437 \pm j1.142940$	.506873	1.370542
6	1.05	18.727	1.126696	$-.789645 \pm j\ .570579$	1.579289	.949099
			1.535284	$-.185878 \pm j\ .993057$	.371755	1.020713
				$-.025910 \pm j1.021602$	.051819	1.044342
	1.10	26.230	1.248053	$-.715339 \pm j\ .476877$	1.430677	.739121
			1.837658	$-.230107 \pm j\ .952094$	.460214	.959433
				$-.040931 \pm j1.029755$	.081861	1.062071
	1.20	36.113	1.501690	$-.646187 \pm j\ .398581$	1.292374	.576425
			2.404505	$-.269710 \pm j\ .902237$	.539420	.886775
				$-.060077 \pm j1.038657$	.120154	1.082417
	1.50	54.202	2.381154	$-.572096 \pm j\ .324068$	1.144191	.432313
			4.230449	$-.308434 \pm j\ .836973$	.616868	.795656
				$-.087368 \pm j1.049425$	.174736	1.108925
	2.00	72.761	4.270072	$-.534998 \pm j\ .290136$	1.069995	.370401
			8.042806	$-.325870 \pm j\ .799822$	.651740	.745907
				$-.104187 \pm j1.055209$	.208374	1.124320
8	1.05	36.268	1.114839	$-.604065 \pm j\ .371382$	1.208130	.502820
			1.259509	$-.257536 \pm j\ .843660$	.515072	.778086
			2.085094	$-.076089 \pm j\ .979591$	.152178	.965388
				$-.015568 \pm j1.010441$	.031137	1.021234
	1.10	46.399	1.229651	$-.541763 \pm j\ .314616$	1.083525	.392490
			1.448643	$-.276093 \pm j\ .780152$	.552186	.684864
			2.608880	$-.023659 \pm j1.014703$	.047318	1.030181
				$-.100633 \pm j\ .962426$	.201267	.936390
	1.20	59.639	1.472198	$-.485227 \pm j\ .267901$	.970453	.307216
			1.819178	$-.287400 \pm j\ .715925$	.574801	.595147
			3.568474	$-.126736 \pm j\ .940478$	.253471	.900562
				$-.033798 \pm j1.019637$	.067597	1.040801
	1.50	83.807	2.319176	$-.426488 \pm j\ .223590$	.852977	.231885
			3.048578	$-.292845 \pm j\ .643036$	.585691	.499254
			6.608405	$-.157688 \pm j\ .909548$	.315377	.852143
				$-.048107 \pm j1.026042$	.096214	1.055076
	2.00	108.575	4.143165	$-.397764 \pm j\ .203316$	.795629	.199554
			5.644525	$-.292952 \pm j\ .605259$	.585904	.452160
			12.912424	$-.174180 \pm j\ .890785$	.348360	.823836
				$-.056895 \pm j1.029716$	.113790	1.063553

(*continued*)

TABLE B-5 (*Continued*)
Case *C* even: 0.1-dB passband ripple

n	ω_s	K_s (dB)	c_i	p_i	a_i	b_i
10	1.05	53.576	1.110055	$-.486272 \pm j\ .273776$	.972543	.311413
			1.186597	$-.272544 \pm j\ .710688$	.545089	.579358
			1.470226	$-.115389 \pm j\ .911607$	.230777	.844342
			2.804853	$-.041773 \pm j\ .983714$	.083546	.969439
				$-.010169 \pm j1.006194$	.020339	1.012529
	1.10	66.262	1.222102	$-.434146 \pm j\ .234898$	.868292	.243660
			1.340720	$-.273555 \pm j\ .645766$	.547110	.491847
			1.754581	$-.136496 \pm j\ .874512$	.272992	.783403
			3.612427	$-.057220 \pm j\ .973594$	.114440	.951160
				$-.015256 \pm j1.008851$	.030511	1.018013
	1.20	82.830	1.459927	$-.387617 \pm j\ .202683$	.775235	.191328
			1.651014	$-.269401 \pm j\ .584238$	.538801	.413911
			2.291423	$-.155607 \pm j.\ 833128$	.311213	.718316
			5.078099	$-.074620 \pm j\ .960707$	.149239	.928526
				$-.021584 \pm j1.011998$	.043168	1.024605
	1.50	113.056	2.293018	$-.339922 \pm j\ .171783$	.679845	.145057
			2.699727	$-.259970 \pm j\ .518096$	.519940	.336007
			4.027356	$-.174590 \pm j\ .781514$	.349180	.641246
			9.688795	$-.096612 \pm j\ .942388$	.193225	.897429
				$-.030465 \pm j1.016193$	.060930	1.033575
	2.00	144.023	4.089232	$-.316795 \pm j\ .157490$	.633589	.125162
			4.929889	$-.253464 \pm j\ .485036$	.506929	.299505
			7.655998	$-.183250 \pm j\ .752817$	.366500	.600315
			19.222417	$-.108980 \pm j\ .931131$	.217960	.878881
				$-.035902 \pm j1.018656$	.071804	1.038948

TABLE B-6
Case C even: 1.0-dB passband ripple

n	ω_s	K_s (dB)	c_i	p_i	a_i	b_i
4	1.05	11.322	1.176045	$-.050129 \pm j1.009723$	.100257	1.022053
				$-.664397 \pm j\ .601548$	1.328793	.803283
	1.10	15.942	1.321589	$-.617627 \pm j\ .535240$	1.235255	.667946
				$-.072023 \pm j1.007836$	.144045	1.020920
	1.20	22.293	1.615455	$-.573215 \pm j\ .472186$	1.146429	.551534
				$-.096950 \pm j1.003197$	.193901	1.015804
	1.50	34.179	2.611679	$-.524307 \pm j\ .404475$	1.048614	.438498
				$-.128327 \pm j\ .994626$	.256654	1.005748
	2.00	46.481	4.733595	$-.499020 \pm j\ .370743$	.998040	.386471
				$-.145844 \pm j\ .988871$	.291689	.999136
6	1.05	29.133	1.126696	$-.479520 \pm j\ .376458$	.959040	.371660
			1.535284	$-.124739 \pm j\ .894415$	.249477	.815537
				$-.018753 \pm j1.000393$	.037507	1.001138
	1.10	36.680	1.248053	$-.437212 \pm j\ .330308$	.874424	.300258
			1.837658	$-.142909 \pm j\ .855552$	.285819	.752392
				$-.027110 \pm j\ .999598$	.054220	.999932
	1.20	46.571	1.501690	$-.397576 \pm j\ .289708$	.795151	.241997
			2.404505	$-.158828 \pm j\ .813561$	.317656	.687107
				$-.037077 \pm j\ .998324$	.074154	.998026
	1.50	64.661	2.381154	$-.174302 \pm j\ .762525$	.348604	.611825
			4.230449	$-.050450 \pm j\ .996191$	.100900	.994942
				$-.355021 \pm j\ .248804$	.710041	.187943
	2.00	83.221	4.270072	$-.333676 \pm j\ .229268$	.667352	.163904
			8.042806	$-.181299 \pm j\ .734590$	.362598	.572492
				$-.058342 \pm j\ .994745$	.116684	.992921
8	1.05	46.727	1.114839	$-.371900 \pm j\ .270505$	.743801	.211483
			1.259509	$-.151848 \pm j\ .763118$	.303695	.605407
			2.085094	$-.047269 \pm j\ .949664$	.094537	.904097
				$-.009859 \pm j\ .999670$	.019717	.999437
	1.10	56.858	1.229651	$-.335499 \pm j\ .237676$	.670998	.169050
			1.448643	$-.158468 \pm j\ .711532$	.316936	.531390
			2.608880	$-.059982 \pm j\ .929136$	.119964	.866892
				$-.014360 \pm j\ .999248$	.025719	.998703
	1.20	70.098	1.472198	$-.302267 \pm j\ .209205$	.604534	.135132
			1.819178	$-.162095 \pm j\ .660152$	.324190	.462075
			3.568474	$-.073157 \pm j\ .905406$	.146315	.825112
				$-.019803 \pm j\ .998642$	.039607	.997679
	1.50	94.266	2.319176	$-.267488 \pm j\ .180775$	.534975	.104229
			3.048578	$-.163159 \pm j\ .602103$	.326319	.389149
			6.608405	$-.088478 \pm j\ .874558$	.176956	.772679
				$-.027224 \pm j\ .997687$	.054449	.996120
	2.00	119.034	4.143165	$-.250369 \pm j\ .167247$	.500739	.090656
			5.644525	$-.162577 \pm j\ .571959$	.325154	.353568
			12.912424	$-.096552 \pm j\ .856791$	.193104	.743414
				$-.031665 \pm j\ .997054$	.063330	.995120

(continued)

TABLE B-6 (*Continued*)

Case C even: 1.0-dB passband ripple

n	ω_s	K_s (dB)	c_i	p_i	a_i	b_i
10	1.05	64.036	1.110055	$-.302228 \pm j\ .211052$	.604457	.135885
			1.186597	$-.154774 \pm j\ .652742$	.309547	.450027
			1.470226	$-.067537 \pm j\ .880131$	.135073	.779193
			2.804853	$-.024923 \pm j\ .970175$	.049846	.941860
				$-.006111 \pm j\ .999653$	.012221	.999343
	1.10	76.722	1.222102	$-.271183 \pm j\ .185925$	.542366	.108108
			1.340720	$-.153648 \pm j\ .599498$	.307296	.383006
			1.754581	$-.078131 \pm j\ .843990$	.156263	.718423
			3.612427	$-.033289 \pm j\ .957369$	.066578	.917664
				$-.008938 \pm j\ .999388$	.017875	.998855
	1.20	93.289	1.459927	$-.243254 \pm j\ .164164$	.486508	.086122
			1.651014	$-.150336 \pm j\ .548673$	.300673	.323643
			2.291423	$-.087630 \pm j.\ 805089$	.175260	.655848
			5.078099	$-.042532 \pm j\ .942201$	.085063	.889552
				$-.012377 \pm j\ .999027$	.024754	.998208
	1.50	123.515	2.293018	$-.214394 \pm j\ .142421$	.428788	.066248
			2.699727	$-.144539 \pm j\ .493399$	.289078	.264334
			4.027356	$-.097026 \pm j\ .757852$	.194052	.583753
			9.688795	$-.054019 \pm j\ .921965$	.108037	.852937
				$-.017097 \pm j\ .998479$	.034194	.997252
	2.00	154.482	4.089232	$-.200312 \pm j\ .132059$	.400624	.057564
			4.929889	$-.140813 \pm j\ .465461$	.281625	.236482
			7.655998	$-.101316 \pm j\ .731995$	.202633	.546082
			19.222417	$-.060405 \pm j\ .910063$	.120810	.831863
				$-.019938 \pm j\ .998124$	.039876	.996650

APPENDIX C

PROPERTIES OF OPERATIONAL AMPLIFIERS

In this appendix we present a brief introduction to the properties of operational amplifiers. An operational amplifier is basically a controlled source in which the forward gain parameter is very large. The most commonly encountered type is the VCVS (voltage-controlled voltage source) operational amplifier. It is represented by the symbol shown in Fig. C-1. The output voltage may be defined as

$$V_o(s) = A_d[V_1(s) - V_2(s)] + A_c \frac{V_1(s) + V_2(s)}{2} \tag{1}$$

where A_d is the *differential-mode gain* and A_c is the *common-mode gain*. These parameters can be used to define a *common-mode rejection ratio* (CMRR) as

$$\text{CMRR} = \frac{A_d}{A_c} \tag{2}$$

Ideally, the input resistance of the operational amplifier is infinite and the output resistance is zero. If the forward gain is sufficiently large so that it may be considered infinite, then the concept of a null port may be used to analyze operational amplifier circuits. A *null port* is simply a port (a pair of terminals) at which both the voltage and the current are simultaneously zero. In general, the input terminal pair of the VCVS operational amplifier realizes a null port as the differential gain A_d approaches infinity.

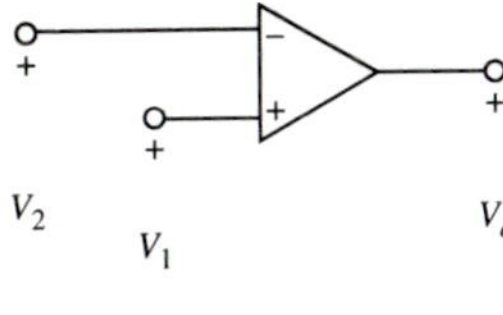

FIGURE C-1
Symbol for VCVS operational amplifier.

As an example, in Fig. C-2(a), a noninverting finite-gain voltage amplifier circuit is shown. The variables $V_i(s)$ and $I_i(s)$ identify the null port. Assuming ideal characteristics, we find that

$$\frac{V_o(s)}{V_s(s)} = \frac{R_1 + R_2}{R_1} \tag{3}$$

Similarly, in Fig. C-2(b) an inverting infinite-gain voltage amplifier circuit is shown. Again $V_i(s)$ and $I_i(s)$ identify the null port. For this circuit we obtain

$$\frac{V_o(s)}{V_s(s)} = -\frac{R_2}{R_1} \tag{4}$$

A higher level of complexity for modeling an operational amplifier occurs when the forward differential-mode gain is not sufficiently large to use the null-port concept. In this case, the general model of the noninverting configuration is modeled as shown in Fig. C-3(a), where the open-loop forward operational amplifier gain is given as

$$V_o(s) = A_d V_i(s) \tag{5}$$

The overall voltage gain $A(s)$ is given as

$$A(s) = \frac{V_o(s)}{V_s(s)} = \frac{A_d}{1 + A_d Z_1/(Z_1 + Z_2)} \tag{6}$$

The (s) notation on Z_1 and Z_2 has been omitted for simplicity. Note that in the limit, as $A_d \to \infty$, $A(s) = (Z_1 + Z_2)/Z_1$. The input and output impedances of the circuit retain their

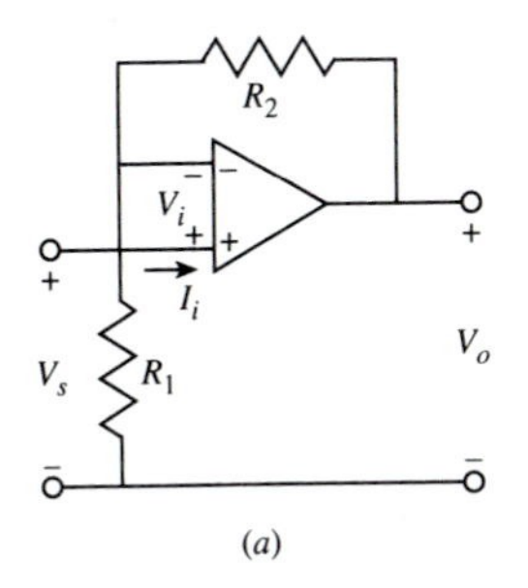

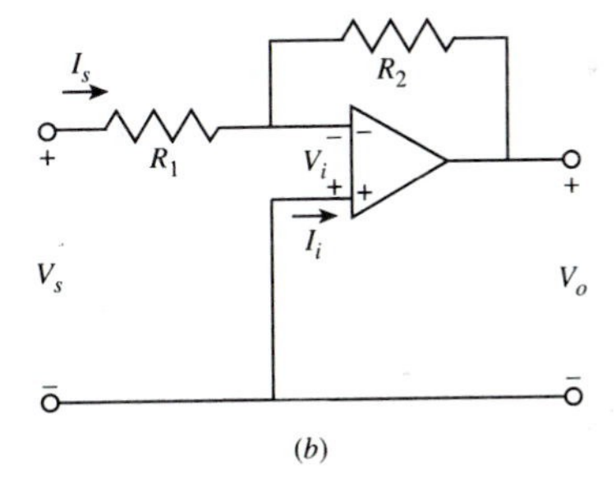

FIGURE C-2
Finite-gain voltage amplifiers: (a) noninverting; (b) inverting.

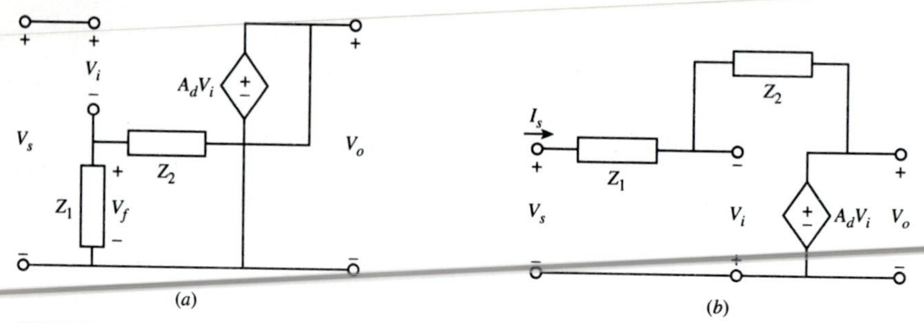

FIGURE C-3
Models for finite-gain amplifiers: (*a*) noninverting; (*b*) inverting.

infinite and zero values, respectively. Similarly, the general model for the inverting configuration is shown in Fig. C-3(*b*). The voltage gain $A(s)$ is

$$A(s) = \frac{V_o(s)}{V_s(s)} = \frac{-A_dZ_2/(Z_1 + Z_2)}{1 + A_dZ_1/(Z_1 + Z_2)} \tag{7}$$

In this expression, as $A_d \to \infty$, $A(s) = -Z_2/Z_1$. For this circuit, the output impedance is zero, but the input impedance is not infinite. It is found to be

$$Z_{\text{in}}(s) = \frac{V_s(s)}{I_s(s)} = Z_1 + \frac{Z_2}{1 + A_d} \tag{8}$$

This finite input impedance causes difficulties in any active *RC* filter applications that use the inverting finite-gain configuration and is one of the reasons why such filters are not widely used.

A still higher level of complexity for the modeling of operational amplifiers occurs when their input and output resistances are considered. For the noninverting finite-gain case, a circuit is shown in Fig. C-4(*a*). For this, the input and output impedances and the

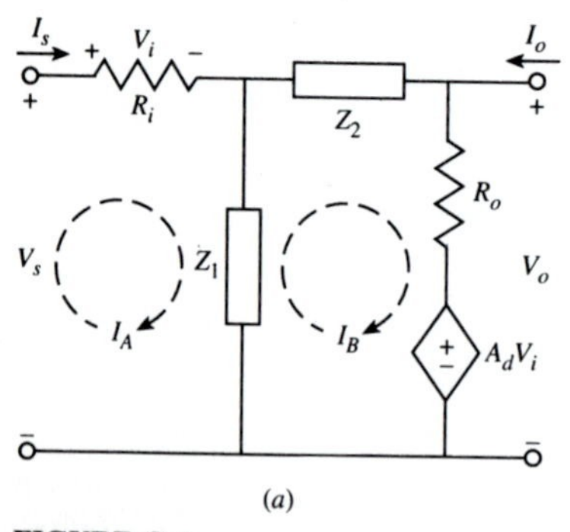

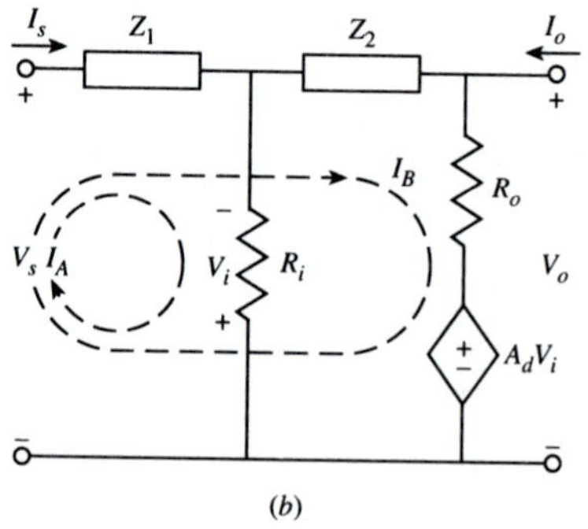

FIGURE C-4
Models for finite-gain amplifiers: (*a*) noninverting; (*b*) inverting.

voltage gain are

$$Z_{\text{in}}(s) = \frac{V_s(s)}{I_s(s)} = \frac{R_i(Z_1 + Z_2 + R_o) + Z_1(Z_2 + R_o) + A_d R_i Z_1}{Z_1 + Z_2 + R_o} \tag{9a}$$

$$Z_{\text{out}}(s) = \frac{V_o(s)}{I_o(s)} = \frac{R_o[R_i(Z_1 + Z_2) + Z_1 Z_2]}{R_i(Z_1 + Z_2 + R_o) + Z_1(Z_2 + R_o) + A_d R_i Z_1} \tag{9b}$$

$$A(s) = \left.\frac{V_o(s)}{V_s(s)}\right|_{I_o(s)=0} = \frac{R_o Z_1 + A_d R_i(Z_1 + Z_2)}{R_i(Z_1 + Z_2 + R_o) + Z_1(Z_2 + R_o) + A_d Z_1 R_i} \tag{9c}$$

A model for the inverting configuration is shown in Fig. C-4(b). For this

$$Z_{\text{in}}(s) = \frac{V_s(s)}{I_s(s)} = Z_1 + \frac{R_i(Z_2 + R_o)}{Z_2 + R_o + R_i + A_d R_i} \tag{10a}$$

$$Z_{\text{out}}(s) = \frac{V_o(s)}{I_o(s)} = \frac{R_o[R_i(Z_1 + Z_2) + Z_1 Z_2]}{R_i(Z_1 + Z_2 + R_o) + Z_1(Z_2 + R_o) + A_d R_i Z_1} \tag{10b}$$

$$A(s) = \left.\frac{V_o(s)}{V_s(s)}\right|_{I_o(s)=0} = \frac{[R_i(R_o - A_d Z_2)]}{R_i(Z_1 + Z_2 + R_o) + Z_1(Z_2 + R_o) + A_d R_i Z_1} \tag{10c}$$

As an example of the above relations, let the parameters of an operational amplifier be $R_i = 10^6\ \Omega$, $R_o = 10^2\ \Omega$, and $A_d = 10^5$. If the feedback impedances in Fig. C-4 have the values $Z_1 = 1\ \text{k}\Omega$ and $Z_2 = 10\ \text{k}\Omega$, then for the noninverting configuration from (9) we find $Z_{\text{in}} = 9.010 \times 10^9\ \Omega$ and $Z_{\text{out}} = 0.011\ \Omega$. Similarly, for the inverting configuration from (10) we obtain $Z_{\text{in}} = 1000.1\ \Omega$ and $Z_{\text{out}} = 0.011\ \Omega$. It should be noted that values of the order of those calculated in the preceding example are not obtainable on a practical basis (except for Z_{in} of the inverting realization). One reason is that the maximum input impedance of an operational amplifier used in a single-input connection cannot exceed the common-mode impedance $R_{i\text{cm}}$ to ground. This impedance is modeled as shown in Fig. C-5. Typically $R_{i\text{cm}}$ is caused by the resistance from the base to the

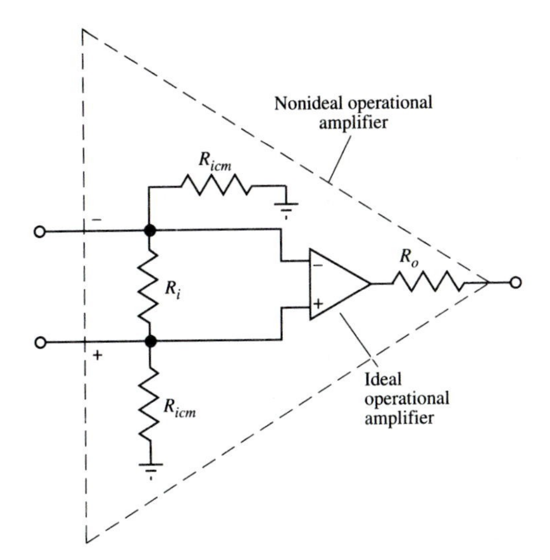

FIGURE C-5
Model for common-mode impedance.

collector of a transistor operating in the normal region. Thus, typical values are in the range of 10 to 100 MΩ depending upon the degree of reverse bias and the value of the collector current. The maximum value of Z_{in} thus is 100 MΩ or less. Similarly the output impedance will be limited to about 1 Ω due to contact resistance and the finite resistance of the conductors used in the operational amplifier.

Most monolithic operational amplifiers use direct (dc) coupling between their various internal stages. Such coupling introduces the possibility of dc errors in the overall amplifier characteristics. There are two sources of such error, namely the bias currents and the offset voltages. The *bias currents* may be modeled by two current sources I_{B1} and I_{B2} as shown in Fig. C-6. In this figure, the solid-line symbol represents an operational amplifier with no dc errors while the dashed-line symbol includes the error sources. The values of these sources represent the bias currents required for proper operation of the operational amplifier. These currents are small, typically in the 100-nA range. For operational amplifiers that are designed using FETs (*f*ield *e*ffect *t*ransistors) in the input, the bias currents are even smaller, typically having values of about 10 pA. The difference between I_{B1} and I_{B2} is called the *offset current*. It is defined as

$$I_{os} = |I_{B1} - I_{B2}| \tag{11}$$

The offset current is typically 5 to 10 percent of the bias current. It is the result of inequalities in the current gain of the two sides of the input differential stage of the operational amplifier.

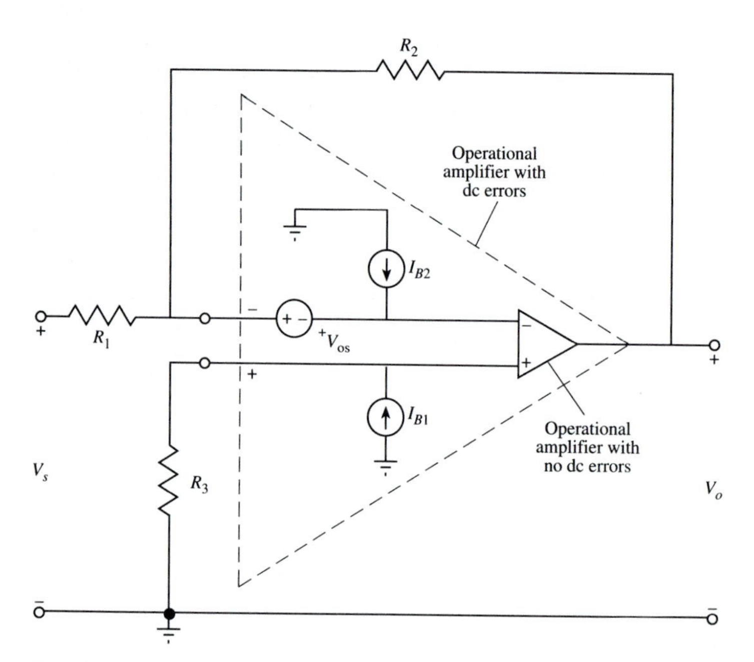

FIGURE C-6
Model for bias currents and offset voltage.

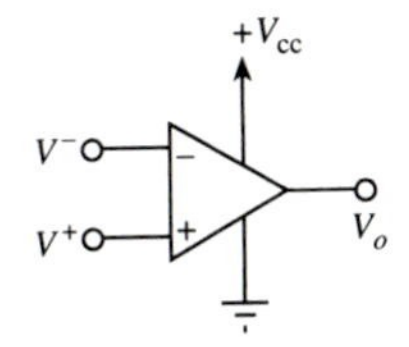

FIGURE C-7
Symbol for single-supply VCVS operational amplifier.

The second source of dc error in an operational amplifier is the *offset voltage*. When an operational amplifier is operated with no differential input, a voltage will appear at the output even though the input voltage is zero. This effect is modeled by the offset voltage source V_{os} in Fig. C-6. The offset voltage accounts for any transconductance variation of the operational amplifier input stages. Typically, values of offset voltage are 1 to 5 mV. The combined effects of the two types of dc errors give

$$V_o(s) = -\frac{R_2}{R_1} V_s(s) \pm \left(1 + \frac{R_2}{R_1}\right) V_{os} + \left(1 + \frac{R_2}{R_1}\right) R_3 I_{B1} - R_2 I_{B2} \tag{12}$$

Since, in general, the polarities to be used with V_{os} and I_{os} are not predictable, (12) must be treated from a worst-case viewpoint. The dc errors will be minimized if the resistances to ground from the inverting and noninverting inputs of the operational amplifier are equal. Since in active filter applications the operational amplifier is always used with feedback, the dc errors do not usually cause serious problems.

Conventional direct-coupled operational amplifiers require the use of a split power supply providing both a positive and a negative polarity output. Recently, however, single-supply operational amplifiers have been introduced. A circuit symbol for the single-supply VCVS is shown in Fig. C-7. The voltages at the inverting and noninverting inputs are shown as V^- and V^+, respectively, while the supply voltage is labeled as V_{cc}. Unlike split-supply operational amplifiers for which dc coupling is readily used, single-supply operational amplifiers must have a biased input. In practice the bias is selected so that under zero signal conditions the output voltage will be midway between the power supply voltage V_{cc} and ground. As an example, consider the use of a single-supply VCVS operational amplifier to realize a unity-gain high-pass realization (see Example 5.3-3). Figure C-8 shows the realization with an additional source V_B being used to provide the bias capability. If V_B is set to $V_{cc}/2$, then the dc value of V_2 will also be

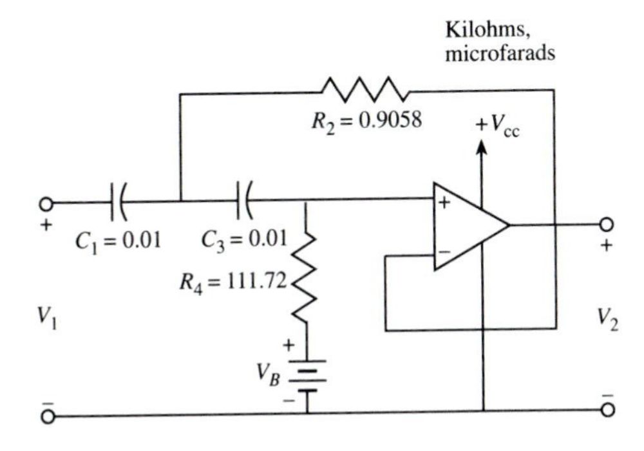

FIGURE C-8
Realization for a unity-gain high-pass filter.

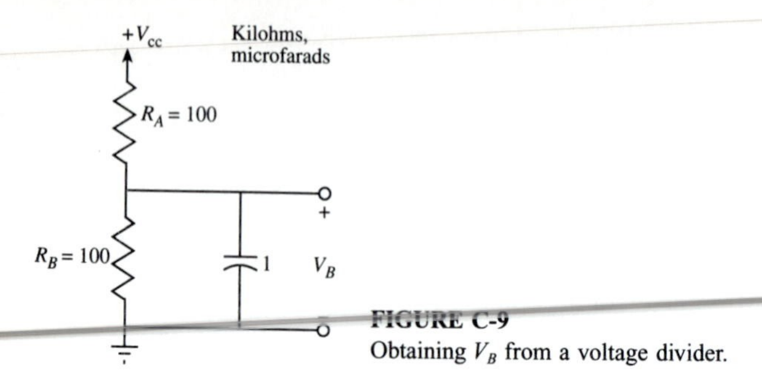

FIGURE C-9
Obtaining V_B from a voltage divider.

$V_{cc}/2$. Voltage V_B can be obtained directly from the power supply V_{cc} by using a voltage divider of the form shown in Fig. C-9. In using such a network, one must make certain that it does not affect the dc characteristics of the circuit. In this case, since $\omega_n \approx 10^4$ rad/s, we calculate from Fig. C-9 that the ac source impedance of V_B is approximately 100 Ω, which is insignificant when placed in series with R_4, which is 111 kΩ.

Now let us consider the ac properties of operational amplifiers. One of the most important characteristics of an operational amplifier, especially with regard to its application in active filters, is its frequency response. As frequency increases, the gain of an operational amplifier decreases due to the bandwidth limitations of its solid-state devices. The more the gain decreases, the more the overall filter realization becomes dependent upon the open-loop gain of the amplifier. In addition, if excessive phase lag is introduced into the open-loop transfer function, a shift in the desired pole locations is produced that can lead to instability.

The actual frequency response of an operational amplifier is a complex expression. A simplified approximation that works well consists of three negative real poles.[1] Such a model predicts that when feedback is applied around it, instability can result. Thus, frequency compensation must be applied to most operational amplifiers before they are usable in active filters or in almost any other application. The objective of most compensation schemes is to achieve an operational amplifier forward-gain function having the form

$$A_d(s) = \frac{A_0 \omega_a}{s + \omega_a} = \frac{\text{GB}}{s + \omega_a} \tag{13}$$

where A_0 is the dc gain, ω_a is the bandwidth, and GB is the gain bandwidth product. A magnitude plot of (13) is given in Fig. C-10 where typical values are $A_0 = 10^5$ and $\omega_a = 10$ rad/s. If we let $s = j\,\text{GB}$ in (13), we see that GB is the radian frequency at which the magnitude of $A_d(j\omega)$ is unity. Discussions of how frequency compensation is

[1] G. E. Tobey, J. D. Graeme, and L. P. Huelsman, *Operational Amplifiers—Design and Applications*, McGraw-Hill Book Company, New York, 1971, chap 5.

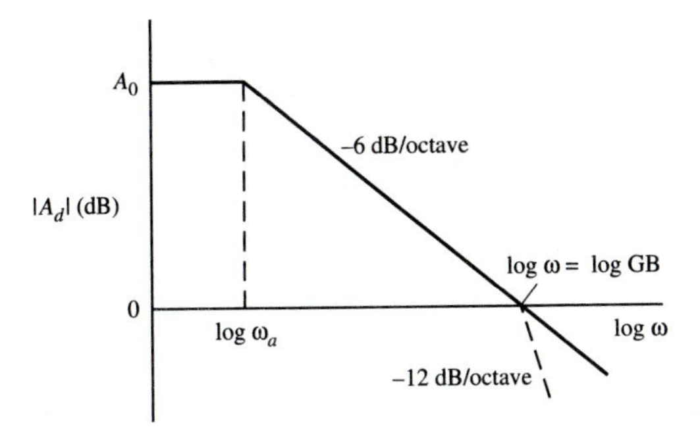

FIGURE C-10
Magnitude of the forward gain of an operational amplifier.

accomplished may be found in the literature.[2] The actual frequency response of most compensated operational amplifiers has a second pole at or above $\omega = \text{GB}$, as shown by the dashed −12-dB/octave line in Fig. C-10. Usually, however, this pole may be ignored.

Another characteristic of an operational amplifier on which active filter performance is dependent is the *slew rate*, i.e., the ability to follow a rapidly changing input signal. This effect is difficult to model since it is a nonlinear one. In active filter applications, it can cause an amplitude-dependent phase lag in the feedback loop. For example, in some filters, as the input signal frequency is swept from low to high values, it is possible for the circuit to break into instability as the frequency approaches the resonant frequency. If, however, the amplitude of the input signal is reduced, a stable sweep can be performed. Such an instability is caused by the fact that while the signal frequency is being increased, the resonance in the filter characteristic also increases the signal amplitude. The operational amplifier's slew-rate limitation thus creates enough phase lag to produce oscillation. To see why the slope of the output voltage of an operational amplifier must be limited to a maximum value (the slew rate), consider the frequency-compensated operational amplifier model shown in Fig. C-11.[3] In this model the source labeled $f(V_i)$ is a VCIS (voltage-controlled current source) whose dependence on an input voltage is given by Fig. C-12. Here SR is the slew rate. The output voltage for Fig. C-11 can be written as

$$V_o(s) = \frac{R\omega_a}{s + \omega_a} f(V_i) \tag{14}$$

where

$$\omega_a = \frac{1}{RC} \tag{15}$$

[2] J. V. Wait, L. P. Huelsman, and G. A. Korn, *Introduction to Operational Amplifier Theory and Applications*, McGraw-Hill, New York, 1992.

[3] J. Solomon, "The Monolithic Op Amp: A Tutorial Study," *IEEE J. Solid-State Circuits*, vol. SC-9, no. 6, December 1974, pp. 314–332.

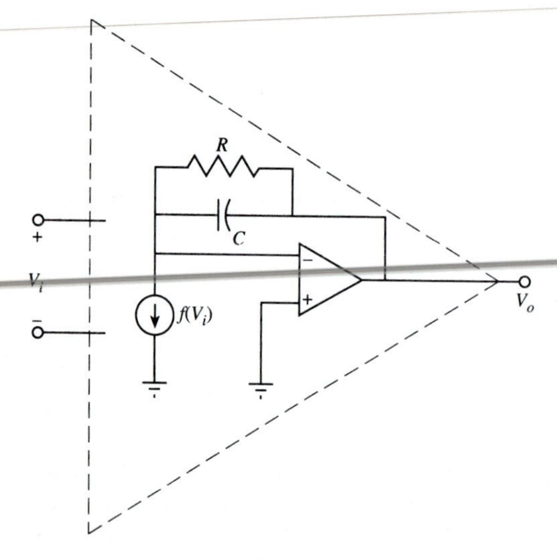

FIGURE C-11
Model of a frequency-compensated operational amplifier.

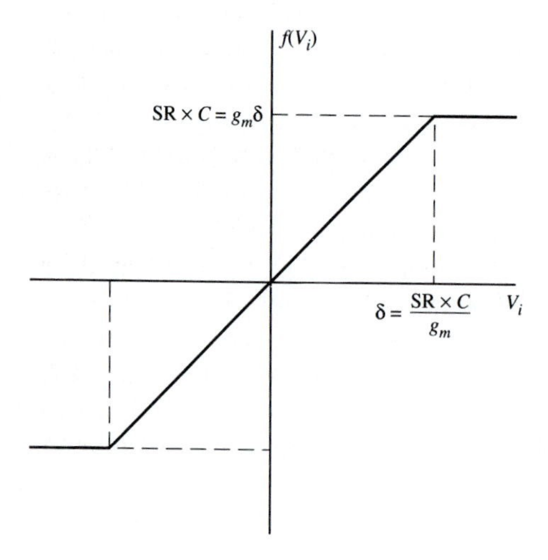

FIGURE C-12
Characteristic for source $f(V_i)$ of Fig. C-11.

For values of V_i less than δ, (14) simplifies to

$$\frac{V_o(s)}{V_i(s)} = g_m R \frac{\omega_a}{s + \omega_a} = \frac{A_0 \omega_a}{s + \omega_a} \tag{16}$$

which is equivalent to (13). Thus $g_m R = A_0$, the small-signal low-frequency (or dc) gain. For frequencies above ω_a, (16) reduces to

$$\frac{V_o(s)}{V_i(s)} \simeq \frac{A_0 \omega_a}{s} = \frac{\mathrm{GB}}{s} \tag{17}$$

If $V_i(s)$ is a step of V volts, then the output time-domain waveform is expressed as

$$v_o(t) = \mathrm{GB} \times V \times t \tag{18}$$

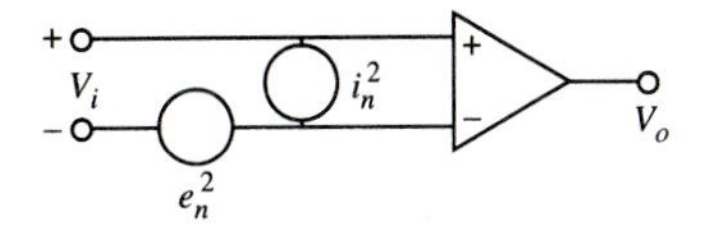

FIGURE C-13
Model for noise characteristics of an operational amplifier.

The slope of the output is seen to be proportional, among other things, to the amplitude of V_i. However, if the magnitude of V_i is greater than δ, then the maximum value of $f(V_i)$ is given as

$$f(V_i)_{\max} = C\frac{dv_o}{dt}\bigg|_{\max} = \mathrm{SR} \times C \tag{19}$$

It is apparent from (19) that the maximum slope of the output voltage is limited to SR, the slew rate of the operational amplifier.

Another characteristic of operational amplifiers that influences active filter performance is noise. This effectively limits the minimum signal level. To obtain large values of signal attenuation in the stopband of a filter characteristic, low noise levels are necessary. The noise characteristics of an operational amplifier can be modeled as shown in Fig. C-13. In this figure the quantities $e_n^2(\omega)$ and $i_n^2(\omega)$ are called *voltage* and *current noise spectral densities*. Their units are mean square volts and mean square amperes, respectively. Their values are also frequently given in rms (root-mean-square) units. Note that, as indicated in the figure, noise sources have no reference polarity. Typical noise spectral densities for an operational amplifier are shown in Fig. C-14. The increase of noise spectral density at decreasing frequencies is usually referred to as *1/f effect*.[4] Another source of noise in active filters is the resistive components. Two equivalent noise models for a resistor of value R are shown in Fig. C-15. For these

$$e_R^2(\omega) = 4kTR \quad \mathrm{V^2/Hz} \qquad i_R^2(\omega) = \frac{4kT}{R} \quad \mathrm{A^2/Hz} \tag{20}$$

where k is the Boltzmann constant of 1.38×10^{-23} Ws/K. Examples of noise calculations may be found in the literature.[5,6]

---

[4] C. D. Motchenbacher and F. C. Fitchen, *Low-Noise Electronic Design*, John Wiley & Sons, Inc., New York, 1973.

[5] F. N. Trofimenkoff, D. H. Treleaven, and L. T. Bruton, "Noise Performance of RC-Active Quadratic Filter Sections," *IEEE Trans. Circuit Theory*, vol. CT-20, no. 5, September 1973, pp. 524–532.

[6] L. T. Bruton and D. H. Treleaven, "Electrical Noise in Low-Pass FDNR Filters," *IEEE Trans. Circuit Theory*, vol. CT-20, no. 2, March 1973, pp. 154–158.

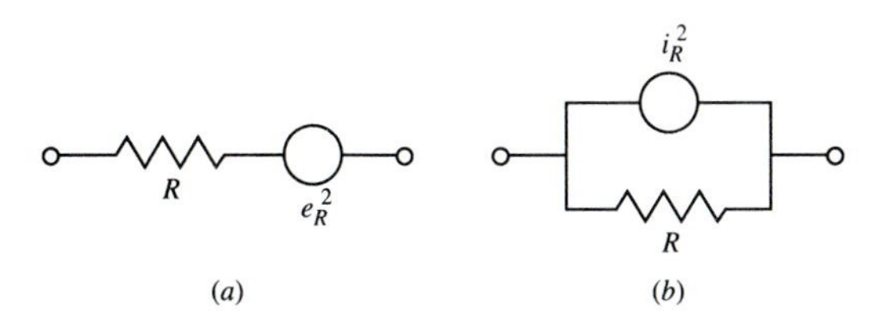

FIGURE C-14
Noise spectral densities for an operational amplifier.

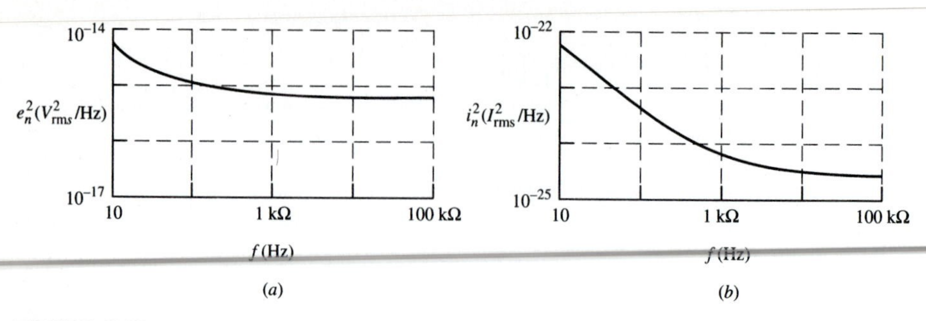

FIGURE C-15
Noise models for a resistor of value R.

Effective design techniques can considerably minimize the effects of noise in active filters. For example, the signal level of the filter should be chosen as large as possible to give the maximum dynamic range. The inherent noise voltage is typically of the order of -100 dBV (100 dB below 1 V) for a 10-Hz bandwidth. For each decade increase in frequency, a 10-dB increase in noise results. For example, if a system has a noise floor of -100 dBV in a 10-Hz bandwidth, then the inherent noise voltage would be -70 dBV over a 10-kHz bandwidth. If we compare the relative contributions of resistors and amplifiers to overall noise, the amplifier is the primary culprit. Thus, the most effective way to decrease output noise levels is to select a low-noise operational amplifier. In most applications $e_n^2(\omega)$ is more important than $i_n^2(\omega)$, so that in practice it is important to select an amplifier with a low value of $e_n^2(\omega)$. If it is important to minimize $1/f$ noise, then a JFET-input operational amplifier should be selected, since the breakpoint in the $e_n^2(\omega)$ curve is less for JFETs than for bipolar junction transistors.

There are some additional miscellaneous parameters of operational amplifiers that deserve mention, since they are important in active filter applications. The first of these concerns the dc error previously characterized by V_{os} and I_{os}. In practice these parameters are dependent upon temperature. Typical values are $dV_{os}/dT \approx 6$ mV/°C and $dI_{os}/dT \approx 2$ pA/°C. Fortunately, the feedback used in active filters will minimize these effects in most realizations. Another temperature-dependent parameter is GB. A typical temperature coefficient of GB is approximately -2000 ppm/°C. A final parameter is the power supply rejection ratio (PSRR). The PSRR is a measure of the ratio of ripple at the output of the amplifier to ripple at the power supply output. In general the PSRR should be greater than A_0 to prevent instability problems.

BIBLIOGRAPHY

Additional information on many of the topics treated in this book can be obtained from the volumes listed below. Tabulations of various approximations and the filters that realize them may be found in Christian and Eisenmann, Craig, Saal, and Zverev. An edited collection of the most significant original research papers that represents benchmarks in the evolution of the active filter field is given in Huelsman (1976).

Baher, H.: *Synthesis of Electrical Networks*, Wiley, New York, 1984.

Balabanian, N., and T. Bickart: *Linear Network Theory: Analysis, Properties, Design, and Synthesis*, Matrix Publishers, Beaverton, OR, 1981.

Blinchikoff, H. J., and A. I. Zverev: *Filtering in the Time and Frequency Domain*, Wiley, New York, 1976.

Brayton, R. K., and R. Spence: *Sensitivity and Optimization*, Elsevier, New York, 1980.

Budak, A.: *Passive and Active Network Analysis and Synthesis*, Houghton Mifflin, Boston, 1974.

Chen, W-K.: *Passive and Active Filters, Theory and Implementations*, Wiley, New York, 1986.

Christian, E.: *LC Filters: Design, Testing, and Manufacturing*, Wiley, New York, 1983.

Christian, E., and E. Eisenmann: *Filter Design Tables and Graphs*, Transmission Networks International, Knightdale, NC, 1977.

Craig, J. W.: *Design of Lossy Filters*, MIT Press, Cambridge, MA, 1970.

Daniels, R. W.: *Approximation Methods for Electronic Filter Design with Applications to Passive, Active, and Digital Networks*, McGraw-Hill, New York, 1974.

Daryanani, G.: *Principles of Active Network Synthesis and Design*, Wiley, New York, 1976.

Huelsman, L. P.: *Active Filters: Lumped, Distributed, Integrated, Digital, and Parametric*, McGraw-Hill, New York, 1970.

——— *Active RC Filters: Theory and Application*, Dowden, Hutchinson and Ross, Stroudsburg, PA, 1976.

Lindquist, C. S.: *Active Network Design*, Steward and Sons, Long Beach, CA, 1977.

Moschytz, G. S.: *Linear Integrated Networks: Design*, Van Nostrand Reinhold, New York, 1975.

——— *Linear Integrated Networks: Fundamentals*, Van Nostrand Reinhold, New York, 1976.

Saal, R.: *The Design of Filters Using the Catalogue of Normalized Low-Pass Filters*, Telefunken G.M.B.H., Backnang/Wurtt., Germany, 1963.

Schaumann, R., M. S. Ghausi, and K. R. Laker: *Design of Analog Filters; Passive, Active RC, and Switched Capacitor*, Prentice-Hall, Englewood Cliffs, NJ, 1990.

Sedra, A. S., and P. O. Brackett: *Filter Theory and Design: Active and Passive*, Matrix Publishers, Champaign, IL, 1977.

Spence, R.: *Linear Active Networks*, Wiley, New York, 1970.

Stephenson, F. W.: *RC Active Filter Design Handbook*, Wiley, New York, 1985.

Temes, G. C., and J. W. LaPatra: *Introduction to Circuit Synthesis and Design*, McGraw-Hill, New York, 1977.

Temes, G. C., and S. K. Mitra: *Modern Filter Theory and Design*, Wiley, New York, 1973.

Van Valkenburg, M. E.: *Introduction to Modern Network Synthesis*, Wiley, 1960.

Wait, J. V., L. P. Huelsman, and G. A. Korn: *Introduction to Operational Amplifier Theory and Applications*, 2nd ed., McGraw-Hill, New York, 1992.

Weinberg, L.: *Network Analysis and Synthesis*, Wiley, New York, 1962; R. E. Krieger Publishing, Huntington, NY, 1975.

Zverev, A. L.: *Handbook of Filter Synthesis*, Wiley, New York, 1967.

INDEX